D1617108

Markov Random Fields

Theory and Application

Markov Random Fields

Theory and Application

Edited by

Rama Chellappa
Department of Electrical Engineering
Center for Automation Research and
Institute for Advanced Computer Studies
University of Maryland
College Park, Maryland

Anil Jain
Computer Science Department
Michigan State University
East Lansing, Michigan

ACADEMIC PRESS, INC.
Harcourt Brace Jovanovich, Publishers

Boston San Diego New York
London Sydney Tokyo Toronto

This book is printed on acid-free paper. ♾

ACADEMIC PRESS, INC.
1250 Sixth Avenue, San Diego, CA 92101-4311

United Kingdom Edition published by
ACADEMIC PRESS LIMITED
24–28 Oval Road, London NW1 7DX

Library of Congress Cataloging-in-Publication Data

Markov random fields : theory and application / Rama Chellappa, Anil Jain, [editors].
p. cm.
Includes bibliographical references and index.
ISBN 0-12-170608-7
1. Markov random fields. I. Chellappa, Rama. II. Jain, Anil K., date.
QA274.45.M37 1991
519.2—dc20 91–31523
CIP

Printed in the United States of America

93 94 95 96 EB 9 8 7 6 5 4 3 2 1

Contents

Contributors

Numbers in parentheses indicate the pages on which the authors' contributions begin.

Barnard, S. T. (245), Artificial Intelligence Center, SRI International, Menlo Park, California 94025

Brown, C. M. (211), Department of Computer Science, University of Rochester, Rochester, New York 14627

Cernuschi-Frias, B. (335), Facultad de Ingenieria, Universidad de Buenos Aires, and CONICET, Argentina

Chatterjee, S. (159), Department of Electrical Engineering, University of California at San Diego, La Jolla, California 92093-0407

Chellappa, R. (69), Department of Electrical Engineering, Center for Automation Research and Insitute for Advanced Computer Studies, University of Maryland, College Park, Maryland 20742

Choe, Y. (273), Hyudai Electronics Industries Co. Ltd, Ichon-Kui, Korea

Chou, P. B. (211), IBM Research Division, T. J. Watson Research Center, P.O. Box 218, Yorktown Heights, New York 10598

Cohen, F. S. (307), Department of Electrical and Computer Engineering, Drexel University, Philadelphia, Pennsylvania 19104

Cooper, D. B. (335), Laboratory for Engineering Man/Machine Systems, Division of Engineering, Brown University, Providence, Rhode Island 02912

Cooper, P. R. (211), Institute for the Learning Sciences and Department of Electrical Engineering and Computer Science, Northwestern University, Evanston, Illinois 60208

Derin, H. (131), Department of Electrical and Computer Engineering, University of Massachusetts, Amherst, Massachusetts 01003

Gelfand, S. B. (499), School of Electrical Engineering, Purdue University, West Lafayette, Indiana 47007

Geman, D. (39), Department of Mathematics and Statistics, University of Massachusetts, Amherst, Massachusetts 01003

Geman, S. (93). Division of Applied Mathematics, Brown University, Providence, Rhode Island 02912

Gidas, B. (471), Division of Applied Mathematics, Brown University, Providence, Rhode Island 02912

Hainsworth, T. J. (409), Department of Statistics, The University of Leeds, Leeds LS2 9JT, United Kingdom

Hung, Y. P. (335), Institute for Information Science, Academia Sinica, Nakang, Taipei, Taiwan

Jain, A. K. (543), Department of Computer Science, Michigan State University, East Lansing, Michigan 48824

Jeng, F. C. (11), Bell Communication Research, 445 South Street, Morristown, New Jersey 07960

Kashyap, R. L. (273), School of Electrical Engineering, Purdue University, West Lafayette, Indiana 47907

Lakshmanan, S. (131), Department of Electrical and Computer Engineering, University of Michigan at Dearborn, Dearborn, Michigan 48128

Manbeck, K. M. (93), Division of Applied Mathematics, Brown University, Providence, Rhode Island 02912

Mardia, K. V. (409), Department of Statistics, The University of Leeds, Leeds LS2 9JT, United Kingdom

Marroquin, J. L. (517), Centro de Investigacion en Matimaticas, A.C., Apartado Postal 402, 36000 Guanajuanto, GTO, Mexico

McClure, D. E. (93), Division of Applied Mathematics, Brown University, Providence, Rhode Island 02912

Mitter, S. K. (499), Department of Electrical Engineering and Computer Science and Laboratory for Information and Decision Systems, Massachusetts Insitute of Technology, Cambridge, Massachusetts 02139

Modestino, J. W. (369), Electrical, Computer, and Systems Engineering Department, Rensselaer Polytechnic Institute, Troy, New York 12180

Nadabar, S. G. (543), Department of Computer Science, Michigan State University, East Lansing, Michigan 48824

Patel, M. A. S. (307), Department of Electrical and Computer Engineering, Drexel University, Philadelphia, Pennsylvania 19104

Poggio, T. (447), Artificial Intelligence Laboratory and Center for Biological Information Processing, Massachusetts Insitute of Technology, Cambridge, Massachusetts 02139

Ramirez, A. (517), Centro de Investigacion in Matimaticas, A.C., Apartado Postal 402, 36000 Guanajuanto, GTO, Mexico

Rangarajan, A. (69), Department of Computer Science and Division of Imaging Science, Department of Diagnostic Radiology, Yale University, New Haven, Connecticut 02650

Rao, T. S. (179), Deparment of Mathematics, University of Manchester Insitute of Science and Technology, Manchester M60 1QE, United Kingdom

Rastogi, S. (11), Department of Electrical, Computer, and Systems Engineering, Rensselaer Polytechnic Institute, Troy, New York 12180

Reynolds, G. (39), VI Corporation, Northampton, Massachusetts 01060

Rosenfeld, A. (1), Center for Automation Research, University of Maryland, College Park, Maryland 20742

Subrahmonia, J. (335), Laboratory for Engineering Man/Machine Systems, Division of Engineering, Brown University, Providence, Rhode Island 02912

Swain, M. J. (211), Department of Computer Science, University of Chicago, Chicago, Illinois 60637

Weinshall, D. (447), IBM Research Division, T. J. Watson Research Center, Yorktown Heights, New York 10598

Wixson, L. E. (211), Department of Computer Science, University of Rochester, Rochester, New York 14627

Woods, J. W. (11), Department of Electrical, Computer, and Systems Engineering, Rensselaer Polytechnic Insitute, Troy, New York 12180

Yang, C. (39), Department of Computer Science, University of Arizona, Tuscon, Arizona 85721

Yuan, J. (179), Department of Mathematics, University of Manchester Insitute of Science and Technology, Manchester M60 1QE, United Kingdom

Zhang, J. (369), Department of Electrical Engineering and Computer Science, University of Wisconsin, Milwaukee, Wisconsin 53201

Preface

Over the last few years, a growing number of researchers from varied disciplines, such as signal and image processing, pattern recognition, image analysis, computer vision, and applied mathematics and statistics, have been utilizing Markov random field (MRF) models for developing optimal, robust algorithms for various problems, such as texture analysis, image synthesis, image restoration, classification and segmentation, surface reconstruction, integration of several low-level vision modules, and sensor fusion. The philosophy of modeling images through the local interaction of pixels or edge elements as specified by Markov models is intuitively very appealing due to the following reasons:

1. One can systematically develop algorithms based on sound principles rather than on some *ad hoc* heuristics for a variety of problems.
2. It is easier to derive quantitative performance measures for characterizing how well the image analysis algorithms work.
3. MRF models can be used to incorporate prior contextual information or constraints in a quantitative way.
4. More importantly, the MRF-based algorithms tend to be local, and tend themselves to parallel hardware implementation in a natural way. The inherent parallel nature of the MRF algorithms implies that MRF algorithms can be suitably mapped onto an artificial neural network architecture.

Due to the multidisciplinary background of the researchers working on MRF models, no single forum currently exists to facilitate an interactive exchange of ideas among these researchers. Considering the growing scientific interest in MRF models, we organized a workshop on this topic in June, 1989 in San Diego, California. The aim of this workshop, sponsored by the Information, Robotics and Intelligent Systems Program of the National Science Foundation, was to bring together active researchers working on theory and applications of MRF models to discuss emerging themes in this field that may play a key role in the development of robust methods for solving a number of problems in computer vision and image processing. Another purpose of the workshop was to discuss successes and failures of MRF models. Invited talks were given by prominent scientists and engineers working in this field. Panel discussions highlighted the limitations and strengths of these models. We were pleasantly surprised to see approximately 100 participants at the workshop. The workshop participants were drawn from academic institutions, research centers, and industry, indicating the broad-based interest in this topic.

This edited volume is an outgrowth of this workshop. The chapters in this book have been written by the invited speakers at the workshop and a few other distinguished researchers who were not able to attend the workshop. The chapters have been arranged in such a way that one can go from representation to applications in image processing, analysis, and computer vision in a natural order. In addition, a chapter on hardware aspects of MRF model-based algorithms has been included and the important related issues on parameter estimation have been addressed. We feel that this book represents the work done by most of the leading researchers in the world and should be a good reference text for engineers, computer scientists, applied statisticians and mathematicians, and physicists who are interested in basic research issues and state-of-the-art in MRF models.

Acknowledgments

We are grateful to all the contributors for their utmost enthusiasm and collaboration that made this edited volume possible. A sincere word of appreciation is due to the Academic Press editors, Ms. Jenifer Swetland and Mr. Brian Miller, who have played a key part in the production of this book. We thank Ms. Delsa Tan for her meticulous technical typing and reformatting, correction of the copyedited version, and the integration of all the manuscripts to complete this volume.

Dr. Ken Laws, former Program Director of the IRIS Program at the National Science Foundation, provided funds to support the Workshop on Markov random fields. The MRF workshop and the preparation of this book were supported by the NSF Grant. No. IRI-8822272.

Image Modeling during the 1980s: A Brief Overview

Azriel Rosenfeld
Center for Automation Research
University of Maryland
College Park, Maryland

The performance of image processing or analysis techniques varies with the nature of the images to which they are applied. Thus it is important for the designer of such techniques to be able to characterize—i.e., to *model*—the class of images that are to be processed or analyzed.

Image models were used by some researchers even in the early years, but the awareness of their importance was enhanced at the end of the 1970s by a workshop on image modeling that was held in Chicago on August 6–7, 1979. The proceedings of this workshop, which were published both in journal and book form [1,2], contained about 20 papers that represented most of the major approaches.

The literature on image modeling deals primarily with probabilistic models for the variations in gray level over a two-dimensional region. Models for the geometries of the regions, or for the decomposition of the image into regions, have been largely ignored. When a region is the image of a three-dimensional surface, models for its gray-level variations should take into account variations in illumination and surface orientation, as well as in reflectivity; but this, too, usually is ignored in the image processing literature. In the following sections, we briefly describe the principal themes in the modeling of image regions that have been treated in the image processing and analysis literature during the 1980s.

1 Random Fields

Two-dimensional time series and random field models probably are the most common class of image models. The different classes of such models suggested in the literature attempt to characterize the correlation among

The author thanks Prof. R. Chellappa for providing input to this section.

ISBN 0-12-170608-7

the neighboring pixels. One of the earliest papers that modeled a 2-D texture using a 1-D time series model is [3]. Applications of 1-D autoregressive time series models to image data compression [4] and 2-D boundary representation and classification may be found in [5–9].

Since images are defined on 2-D grids, it is more appropriate to model them as realizations of 2-D time series or random field models. A straightforward generalization of the 1-D time series AR model to 2-D is obtained by assuming a separable correlation function for the 2-D model. This simple model and its extensions have found many applications in image restoration [10], image compression [11], and texture classification and segmentation [12].

The 1979 image modeling workshop brought together a number of leading researchers in image modeling. Since then, the research activities in the image modeling area have increased considerably. Researchers in the image processing and analysis area became aware of significant work in the statistics literature [13–19]. Also, pioneering work reported in the engineering literature [20–21] started receiving attention. Owing to the cross-fertilization of ideas from image processing, spatial statistics, and statistical physics, the last decade has witnessed a significant amount of research activity in image modeling with special emphasis on Markov random fields (MRF), the topic of this book. The strength of MRF lies in the fact that more general image structures can be modeled. Whether the interactions are among neighboring pixels or regions, one can use similar graph structures for representation.

Since 1979, a plethora of papers has appeared on various aspects of MRF models with applications to texture synthesis [22–24], classification [25–27], segmentation of visible and synthetic aperture radar images [28–39], image restoration [40–46], integration of early vision modules [47], and artificial neural networks [32, 48]. Issues such as parameter estimation and hypothesis testing in MRF models also have been addressed [18, 49–58]. As most of the researchers who have contributed to the general MRF area also have written chapters for this book, it seems unnecessary to expand on the vast number of papers published in this area. Useful reviews on MRF models may be found in [59–62].

2 Random Geometric Processes

Geometric probabilists have studied many types of random planar patterns, including random point and line patterns, *bombing* patterns, and random tessellations [63]. Such patterns can serve as a basis for defining classes of *mosaic models* for images [64–68]. A general theory of patterns has been developed by Grenander and his colleagues; for a comprehensive treatment, see [69].

3 Fractals

Many types of natural patterns have hierarchical structures that are (approximately) self-similar under scale change, and can be modeled by so-called fractal processes, which have been studied extensively by Mandelbrot [70] (also discussed in [71,72]); on their application to natural scenes, see [73–76].

4 Texture

Techniques for texture classification, segmentation, and synthesis make extensive use of image models. For surveys of the literature on texture, see [77–80].

5 Image Formation Models

Models that relate the surface orientation and observed image intensity through nonlinear reflection mappings have been of concern to computer vision researchers. These models have been used for the inverse problem of extracting surface shape from intensity, the so-called shape from shading problem. Notable papers on this topic have been reprinted or referred to in [81]. Image processing researchers have not paid much attention to modeling images this way. Much work needs to be done in integrating the traditional random field models discussed in Section 1 and the deterministic models of image formation.

Only selected references are cited here; for a more extensive bibliography, see the section on texture models in the author's annual bibliography on "Picture Processing" (since 1987, "Image Analysis and Computer Vision"), which appears each year in *Computer Vision, Graphics, and Image Processing* (since 1991, *CVGIP: Image Understanding*). For survey papers that cover many aspects of image modeling, see [82–85], as well as the book [86].

Bibliography

[1] *Comput. Graphics Image Processing*, vol. 12, 1980, 1–406.

[2] A. Rosenfeld, (ed.), *Image Modelling*, Academic Press, 1981.

[3] B.H. McCormick and S.N. Jayaramamurhy, Time Series Model for Texture Synthesis, *Intl. Jl. Comput. Information Science*, vol. 3, 1974, 329–343.

[4] E.J. Delp, R.L. Kashyap, and O.R. Mitchell, Image Data Compression Using Autoregressive Time Series Models, *Pattern Recognition*, vol. 11, 1979, 313–323.

[5] D.B. Cooper, Maximum Likelihood of Markov Process Blob Boundaries in Noisy Images, *IEEE Trans. Patt Anal. Mach. Intell.,* vol. 1, Oct. 1979, 373–383.

[6] R.L. Kashyap and R. Chellappa, Stochastic Models for Closed Boundary Analysis: Representation and Reconstruction, *IEEE Trans. Information Theory*, vol. 27, Sept. 1981, 627–637.

[7] P.F. Singer and R. Chellappa, Classification of 2-D Closed Boundaries Using Stochastic Models, *Proc. IEEE Computer Society Conf. on Comput. Vision and Patt. Recognition*, Washington D.C., June 1983, 146–147.

[8] S.R. Dubois and F.H. Glanz, An Autoregressive Model Approach to 2-D Shape Classification, *IEEE Trans. Patt. Anal. Mach. Intell.*, vol. 7, Jan. 1986, 55–66.

[9] D. Anastassiou and D.J. Sakrison, A Probability Model for Simple Closed Random Curves, *IEEE Trans. Information Theory*, vol. 27, May 1981, 376–381.

[10] J.W. Woods and C.H. Radewan, Kalman Filtering in Two Dimensions, *IEEE Trans. Information Theory*, vol. 23, July 1977, 473–482.

[11] P.A. Maragos, R.W. Schafer, and R.M. Mersereau, Two-Dimensional Linear Prediction and Its Application to Adaptive Predictive Coding of Images, *IEEE Trans. Acoust., Speech and Signal Proc.*, vol. 32, Dec. 1984, 1213–1229.

[12] C.W. Therrien, An Estimation Theoretic Approach to Terrain Image Segmentation, *Comput. Vision, Graphics, Image Processing*, vol. 22, June 1983, 313–326.

[13] P. Whittle, On Stationary Processes in the Plane, *Biometrika*, vol. 41, 1954, 434–449.

[14] P. Levy, *Processes Stochastiques et Movement Brownien*, Gauthier-Villars, Paris, 1948.

[15] P. Levy, A Special Problem of Brownian Motion and a General Theory of Gaussian Random Functions, in *Proc. 3rd Berkeley Symp. on Mathematical Statistics and Probability*, vol. 2, University of California Press, Berkeley, 1956.

[16] Y.A. Rosanov, On Gaussian Fields with Given Conditional Distributions, *Theory of Probability and Applications*, vol. XII, 1967, 381–391.

[17] P.A.P. Moran, Necessary Conditions for Markovian Processes on a Lattice, *Jl. Appl. Probability*, vol. 10, Sept. 1973, 605–612.

[18] P.A.P. Moran, A Gaussian-Markovian Process on a Square Lattice, *Jl. Appl. Probability*, vol. 10, March 1973, 54–62.

[19] J.E. Besag, Spatial Interaction and the Statistical Analysis of Lattice Systems, *Jl. Royal Stat. Soc.*, Ser. B, vol. 36, 1974, 192–236.

[20] K. Abend, T.J. Harley and L.N. Kanal, Classification of Binary Random Patterns, *IEEE Trans. Information Theory*, vol. 11, Oct. 1965, 538–544.

[21] J.W. Woods, Two-Dimensional Discrete Markovian Fields, *IEEE Trans. Information Theory*, vol. 18, March 1972, 232–240.

[22] M. Hassner and J. Sklansky, Markov Random Fields of Digitized Image Texture, *Proc. IEEE Computer Society Conf. on Patt. Recognition and Image Processing*, June 1978, 346–351.

[23] G.R. Cross and A.K. Jain, Markov Random Field Texture Models, *IEEE Trans. Patt. Anal. Mach. Intell.*, vol. 5, Jan. 1983, 25–39.

[24] R. Chellappa, S. Chatterjee, and R. Bagdazian, Texture Synthesis and Coding Using Gaussian Markov Random Field Models, *IEEE Trans. Systems, Man, and Cybernetics*, vol. 15, March 1985, 293–303.

[25] R.L. Kashyap, R. Chellappa, and A. Khotanzad, Texture Classification Using Features Derived from Random Field Models, *Patt. Recognition Letters*, vol. 1, Oct. 1982, 43–50.

[26] R. Chellappa and S. Chatterjee, Classification of Textures Using Gaussian Markov Random Field Models, *IEEE Trans. Acoust., Speech and Signal Proc.*, vol. 33, Aug. 1985, 959–963.

[27] R.L. Kashyap and A. Khotanzad, A Model-Based Method for Rotation Invariant Texture Classification, *IEEE Trans. Patt. Anal. Mach. Intell.*, vol. 8, July 1986, 472–481.

[28] H. Derin and H. Elliott, Modeling and Segmentation of Noisy and Textured Images Using Gibbs Random Fields, *IEEE Trans. Patt. Anal. Mach. Intell.*, vol. 9, Jan. 1987, 38–55.

[29] J.E. Besag, On the Statistical Analysis of Dirty Pictures (with Discussion), *Jl. Royal Stat. Soc.*, Ser. B, vol. 48, 1986, 259–302.

[30] F.S. Cohen and D.B. Cooper, Simple Parallel Hierarchical and Relaxation Algorithms for Segmenting Noncausal Markovian Random Fields, *IEEE Trans. Patt. Anal. Mach. Intell.*, vol. 9, March 1987, 195–219.

[31] H. Derin and W.S. Cole, Segmentation of Textured Images Using Gibbs Random Fields, *Comput. Vision, Graphics, Image Processing*, vol. 35, July 1986, 72–98.

[32] J.L. Marroquin, S. Mitter, and T. Poggio, Probabilistic Solution of Ill-Posed Problems in Computational Vision, *Jl. American Stat. Assn.*, vol. 82, March 1987, 76–89.

[33] B.S. Manjunath, T. Simchony and R. Chellappa, Stochastic and Deterministic Networks for Texture Segmentation, *IEEE Trans. Acoust., Speech, Signal Proc.*, vol. 38, June 1990, 1039–1049.

[34] D. Geman, S. Geman, C. Graffigne, and P. Dong, Boundary Detection by Constrained Optimization, *IEEE Trans. Patt. Anal. Mach. Intell.*, vol. 12, July 1990, 609–628.

[35] B.D. Ripley, Statistics, Images and Pattern Recognition, *Canadian Jl. Statist.*, vol. 14, 1986, 83–111.

[36] J.W. Modestino and J. Zhang, A Markov Random Field Model Based Approach to Image Interpretation, *Proc. IEEE Computer Society Conf. on Comput. Vision and Patt. Recognition*, June 1990, 458–465.

[37] Z. Fan and F.S. Cohen, Textured Image Segmentation as a Multiple Hypothesis Test, *IEEE Trans. Circuits and Systems,* vol. 35, June 1988, 691–702.

[38] H. Derin, P.A. Kelly, G. Vézina, and S.G. Labitt, Modeling and Segmentation of Speckled Images Using Complex Data, *IEEE Trans. Geoscience and Remote Sensing*, vol. 28, Jan. 1990, 76–87.

[39] E. Rignot and R. Chellappa, Segmentation of Synthetic Aperture Radar Complex Data, *Jl. Optical Society of America*, A, Vol. 8, Sept. 1981, 1499–1505.

[40] A.K. Jain and J.R. Jain, Partial Differential Equations and Finite Difference Methods in Image Processing: Part II: Image Restoration, *IEEE Trans. Automatic Control*, vol. 23, Oct. 1978, 817–834.

[41] R. Chellappa, and R.L. Kashyap, Digital Image Restoration Using Spatial Interaction Models, *IEEE Trans. Acoust., Speech and Signal Proc.*, vol. 30, June 1982, 461–472.

[42] S. Geman and D. Geman, Stochastic Relaxation, Gibbs Distributions and Bayesian Restoration of Images, *IEEE Trans. Patt. Anal. Mach. Intell.*, vol. 6, Nov. 1984, 721–741.

[43] B. Chalmond, Image Restoration Using an Estimated Markov Model, Dept. Math., Univ. Paris, Orsay, France, 1987.

[44] T. Simchony, R. Chellappa and Z. Lichtenstein, Relaxation for MAP Estimation of Gray Level Images Corrupted by Multiplicative Noise, *IEEE Trans. Information Theory*, vol. 36, May 1990, 608–613.

[45] F.C. Jeng and J.W. Woods, Simulated Annealing in Compound Gaussian Markov Random Fields, *IEEE Trans. Information Theory*, vol. 36, Jan. 1990, 94–107.

[46] T. Simchony, R. Chellappa and Z. Lichtenstein, Pyramid Implementation of Optimal Step Conjugate Gradient Algorithms for Some Computer Vision Problems, *IEEE Trans. Systems, Man, and Cybernetics*, vol. 19, Nov. 1989, 1408–1425.

[47] T. Poggio, E.B. Gamble and J.J. Little, Parallel Integration of Vision Modules, *Science*, vol. 242, Oct. 1988, 436–440.

[48] A. Ragarajan, R. Chellappa, and B.S. Manjunath, Markov Random Fields and Neural Networks with Applications to Early Vision Problems, in *Artificial Neural Networks and Statistical Pattern Recognition: Old and New Connections*, I.K. Sethi and A.K. Jain (eds.), Elsevier, 1991.

[49] J.E. Besag and P.A.P. Moran, On the Estimation and Testing of Spatial Interaction in Gaussian Lattice, *Biometrika*, vol. 62, Dec. 1975, 555–562.

[50] H. Künsch, Thermodynamics and Statistical Analysis of Gaussian Random Fields, *Z. Wahr. Ver. Geb*, vol. 58, Nov. 1981, 407–421.

[51] K. Ord, Estimation Methods for Models of Spatial Interaction, *Jl. Amer. Stat. Assn.*, vol. 70, March 1975, 120–126.

[52] J.E. Besag, Statistical Analysis of Non-Lattice Data, *The Statistician*, vol. 24, 1975, 179–195.

[53] J.E. Besag, Efficiency of Pseudo-likelihood Estimation for Simple Gaussian Fields, *Biometrika*, vol. 64, Dec. 1977, 616–618.

[54] J.E. Besag, Errors-in-Variable Estimation for Gaussian Lattice Scheme, *J. Royal Stat. Soc.*, Ser. B, vol. 39, 1977, 673–678.

[55] R.L. Kashyap and R. Chellappa, Estimation and Choice of Neighbors in Spatial Interaction Models of Images, *IEEE Trans. Information Theory*, vol. 29, Jan. 1983, 60–72.

[56] G. Sharma and R. Chellappa, A Model Based Approach for the Estimation of 2-D Maximum Entropy Power Spectra, *IEEE Trans. Information Theory*, vol. 31, Jan. 1985, pp. 90–99.

[57] S. Geman and C. Graffigne, Markov Random Field Image Models and Their Applications to Computer Vision, *Proc. Intl. Congress of Mathematics 1986*, A.M. Gleason (ed.), American Mathematical Society, 1987.

[58] B. Gidas, Consistency of Maximum Likelihood and Pseudo-Likelihood Estimators for Gibbs Distributions, *Proc. Workshop on Stochastic Differential Systems in Applications in Electrical and Computer Engineering, Control Theory and Operations Research*, IMA, University of Minnesota, 1986.

[59] R.L. Kashyap, Analysis and Synthesis of Image Patterns by Spatial Interaction Models, in *Progress in Pattern Recognition*, vol. 1, L.N. Kanal and A. Rosenfeld (eds.), North-Holland, 1981.

[60] R. Chellappa, Two-Dimensional Discrete Gauss Markovian Random Field Models for Image Processing, in *Progress in Pattern Recognition*, vol. 2, L.N. Kanal and A. Rosenfeld (eds.), North-Holland, 1985.

[61] H. Derin and P.A. Kelly, Discrete-index-Markov-type Random Processes, *Proc. IEEE*, vol. 77, Oct. 1989, 1485–1510.

[62] R.C. Dubes and A.K. Jain, Random Field Models in Image Analysis, *Jl. Appl. Stat.*, vol. 16, 1989, 131–164.

[63] D. Stoyan, W.S. Kendall, and J. Mecke, *Stochastic Geometry and Its Applications*, Wiley, New York, 1987.

[64] N. Ahuja, Mosaic Models for Images—I. Geometric Properties of Components in Cell-Structure Mosaics, *Information Sciences*, vol. 23, 1981, 69–104.

[65] N. Ahuja, Mosaic Models for Images—II. Geometric Properties of Components in Coverage Mosaics, *Information Sciences*, vol. 23, 159–200.

[66] N. Ahuja, Mosaic Models for Images—III. Spatial Correlation in Mosaics, *Information Sciences*, vol. 24, 43–69.

[67] N. Ahuja and A. Rosenfeld, Mosaic Models for Textures, *IEEE Trans. Patt. Anal. Mach. Intell.*, vol. 3, Jan. 1981, 1–11.

[68] N. Ahuja, T. Dubitzki, and A. Rosenfeld, Some Experiments with Mosaic Models for Images, *IEEE Trans. Systems, Man, and Cybernetics*, vol. 10, Nov. 1980, 744–749.

[69] U. Grenander, *Lectures in Pattern Theory* (3 vols.), Springer, New York, 1976–81.

[70] B.B. Mandelbrot, *Fractals and a New Geometry of Nature*, Freeman, San Francisco, CA, 1981.

[71] K.J. Falconer, *The Geometry of Fractal Sets*, Cambridge University Press, Cambridge, UK, 1985.

[72] H.O. Peitgen and D. Saupe, *The Science of Fractal Images*, Springer, Berlin, 1988.

[73] M. Barnsley, *Fractals Everywhere*, Academic Press, Boston, 1988.

[74] A.P. Pentland, Fractal-Based Description of Natural Scenes, *IEEE Trans. Patt. Anal. Mach. Intell.*, vol. 6, Nov. 1984, 661–674.

[75] J.M. Keller, R.M. Crownover, and R.Y. Chen, Characteristics of Natural Scenes Related to the Fractal Dimension, *IEEE Trans. Patt. Anal. Mach. Intell.*, vol. 9, Sept. 1987, 621–627.

[76] P. Kube and A. Pentland, On the Imaging of Fractal Surfaces, *IEEE Trans. Patt. Anal. Mach. Intell.*, vol. 10, Sept. 1988, 704–707.

[77] R.M. Haralick, Image Texture Survey, in *Fundamentals in Computer Vision*, O.D. Faugeras (ed.), Cambridge University Press, Cambridge, UK, 1983, 145–172; also in *Handbook of Statistics, Vol. 2: Classification, Pattern Recognition, and Reduction of Dimensionality*, P.R. Krishnaiah and L.N. Kanal (eds.), North-Holland, Amsterdam, 1982, 399–415.

[78] H. Wechsler, Texture Analysis—A Survey, *Signal Processing*, vol. 2, 1980, 271–282.

[79] L. Van Gool, P. Dewaele, and A. Oosterlinck, Texture Analysis Anno 1983, *Comput. Vision, Graphics, Image Processing*, vol. 29, March 1985, 336–357.

[80] R.M. Haralick, Statistical Image Texture Analysis, in *Handbook of Pattern Recognition and Image Processing*, T.Y. Young and K.S. Fu (eds.), Academic Press, Orlando, FL, 1986, 247–279.

[81] B.K.P. Horn and M.J. Brooks (eds.), *Shape from Shading*, M.I.T. Press, Cambridge, MA, 1989.

[82] A.K. Jain, Advances in Mathematical Models for Image Processing, *Proc. IEEE*, vol. 69, May 1981, 502–528.

[83] N. Ahuja and B. Schachter, Image Models, *Computing Surveys*, vol. 13, Dec. 1981, 373–397.

[84] N. Ahuja and A. Rosenfeld, Image Models, in *Handbook of Statistics Vol. 2: Classification, Pattern Recognition, and Reduction of Dimensionality*, P.R. Krishnaiah and L.N. Kanal (eds.), North-Holland, Amsterdam, 1982, 383–397.

[85] R.L. Kashyap, Image Models, in *Handbook of Pattern Recognition and Image Processing*, T.Y. Young and K.S. Fu (eds.), Academic Press, Orlando, FL, 1986, 281–310.

[86] N. Ahuja and B.J. Schachter, *Pattern Models*, Wiley, New York, 1983.

Compound Gauss-Markov Random Fields for Parallel Image Processing

Fure-Ching Jeng[†], John W. Woods[‡], and Sanjeev Rastogi[‡]

[†]Bell Communications Research
Morristown, New Jersey

[‡]Department of Electrical, Computer, and Systems Engineering
Rensselaer Polytechnic Institute
Troy, New York

1 Introduction

Due to limited computing power, linear-shift invariant (LSI) models have been generally used for image estimation and restoration. However, an LSI model is not capable of fully describing the complex nature of an image. The major problem is that the resulting LSI image estimate suffers from over-smoothing of the edges which are quite important to the human visual system (HVS). Partly as a result, various space-variant and adaptive filters have been proposed [1, 2, 3, 4] wherein each pixel (or block of pixels) is filtered by a switched linear filter where the switching is governed by a visibility function. This visibility function can be based on either a local variance estimate [1] or an average estimate of the isotropic gradient [2] of the image. A more complex visibility function which depends upon both a mix of the directional components and an isotropic high frequency component was used in [3]. One of the problems with all these approaches is that the filter switching is rather *ad hoc* and cannot be justified on a theoretical basis. It would be better if the filter transitions could be related to an image model.

One way to embed a mathematical approach into the above adaptive methods is to have the various models' switching controlled by a hidden random field. Toward this end, a class of models called compound Gauss-Markov models has been proposed [5, 6]. A compound random field consists

ISBN 0-12-170608-7

of two levels, called upper and lower. The upper level random field is the observed image, which is composed of several LSI submodels representing a variety of local characteristics of images, e.g., edge orientation or texture. The lower or hidden level is a random field with a finite and discrete range space whose purpose is to govern transitions between the observed LSI submodels.

The doubly stochastic Gaussian (DSG) field [5] possesses a conditional Gaussian, autoregressive (AR) upper or observed level, whose model coefficients have a causal support. They are switched by a hidden (hence lower level) 2-D causal Markov chain to generate the required local edge structure. The DSG random field is a Markov random field but the observed upper level component, by itself, is not, i.e., the image data component of the DSG, by itself, is not Markov. These models are discussed in [11].

Geman and Geman [6], on the other hand, used a noncausal neighborhood system in their image model. In their nomenclature, our lower level random field becomes their line process which was defined on an interpixel grid system. Their upper level random field is a noncausal conditional Markov random field also, but with a finite discrete range, i.e., a noncausal 2-D Markov chain. Since they modeled images on a finite range space, they thereby excluded Gaussian models, which are very widely used in image estimation and restoration. Generalizing their model, we have introduced an extension of the Gauss-Markov random field [7, 8]. Our compound Gauss-Markov (CGM) model [9] has continuous grey levels as an upper level observations model but retains the Gemans' lower level line process (field) as the structural model.

In this chapter, CGM models will be used for image restoration. The MAP estimator and MMSE estimator will be developed by means of simulated annealing (stochastic relaxation) and the Gibbs sampler respectively. The MAP estimator was first introduced by Geman and Geman [6] where they presented a convergence proof for this iterative method. Their method of proof is confined to images that are modeled on a finite range space which, as mentioned above, unfortunately excludes the case of CGM models. When applying simulated annealing to our models, a proof for the convergence of simulated annealing to the MAP estimate was thus lacking. We have obtained such a proof for a broad subclass of CGM models [10]. This chapter will concentrate on the algorithmic aspects and experimental results of applying parallel simulated annealing and the Gibbs sampler with CGM models to image restoration problems. Those interested in the more theoretical aspects of this work are referred to [10] and [11].

It is well known that simulated annealing and the Gibbs sampler are computationally demanding methods. A sequential implementation which will take several CPU hours is not very appealing in practice. Fortunately, due to the Markov property of our image model, computing the *a posteriori* probability by simulated annealing and the Gibbs sampler only involves neighboring pixels which are contained in a relatively small region. Hence the MAP and MMSE estimators can be implemented in a parallel fashion to reduce processing time [16]. The essence of the parallel implementation is to partition an image into several disjoint regions such that any two pixels in a given disjoint region are not neighboring each other. Hence, we can compute the *a posteriori* probability for all pixels in the same region simultaneously. Therefore, each processor in a massively parallel computer is assigned to each pixel and all processors belonging to the same region perform the same computation simultaneously. Hence, the employment of a massively parallel computer can cut down processing time so that simulated annealing becomes practically usable.

A massively parallel computer can be classified into either single instruction multiple datastream (SIMD) or multiple instruction multiple datastream (MIMD). All processors in an SIMD machine perform the same instruction while each processor in an MIMD machine can execute a different instruction. Although an MIMD machine is more flexible, it is more difficult to implement. For example, in [6], an MIMD machine was proposed for parallel implementation of simulated annealing with the constraint that two neighboring processors can not update the *a posteriori* probability simultaneously. As discussed previously, the operation on each pixel is the same for all pixels, so an SIMD machine is more suitable for the parallel implementation of simulated annealing [16]. In this chapter, an experiment on the parallel implementation of simulated annealing on an SIMD machine—the DAP 510 computer—will be described in detail.

One important issue which will not be addressed much in this chapter is parameter identification. The parameter identification problem can be classified into two categories: *supervised* and *unsupervised.* In the supervised problem, a training (or original) image is available and the parameters can be obtained directly from this training image. This approach makes sense only if the given image in the problem possesses the same statistics as the training image. In the other category, *unsupervised*, we must obtain the model parameters from the degraded image itself. The solution to this unsupervised problem usually leads to an EM (expectation and maximization) type of method [14], which to the authors' knowledge, has not yet been extended to CGM models. Thus in this chapter, we use a simple and somewhat *ad hoc* parameter estimation approach to be given later in Section 5.

Here we start out by first defining the CGM model. Then the estimation problems are formulated and the parallel estimators are described. Next, a section devoted to the parallel implementation of simulated annealing on the DAP 510 computer will be presented. Finally we give simulation results that provide comparisons between the performance of compound Gaussian models with simulated annealing versus simple Gaussian models with conventional Wiener filtering. The intent here will be to see which algorithms result in subjectively pleasing images. We then draw conclusions about which mathematical models best match the human visual system (HVS).

2 Compound Markov Random Fields

As described in the Introduction, a simple shift-invariant Gaussian AR model can often lead to oversmoothing of image data. In this section, we will present a compound Markov model, the noncausal CGM image model, which can solve this problem.

The Gemans' compound model can be generalized by incorporating a conditional Gauss-Markov observational model [7, 8] together with their line field, as indicated following. A simple Gauss-Markov (GM) model can be described as follows:

$$s(m,n) = \sum_{kl \in \Re} c_{kl} s(m-k, n-l) + w(m,n), \tag{1}$$

where $w(m,n)$ is a Gaussian random field satisfying the following covariance constraint:

$$E[w(m,n)w(k,l)] = \begin{cases} \sigma_w^2 & \text{if } (m,n) = (k,l) \\ -c_{m-k,n-l}\sigma_w^2 & \text{if } (m-k, n-l) \in \Re \\ 0 & \text{otherwise,} \end{cases} \tag{2}$$

and $\Re$ is shown in Fig. 1. A random field described by the above equation is a Markov random field with a neighborhood support $\Re$ [8]. By the equivalence of Markov random fields and Gibbs distributions, we can write the joint probability density function (pdf) $p(S)$ as a Gibbsian pdf:

$$p(S) = \frac{1}{Z_1} e^{-U_s(S)}$$

and

$$U_s(S) \triangleq \sum_{c_s \in C_s} V_{c_s}(S), \tag{3}$$

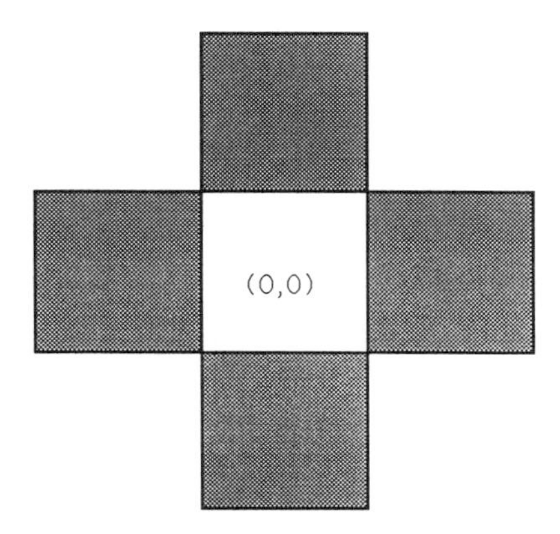

Figure 1: Coefficient support region $\Re$ of the first order model.

where c_s is a clique, C_s denotes the clique system for the given Markov neighborhood system, and Z_1 is a normalizing constant which is functionally independent of S, where the matrix

$$S \triangleq \begin{pmatrix} s_{11} & s_{12} & \cdots & s_{1N} \\ s_{21} & s_{22} & \cdots & s_{2N} \\ \vdots & \vdots & \vdots & \vdots \\ s_{N1} & s_{N2} & \cdots & s_{NN} \end{pmatrix}.$$

Each $V_{c_s}(S)$ is a function on the sample space with the property that $V_{c_s}(S)$ involves only those $s(m,n)$ of S for which $(m,n) \in c_s$. In the above GM model, we have

$$V_{c_s}(S) = \begin{cases} \frac{s^2(m,n)}{2\sigma_w^2} & \\ -\frac{c_{m-k,n-l}s(m,n)s(k,l)}{\sigma_w^2} & \text{if } (m-k, n-l) \in \Re. \end{cases}$$

A compound Gauss-Markov or CGM model consists of several conditionally Gauss-Markov submodels with an underlying structure or line field. Here

$$s(m,n) = \sum_{kl \in \Re} c_{kl}^{\underline{\ell}(m,n)} s(m-k, n-l) + w^{\underline{\ell}(m,n)}(m,n), \tag{4}$$

where $w^{\underline{\ell}(m,n)}(m,n)$ is a conditionally Gaussian noise whose variance controlled by the $\underline{l}(m,n)$ and $\underline{\ell}(m,n)$ is a vector which consists of the four nearest neighbors of the line field surrounding the pixel $s(m,n)$, with $\ell(m,n)$ denoting the Gemans' line field. This line field takes on two values indicating whether a *bond* is broken or not. If a bond is broken between adjacent pixels, then there is weak covariance between them, otherwise a strong covariance exists.

The joint mixed probability density function (mpdf) of S, L is the Gibbs distribution

$$p(S,L) = \frac{1}{Z_2} e^{-\frac{U_s(S|L)+U_l(L)}{T}}, \tag{5}$$

where the constant Z_2 is independent of the matrices S, L. The parameter T is a so-called temperature parameter, U_s is defined in (3), and U_l is of the form

$$U_l(L) = \sum_{c_l \in C_l} V_{c_l}(L),$$

where the matrix

$$L \triangleq \begin{pmatrix} l_{11} & l_{12} & \cdots & l_{1N'} \\ l_{21} & l_{22} & \cdots & l_{2N'} \\ \vdots & \vdots & \vdots & \vdots \\ l_{N'1} & l_{N'2} & \cdots & l_{N'N'} \end{pmatrix}$$

with N'^2 is the total number of points in the line field over the finite observation region.

The clique system c_l used in this chapter is the same system used in [6] and is shown in Fig. 2.

For a CGM model to have a valid conditional Markov random field given a realization of the line field L, there are covariance constraints on $w^{\underline{\ell}(m,n)}(m,n)$:

$$E[w^{\underline{\ell}(m,n)}(m,n)w^{\underline{\ell}(k,l)}(k,l)]$$

$$= \begin{cases} \sigma^2_{w_{\underline{\ell}(m,n)}(m,n)} & \text{if } (m,n) = (k,l) \\ -c^{\underline{\ell}(k,l)}_{m-k,n-l}\sigma^2_{w_{\underline{\ell}(m,n)}(m,n)} & \text{if } (m-k,n-l) \in \Re \\ 0 & \text{otherwise.} \end{cases} \tag{6}$$

By commutativity of the covariance of two random variables, we have the following constraint on the submodel coefficients:

$$c^{\underline{\ell}(k,l)}_{m-k,n-l}\sigma^2_{w_{\underline{\ell}(m,n)}(m,n)} = c^{\underline{\ell}(m,n)}_{k-m,l-n}\sigma^2_{w_{\underline{\ell}(k,l)}(k,l)} \quad \text{if } (m-k,n-l) \in \Re \tag{7}$$

This constraint reduces the total number of free parameters that can be determined by parameter identification. Consequently, a classical unconstrained least-squares approach is not appropriate here.

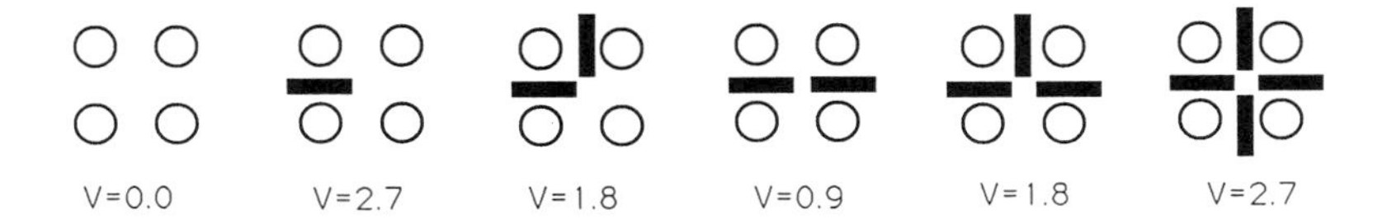

Figure 2: The clique system c_l.

Since the CGM model has a Gibbs probability distribution, it is a Markov random field. Depending on the functional $U_s(S|L)$, the hidden random field L, by itself, may or may not be Markov, as can be seen from the following expression for the probability function $P(L)$,

$$\begin{aligned} P(L) &= \frac{1}{Z_2} \int_S e^{-\frac{U_s(S|L)+U_l(L)}{T}} dS, \\ &= \frac{e^{-\frac{U_l(L)}{T}}}{Z_2} \int_S e^{-\frac{U_s(S|L)}{T}} dS. \end{aligned}$$

If $\int_S e^{-\frac{U_s(S|L)}{T}} dS$ has the form of a Gibbs distribution for L, then L is a Markov random field, otherwise it is not. In general, $\int_S e^{-\frac{U_s(S|L)}{T}} dS$ will not be a Gibbs distribution for L.

3 Problem Formulations and Estimator Solutions

We wish to estimate the original unblurred image $s(m,n)$ from the degraded observations $r(m,n)$:

$$r(m,n) = \sum_{k,l \in \Re_h} h(k,l)s(m-k,n-l) + v(m,n), \tag{8}$$

where $h(k,l)$ is the blur point spread function (psf) with support region $\Re_h$ and $v(m,n)$ is a zero-mean, white noise independent of $s(m,n)$ and having variance σ_v^2. Without loss of generality, we assume that the image size is N $\times$ N. We discuss the MAP estimator in the next two sections and the MMSE estimator in Sections 3.3 and 3.4.

3.1 MAP Estimate: Problem Formulation

The *a posteriori* joint MPDF for S and L given R can be decomposed as

$$p(S, L \mid R) = \frac{p(R|S, L)p(S, L)}{p(R)} = \frac{p(R|S)p(S \mid L)P(L)}{p(R)} \tag{9}$$

where the matrices S, L have been previously defined and R is defined as

$$R \triangleq \begin{pmatrix} r_{11} & r_{12} & \cdots & r_{1N} \\ r_{21} & r_{22} & \cdots & r_{2N} \\ \vdots & \vdots & \vdots & \vdots \\ r_{N1} & r_{N2} & \cdots & r_{NN} \end{pmatrix}.$$

Our goal is to find the joint matrix estimate $\hat{S}$ and $\hat{L}$ such that

$$p(\hat{S}, \hat{L} \mid R) = \max_{S,L} p(S, L \mid R). \tag{10}$$

It can be seen that the MAP criteria lead to the need for a frame store to hold the entire image during processing. The criteria also indirectly lead to iterative algorithms requiring massive parallelism for their efficient solution.

3.2 MAP Estimator: Simulated Annealing Approach

Since the criterion function $p(S, L \mid R)$ is nonlinear, it is extremely difficult to find the optimum solution $\hat{S}, \hat{L}$ by any conventional method. In [6], a relaxation technique, called stochastic relaxation or simulated annealing, is developed to search for MAP estimates from degraded observations. In the following, we will present a version of simulated annealing appropriate for CGM models.

Assume that $s(m, n)$ is a CGM random field and the observations $r(m, n)$ satisfy (8). Some notation is first defined for the sake of conciseness:

$$S_{(m,n)} \triangleq \{s(k, l) : \ \forall (k, l) \neq (m, n)\},$$

$$L_{(m,n)} \triangleq \{l(k, l) : \ \forall (k, l) \neq (m, n)\}.$$

The conditional *a posteriori* density function $\pi^s(m, n)$ for $s(m, n)$ given $S_{(m,n)}$, L and R is

$$\begin{aligned}
\pi^{s}(m,n) &\triangleq p[s(m,n) \mid S_{(m,n)}, L, R] = \frac{p(S,L,R)}{\int p(S,L,R)ds(m,n)} \\
&= \frac{p(R|S)p(S|L)P(L)}{\int p(R|S)p(S|L)P(L)ds(m,n)} \\
&= \frac{1}{Z_3} exp\left(-\frac{(s(m,n) - \sum_{kl} c_{kl}^{l(m,n)} s(m-k,n-l))^2}{2T\sigma^2_{w_{\underline{l}(m,n)}}} \right. \\
&\quad \left. - \frac{\sum_{ij\in\Re_h}(r(m+i,n+j) - \sum_{kl\in\Re_h} h_{kl}s(m+i-k,n+j-l))^2}{2T\sigma_v^2} \right), \qquad (11)
\end{aligned}$$

where Z_3 is a normalizing constant. The conditional *a posteriori* probability function $\pi^l(m,n)$ for $l(m,n)$ given $L_{(m,n)}$, S and R is

$$\begin{aligned}
\pi^{l}(m,n) &\triangleq P[l(m,n) \mid L_{(m,n)}, S, R] = \frac{p(S,L,R)}{\sum_{l(m,n)} p(S,L,R)} \\
&= \frac{p(S,L)p(R|S)}{\sum_{l(m,n)} p(S,L)p(R|S)} \\
&= \frac{p(S,L)}{\sum_{l(m,n)} p(S,L)} \\
&= \frac{p(s(i,j),s(i',j') \mid L)p(S_{(i,j),(i',j')}|s(i,j),s(i',j'),L)P(L)}{\sum_{l(m,n)} p(s(i,j),s(i',j') \mid L)p(S_{(i,j),(i',j')}|s(i,j),s(i',j'),L)P(L)} \\
&= \frac{p(s(i,j),s(i',j') \mid L)P(L)}{\sum_{l(m,n)} p(s(i,j),s(i',j') \mid L)P(L)}, \qquad (12)
\end{aligned}$$

where $l(m,n)$ is located between the two neighboring pixels $s(i,j)$, $s(i',j')$ and

$$S_{(i,j),(i',j')} \triangleq \{s(k,l) : \forall (k,l) \neq (i,j) or (i',j')\}.$$

The dependence of $p(s(i,j), s(i',j') \mid L)P(L)$ on $l(m,n)$ can be determined from Eq. (5). In fact, $\pi^l(m,n)$ can be explicitly written as follows:

$$\pi^l(m,n) = \frac{1}{Z_4} exp \left(- \frac{\frac{s^2(i,j)}{2\sigma^2_{w_{\underline{l}(i,j)}}} - \frac{c^{\underline{l}(i',j')}_{i-i',j-j'} s(i,j)s(i',j')}{\sigma^2_{w_{\underline{l}(i,j)}}} + \frac{s^2(i',j')}{2\sigma^2_{w_{\underline{l}(i',j')}}} + \sum_{(m,n)\in c_l} V_{c_l}(L)}{T} \right),$$

where Z_4 is a normalizing constant.

The simulated annealing method can be implemented in a sequential or parallel manner [12]. We describe parallel simulated annealing here. We keep the temperature constant for each sweep of the image and reduce the temperature only after the complete sweep (or iteration) on the image. Since $\pi^s(m,n)$ and $\pi^l(m,n)$ both depend on $T(t)$, we henceforth denote them by $\pi^s_t(m,n)$ and $\pi^l_t(m,n)$ respectively.

3.2.1 Parallel Simulated Annealing Algorithm

A general structure of parallel simulated annealing algorithm is presented here. An implementation on a particular machine, the DAP 510 computer, will be described in the next section. As pointed out in [6, 15], stochastic relaxation converges very slowly to the MAP estimate because of the logarithmic temperature cooling schedule. Now, in [6], it was suggested that simulated annealing can be implemented on an asynchronous parallel machine to speed up the processing. However, an asynchronous machine may be rather difficult to implement and use compared to a single-instruction multiple-data (SIMD) machine. Here we present a coding scheme for implementation of simulated annealing on an SIMD machine as has been suggested by Murray, Kashko and Buxton in [16]. A parallel segmentation algorithm using relaxation and its mapping onto array processors is also discussed in [17]. The total speed-up depends on the size of the neighborhood of the Markov random field. In theory, we can have $2N^2/k$ speed-up if we have enough processors where N^2 is the total number of pixels in the image and k is the size of the neighborhood of the Markov random field. We describe this maximally parallel "coding scheme" next.

Instead of updating one pixel at each iteration, we can partition the entire image into disjoint regions (called coding regions) such that pixels which belong to the same disjoint region are conditionally independent given the data of all the other disjoint regions. The total number of coding regions depends on the size of the neighborhood support of the image model, hence the speed-up factor given above. Since the pixels in the same coding region are conditionally independent, we can update them simultaneously at each iteration by using the previous iteration result in the simulated annealing procedure. A parallel version of our simulated annealing procedure can thus be described as:

1. Set $t = 0$ and assign an initial configuration denoted as S_{-1}, L_{-1}, an initial temperature $T(0) = 1$ and the disjoint coding regions for S denoted as $\Gamma_1^s, \Gamma_2^s, ..., \Gamma_{\alpha_1}^s$ and the disjoint coding regions for L denoted as $\Gamma_1^l, \Gamma_2^l, ..., \Gamma_{\alpha_2}^l$. Here, we assume that α_1 and α_2 are the total number of coding regions for S and L respectively.

2. The evolution $L_{t-1} \longrightarrow L_t$ of the line field can be obtained by sampling the new value of the coding region $\Gamma_{i_t}^l$ of the structure image L_t based on the conditional probability mass function $\pi_t^l(m,n)$ $\forall(m,n) \in \Gamma_{i_t}^l$, where $\Gamma_{i_t}^l$ is the coding region visited at time t.

3. Set $t = t + 1$. Go back to Step 2 until the whole line field is finished.

4. The evolution $S_{t-1} \longrightarrow S_t$ of the image can be obtained by sampling the new value of the coding region $\Gamma_{i_t}^s$ of the image S_t based on the conditional density function $\pi_t^s(m,n)$ $\forall(m,n) \in \Gamma_{i_t}^s$. where $\Gamma_{i_t}^s$ is the coding region visited at time t.

5. Set $t = t + 1$ and assign a new value to $T(t+1)$ after each complete sweep of the image and go forward to the next step, otherwise keep the temperature unchanged and go back to the previous step.

6. Go to step 2 until $t > t_f$, where t_f is a specified integer.

The following theorem from [18] guarantees that this parallel processing version of simulated annealing will converge to the MAP estimate under the stated conditions.

Theorem 1 *If the following conditions are satisfied:*
(a) $\sum_{kl} \mid c_{kl}^{l} \mid = \rho < 1 \qquad \forall \underline{l}$,
(b) $T(t) \longrightarrow 0$ *as* $t \longrightarrow \infty$, *and*
(c) $T(t) \geq C/log(1 + k(t))$,
then for any starting configuration S_{-1}, L_{-1}, *we have*

$$p(S_t, L_t \mid S_{-1}, L_{-1}, R) \longrightarrow \pi_0(S, L), \text{ as } t \longrightarrow \infty. \tag{13}$$

3.3 MMSE Estimate: Problem Formulation

The object is to find the minimum mean-squared error (MMSE) estimate of S given the observations R, i. e., $\hat{S}$ such that

$$E[\|\hat{S} - S\|^2 | R] = \min_{S'} E[\|S' - S\|^2 | R],$$

where

$$\|S\|^2 \triangleq \sum_{m,n} |s(m,n)|^2.$$

It is well known that the optimal MSE estimate is the conditional mean, i.e.,

$$\hat{S} = E[S|R].$$

The conditional mean can be obtained easily if S is a simple AR model without the hidden line field L. The optimal estimator is the Wiener filter. Unfortunately, due to the complex nature of the compound field constituting the CGM model, the analytical solution for the condition mean is almost impossible to find. Although a closed form solution for optimal estimator cannot be obtained, nevertheless a simulation scheme can be employed to realize a random field with the *a posteriori* probability density based on the given observations. Then the sample average from this simulation will approximate the desired conditional mean.

3.4 MMSE Estimator: Gibbs Sampler Approach

The classical Monte-Carlo simulation [20] has been used for evaluating the mean of a function of a vector-valued random variable. It can be used to obtain the MMSE estimate or conditional mean, by sampling the *a posteriori* pdf. The sample mean converges to the ensemble mean due to the *law of large numbers* from *ergodic theory.* In [6], Geman and Geman proposed an algorithm called the Gibbs sampler to simulate a conditional random field. The Gibbs sampler is the same as simulated annealing except that the temperature T is held constant and set $T = 1$ during all the iterations. It can be shown that probability function of the random field generated by the Gibbs sampler is the *a posteriori* probability $P(S, L|R)$ [6, 18]. By averaging all sample values for each pixel separately, the approximate conditional mean can be obtained due to the *law of large numbers.*

The parallel version of the Gibbs sampler can thus be described as:

1. Set $t = 0$ and assign an initial configuration denoted as S_{-1}, L_{-1}, an initial temperature $T(0) = 1$ and the disjoint coding regions for S denoted as $\Gamma_1^s, \Gamma_2^s, \ldots, \Gamma_{\alpha_1}^s$ and the disjoint coding regions for L

denoted as $\Gamma^l_1, \Gamma^l_2, ..., \Gamma^l_{\alpha_2}$. Here, we assume that α_1 and α_2 are the total number of coding regions for S and L respectively.

2. The evolution $L_{t-1} \longrightarrow L_t$ of the line field can be obtained by sampling the new value of the coding region $\Gamma^l_{i_t}$ of the structure image L_t based on the conditional probability mass function $\pi^l_t(m,n)\ \forall(m,n) \in \Gamma^l_{i_t}$, where $\Gamma^l_{i_t}$ is the coding region visited at time t.

3. Set $t = t + 1$. Go back to Step 2 until the whole line field is finished.

4. The evolution $S_{t-1} \longrightarrow S_t$ of the image can be obtained by sampling the new value of the coding region $\Gamma^s_{i_t}$ of the image S_t based on the conditional density function $\pi^s_t(m,n)\ \forall(m,n) \in \Gamma^s_{i_t}$. where $\Gamma^s_{i_t}$ is the coding region visited at time t.

5. Set $t = t + 1$ after each complete sweep of the image and go to step 2 until $t > t_f$, where t_f is a specified integer.

6. Obtain an approximation to the conditional mean $\hat{s}(m,n)$ by computing the sample average $\frac{1}{t}\sum_i s_i(m,n)$.

4 Implementation of Parallel Simulated Annealing on the DAP 510 Machine

In this section, we will describe our implementation of the parallel simulated annealing (SA) algorithm on the DAP 510 machine.

4.1 Massively Parallel Computers

There are a number of fine grain, massively parallel computers that are currently available for applications in areas such as image processing, graphics, and computer-aided design. Applications in these areas involve manipulation of large volumes of data, and algorithms in which parallelism is often inherent. This is due to the fact that although a typical image has large dimensions, many algorithms involve only local operations on each pixel, which can be performed by using a window-based approach. Such algorithms are especially suited to massive parallelism, since processors assigned to individual pixels can execute the same instructions simultaneously. This is known as the Single Instruction Multiple Datastream (SIMD) principle (Fig. 3).

In these parallel machines, the processors (known as "processing elements" or PEs) are arranged in a matrix. In most machines, each PE

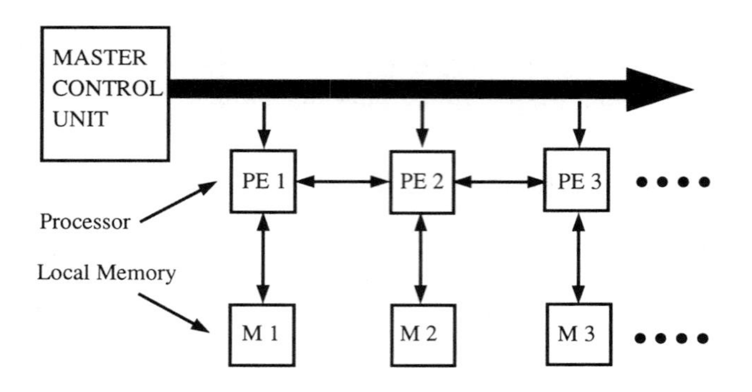

Figure 3: A Single Instruction Multiple Datastream Machine

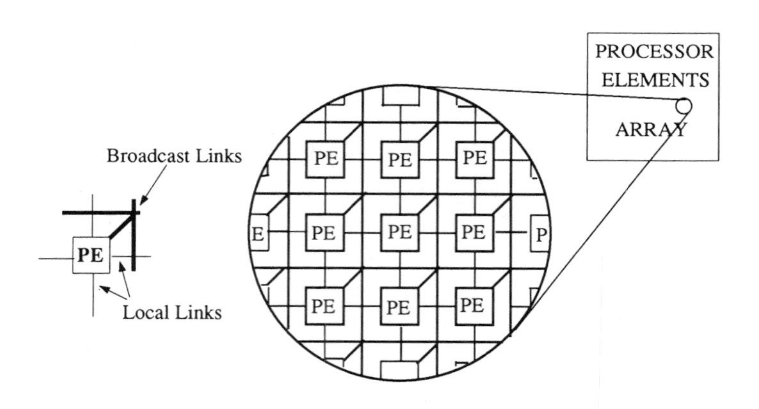

Figure 4: Nearest Four Neighbor Connected Distributed Processor

is provided with connections to at least its four nearest neighbors, which yields a very high level of connectivity. In addition, a bus system often connects all the PEs in each row and all the PEs in each column. These row and column data highways, present in the DAP architecture, provide a rapid instruction/data broadcasting or fetching capability. Each PE is also connected to its own local memory, of limited size. The processor array is controlled by a master control unit (MCU), which acts like a conventional CPU except that it does not itself execute all the instructions. Parallel instructions are only decoded by the MCU and then broadcast to the entire array of processors, where they are executed by all the PEs simultaneously, each operating on the data in its own local memory (Fig. 4).

The parallel instruction codes are generally written in a high level parallel language which can manipulate vectors and matrices in the same way as conventional languages handle scalars. The code normally resides in the code memory of the MCU, and the machine may be accessed through a host workstation.

4.2 DAP 510 Architecture

The DAP 510 (distributed array processor) is a massively parallel computer located in the Center for Image Processing Research at Rensselaer. In the DAP 510, the PE array is square and of size 32 x 32, or 1024 PEs total. Each PE can communicate directly with its north, east, south and west neighbor, and with any other PE through the aforementioned row and column highways. Also, the last PE in a row or column is directly connected to the first PE in that row or column. The local memory size for each PE is four kilobytes, and the operating clock rate is 10MHz. Connections are provided for fast I/O up to 50MB/sec between the internal bus and host peripherals. Thus intermediate images can also be displayed on a monitor in real time during the iterative process, thus implementing the concept of scientific visualization. Figure 5 gives a pictorial representation of the DAP architecture. The DAP programming environment is FORTRAN-PLUS which is a parallel extension to conventional FORTRAN IV. The MCU of the DAP can be accessed through a Sun or Vax workstation which acts as the host machine.

4.3 Parallel Simulated Annealing Algorithm

The conditional independence property of the Markov random field can be exploited for the parallel implementation of the SA algorithm. Equations (11) and (12) imply that the updating of pixel $s(m, n)$ requires only the pixels within a neighborhood which is twice the size of blur support region

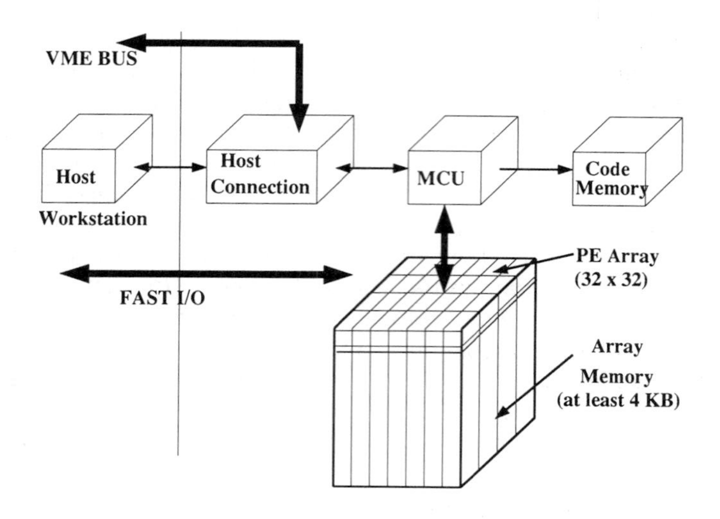

Figure 5: DAP 510 Architecture

$\Re_h$. Also, for an update of the line field site $l(m,n)$, only its four nearest line sites are required. The whole image can thus be divided into conditionally independent and disjoint *coding regions* for their parallel update. These coding regions are then mapped onto the individual processors of the DAP. For synchronous operation of all processors, the temperature is kept constant during one update of the field. It is then reduced at the fastest rate possible, before starting the next iteration. For the Gaussian case, a lower bound on the annealing rate, i.e., the permitted rate of temperature decrease, is given by

$$T(k) \geq \frac{C}{ln(1+k)}. \tag{14}$$

The parallel SA algorithm to obtain the MAP image estimate is given below, assuming the conditional independence of coding blocks.

- DIVIDE the image into coding blocks matching the size of the PE array.
- CRINKLE MAP the blocks onto individual processor elements.
- SET the starting temperature T_0 at a high value.

- **begin** {DO IN PARALLEL} for each PE:
- In a raster-scan fashion, repeat the following steps, until an entire field update.
- UPDATE the horizontal line sites, using Gibbs sampler and (12).
- UPDATE the vertical line sites similarly.
- COMPUTE the local models for the pixels from the current line field.
- GENERATE new samples for pixels using the Gibbs' sampler and (11).
- **end** {DO IN PARALLEL}.
- INCREASE the iteration count k.
- DECREASE the temperature T_k using (14).
- If NOT CONVERGED, GO BACK to the parallel DO loop.

The algorithm to obtain the MMSE estimate is the same as above except that the temperature is kept constant at $T = 1$, and (11) is used to update the current MMSE estimate, after a new image sample is generated.

4.4 Implementation of Parallel SA Algorithm

Since the image size is larger than the PE array edge size of the DAP, we first must map the image onto the local PE memory in a *crinkled* format (Fig. 6). Thus for a n x n processor machine, and an image of size N x N, each block of size (N/n) x (N/n) is mapped onto an individual processor. Each processor updates a site only within its own block, keeping a constant distance from the sites being updated in its neighboring blocks. Care has to be taken when a processor updates a site near the boundary of its own block. In order to process these pixels, either the PE can obtain needed data through its direct link with an adjacent PE, or else the whole image can be shifted in the proper direction through highly efficient *parallel data transforms.*

Figures 7 and 8 show that full efficiency of the machine can be maintained by selecting a judicious processing schedule for implementing parallel SA. If the block size (K) is greater than 2 x 2, then all processors are kept busy for the line field update. Due to the presence of both horizontal and vertical lines, the entire line field update requires $2K^2$ cycles, a saving of

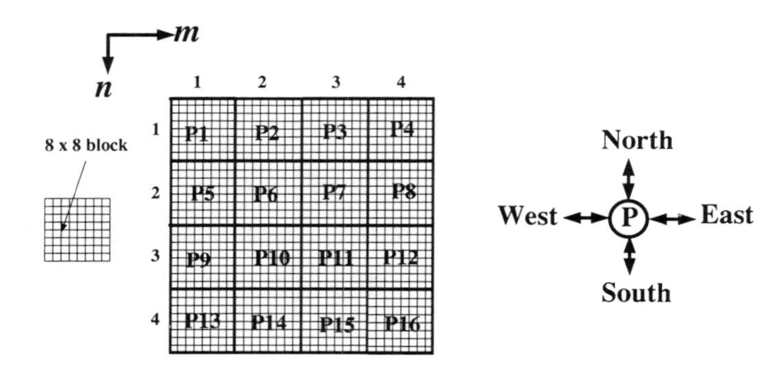

Figure 6: Image Mapping Using Crinkled Format(Image: 32 x 32, Processor: 4 x 4)

$O(2n^2)$. It is also noted, that for a blur support of M x M and $K > M$, the image field update is accomplished in only K^2 cycles.

Parallel SA was implemented on a distributed array processor of size 32 x 32. We considered the monochrome *Lenna* image of size 256 x 256 pixels. The image was artificially blurred and then corrupted by independent and additive white Gaussian noise. For this parallel SA implementation, the parameters for the model, and clique potentials were taken from [6, 10].

4.5 Future Research Issues

To restore images blurred by a PSF of much bigger size, say twice as large as the blurs considered here, alternative strategies will have to be used to preserve the conditional independence of different coding regions. One could process alternate coding regions at the same time. This would reduce the machine efficiency by one half, but it would enable us to process images with a blur support region twice as large. Another alternative is to process a subsampled image, say subsampled by 2×2. The new smaller image would have a blur extent of roughly half the size of the original PSF. This approach may be quite practical since in reality, many images are oversampled.

To further speedup the SA convergence, use of non-Gaussian Markov fields is currently under investigation. As discussed is Szu [19] for deterministic optimization, it is possible to decrease the temperature more rapidly if

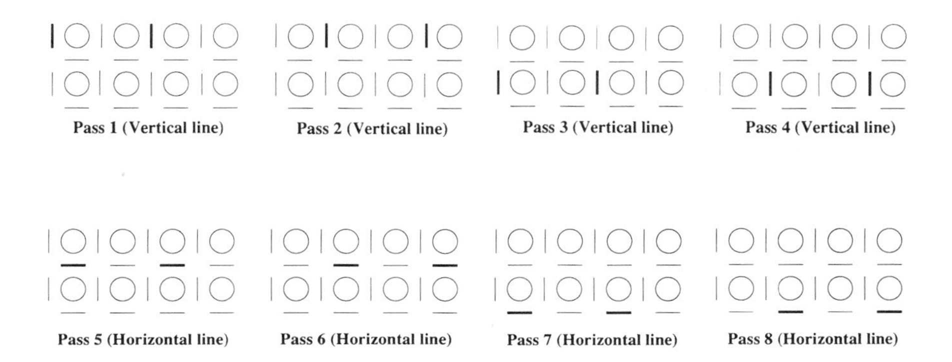

Figure 7: Line Field Parallel Update (Block Size: 2 x 2)

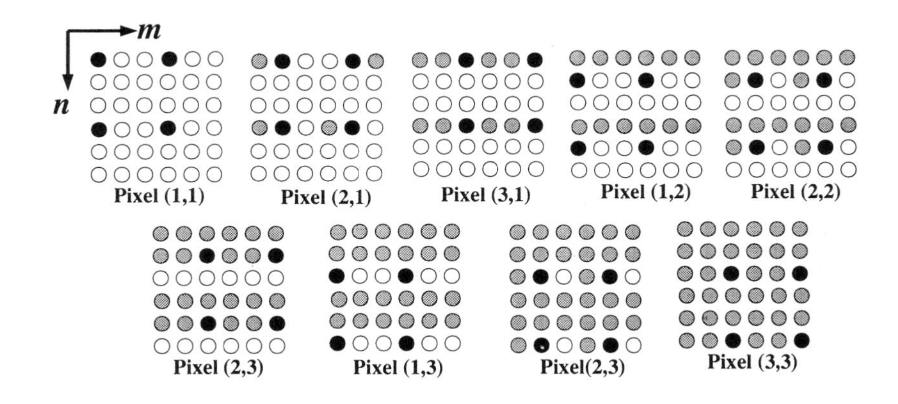

Figure 8: Image Field Parallel Update (Block size: 3 x 3, Blur support: 3 x 3)

the distributions have peaks similar to the Gaussian case but have broader tails. In particular, use of Laplacian distributed interpolation residuals seems quite appropriate for image models.

5 Parameter Identification

In this section we will describe a practical way to obtain parameters which we have employed in our simulations. Various parameters, such as submodel coefficients and transition probabilities, are needed before we can implement the various filtering algorithms. Since these parameters are not often available, we must estimate them from the data. Here we take the *supervised* approach of estimating these parameters from training data that may be prototypical of the expected noisy images. Using four directional edge operators scanning through the training image (the original unblurred image is used here), we can partition this image into six disjoint regions. Each region corresponds to a different submodel which will be used to model similar parts of the image. After obtaining a model configuration, i.e., the hidden structure of the whole image, we can use *maximum likelihood* (ML) identification techniques to estimate the various submodels' coefficients.

Due to the noncausal structure of the CGM model, the ML parameter estimation is quite complicated. The difficulty is twofold. First, ML identification for a noncausal Gauss-Markov model is not the classical least-squares problem [7]. A method called the *coding scheme* has been widely used. However, this method is not efficient in that only 50% of the data is used. Also the result is not unique since many different coding schemes exist. In [7], a consistent estimation method was proposed for the noncausal Gauss-Markov model. This scheme is quite similar to least-square error identification. We adopt this method for parameter identification in the CGM case. Second, due to the covariance constraints on the submodel coefficients (7), the set of coefficients obtained through the above separate submodel identification procedures will be usually not consistent. There are two ways around this difficulty. One of them is to modify the identification results from the least-squares method to make the solutions consistent. Another way is to heuristically assign a set of consistent model coefficients that seems reasonable, much as was done in [6]. Here we use this latter *ad hoc* approach to obtain a consistent set of model coefficients as needed for our simulations. For the line field in our CGM model, we use the same Gibbs distribution as in [6].

6 Experimental Results

In this section, we present the performance of parallel implementation on the DAP 510. A comparison of the processing time between sequential processing and the parallel processing is given. We also we show some restoration results based on the CGM model and compare them with Wiener filtering results. All simulations use first order image models. The blurring function was the 5×5 uniform blur with blurred-signal to noise ratio (BSNR) = 40dB.

In the sequential implementation of SA, the image boundary pixels were not processed, giving rise to a visible distortion near the image boundaries in the restored image. However, for the parallel implementation on the DAP, we assumed the image to be periodic over two-dimensional space. This gave us a more satisfactory results within about five rows or columns of the image boundary.

Table 1 gives performance of the parallel implementation of SA, as compared to the sequential implementation in terms of the improvement in signal-to-noise ratio defined by

$$\text{Improvement (in dB)} = 10 \log_{10} \left(\frac{\| R - S \|^2}{\| \hat{S} - S \|^2} \right),$$

$$\| E \|^2 \triangleq \sum_{m,n \in \mathcal{Z}} e^2(m,n) \tag{15}$$

Computed Image estimate	exclude 4 pixel borders			exclude 15 pixel borders		
	initial	final	gain	initial	final	gain
Sequential MAP (SA)	22.80	17.47	5.33	23.10	16.82	6.28
Parallel MAP (SA)	22.80	16.44	6.36	23.10	16.73	6.37
Parallel MMSE (SA)	22.80	16.83	6.96	23.10	16.11	6.99
Wiener (known LSI model)	22.80	16.69	6.11	23.10	16.96	6.14

Table 1: Mean Square error (in dB) for 5 x 5 uniform blur: BSNR = 40 dB.

Implementation	Machine type	Computational time	
		200 iterations	each iteration
Sequential SA	VAX 11/780	$\simeq$ 20 hr	$\simeq$ 6 min
Parallel SA	DAP 510	12 min 10 sec	3.65 sec

Table 2: Computational time for 5 x 5 uniform blur. Image size : 256 x 256.

Table 2 gives the computational time required for SA algorithm execution. The sequential machine used was a VAX 11/780 with floating point

Figure 9: Original image.

Figure 10: Blurred and Noisy image with BSNR =40 dB.

Figure 11: Wiener filter result for the Gauss-Markov model with BSNR =40 dB.

Figure 12: Parallel MAP estimate for the compound Gauss-Markov model with BSNR =40 dB.

Figure 13: Parallel MMSE estimate for the compound Gauss-Markov model with BSNR =40 dB.

Figure 14: Sequential MAP estimate for the compound Gauss-Markov model with BSNR =40 dB.

accelerator. Although not truly comparable due to use of different machines, it is seen that the time per iteration has been drastically reduced by a factor of about 100.

Figs. 9 and 10 show the original image and the degraded (blurred and noisy) image respectively. A Wiener filtering result based on an AR model is shown in Fig. 11. The parallel MAP and MMSE results based on the CGM model from the same blurred and noisy image are shown in Fig. 12 and Fig. 13 respectively. For comparison purposes, a sequential MAP restoration result is also shown in Fig. 14.

From Figs. 11-14, it is clearly seen that the compound models provide sharper images without the noise amplification that results from LSI Gauss-Markov based estimates. The compound model allows the restoration to proceed with strong noise reduction in fairly flat regions of the image, yet still retain sharp edge restoration. The Wiener filtering result clearly suffers from an inability to deal with this problem. Comparing the parallel MAP and parallel MMSE results shown in Fig. 12 and Fig. 13, we find that any visual difference between these two is too small to see in the printed pictures. However, viewing on the computer monitor indicates that the MMSE estimate is a bit cleaner than the MAP estimate.

7 Conclusions

In this chapter, a compound Gauss-Markov field, i.e. the CGM model, has been employed for modeling images and parallel MAP and MMSE estimators have been developed for image restoration. The nice feature of compound random fields is that the upper level field can provide a description of the local characteristics for each pixel while the lower level field provides the mechanism for the model transition to represent local structure. Therefore, compound models can more faithfully reflect this structure, which is apparent in real images. We would then expect that processing results based on compound models can be much better. In fact, our simulation results have shown that compound Gauss-Markov models do provide much improved visual results by preserving edge sharpness with much less noise in the flatter regions of the image.

Due to the complexities of the compound Gauss-Markov models, it is in general hard to find an analytic solution for MAP and MMSE estimates. Appropriate versions of simulated annealing and the Gibbs sampler for the CGM model were developed and applied to obtain these estimates. Due to the extensive computation of simulated annealing and the Gibbs sampler, a "coding" version for parallel implementation was developed to speed up the processing. Parallel implementation of simulated annealing and the Gibbs sampler on a massively parallel computer, the DAP 510 machine, has been conducted. A processing time speedup by two orders of magnitude has been obtained while the quality of the resulting image is still maintained.

Acknowledgment

Partial support for this work was provided by the Directorate for Computer and Information Sciences and Engineering of the National Science Foundation under Institutional Infrastructure Grant No. CDA-8805910 and also Grant No. MIP-8703021.

Bibliography

[1] F. C. Jeng and J. W. Woods, "Inhomogeneous Gaussian Image Models for Image Estimation and Restoration," *IEEE Trans. Acoust., Speech, Signal Processing,* vol. 36, pp. 1305-1312, August 1988.

[2] J. F. Abramatic and L. M. Silverman, "Nonlinear Restoration of Noisy Images," *IEEE Trans. Pattern Anal. Machine Intell.*, vol. 4, pp. 141-149, March 1982.

[3] H. E. Knutsson, R. Wilson, and G. H. Granlund, "Anisotropic Nonstationary Image Estimation and Its Applications: Part I-Restoration of Noisy Images," *IEEE Trans. Commun.*, vol. 31, pp. 388-397, March 1983.

[4] S. A. Rajala and R. J. P. DeFigueiredo, "Adaptive Nonlinear Image Restoration by a Modified Kalman Filtering Approach," *IEEE Trans. Acoust., Speech, and Signal Process.*, vol. 29, pp. 1033-1042, October 1981.

[5] J. W. Woods, S. Dravida, and R. Mediavilla, "Image Estimation Using Doubly Stochastic Gaussian Random Field Models," *IEEE Trans. Pattern Anal. and Machine Intell.,* vol. 9, pp.245-253, March 1987.

[6] S. Geman and D. Geman, "Stochastic Relaxation, Gibbs Distributions, and the Bayesian Restoration of Images," *IEEE Trans. Pattern Analysis and Machine Intelligence,* vol. 6, pp. 721-741, Nov. 1984.

[7] R. L. Kashyap and R. Chellappa, "Estimation and Choice of Neighbors in Spatial-Interaction Models of Images," *IEEE Trans. Inform. Theory,* vol. 29, pp. 60-72, Jan. 1983.

[8] J. W. Woods, "Two-dimensional Discrete Markovian Fields," *IEEE Trans. Inform. Theory,* vol. 18, pp. 232-240, March 1972.

[9] F.C. Jeng and J. W. Woods, "Image Estimation by Stochastic Relaxation in the Compound Gaussian Case," *Proceedings ICASSP 1988,* pp. 1016-1019, New York, NY, April 1988.

[10] F. C. Jeng and J. W. Woods, "Simulated Annealing in Compound Gaussian Random Fields," *IEEE Trans. Inform. Theory,* vol. 36, pp. 94-107, Jan. 1990.

[11] F.C. Jeng and J. W. Woods, "Compound Gauss-Markov Random Fields for Image Estimation," *IEEE Trans. Signal Processing,* vol. 39, pp. 683-691, March 1991.

[12] F.C. Jeng and J. W. Woods, "Compound Gauss-Markov Random Fields for Image Processing," Chapter 5 in *Image Restoration,* Editor A. K. Katsaggelos, Springer-Verlag, Berlin, 1991, pp. 89–108.

[13] J. Besag, "On the Statistical Analysis of Dirty Pictures," *Journal of Royal Statistics Society B,* vol. 48, pp. 259-302, 1986.

[14] A. P. Dempster, N. M. Laird, and D. B. Rubin, "Maximum Likelihood from Incomplete Data via the EM Algorithm," *Annals of the Royal Statistical Society,* pp. 1-38, 1978.

[15] S. Kirkpatrick, C. D. Gelatt and M. P. Vecchi, "Optimization by Simulated Annealing," *Science,* vol. 220, pp. 671-680, May 1983.

[16] D. W. Murray, A. Kashko and H. Buxton, "A Parallel Approach to the Picture Restoration Algorithm of Geman and Geman on an SIMD Machine," *Image and Vision Comput.,* vol. 4, pp. 133-142, August 1986.

[17] H. Derin and C. S. Won, "A Parallel Image Segmentation Algorithm Using Relaxation with Varying Neighborhoods and its Mapping to Array Processors, " *Computer Vision, Graphics, and Image Process.,* vol. 40, pp. 54-78, 1987.

[18] F. C. Jeng, "Compound Gauss-Markov Random Fields for Image Estimation and Restoration," *Ph.D. Thesis,* Rensselaer Polytechnic Institute, Troy, N.Y., June, 1988.

[19] H. H. Szu, "Non-Convex Optimization", *Proc. SPIE Int. Conf. on Real Time Signal Process.,* pp. 59-65, 1986.

[20] A.J.M. Hammersley and A.D.C. Handscomb, *Monte Carlo Methods*, Methuen Co. Ltd., London, pp. 113-126, 1965.

Stochastic Algorithms for Restricted Image Spaces and Experiments in Deblurring

Donald Geman[†], George Reynolds[‡], and Chengda Yang[§]

[†]Department of Mathematics and Statistics
University of Massachusetts at Amherst
Amherst, Massachusetts

[‡]VI Corporation
Northampton, Massachusetts

[§]Department of Computer Science
University of Arizona
Tuscon, Arizona

1 Introduction

The computational demand of stochastic relaxation algorithms used in image processing, such as the Metropolis algorithm and the Gibbs Sampler, is heavily dependent on both the spatial and brightness resolution. In particular, at a given spatial resolution, i.e., given the number of pixels is fixed, the amount of computation required to (approximately) reach equilibrium in simulations or the extremal states in optimization increases roughly linearly with the number of allowed grey levels and may be quite substantial for a large dynamic range for even moderately sized pixel lattices. For this reason, it is commonplace to introduce computational shortcuts, such as altering the recipe for updating individual pixels by exploring only a portion of the full dynamic range, say those grey levels lying in an interval about the current intensity.

Geman and Reynolds [6] suggested a related modification in order to reduce the computational burden. Let $\Gamma = \{0, 1, ..., L-1\}$ denote the grey scale range and let S denote the pixel lattice. In the Gibbs Sampler (see Section 2), pixels are visited in either a random or cyclic order; each time

ISBN 0-12-170608-7

the grey level value x_s at a pixel $s \in S$ is updated, a sample is taken from the (discrete) probability measure

$$\nu_s(d\gamma) = \Pi(x_s \in d\gamma | x_t, t \neq s),$$

which is the conditional distribution of x_s given the current values x_t at all other pixels $t \neq s$. If all grey level configurations are allowed (i.e., with the usual product configuration space Γ^S), then each probability measure ν_s puts positive mass over the entire dynamic range; in particular, L many weights must be computed in order to generate a random sample from ν_s. In [6], the mass of ν_s is (proportionally) redistributed over the union of six integer intervals whose centers are current grey level x_s, the grey levels at the four neighbors of s, and the grey level of the data at s. The purpose of this truncation is to approximate ν_s by ignoring grey level values with extremely low probabilities. Specifically, in the case of a full dynamic range with $L = 256$ grey levels, and an interval of radius five about each of the resulting six values, then the modified sampling distribution has, on the average, 15 to 25 weights rather than 256. This yields about an order of magnitude decrease in the number of operations performed with no apparent change since the true distribution ν_s places virtually zero mass on the complement of the reduced support.

1.1 Restricted Convergence

However, the theoretical behavior of this algorithm was not determined in [6]; in particular, it is unclear what is the asymptotic distribution of the associated Markov chain on Γ^S. Basically, this question was settled in [50]: by restricting the space of allowable images to certain subsets $\Omega \subset \Gamma^S$, and by modifying the truncation procedure, the reduced support at pixel s may be identified with a one-dimensional "section" at s, i.e., with the set of values $\gamma \in \Gamma$ for which the configuration obtained from $x \in \Omega$ by putting $x_s = \gamma$ (keeping x_t for $t \neq s$) will remain in Ω. The standard relaxation algorithms retain their convergence properties (both for simulation and annealing) *relative to* Ω. The proof follows along the lines in [5], although modifications are necessary to account for the fact that Ω is no longer a product space, which alters the usual setting for Markov random fields. The main features of the proof are presented in Section 4 for the case of annealing with the Gibbs Sampler; the reader is referred to [50] for analogous results for the Metropolis algorithm.

The restricted image spaces are nested relative to an "isolation index," which determines the extent to which an individual pixel $s = (i, j)$ may be separated (in intensity) from *all* its "neighbors"; for convenience, the neighbors are the four nearest ones, namely $(i, j+1), (i, j-1), (i+1, j)$, and $(i-1, j)$. *There is no restriction on the magnitude of the gradients allowed, and hence none on the nature of boundaries.* Indeed, most images encountered in practice satisfy this condition for a relatively small isolation index. The computational requirement of the new algorithms is proportional to the degree of isolation; the standard algorithms correspond to no restrictions, i.e., $\Omega = \Gamma^S$. However, under mild assumptions (for example specifying that no pixel may be separated from all its neighbors by more than five grey levels) the computational cost is, in fact, substantially reduced.

1.2 An Application to Image Deblurring

The problem considered in [6] is image deconvolution or deblurring; it is typical of ill-conditioned inverse problems that frequently arise in low-level computer vision and other fields. The simplest degradation model accounting for blur and noise is the familiar linear one:

$$y_s = (\mathcal{K}x^0)_s + \eta_s, \quad s \in S', \tag{1}$$

where y is the data, S' is the measurement lattice, $x^0 = \{x^0_s, \ s \in S\}$ is the uncorrupted "true" image, η is taken as white noise, and $\mathcal{K}$ is an operator representing the point spread function (PSF). (In general, $S' \subset S$ due to the nature of optical blurring and data acquisition; we must choose S sufficiently large that the "blur mask" falls entirely within S when centered at any pixel within S'.) The work in [6] is extended here to experiments that involve two-dimensional uniform blur and one-dimensional uniform (i.e., motion) blur and that compare the new, "exact" algorithm in Ω with the ad hoc, "union" algorithm in [6]. Fortunately, as the theory predicts, there is little difference, which accounts for the earlier observations that the quality of the reconstructions was virtually uncompromised by reducing the support of ν_s. The experimental results are described in Section 7.

Many image restoration methods employ *a priori* smoothness constraints in addition to those derived from the modeling the image formation process. The purpose is to approximate x^0 by functions that are locally smooth away from visual boundaries, converting the ill-posed inverse problem (1) into a well-formulated (and hopefully well-conditioned) optimization problem. Our method belongs to this category: the restoration $\hat{x}$ is defined as any (global) minimum of a function

$$H(x) = \Phi(x) + \lambda \sum_{s \in S'} (y_s - (\mathcal{K}x)_s)^2, \quad x \in \Omega.$$

Here λ is a positive ("smoothing") parameter, and the "regularization term" $\Phi(x)$ imposes a first, second, or third order continuity condition on x. For example, in the first order case,

$$\Phi(x) = \sum_{<s,t>} \phi(\frac{x_s - x_t}{\Delta}),$$

where Δ is a positive (scale) parameter and the summation is over nearest neighbor pairs. The effect is to emphasize images that are approximately locally constant and whose blurred values resemble the data. In the higher order cases, the corresponding summands involve discrete (linear) approximations to the differences between elements of the gradient vector (second order) and Hessian matrix (third order) at adjacent pixels; see Section 6. These higher order models support reconstructions exhibiting planar and quadric patches in addition to constant regions. In all cases,

$$\phi(u) = \frac{-1}{1 + |u|}. \tag{2}$$

The traditional choice for ϕ in constrained least-squares restoration is the quadratic function $\phi(u) = u^2$. Despite its computational advantages (resulting in a *linear* estimate $\hat{x} = \hat{x}(y)$), we find it ill-suited to image deblurring, mainly because the rapid growth as $u \to \infty$ inhibits the recovery of large intensity gradients. (Similarly, boundary formation may be inhibited for other ϕ's for which $\phi(\infty) = \infty$.) In addition, the fact that $\phi'(0) = 0$ also contributes to "over-smoothing." In contrast, the reconstructions obtained with the function ϕ in Eq. (2) are more accurate in the vicinity of discontinuities due to *concavity* on $[0, \infty)$ and the *finite asymptotic behavior* ($\phi(\infty) = 0$). In fact, such functions ϕ have a *strictly non-interpolating property*, which will be explained in Section 6.1 and which may be especially important if sharp transitions represent much of the information content of the image. Other nonquadratic stabilizers appear in [2], [3], [4], [11], and [16].

The attributes of ϕ provide another advantage. One is able to choose λ as a function of the other parameters (namely, $\Delta, \mathcal{K}$, and σ^2, the noise variance) and a "confidence level" $\alpha \approx 1$, such that H has the following property: if x^0 belongs to a certain "ideal" image class, then, with probability at least α, x^0 is a *coordinate-wise* minimum of H; see Section 6.3. (Arranging for x^0 to be a highly probable *global minimum* of H is a more delicate matter.) Our persistent experience has been that this formula for λ yields results comparable to those obtained by extensive trial-and-error.

Finally, regularization (or "prior") models for which $\phi(\infty) < \infty$ were first introduced in [9] and subsequently applied to image restoration problems in [6] and [10]. We shall refer to them as *implicit discontinuity models* because they support image discontinuities *without* the use of a "line process" ([8]) or other auxiliary device for marking the location of jumps and suspending the smoothness constraints in their vicinity. In fact, it can be shown that the cost functionals H have a "dual representation" in terms of a *coupled* cost functional, which involves a continuous-valued and non-interacting line process [6]. This correspondence is summarized in Section 6.2.

2 Stochastic Relaxation Algorithms

Recall that the total image space is $\Gamma^S = \{x : \quad x_s \in \Gamma \quad \forall s \in S\}$, where $S = \{(i,j) : \ i,j = 1,2,....,N\}$. In some studies, the estimated image $\hat{x}$ is defined as a property of the Gibbs distribution

$$\Pi_\tau(x) = e^{-H(x)/\tau} / \sum_{x \in \Gamma^S} e^{-H(x)/\tau}, \tag{3}$$

where τ is a positive parameter and $H : \Gamma^S \rightarrow R$ is as previously stated. For example, we might define $\hat{x}$ as the *mean* of Π_τ, which is analytically intractable but may be estimated from empirical averages by sampling from Π_τ. (In the Bayesian formulation, the regularization term is associated with the log likelihood of a "prior distribution" on Γ^S and the data term is associated with the log likelihood of the (conditional) distribution of the data, y, given the true image; thus Eq. (3) represents the "posterior distribution.")

Here we investigate another estimator, namely any value x_{min} that minimizes H. (This estimator may be interpreted as the *mode* of Π_τ, but is not a property of the distribution *per se*.) Due to the customary nonconvexity of H and large size of the image space, deterministic algorithms for finding x_{min} are computationally unfeasible, especially when L is large. One alternative is to employ suboptimal deterministic algorithms, for instance graduated nonconvexity [3] and ICM [1]. Another alternative is to use stochastic algorithms to find an approximation to x_{min}; two well-known types are the Metropolis algorithm (originally introduced in [17]) and the Gibbs Sampler (see [7], [8], [12], and the physics literature on the "heat-bath algorithm").

Stochastic relaxation is a Monte Carlo method designed for sampling from probability distributions of the form Π_τ. When the goal is to minimize a function H, stochastic relaxation is combined with annealing by varying the control parameter τ (which corresponds to temperature in a real physical system) during the sampling process; this increasingly concentrates the mass of Π_τ in the vicinity of x_{min}. The simulated annealing algorithm is computationally demanding but has the desirable feature of converging to a global minimum of H. However, this is (by definition) an *asymptotic* statement and is usually impossible to strictly realize in practice. Nonetheless, we have found the estimates so provided more accurate for deblurring than those resulting from other optimization methods.

Let us review the ingredients of stochastic relaxation. One generates a Markov chain $\{X(k),\ k = 0, 1, ...\}$ with state space Γ^S, which represents, in our case, successive restorations. The initial value $X(0)$ is arbitrary in principle, although, in practice, this choice can be quite important. If the transition dynamics are suitably chosen, then the asymptotic distribution of the Markov chain is Π_τ itself in simulation (i.e., fixed temperature sampling) and is the uniform probability measure μ on the set Γ^S_{min} of global minima of H in optimization by annealing; in particular, in the latter case, $Pr(X(k) \in \Gamma^S_{min}) \to 1$ as $k \to \infty$.

At each stage k of the algorithm, $k = 1, 2, ...$, one updates the pending restoration at a single, predetermined pixel a_k. Let $\{a_k : k = 1, 2, ...\}$ denote this site visitation schedule; usually, the sites are simply visited in a raster scan, the only technical requirement being that every site be visited infinitely often, i.e., there be a sequence of integers $0 = K_0 < K_1 < K_2 < ...$ such that, for each $m = 0, 1, , 2...,\ S \subseteq \{a_{K_m+1}, a_{K_m+2}, ..., a_{K_{m+1}}\}$. Let $\{T_1, T_2, ...\}$ denote the corresponding "annealing schedule"; thus, T_k is the temperature to be used at the kth update. Let $\Pi(x) = \Pi_1(x)$, and, for each $k = 1, 2, ...$, let $\Pi^{(k)}(x) = \Pi_{T_k}(x)$ and let $P^{(k)}$ be the probability transition matrix on $\Gamma^S \times \Gamma^S$ defined by

$$P^{(k)}_{xz} = \begin{cases} \Pi^{(k)}(z_{a_k}|x_{(a_k)}) & \text{if } x_{(a_k)} = z_{(a_k)} \\ 0 & \text{otherwise.} \end{cases}$$

In the above $x_{(s)}$ refers to the configuration x with x_s removed, and $\Pi^{(k)}(x_s|x_{(s)}) = \Pi^{(k)}(X_s = x_s|X_t = x_t,\ t \neq s)$, the local conditional distribution of the Markov random field associated with $\Pi^{(k)}$. Finally, $\{X(1), X(2), ...\}$ is the Markov chain with transitions $\{P^{(1)}, P^{(2)}, ...\}$. Then

Simulation If $T_k \equiv 1$, then $\lim_{k\to\infty} Pr(X(k) = x) = \Pi(x)\ \ \forall x \in \Gamma^S$;

Optimization If $T_k = \dfrac{c}{log(k_0 + k)}$ for certain constants c and k_0, then $\lim_{k\to\infty} Pr(X(k) = x) = \mu(x)\ \ \forall x \in \Gamma^S$.

When L is large, for instance $L = 256$, practical problems are encountered. The exact recipe becomes quite demanding because each time a site a_k is visited, L many weights must be computed in order to correctly generate a random number that follows the distribution $\Pi^{(k)}(X_{a_k} \in d\lambda|X_{(a_k)} =$

$X_{(a_k)}(k-1)$). Thus, the time required to generate $X(k)$ from $X(k-1)$ is proportional to L.

Another popular Monte Carlo relaxation method is the Metropolis algorithm, which involves a "state-generating" matrix Q. Whereas the transition time from $X(k-1)$ to $X(k)$ is independent of L, there is still a problem when L is large. For example, for one common choice of Q, a new candidate for x_{a_k} is uniformly generated over the entire grey level range; consequently, the chance to *accept* the new value is very low and usually $X(k-1) = X(k)$. For other choices of Q, many steps are required before large intensity changes are reasonably likely. As a result, image boundaries are difficult to alter, and erroneously placed boundaries represent "deep" local minima. From here on we shall restrict attention to the Gibbs Sampler; we refer the reader to [50] for results of the Metropolis algorithm analogous to those in Section 4.

3 Restricted Image Spaces

For each $s = (i,j) \in S$, let ∂s denote the set of neighbors of s, i.e., $\partial s = S \bigcap \{(i-1,j),(i+1,j),(i,j-1),(i,j+1)\}$.

Definition: An image $x \in \Gamma^S$ has no isolated pixels at level $\gamma_0 \in \Gamma$ if $\min_{t \in \partial s} |x_s - x_t| \leq \gamma_0, \ \ \forall s \in S$. We shall refer to γ_0 as the allowed isolation level and write Ω_{γ_0} for the subset of Γ^S consisting of all such images.

Notice that the definition does not restrict image gradients; it simply requires that each pixel be associated with at least one of its four neighbors in the sense that its intensity not differ from each of theirs by more than γ_0. This is a reasonable assumption, especially when the image size is large, because the true image x^0 is likely to be close to one in Ω_{γ_0} for some $\gamma_0 << L$, both visually and as vectors. If we are willing to assume that the true image belongs to some Ω_{γ_0}, then obviously we needn't search for it within Γ^S, but only within Ω_{γ_0}.

From here on, let us assume that γ_0 is fixed (we shall take $\gamma_0 = 5$ in our experiments) and simply write Ω for Ω_{γ_0}. Let $x \in \Omega$ and $s \in S$. We wish to compute the set of grey levels that may be assumed by x_s in order to keep $x \in \Omega$; this set will be the support of the local conditional distribution in the modified Gibbs Sampler.

Let $t \in \partial s$. We say that x_t *is exclusively connected to* x_s *if* $\forall u \in \partial t$, $u \neq s$, $|x_t - x_u| > \gamma_0$.

Observe that $x \in \Omega$ and x_t exclusively connected to x_s implies that

$|x_t - x_s| \le \gamma_0$, and that x_t is exclusively connected to x_s does not imply that x_s is exclusively connected to x_t. Let

$$J_s(x) = \{t \in \partial s : \; x_t \;\; is \;\; \text{exclusively connected to} \;\; x_s\}.$$

For each $l \in \Gamma$, let $I(l) = \{l - \gamma_0, l - \gamma_0 + 1, ..., l-1, l, l+1, ..., l+\gamma_0 - 1, l+\gamma_0\} \bigcap \Gamma$ and set

$$E_s(x) = \begin{cases} \displaystyle\bigcap_{t \in J_s(x)} I(x_t) & \text{if } J_s(x) \text{ is not empty} \\ \displaystyle\bigcup_{t \in \partial s} I(x_t) & \text{if } J_s(x) \text{ is empty.} \end{cases}$$

Now, $x \in \Omega$ implies $x_s \in E_s(x)$, so that $E_s(x)$ is never empty. It is not difficult to show that $E_s(x) = E_s(z)$ for any two configurations $x, z \in \Omega$ that agree everywhere except possibly at pixel s. (Just observe that $J_s(x) = J_s(z)$.) Moreover,

If $x \in \Omega$, $z \in \Gamma^S$, and $\forall t \in S$, $t \neq s$, $x_t = z_t$ then $z \in \Omega$ iff $z_s \in E_s(x)$.

To see this, first notice that, since $z_t = x_t$, $\forall t \neq s$, we can define $E_s(z) = E_s(x)$, and since only pixels in $\{s\} \cup \partial s$ are affected, we need only check these.

If $z_s \in E_s(x)$, then $z_s \in I(x_t)$ for some $t \in \partial s$, which implies that $\min_{t \in \partial s} |z_s - z_t| \le \gamma_0$. Meanwhile, if an x_t with $t \in \partial s$ is not exclusively connected to x_s, then x_t is still connected to another of its neighbors. If an x_t with $t \in \partial s$ is exclusively connected to x_s, then the x_t is still connected to z_s because $z_s \in \bigcap_{u \in J_s(x)} I(x_u) \subseteq I(x_t)$, which implies $z \in \Omega$.

If $z_s \notin E_s(x)$, then in the case $J_s(x)$ is nonempty, this implies $z_s \notin I(x_t)$ for some x_t exclusively connected to x_s, and hence x_t becomes isolated above the allowed level; in the case $J_s(x)$ is empty, this implies z_s is isolated because $|z_s - z_t| > \gamma_0$ for all $t \in \partial s$ since $z_s \notin \bigcup_{t \in \partial s} I(x_t)$. Thus, $z \notin \Omega$.

The important computational feature of Ω is that, for any x, $|E_s(x)| \ll L$ when $\gamma_0 \ll L$. Hence the amount of computation for each step in the Gibbs Sampler is correspondingly reduced if we are willing to restrict H to Ω and define our reconstruction $\hat{x}$ as any member of $\Omega_{min} = \{x : \; H(x) \le H(z) \;\; \forall z \in \Omega\}$. As seen from these arguments, although the Ω is a proper subset of Γ^S, the definition of Ω is based only on a local property, sufficiently simple that "sections" of the set Ω can be rapidly determined.

4 Restricted Sampling and Optimization

In the standard proofs of convergence of the Markov chain generated by the Gibbs Sampler it is assumed that the distribution Π is defined on the product space Γ^S; see, e.g., [5]. As a result, we cannot directly apply those results to Ω, but fortunately they may be extended to a class of subsets of Γ^S that include Ω.

Let us say that a subset Θ is *coordinatewise linked* if $\forall z, x \in \Theta, \ z \neq x, \ \exists n \geq 1$ and a sequence of distinct states $z = z^0, z^1, ..., z^n = x$ in Θ such that z^k and z^{k+1} differ at only one site, i.e.,

$$\left|\{s : z_s^k \neq z_s^{k+1}\}\right| = 1 \ \forall k = 0, 1, ..., n-1.$$

Then it can be shown (see [50]) that, if $\gamma_0 \geq 1$,

Proposition: Each Ω is coordinatewise linked.

The following theorem extends the results stated in Section 2 for the Gibbs Sampler algorithm to any coordinatewise linked subset Θ of Γ^S. Let H, $\Pi(x)$ be defined as previously, but relative to Θ, and let $\Theta_{min} = \{x \in \Theta : H(x) \leq H(z), z \in \Theta\}$. In addition, let $\{a_k\}$, $\{T_k\}$, and $\{K_j\}$ be as in Section 2, as well as $\Pi^{(k)}(x), P^{(k)}$ and the Markov chain $\{X(k)\}$, all relative to Θ instead of Γ^S. Finally, let μ be uniform distribution on Θ_{min}.

Theorem. Suppose $\exists N^* < \infty$ such that, $\forall m = 0, 1, 2, ..., \ K_{m+1} - K_m \leq N^*$. Then
1. If $T_k \equiv 1$, then $\lim_{k \to \infty} Pr(X(k) = x) = \Pi(x) \ \ \forall x \in \Theta$;
2. If $T_k = \dfrac{c}{log(k_0 + k)}$ for certain c and k_0, then $\lim_{k \to \infty} Pr(X(k) = x) = \mu(x) \ \ \forall x \in \Theta$.

Proof:
The proof is essentially the same as in [5], although some modifications are necessary. Notice that the only added requirement is that the site visitation schedule must cover S with a *bounded* period. This condition is satisfied in all practical implementations but eliminates certain difficulties in the proof when Θ is not a product space. We shall only give the proof for annealing; the proof in the simulation case is virtually identical.

Claim 1: Let $P_{xz}^{(k_1,k_2)} = Pr(X(k_2) = z | X(k_1) = x)$, $P^{(k_1,k_2)}$ be the corresponding transition matrix, and let $\alpha(P^{(k_1,k_2)})$ be the ergodic coefficient of $P^{(k_1,k_2)}$ (see [14] and the following). Then

$\exists 0 = \tau_0 < \tau_1 < \tau_2 < \ldots$ such that $\sum_{n=1}^{\infty} \alpha(P^{(\tau_{n-1},\tau_n)}) = \infty.$

Proof of Claim 1:

Let $M = \max_{x,z\in\Theta} \max\{m: \ x, z \text{ are coordinatewise linked in } m \text{ steps}\}$.

Let $\tau_n = K_{nM}, \quad n = 0, 1, 2, \ldots$. Let $x, z \in \Theta$. Then, by the definition of a coordinatewise linked set, $\exists m' \leq M$, and $x = x^0, x^1, \ldots, x^{m'} = z$, all in Θ, such that $\forall m = 0, 1, 2, \ldots, m' - 1, \quad \left|\{s: \ x_s^m \neq x_s^{m+1}\}\right| = 1$. Set $x^{m'+1} = x^{m'+2} = \ldots = x^M = z$. It is easy to show that for fixed s and $x_{(s)}$,

$$\min_{x_s} \Pi^{(k)}(x_s|x_{(s)}) \geq \min_{x_s} \Pi^{(k+1)}(x_s|x_{(s)}),$$

and consequently,

$$\min_{s,x} \Pi^{(k)}(x_s|x_{(s)}) \geq \min_{s,x} \Pi^{(k+1)}(x_s|x_{(s)}).$$

Then

$$
\begin{aligned}
P_{xz}^{(\tau_{n-1},\tau_n)} &= Pr\left(X\left(K_{nM}\right) = z | X\left(K_{(n-1)M}\right) = x\right) \\
&\geq \prod_{m=1}^{M} Pr\left(X\left(K_{(n-1)M+m}\right) = x^m | X\left(K_{(n-1)M+m-1}\right) = x^{m-1}\right) \\
&\geq \prod_{m=1}^{M} \left(\min_{u\in\Theta} \min_{s\in S} \Pi^{(K_{(n-1)M+m})}\left(u_s|u_{(s)}\right)\right)^{K_{(n-1)M+m} - K_{(n-1)M+m-1}} \\
&\geq \prod_{m=1}^{M} \left(\min_{u\in\Theta} \min_{s\in S} \Pi^{(K_{(n-1)M+m})}\left(u_s|u_{(s)}\right)\right)^{N^*} \\
&\geq \left(\min_{u\in\Theta} \min_{s\in S} \Pi^{(K_{nM})}\left(u_s|u_{(s)}\right)\right)^{N^* M} \\
&= \left(\min_{u\in\Theta} \min_{s\in S} \frac{\Pi^{(K_{nM})}(u)}{\Pi^{(K_{nM})}\left(u_{(s)}\right)}\right)^{N^* M} \\
&\geq \left(\min_{u\in\Theta} \Pi^{(K_{nM})}(u)\right)^{N^* M} \\
&= \left(\min_{u\in\Theta} \frac{e^{-H(u)/T_{K_{nM}}}}{\sum_{v\in\Theta} e^{-H(v)/T_{K_{nM}}}}\right)^{N^* M} \\
&\geq \left(\frac{e^{-(H_{max} - H_{min})/T_{K_{nM}}}}{|\Theta|}\right)^{N^* M} \quad \text{where } H_{max} = \max_{x\in\Theta} H(x) \text{ and } H_{min} \\
&= \min_{x\in\Theta} H(x) \\
&= |\Theta|^{-N^* M} e^{-C/T_{K_{nM}}} \qquad \text{where } c = (H_{max} - H_{min})N^* M
\end{aligned}
$$

$$= |\Theta|^{-N^*M} \frac{1}{k_0 + K_{nM}}$$
$$\geq |\Theta|^{-N^*M} \frac{1}{k_0 + nN^*M}.$$

Therefore, by Lemma V.2.2. in [14],

$$\sum_{n=1}^{\infty} \alpha(P^{(\tau_{n-1},\tau_n)}) = \sum_{n=1}^{\infty} \left(\min_{x,z\in\Theta} \sum_{u\in\Theta} \min\left(P_{xu}^{(\tau_{n-1},\tau_n)}, P_{zu}^{(\tau_{n-1},\tau_n)}\right)\right)$$
$$\geq \sum_{n=1}^{\infty} \sum_{u\in\Theta} \min_{x,z\in\Theta} \min\left(P_{xu}^{(\tau_{n-1},\tau_n)}, P_{zu}^{(\tau_{n-1},\tau_n)}\right)$$
$$= \sum_{n=1}^{\infty} \sum_{u\in\Theta} \min_{x\in\Theta} P_{xu}^{(\tau_{n-1},\tau_n)} \geq |\Theta| \sum_{n=1}^{\infty} \min_{x,u\in\Theta} P_{xu}^{(\tau_{n-1},\tau_n)}$$
$$\geq |\Theta|^{1-N^*M} \sum_{n=1}^{\infty} \frac{1}{k_0 + nN^*M}$$
$$= \infty.$$

Claim 2: For $k = 1, 2, ...,$ $P^{(k)}$ and $\Pi^{(k)}$ satisfy the invariant (equilibrium) relation

$$\forall \ x \in \Theta, \qquad ((P^{(k)})'\Pi^{(k)})(x) = \Pi^{(k)}(x).$$

Proof of Claim 2: Let $E_s(x)$ again denote the support of the local conditional distribution $\Pi^{(k)}(x_s|x_{(s)})$, $x \in \Theta$.

$$\begin{aligned}
((P^{(k)})'\Pi^{(k)})(x) &= \sum_{z\in\Theta} \Pi^{(k)}(z) P_{z,x}^{(k)} = \sum_{z:\ z_{(a_k)} = x_{(a_k)}} \Pi^{(k)}(z) P_{z,x}^{(k)} \\
&= \sum_{z:\ z_{(a_k)} = x_{(a_k)}} \Pi^{(k)}((z_{a_k}, x_{(a_k)}))\Pi^{(k)}(x_{a_k}|z_{(a_k)}) \\
&= \Pi^{(k)}(x_{a_k}|x_{(a_k)}) \sum_{z_{a_k}\in E_{a_k}(x)} \Pi^{(k)}((z_{a_k}, x_{(a_k)})) \\
&= \Pi^{(k)}(x_{a_k}|x_{(a_k)})\Pi^{(k)}(x_{(a_k)}) = \Pi^{(k)}(x).
\end{aligned}$$

Claim 3: $\sum_{k=1}^{\infty} \sum_{x\in\Theta} |\Pi^{(k+1)}(x) - \Pi^{(k)}(x)| \ < \infty.$

Proof of Claim 3: The proof is the same as in [5].

To finish the proof of the theorem, by Claim 1 and Theorem V.3.2 in [14], $\{X(k)\}$ is weakly ergodic, and using Claim 2, Claim 3, and Theorem V.4.3 in [14], we see that in fact $\{X(k)\}$ is strongly ergodic. Clearly, then,

$\lim_{k\to\infty} \Pi^{(k)}(x) = \mu(x)$. Since Θ is finite, these facts, together with strong ergodicity, imply the conclusions of this theorem. ⊘

Returning now to the case $\Theta = \Omega$, $E_s(x)$ is the support of the local conditional distribution $\Pi^{(k)}(\cdot|x_{(s)})$. Therefore, to implement the Gibbs Sampler relative to Θ, the number of conditional probability values that must be calculated in order to generate $X(k)$ from $X(k-1)$ is only $|E_{a_k}(X(k-1))|$ instead of L.

5 The Image Deblurring Problem

The image restoration problem is to recover an ideal brightness pattern $x^0 = \{x^0(\vec{u}) : \vec{u} \in D\}$, where D is a planar domain, from the light measurements actually recorded by the sensor. For example, light in the visible range of the electromagnetic spectrum is sensed by a CCD camera, and the continuous distribution x^0 is "digitized" into samples y_s, where $s \in S'$ (an $M \times M$ rectangular lattice) and the values assumed by y_s are "quantized" to grey level values in $\Gamma = \{0, 1, ..., L-1\}$; often $M = 256$ or 512, and $L = 256$. For electro-optical and other devices, the transformation from x^0 to y involves the degradation of the signal by the transport medium, optical blurring, radiometric distortion, and various sources of noise, such as quantum and thermal fluctuations; in addition, there is a loss of information in the discretization process itself, namely digitization and quantization.

In many situations the dominant effect is blurring, which may be due to defocusing, motion, atmospheric turbulence, or other factors. The most elementary model that properly accounts for blur is the fully discrete, linear one given in Section 1, in which the domain D is converted into a discrete $N \times N$ array of pixels S. In matrix notation,

$$y = \mathcal{K}x^0 + \eta, \tag{4}$$

wherein we regard x^0 as an $N^2 \times 1$ vector, y and η as $M^2 \times 1$ vectors, and $\mathcal{K}$ as an $M^2 \times N^2$ matrix.

In general, $M < N$. There are exceptional cases in which one may assume $M = N$ (e.g., sometimes the values of x^0 are known on the "boundary" $S \setminus S'$), but in the more realistic formulation the system is underdetermined. Moreover, even if $\mathcal{K}$ were invertible, the "inverse problem" is usually ill-conditioned because the matrix $\mathcal{K}$ is nearly singular.

Finally, we shall assume that the "forward" (image formation) problem is completely specified, i.e., the distribution of y given x^0 is known. More specifically, our experiments involve (known) uniform blurs over square regions and linear segments (uniform motion blur), and the process η consists of white, centered Gaussian noise with (known) variance σ^2.

6 Implicit Discontinuity Models

Our estimated image is any (global) minimum of the functional $H^m : \Omega \to [0, \infty)$ given by

$$\begin{aligned} H^m(x) &= \Phi^m(x) + \lambda \parallel y - \mathcal{K}x \parallel^2 &(5)\\ &= \sum_C \phi(D_C^m(x)/\Delta) + \lambda \sum_{s \in S'} (y_s - (\mathcal{K}x)_s)^2. &(6)\end{aligned}$$

Recall that $\phi(u) = -1/(1+|u|)$ and Δ,λ are positive parameters; Φ^m refers to a smoothness constraint of order $m = 1, 2$, or 3, and there is a class of "cliques" C associated with each m. In the first order case, a clique C is any pair (s, t) of adjacent horizontal or vertical pixels and

$$D_C^1(x) = x_s - x_t, \quad C = (s, t).$$

For the planar case, looking at second differences, i.e. differences between components of the gradient at adjacent pixels, yields cliques of three types, each involving three or four pixels:

$$(1)\ \begin{matrix} s\bullet \\ t\bullet \\ u\bullet \end{matrix} \qquad (2)\ \begin{matrix} s\bullet & t\bullet \\ u\bullet & v\bullet \end{matrix} \qquad (3)\ s\bullet \quad t\bullet \quad u\bullet .$$

Now define

$$D_C^2(x) = \begin{cases} x_s - 2x_t + x_u & \text{if } C \text{ is of type (1) or (3)} \\ x_s - x_t - x_u + x_v & \text{if } C \text{ is of type (2).} \end{cases}$$

Finally, for the quadric case, looking at third differences, i.e., differences between components of the (discrete) Hessian matrix at adjacent pixels, yields cliques involving either four or six pixels:

$$
(1)\ \begin{matrix} s\bullet \\ t\bullet \\ u\bullet \\ v\bullet \end{matrix} \qquad (2)\ \begin{matrix} p\bullet & q\bullet & r\bullet \\ s\bullet & t\bullet & u\bullet \end{matrix} \qquad (3)\ \begin{matrix} p\bullet & s\bullet \\ q\bullet & t\bullet \\ r\bullet & u\bullet \end{matrix} \qquad (4)\ s\bullet\ t\bullet\ u\bullet\ v\bullet .
$$

In this case we define

$$
D_C^3(x) = \begin{cases} x_s - 3x_t + 3x_u - x_v & \text{if } C \text{ is of type (1) or (4)} \\ x_p - 2x_q + x_r - x_s + 2x_t - x_u & \text{if } C \text{ is of type (2) or (3).} \end{cases}
$$

Let us say that x is planar on a subregion $T \subset S$ if there are constants A, B, C such that $x_{i,j} = Ai + Bj + C$ for all $(i,j) \in T$ and that x is quadric if there are constants A, B, C, D, E, F such that $x(i,j) = Ai^2 + Bj^2 + Cij + Di + Ej + F$ for all $(i,j) \in T$. It is then easy to check that x *is constant, planar, or quadric on* S *if and only if* $D_C^m(x) = 0$ *for every* C *for* $m = 1, 2$ *or* 3.

6.1 A Noninterpolating Property

One motivation for the choice of ϕ is the following observation from [6]: Consider just a one-dimensional discrete signal and the class J_δ of real-valued functions defined on the integers from 0 to M that have the property that $x(0) = 0$ and $x(M) = \delta$. *Then, provided* ϕ *is even, concave and increasing, and for any choice of* M *and* Δ, *the function* $\Phi(x) = \sum_{i=1}^{M} \phi((x(i) - x(i-1))/\Delta)$ *is minimized over* J_δ *by those functions in* J_δ *with a single jump.* Moreover, it is not difficult to extend this property of minimizing the number of discontinuities to higher order derivatives. For example, again in one dimension, functionals of the form $\sum \phi(x(i+1) + x(i-1) - 2x(i))$ are minimized by curves displaying the fewest number of linear segments subject to boundary conditions on $x(0), x(1), x(M-1)$, and $x(M)$. As a result, reconstructions with such models are quite faithful to the original image in the neighborhood of abrupt transitions.

6.2 Explicit Discontinuities

There is a natural correspondence between the type of cost functionals we are using and those involving a *noninteracting* line process. Consider a process $b = (b_C)$ indexed by the appropriate cliques C (depending on the order of the model) where each b_C assumes *continuous* values on $[0, \infty]$ and represents the strength of the constraint associated with C. In the first-order case, the relationship between b and a binary line process l is simply $b_{s,t} = 1 - l_{s,t}$.

For the moment, suppose ϕ is unspecified, but H is of the form in Eq. (6). Then [6] one can identify conditions on ϕ such that there exists a "dual functional"

$$H^*(x,b) = \sum_C (b_C(D_C(x)/\Delta)^2 + \phi^*(b_C)) + \lambda \sum_{s \in S'} (y_s - (\mathcal{K}x)_s)^2,$$

such that $H(x) = \inf_b H^*(x,b)$, in which case the problems of minimizing H and H^* are equivalent. Here, ϕ^* is a strictly *decreasing* function on $[0,\infty)$, $\phi^*(0) = 0$ and represents the "penalty" incurred by enforcing a smoothness constraint at level b_C. Since there are no interactions among the b-variables, we seek conditions on ϕ for which there exists a function ϕ^* with

$$\phi(u) = \inf_{0 \leq z}(zu^2 + \phi^*(z)).$$

This has the simple geometric interpretation that ϕ is the lower envelope of a family of quadratic functions. In [6] it is shown that (some details aside) such a dual functional exists whenever $\phi(\sqrt{u})$ is *concave* and $lim_{u \to +\infty}\phi(u) = 0$. This covers our choice (2) as well as those appearing in [3], [9], and elsewhere. Indeed, an example of this correspondence (involving a binary-valued line process) was first noticed in [3].

It would be interesting to explore how the computational difficulties of minimizing H might be reduced by reformulating the optimization problem using H^*. Notice that the term $(D_C(x)/\Delta)^2$ is *quadratic* in x because $D_C(x)$ is linear in x. Consider the Markov random field (X, B) with (joint) distribution

$$\Pi(x,b) = e^{-H^*(x,b)} / \sum_x \int e^{-H^*(x,b)} db.$$

(Actually, some adjustments must be made to achieve integrability since the density is improper as it stands.) Then under this probability law the process X is *conditionally ("intrinsic") Gaussian* given B and the variables B_C are *conditionally independent* given X, with the same density up to a single parameter depending on $D_C(X)$. Consequently, stochastic relaxation with the dual process is much faster than with the original one.

6.3 Parameter Selection

The problem of choosing λ (as well as other parameters) has drawn considerable attention (see e.g., [15]), and there is certainly no consensus methodology. One approach is to select Δ on an ad hoc basis and then to choose

λ to satisfy the constraint $\| y - \mathcal{K}\hat{x} \|^2 \approx M^2\sigma^2$, where $\hat{x}$ minimizes H^m. Of course this is motivated by the simple observation that, by the law of large numbers, this constraint is satisfied by x^0. A more common approach is to regard one or both of Δ and λ as unknown (hyper-)parameters of the posterior probability distribution associated with H^m and attempt to *estimate* Δ and λ from the data using standard statistical procedures such as maximum likelihood and method of moments; see, e.g., [2], [9], [15].

We prefer another strategy. Let us say that x is a *coordinate-wise minimum* for a function H if any change in x at a single coordinate (i.e., pixel) increases the value of H. Due to the noise, the set of such values is actually a *random set*, call it $\mathcal{W}_y$. We then seek conditions on Δ and λ such that

$$Pr\{x^0 \in \mathcal{W}_y\} \approx 1$$

under certain (highly idealized) assumptions about x^0. In particular, any coordinate-wise descent algorithm will (likely) remain at x^0 if it arrives there. (Ideally, our cost function would be *globally* minimized (over Γ^S or Ω) by the true image, at least again in very simple cases, but these results appear elusive.)

The formal results on coordinate-wise minima appear in [6]. One is then able to specify λ as a function of the other parameters, namely Δ, σ and $\mathcal{K}$. We emphasize that this approach is different from most of the work in statistical image reconstruction, particularly Bayesian methods, in which λ is regarded as a model parameter that should be *estimated from the data.* Our choice is *independent of the data.*

Specifically, the formula that results from this analysis is the following; we refer the reader to [6] for the missing information about how to determine the "confidence level" α mentioned in Section 1 and the idealized image class. Let ζ denote the sum of squares of the blur coefficients and let $c(m)$ a constant depending on the model order: $c(1) = 2, c(2) = 5, c(3) = 14$. Recall that σ^2 is the noise variance. Then

$$\lambda = \begin{cases} c(m)/(6\Delta\sqrt{\zeta}\sigma) & \text{if } \sigma \leq \sqrt{\zeta}\Delta/6 \\ c(m)/(\sqrt{\zeta}\Delta/2 + 3\sigma)^2 & \text{if } \sigma \geq \sqrt{\zeta}\Delta/6. \end{cases} \tag{7}$$

All the experiments use Eq. (7) for choosing λ.

7 Experiments

Depending on the situation, we choose a model order m and $\hat{x}$ is defined by

$$H^m(\hat{x}) = min_{x\in\Omega} H^m(x). \tag{8}$$

The optimization problem involved in minimizing H^m is formidable, especially for the higher order models. Some of the practical implementation issues are now addressed, after which the individual experiments are described.

We present experiments on five images of varying difficulty. We used two different algorithms: The first algorithm operates in the restricted image space Ω_{γ_0} and uses the "correct" recipe for $E_s(x)$ given in Section 3; we shall refer to this as the "exact" algorithm. The second algorithm operates in the product space Γ^S and is the "union" algorithm from [6].

A pseudo-code description of the "exact" algorithm is as follows:

Select an integer radius γ_0 *(equal to 5 in all our experiments.) For each site* $s = (i, j)$ *consider the neighborhood about* s*:*

$$\begin{array}{ccccc} & & \bullet x_{(i-2,j)} & & \\ & \bullet x_{(i-1,j-1)} & \bullet x_{(i-1,j)} & \bullet x_{(i-1,j+1)} & \\ \bullet x_{(i,j-2)} & \bullet x_{(i,j-1)} & \bullet x_{(i,j)} & \bullet x_{(i,j+1)} & \bullet x_{(i,j+2)} \\ & \bullet x_{(i+1,j-1)} & \bullet x_{(i+1,j)} & \bullet x_{(i+1,j+1)} & \\ & & \bullet x_{(i+2,j)} & & \end{array}$$

If $| x_{i,j-1} - x_{i,j-2} |> \gamma_0$ *and* $| x_{i,j-1} - x_{i-1,j-1} |> \gamma_0$ *and* $| x_{i,j-1} - x_{i+1,j-1} |> \gamma_0$*, then* $a_1 = 0$*; else* $a_1 = 1$*.*

If $| x_{i,j+1} - x_{i,j+2} |> \gamma_0$ *and* $| x_{i,j+1} - x_{i-1,j+1} |> \gamma_0$ *and* $| x_{i,j+1} - x_{i+1,j+1} |> \gamma_0$*, then* $a_2 = 0$*; else* $a_2 = 1$*.*

If $| x_{i-1,j} - x_{i-2,j} |> \gamma_0$ *and* $| x_{i-1,j} - x_{i-1,j-1} |> \gamma_0$ *and* $| x_{i-1,j} - x_{i-1,j+1} |> \gamma_0$*, then* $a_3 = 0$*; else* $a_3 = 1$*.*

If $| x_{i+1,j} - x_{i+2,j} |> \gamma_0$ *and* $| x_{i+1,j} - x_{i+1,j-1} |> \gamma_0$ *and* $| x_{i+1,j} - x_{i+1,j+1} |> \gamma_0$*, then* $a_4 = 0$*; else* $a_4 = 1$*.*

If $a_1a_2a_3a_4 = 1$*, then the local section is formed by the union of the four intervals* $I_k, k = 1, 2, 3, 4$*, of radius* γ_0 *centered at* $x_{i,j-1}$*,* $x_{i-1,j}$*,* $x_{i+1,j}$ *and* $x_{i,j+1}$*, respectively; else the local section is the intersection* $\bigcap_{k:a_k=0} I_k$ *of the intervals of radius* γ_0 *for which* $a_k = 0$*.*

Recall that the algorithm in [6] takes the union of *six* intervals (of equal radius), four about the neighbors of s, one about the current value at s, and one about the data y_s. (A slight modification, resulting in no perceptible difference, is to drop the interval about the data.) We shall compare the exact and union algorithms, using the same annealing protocol. In general, the only differences we noticed are that (naturally) occasional isolated pixels will occur with the union algorithm (but are barely observable) and that minor artifacts sometimes occur using the exact algorithm, mainly in the form of isolated *pairs*. Quantitatively, the differences are also minor; the estimates provided by the two algorithms achieve approximately the same value of H and are approximately the same distance (in the L^1 norm) from x^0.

The first image (see Fig. 1) is a simple locally constant ("Mondrian") image and the second (see Fig. 2) is locally planar. Both of these are 64×64 synthetic images. The third (Fig. 3) is a text image generated using X windows, the original having two grey levels 230 and 102 but blurred and reconstructed in the dynamic range 0 to 255, as are all the images. The fourth (Fig. 4) is the image of a soccer ball, and the fifth (see Fig. 5) is the image of a face, both obtained from a standard vidicon camera.

After blurring, white Gaussian noise was added with means zero and variance σ^2 determined by first specifying the dB level. Recall that dB= $10 \log_{10}(SNR)$ in which SNR denotes the signal-to-noise ratio, defined by

$$\begin{aligned} SNR &= \hat{\sigma}^2(y)/\sigma^2 \\ &= \frac{1}{|S|} \sum_s (y_s - \bar{y})^2/\sigma^2, \end{aligned}$$

where $\bar{y}$ is the mean of the data (=signal). The value of SNR is essentially unchanged if $\hat{\sigma}^2(y)$ is replaced by $\hat{\sigma}^2(\mathcal{K}x^0)$ since the difference between these is of much smaller order than $\hat{\sigma}^2(\mathcal{K}x^0)$ (unless σ^2 is absurdly large). Consequently, given dB, the data is obtained by adding noise with variance $\sigma^2 = \hat{\sigma}^2(\mathcal{K}x^0)/(10^{dB/10})$.

In every experiment, λ was chosen according to the value given in Eq. (7). The choice of Δ is ad hoc. In a standard image with 256 grey levels, it seems reasonable that an edge of 20 to 30 grey levels is significant. On the other hand, a change of 2 or 3 in the *slope* of a planar surface is visually significant. In fact, in a large number of experiments on these images, we have found that setting $\Delta \approx 10$ in the first order model, and $\Delta \approx 5$ in the second and third order models, yields consistently good results. Finally, all the experiments were run using 300 sweeps (cycles of the pixel lattice) for each order and dropping temperature linearly from an initial value $\tau = 0.5$ to a final value $\tau = .01$.

7.1 Locally Constant Image

See Fig. 1 and Fig. 6. This result clearly indicates the utility of the first order model *when the original is indeed a Mondrian.* The image was blurred with a 1×20 horizontal motion blur and 40dB noise was added. In this case the noise standard deviation is $\sigma = 1.6$ grey levels. Δ was chosen to be 10 and an application of Eq. (7) yields $\lambda = 0.037$. Observe the qualitative similarity between the results of the exact and the union algorithms.

Figure 1: Top: Locally constant image. Bottom: 1×20 motion blur with 40dB noise.

Figure 2: Top: Locally planar image. Bottom: 7×7 uniform blur with 40dB noise.

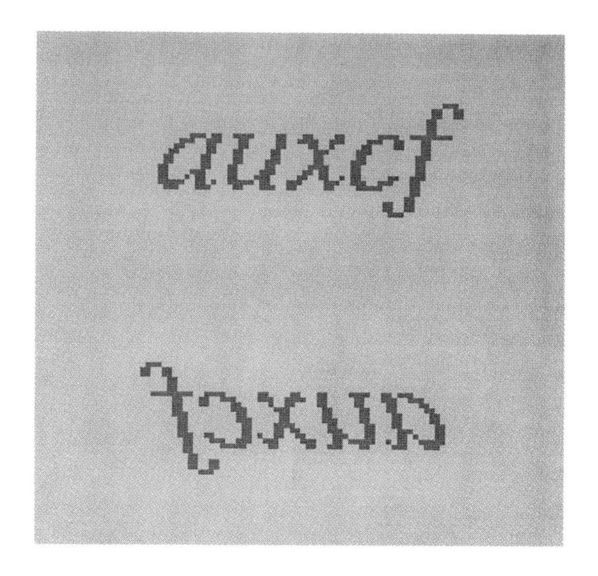

Figure 3: Top: Text image. Bottom: 1×20 motion blur with 25dB noise.

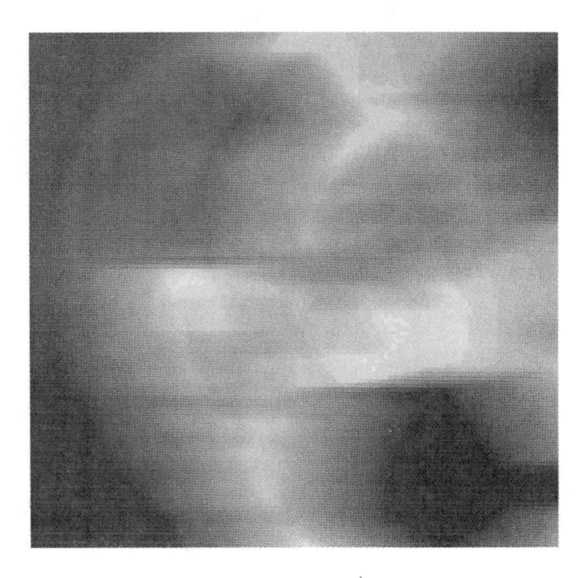

Figure 4: Top: Soccer ball image. Bottom: 1×30 motion blur with 60dB noise.

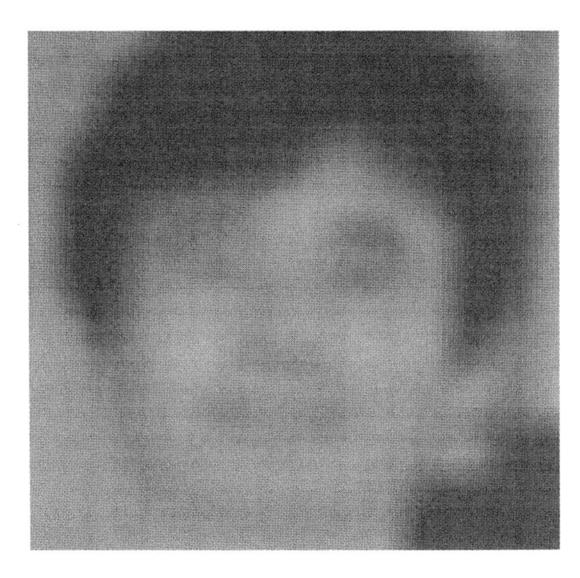

Figure 5: Top: Face image. Bottom: 7×7 uniform blur with 40dB noise.

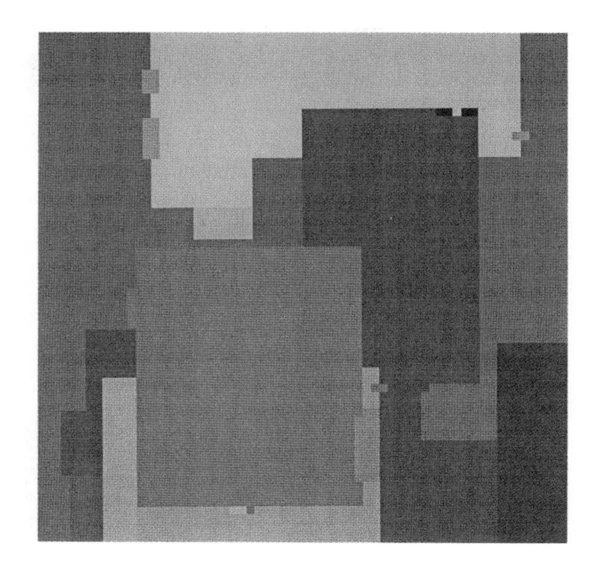

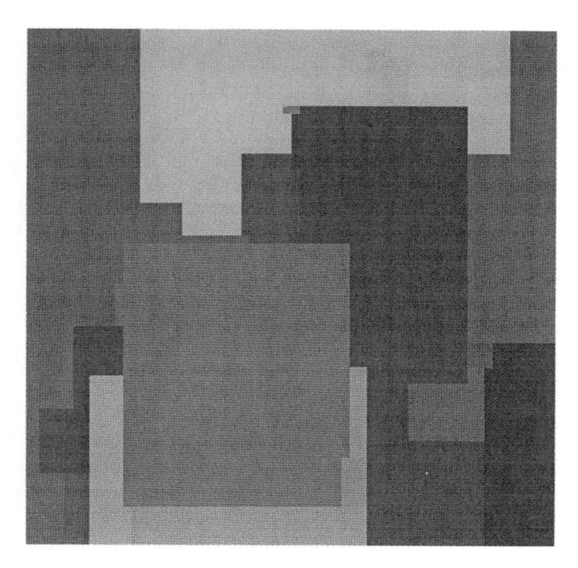

Figure 6: Restorations. Top: "Exact" algorithm. Bottom: "Union" algorithm.

7.2 Locally Planar Image

See Fig. 2 and Fig. 7. In this experiment, the output of the first order model is used as as a *starting point* for the second order model. This *coarse-to-fine* approach is discussed in [6], where results of the first order model for Gaussian blur are shown. The results here (not shown) are the same: the first order discontinuities are found by the first order model but the linear ramps are terraced; however, the planar facets then emerge with the second order model. This image was blurred with a a 7×7 uniform blur; the noise is at 40dB ($\sigma = 0.6$). $\Delta = 10$ in both the first and second order models; $\lambda = 0.32$ for the first order model and $\lambda = 0.8$ for the second order model. Again, notice the similarity between the results.

7.3 Letter Image

See Fig. 3 and Fig. 8. This image is 100×100 and was degraded with 1×20 horizontal uniform motion blur and 25dB noise ($\sigma = 0.8$). Here again the two algorithms yield comparable results, and we only show those from the "union" in the next two experiments.

7.4 Soccer Ball

See Fig. 4 and Fig. 9. This image is 128×128. The blur is 1×30 horizontal motion blur and the noise is very small, just 60dB ($\sigma = 0.07$). This is a real image taken with a standard vidicon camera and the result is typical of those possible in high signal-to-noise situations. Fig. 9 compares the output of the first and second order models (both starting at the data) using the "union" algorithm. In this case $\Delta = 10$ was chosen for both models and $\lambda \approx 1.0$ for the first order model and $\lambda \approx 2.5$ for the second order model. The output of the second order model is slightly more "realistic" in the sense that extended gradients are more faithful to the original (e.g., less terraced) with the second order model.

7.5 Face Image

See Fig. 5 and Fig. 10. This image was also taken with a vidicon camera, and is 100×100 pixels. It was blurred using a 7×7 uniform mask and 40dB noise ($\sigma = 0.6$) was added. The restoration was obtained using the *third* order model (starting at the data) and the "union" algorithm.

Figure 7: Restorations. Top: "Exact" algorithm. Bottom: "Union" algorithm.

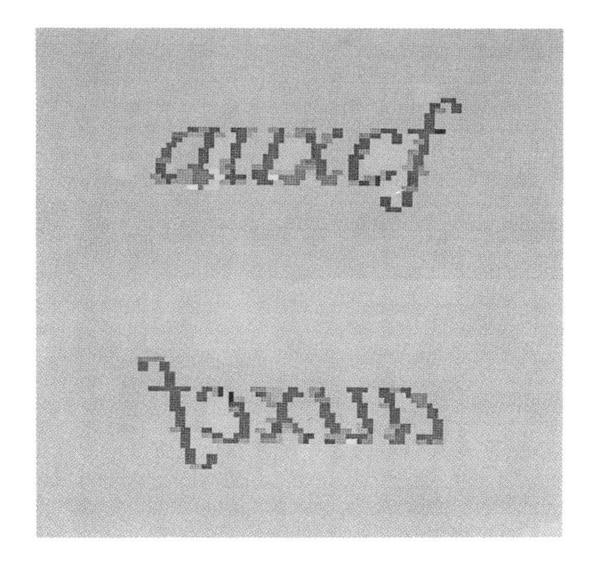

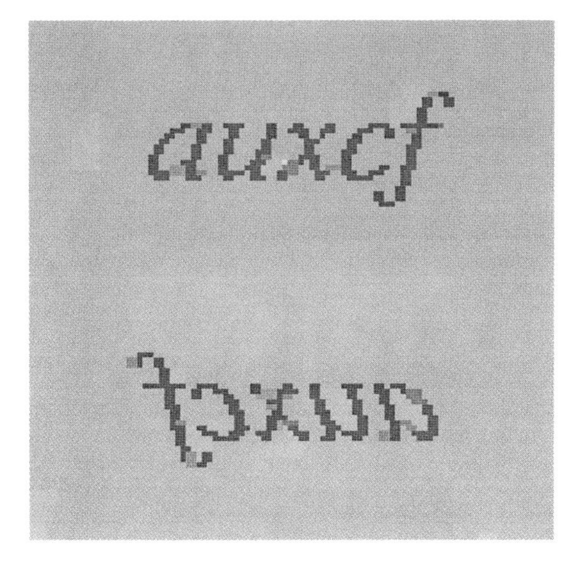

Figure 8: Restorations. Top: "Exact" algorithm. Bottom: "Union" algorithm.

Figure 9: Restorations. Top: first order, "Union" algorithm. Bottom: second order, "Union" algorithm.

Figure 10: Restoration. Third order, "Union" algorithm.

Bibliography

[1] J. Besag, (1986). "On the statistical analysis of dirty pictures," (with discussion), *J. Royal Statist. Soc.*, Ser. B, 48, pp. 259–302.

[2] J. Besag, (1989). "Towards Bayesian image analysis," *J. Appl. Statistics*, 16.

[3] A. Blake and A. Zisserman, (1987), *Visual Reconstruction*, MIT Press, Cambridge, Massachusetts.

[4] D. Geiger and F. Girosi, (1989). "Mean field theory for surface reconstruction and visual integration," A.I. Memo No. 1114, Artificial Intelligence Laboratory, M.I.T.

[5] D. Geman and S. Geman, (1987). "Relaxation and annealing with constraints," Complex Systems Technical Report 35, Division of Applied Mathematics, Brown University.

[6] D. Geman and G. Reynolds, (1992). "Constrained restoration and the recovery of discontinuities," *IEEE Trans. Pattern Anal. Machine Intell.*, 14, pp. 367–383.

[7] S. Geman, (1985). "Stochastic relaxation methods for image restoration and expert systems," in *Automated Image Analysis*, D.B. Cooper, R.L. Launer, and D.E. McClure, eds., Academic Press, New York.

[8] S. Geman and D. Geman, (1984). "Stochastic relaxation, Gibbs distributions, and the Bayesian restoration of images", *IEEE Trans. Pattern Anal. Machine Intell.*, 6, pp. 721-741.

[9] S. Geman and D. E. McClure, (1987). "Statistical methods for tomographic image reconstruction," in *Proceedings of the 46th Session of the International Statistical Institute*, Bulletin of the ISI, Vol. 52.

[10] S. Geman, D.E. Mclure, and D. Geman, (1992). "A nonlinear filter for film restoration and other problems in image processing," CVGIP: Graphical Models and Image Processing, 54, pp. 281–289.

[11] P. J. Green, (1990). "Bayesian reconstruction from emission tomography data using a modified EM algorithm," *IEEE Trans. Medical Imaging*, 9, pp. 84–93.

[12] U. Grenander, (1983). "Tutorial in pattern theory," Technical Report, Div. of Applied Mathematics, Brown University.

[13] J.M. Hammersley and D.C. Handscomb, (1964). *Monte Carlo Methods.* Methuen and Company, London.

[14] D. Issacson and R. Madsen, (1976). *Markov Chains: Theory and Applications.* John Wiley and Sons, New York.

[15] J. W. Kay, (1988). "On the choice of regularization parameter in image restoration," *Springer Lecture Notes in Computer Science*, 301, pp. 587–596.

[16] Y.G. Leclerc (1989) "Constructing simple stable descriptions for image partitioning," *International Journal of Computer Vision*, 3, pp. 73-102.

[17] N. Metropolis, A.W. Rosenbluth, M.N. Rosenbluth, A. H. Teller, and E. Teller, (1953). "Equations of state calculations by fast computing machines," *J. Chemical Physics*, 21, pp. 1087–1091.

[18] C. Yang, (1991), "Stochastic methods in image restoration," Ph.D. thesis, University of Massachusetts.

A Continuation Method for Image Estimation Using the Adiabatic Approximation

A. Rangarajan[‡§] and Rama Chellappa[†]

[‡]Department of Computer Science
Yale University
New Haven, Connecticut

[§]Division of Imaging Science
Department of Diagnostic Radiology
Yale University
New Haven, Connecticut

[†]Department of Electrical Engineering
Institute for Advanced Computer Studies
Center for Automation Research
University of Maryland
College Park, Maryland

1 Introduction

Global optimization techniques for image estimation have become very popular in recent years [1]. The power of global optimization lies in its iterative and nonlinear nature. Iterative algorithms make global transfer of information possible. The most important problem in global optimization is overcoming non-convexity, which often traps algorithms in local minima. In the past decade, stochastic relaxation algorithms [1] have been used to overcome the problem of non-convexity. Specifically, simulated annealing uses noise to avoid getting trapped in local minima. Simulated annealing iteratively samples from the conditional distribution of each variable, while a control parameter, the temperature, is varied from high to low values. At low values of the temperature, the algorithm samples from the most probable configurations.

Partially supported by the Joint Services electronics program under the contract F49620-88-C-0067 and by the NSF grant MIP-84-51010.

ISBN 0-12-170608-7

The deterministic analog to stochastic relaxation also involves the variation of a control parameter during the iterations. These deterministic annealing procedures are called *continuation* methods [2], since they involve tracking the minima as the control parameter is varied. The different values of the control parameter generate a sequence of cost functions. The original cost function is increasingly, closely approximated and is asymptotically reached at the final values of the control parameter.

We are interested in the problem of image estimation from a noisy scene while simultaneously preserving the edges. This requires feedback from the edge detection process to the estimation process and vice versa. This can be accomplished by formulating the problem as one of joint estimation of the image and the edges (discontinuities, line processes) by global minimization of a cost function. We are interested in continuation schemes due to the computationally intensive nature of simulated annealing and the non-convexity of the problem. The cost function we use is completely general, with the popular *weak membrane* [3] being a special case. Specifically, we seek to minimize a cost function that encourages the formation of unbroken contours (hysteresis) [4] and discourages multiple responses to a single discontinuity (non-maximum suppression) [5]. We would like these additional constraints to involve the discontinuities alone, since that would aid in the export of these algorithms to other modules involving discontinuities [6]. These constraints are missing in the weak membrane, leading to a cost function that has no interactions between the discontinuities.

Continuation methods already exist for the weak membrane [3, 7, 8]. We are interested in relating continuation methods, which incorporate the additional constraints, to the weak membrane-based continuation methods. The *adiabatic approximation* [9] plays an important role in helping us retain the properties of the weak membrane continuation methods in the presence of additional constraints. The adiabatic approximation also helps us relate our algorithm to Mean Field Theory (MFT) [7], which is a generic, deterministic annealing procedure. We also show a relationship between the adiabatic approximation and the Iterated Conditional Modes (ICM) [10] algorithm. Our resulting algorithm uses a combination of the Conjugate Gradient (CG) and ICM algorithms. We present experimental results on an aerial image and demonstrate the realm of validity of the adiabatic approximation.

Section 2 introduces the adiabatic approximation and traces a relationship to continuation methods in general. In Section 3, we use the adiabatic approximation to derive a continuation method that incorporates additional constraints but still retains ties to the weak membrane continuation methods. Section 4 provides solutions for arbitrary interaction terms that can be used in the ICM algorithm. In Section 5, we specialize to the Graduated Non-Convexity (GNC) sequence and show that our algorithm retains the properties of the GNC sequence while incorporating further constraints. Section 6 discusses the results. Conclusions are presented in Section 7.

2 Continuation Methods and Adiabatic Approximation

The relationship between Gibbs distributions [1], Bayesian inference, and cost functions now is well known. If the prior and degradation models are Gibbsian, the posterior also is Gibbsian and can be specified completely by a Hamiltonian (cost function). The maximum *a posteriori* (MAP) estimate then reduces to finding the global minimum of the cost function. The specific form of the assumed Hamiltonian is written as follows:

$$\begin{aligned}\mathcal{H}(\mathbf{f},\mathbf{v},\mathbf{h}) = &\sum_{\{i,j\}} (f(i,j)-d(i,j))^2 \\ &+\sum_{\{i,j\}} \left(\lambda^2 f_x^2(i,j)\,(1-v(i,j)) + \alpha\ v(i,j)\right) \\ &+\sum_{\{i,j\}} \left(\lambda^2 f_y^2(i,j)\,(1-h(i,j)) + \alpha\ h(i,j)\right) \\ &+\mathcal{H}_c(\mathbf{v},\mathbf{h}),\end{aligned} \tag{1}$$

where $\mathbf{d}$ is the data and $\mathbf{f}$ is the required intensity image estimate. The line processes $\mathbf{v}$ and $\mathbf{h}$ are located between adjacent data variables and denote the presence (or absence) of a discontinuity. We seek the minimum of the Hamiltonian ($\mathcal{H}$) or energy function in the space of $\mathbf{f}$, $\mathbf{v}$, and $\mathbf{h}$, where $v(i,j),\ h(i,j) \in \{0,1\}$, and $f_x(i,j) \stackrel{\text{def}}{=} f(i+1,j)-f(i,j)$ and $f_y(i,j) \stackrel{\text{def}}{=} f(i,j+1)-f(i,j)$.

The first term in Eq. (1) attempts to keep the restoration of $\mathbf{f}$ close to the observed data in a least-squares sense. This term is due to the additive noise assumption. The second term encodes our assumption that the data is smooth everywhere except at the discontinuities. The third term enforces a penalty on incorporating a discontinuity in the restoration. The last term ($\mathcal{H}_c(.)$) can be used to incorporate prior assumptions on the nature of discontinuities. Edges in images tend to be connected and form contours. This term can reflect these beliefs.

When the $\mathcal{H}_c(.)$ term is absent, the line processes become independent of each other and can be eliminated at the outset [3].

$$z = \begin{cases} 0 & \lambda^2 f_p^2 < \alpha \\ 1 & \lambda^2 f_p^2 \geq \alpha \end{cases}, \tag{2}$$

where z and f_p are the generic symbols used for the line process and the data gradient in this chapter.

It is possible to eliminate the line processes because they are independent of each other. The line process is an *unobservable* process, as it is not directly linked to any observed data. We observe and record degraded data only, not degraded edges. Substituting Eq. (2) in Eq. (1) with the $\mathcal{H}_c(.)$ term removed, we get

$$\begin{aligned} \mathcal{H}_b(\mathbf{f}, \mathbf{v}, \mathbf{h}) &= \sum_{\{i,j\}} (f(i,j) - d(i,j))^2 \\ &+ \sum_{\{i,j\}} \Big(g^*(f_x(i,j)) + g^*(f_y(i,j)) \Big). \end{aligned} \tag{3}$$

where

$$g^*(f_p) = \begin{cases} \lambda^2 f_p^2 & \lambda^2 f_p^2 < \alpha \\ \alpha & \lambda^2 f_p^2 \geq \alpha \end{cases}. \tag{4}$$

This energy function, Eq. (3), is the popular *weak membrane.* The term weak membrane arises from the physical nature of this Hamiltonian. If discontinuities are absent, the reconstruction would be like a membrane that is continuous everywhere. When discontinuities are present, the membrane no longer is continuous. The parameters λ and α determine the smoothness of the weak membrane and the number of discontinuities obtained.

The energy function as it stands Eq. (3) is non-convex due to the nature of the g^* function. Several researchers [3, 7, 8, 11, 12] have proposed a variety of continuation methods or convex formulations to deal with this problem. A continuation method essentially tracks minima through the variation of a control parameter: the original energy function increasingly is closely approximated during this variation. All these approaches can be synthesized conceptually by replacing the g^* function by either a solitary g function or by a sequence of $g^{(k)}$ functions: the integer k is the index of the sequence. Henceforth, we will refer to this sequence simply by the g function since our generalized framework is valid for all members of a given sequence.

$$\begin{aligned} \mathcal{H}_b(\mathbf{f}, \mathbf{v}, \mathbf{h}) &= \sum_{\{i,j\}} (f(i,j) - d(i,j))^2 \\ &+ \sum_{\{i,j\}} \Big(g(f_x(i,j)) + g(f_y(i,j)) \Big) \end{aligned} \tag{5}$$

At this point, we are interested in the *GNC algorithm* [3]. The GNC algorithm is a continuation method designed for the weak membrane, wherein a sequence of cost functions is constructed such that the first function in the sequence is convex. The original cost function is reached asymptotically by varying the continuation parameter. Convexity of the first stage automatically guarantees the positive definiteness of the Hessian matrix. When the control parameter is changed, the Hessian (evaluated at the minimum of the first stage) no longer is convex. This is reflected by several eigenvalues changing sign. The negative eigenvalues reflect loss of stability. Also, the GNC cost function is a function of the intensities only. We now relate the GNC algorithm with the more specific methods of synergetics.

A basic idea in the methods of synergetics [9, 13] is to obtain a new cost function that is a function of the eigenvectors corresponding *only* to the negative (unstable) eigenvalues. To accomplish this, the eigenvectors corresponding to the positive (stable) eigenvalues should be replaced by their solution with respect to the other eigenvectors. We elucidate this point with a simple example.* The following system of equations does not correspond to gradient descent on a cost function. However, the equations illustrate our point with respect to positive and negative eigenvalues.

$$\frac{du}{dt} = \alpha u - us, \tag{6}$$

$$\frac{ds}{dt} = -\beta s + u^2, \tag{7}$$

where $\alpha,\ \beta > 0$. Linearizing this system around the point $(u, s) = (u^0, s^0)$, we get

$$\begin{aligned} \frac{du}{dt} &= \alpha u - u^0 s - s^0 u, \\ \frac{ds}{dt} &= -\beta s + 2uu^0. \end{aligned} \tag{8}$$

When gradient descent is used on a cost function $V(x)$, we get the system of equations, $\frac{dx}{dt} = -\frac{\partial V}{\partial x}$. The Hessian (the matrix of second partials) is evaluated at x^0. The corresponding Hessian for (8) is

$$H = \begin{bmatrix} -\alpha + s^0 & u^0 \\ -2u^0 & \beta \end{bmatrix}. \tag{9}$$

Assuming the simplest initial conditions, $(u^0, s^0) = (0, 0)$, we get

$$H = \begin{bmatrix} -\alpha & 0 \\ 0 & \beta \end{bmatrix}, \tag{10}$$

*Reprinted with permission from *Advanced Synergetics* by H. Haken, Springer-Verlag, New York, 1987.

which is diagonal. Since α and β both are greater than zero, it follows that the linear system is unstable and that u corresponds to the unstable eigenvalue and s to the stable eigenvalue. As mentioned previously, one of the basic ideas in synergetics is *reparametrization about the unstable modes*, where by unstable modes, we mean the eigenvectors corresponding to the unstable eigenvalues. In our example, this involves solving for s in terms of u,

$$\frac{ds}{dt} = 0 \Rightarrow \ s = \frac{u^2}{\beta}, \tag{11}$$

which corresponds to keeping u fixed and finding the equilibrium state of s with respect to u. When this is substituted back in Eq. (6), we get

$$\frac{du}{dt} = \alpha u - \frac{u^3}{\beta}, \tag{12}$$

which can be obtained by gradient descent on the cost function,

$$V(u) = \frac{u^4}{4\beta} - \frac{\alpha u^2}{2}. \tag{13}$$

$V(u)$ is shown in Fig. 1 for different values of α. The minimum of the cost function is at $u = \pm\sqrt{\alpha\beta}$, which also gives $s = \alpha$. Note that the equilibrium values of u and s could have been obtained easily from Eqs. (6) and (7) by setting $\frac{du}{dt}$ and $\frac{ds}{dt}$ to zero. However, the approach to equilibrium is dictated by the dynamics of u and not s. To understand this clearly, we vary α from negative to positive. Only when α is greater than zero do we get a nonzero minimum for u and s. A bifurcation occurs when α crosses over from negative to positive. This clearly indicates that the sign of α determines the equilibrium state. The *effective* energy function $V(u)$ dictates the approach to equilibrium. The example illustrates the following points. The eigenvalues of the linearized system give us information about the stable and unstable modes of the nonlinear system. Eliminating the stable modes gives us effective energy functions in terms of the unstable modes. Variation of a control parameter α acts as a control on convexity. As α varies from zero to positive values, the effective energy function becomes non-convex (Fig. 1).

We now begin to see a close relationship between the methods of synergetics and the GNC algorithm. The GNC algorithm begins by eliminating the discontinuities from the energy function. The effective energy function is approximated by a sequence of energy functions depending on a continuation parameter. As the continuation parameter is varied, the energy function that initially is convex becomes increasingly non-convex. There

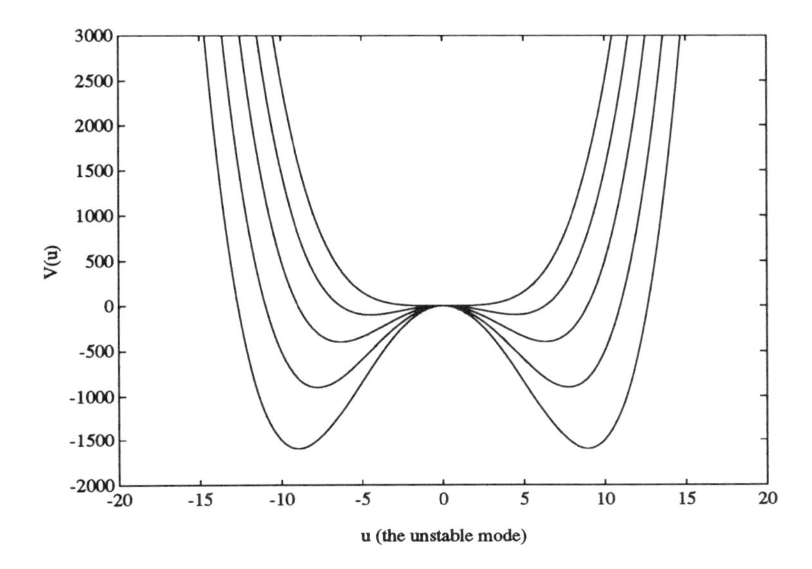

Figure 1: The cost function $V(u)$ for different values of α and $\beta = 1$

is one crucial difference between the GNC algorithm and synergetics. In GNC, the discontinuities are eliminated before the introduction of the continuation parameter. However, we can show that the GNC algorithm also can be viewed in terms of a general energy function sequence where elimination of discontinuities generates the GNC sequence. The elimination of the discontinuities would be similar to the elimination of the stable mode in Eq. (11). Another important distinction is that there appear to be no equivalents of the stable and unstable eigenvalues in the GNC sequence. Rather, the discontinuities were eliminated because there were no interactions between them. Once we identify the discontinuities with the stable modes and the data with the unstable modes, a relation appears.

Elimination of the stable modes requires identification of the positive and negative eigenvalues followed by reparametrization. The calculation of eigenvalues is impossible for systems with a large number of variables (as is our case). However, note that elimination of the *dynamics* of s necessitates the assumption that s is close to its equilibrium state, given u. In other words, $\frac{ds}{dt}$ is assumed to be close to zero. Elimination of the dynamics of the stable modes based on this assumption is known as the *adiabatic approximation.*

The adiabatic approximation assumes that s always is close to its equilibrium state given u $(-\beta s + u^2 = 0)$. We have been concerned chiefly concerned with gradient descent dynamics and the adiabatic approximation. The principle remains the same when other dynamics are considered. It states: *When a variable always is close to its equilibrium state with the other variables held fixed, the dynamics of that variable can be eliminated.* The eliminated variables are termed *fast relaxing*, since they always are close to their equilibrium states given the states of the other variables.

The dynamical system with respect to the slow variables remains unchanged except that the discontinuities now are set to their equilibrium states. We noted earlier that the adiabatic approximation meant that the discontinuities always were close to their equilibrium states. To obtain an effective energy function, the closed-form solution of the discontinuities w.r.t the intensities is necessary. This is difficult to obtain when there are significant interactions among the discontinuities. The best alternative is to use a descent algorithm that continuously sets the discontinuities to their equilibrium states, keeping the other variables fixed. Such an algorithm already exists; the ICM algorithm. To our knowledge, a relationship between ICM and the adiabatic approximation has not been pointed out.

The ICM algorithm was first suggested by Besag in [10]. Though ICM was set in a probabilistic context, the algorithm is strictly deterministic. ICM maximizes the conditional probability with respect to the variable at (i, j), which is identical to setting it to its equilibrium state, keeping the neighbors (other variables) fixed. Once this is done, we move to a new site that can be picked at random. Since the conditional probability structure does not extend beyond a few neighbors, the sites can be updated in a quasi-parallel fashion; update those sites that are not neighbors of each other. This procedure is guaranteed to find a local minimum of the cost function.

When the line processes are independent of each other, the updating on the line processes is completely parallel and a closed-form solution (given the data) can be found. ICM then leads to a parallel updating of the line processes and a suitable descent algorithm for the data.

When the line processes are not independent of each other, ICM iteratively sets them to their equilibrium states until a minimum is reached (for all line processes). This is equivalent to the adiabatic approximation in the presence of interactions (where closed-form solutions are difficult). If the adiabatic approximation is valid, ICM should converge very quickly. We can verify, therefore, the adiabatic approximation by this self-consistent argument. If ICM does not converge quickly (two or three iterations), then the effective energy function obtained by the adiabatic approximation does not provide an accurate picture of the dynamics. This would indicate that the line processes interact with each other over a much larger range than previously suspected, which is why they do not converge quickly.[†]

A recent, popular approach to combinatorial optimization is the method of *Hopfield networks* [14]. The method constructs a Lyapunov function whose minimum yields the expected value of the line processes. If there are no interactions between the vertical and horizontal line processes, we can use the one-dimensional cost function $\mathcal{H}(\mathbf{f}, \mathbf{l})$, where $\mathbf{l}$ is the digital line process. The Hopfield network minimizes the free energy F,

$$F = \mathcal{E}_l V(\mathbf{f}, \mathbf{l}) - \frac{1}{\beta} S = -\frac{1}{\beta} \log Z, \tag{14}$$

where $\mathcal{E}_l$ is the expectation operator with respect to l, and S is the entropy, defined as

$$\sum_{\{i,j\}} \left(-p_{ij} \log p_{ij} - (1 - p_{ij}) \log(1 - p_{ij})\right),$$

where p_{ij} is the probability that $l(i,j) = 1$. Hopfield identifies $\text{Prob}(l(i,j) = 1)$ with the expected value of $l(i,j)$, which is $z(i,j)$ in our notation. Therefore,

$$\begin{aligned} F &= V(\mathbf{f}, \mathbf{z}) \\ &+ \frac{1}{\beta} \sum_{\{i,j\}} \Big(z(i,j) \log z(i,j) + (1 - z(i,j)) \log(1 - z(i,j)) \Big). \end{aligned} \tag{15}$$

The first term arises by substitution of $z(i,j)$ for the expectation of $l(i,j)$. The continuation parameter for the sequence of energy functions is β, which is the inverse of the temperature. For low β, the free energy is dominated by the entropy and at high β, the free energy is identical to the cost function. The following dynamical system performs gradient descent on the free energy:

$$\begin{aligned} \frac{df(i,j)}{dt} &= -\frac{\partial V(\mathbf{f}, \mathbf{z})}{\partial f(i,j)}, \\ \frac{dw(i,j)}{dt} &= -w(i,j) - \frac{\partial V(\mathbf{f}, \mathbf{z})}{\partial z(i,j)}, \end{aligned} \tag{16}$$

where $z = \frac{1}{1+\exp(-\beta w)}$.

These equations converge to the mean field solutions [15]. In MFT, each variable is set to its mean value given the other variables. However, in Eq. (16), this solution also is the minimum of Eq. (15).

[†] In that case, an effective energy function where the adiabatic approximation now is applied to the data, might yield a better picture.

What happens if we assume that the $w(i,j)$ can be eliminated using the adiabatic approximation? $w(i,j)$ can be set to its equilibrium state given the other variables, which is:

$$w(i,j) = -\frac{\partial V(\mathbf{f},\mathbf{z})}{\partial z(i,j)}$$

The main point is that application of the adiabatic approximation in Eq. (16) yields a free energy that is a function of $\mathbf{f}$ and the "frozen" values of $\mathbf{z}$. Moreover, the frozen values of $z(i,j)$ are the mean field values. If interactions are absent, the free energy bears strong physical ties to the GNC sequence [7]. If interactions are present, the validity of the adiabatic approximation ensures that the modified free energy is an improvement over the no interactions case.

To summarize, we can extend the continuation methods developed in the absence of interaction terms. First, we assume that the discontinuities are *fast relaxing*. Then we derive cost functions in the space of the intensities *and* line processes that are equivalent to the previous continuation methods which are functions only of the intensities. Interactions now are added and we test if the adiabatic approximation is still valid; ICM should converge quickly. If it does, we can claim that the new constraints are locally active, and that they do not introduce any spurious global effects. If it does not, then the new constraints are more dominant, leading to a completely different cost function. The properties of the old continuation method no longer hold.

3 Framework for Addition of Constraints

In this section, we first obtain an equivalent cost function to Eq. (5). Then we add further constraints and solve for the line process. The first objective is to reintroduce the line processes in g function formalism. This objective is driven by the overall purpose of adding arbitrary interaction terms to the weak membrane while retaining the continuation method. From this viewpoint, examine the new energy function introduced here:

$$\begin{aligned}\mathcal{H}_b(\mathbf{f},\mathbf{t}_v,\mathbf{t}_h) &= \sum_{\{i,j\}} (f(i,j)-d(i,j))^2 \\ &+ \sum_{\{i,j\}} g_t(f_x(i,j),t_v(i,j)) \\ &+ \sum_{\{i,j\}} g_t(f_y(i,j),t_v(i,j)),\end{aligned} \tag{17}$$

where the nature of the new process $\mathbf{t}$ will be specified. Equation (17) is obtained from Eq. (5) by replacing $g(f_p)$ with $g_t(f_p, t)$. Consider the following choice for g_t,

$$g_t(f_p, t) = g(t) + \frac{g^{'}(t)}{2t}(f_p^2 - t^2). \tag{18}$$

This choice of g_t guarantees $t = f_p$ to be a minimum of Eq. (18). When $t = f_p$ is substituted back in Eq. (18), $g_t(f_p, f_p) = g(f_p)$ and we retrieve the original sequence. For $t = f_p$ to be a minimum of Eq. (18), we also require:

$$\frac{\partial^2 g_t(f_p, t)}{\partial t^2} > 0 \text{ at } t = f_p. \tag{19}$$

These conditions translate in turn to

$$\begin{aligned} \frac{\partial g_t(f_p, t)}{\partial t} &= \frac{1}{2}(1 - \frac{f_p^2}{t^2})(g^{'}(t) - tg^{''}(t)) = 0, \\ \frac{\partial^2 g_t(f_p, t)}{\partial t^2} &= \frac{1}{2}(1 - \frac{f_p^2}{t^2})(-tg^{'''}(t)) \\ &+ \frac{f_p^2}{t^3}(g^{'}(t) - tg^{''}(t)) > 0, \end{aligned}$$

where $g^{''}$ and $g^{'''}$ are the second and third derivatives w.r.t t.

The general conditions we obtain are:

$$\begin{aligned} \text{For } t = f_p, \ (g^{'}(t) - tg^{''}(t)) &> 0, \\ \text{and for } t = -f_p, \ (g^{'}(t) - tg^{''}(t)) &< 0. \end{aligned} \tag{20}$$

It is useful to switch to a *scale-invariant* system. First, note that the sequence of energy functions can be rewritten as follows:

$$\begin{aligned} \mathcal{H}_b(\mathbf{f}, \mathbf{t}_v, \mathbf{t}_h) &= \sum_{\{i,j\}} (f(i,j) - d(i,j))^2 \\ &+ \alpha \sum_{\{i,j\}} \left(g_t(\frac{f_x(i,j)}{f_p^0}, t_v(i,j)) \right) \\ &+ \alpha \sum_{\{i,j\}} \left(g_t(\frac{f_y(i,j)}{f_p^0}, t_h(i,j)) \right), \end{aligned} \tag{21}$$

where $f_p^0 = \sqrt{\frac{\alpha}{\lambda^2}}$. $g_t(f_p, t)$ is scaled down by a factor of α, and f_p is scaled down by the factor f_p^0. Finally, we can rewrite g_t as follows:

$$g_s(f_p, s) = g_t(f_p, \sqrt{s}) = g_2(s) + g_2^{'}(s)(f_p^2 - s), \tag{22}$$

where $g_2(s) = g(\sqrt{s})$. The process s is the gradient-magnitude squared process; $s \geq 0$.

Equation (22) is the simplest expression that allows us to introduce a new process s such that the infimum with respect to s gives us the $g(f_p)$ sequence. The rescaling (of f_p and g_s) is not necessary for the preceding statement to be valid.

$$g(f_p) = inf_{(s \geq 0)} g_s(f_p, s) \Rightarrow s = f_p^2. \tag{23}$$

The sufficient condition for a minimum at $s = f_p^2$ is

$$g_2^{''}(s) \leq 0, \forall s \geq 0. \tag{24}$$

An alternative proof of this derivation can be found in [16, 17]. The authors in [17] also have pointed out that the concavity of $g_2(s)$ is necessary for the process s to exist.

3.1 Recovery of the line process

So far, we have derived an equivalent energy function and introduced an attribute process s. The weak membrane energy function initially was formulated with a digital line process. It is possible to define a monotonic transformation from s to an analog line process z ($z \in [0, 1]$). However, there are numerous choices and a lack of a suitable criterion that allows us to select a single analog line process. Intuitively, the line process z should equal zero at $s = 0$ and equal one at $s = \infty$.

The transformation from $g(t)$ to $g_2(s)$ does not change the nature of the g function. The derivative $g_2^{'}(s)$ decreases with s since $g_2^{''}(s)$ is necessarily negative.

Consider a $g_2(s)$ function that, in addition to being concave, also is restricted by two further criteria: $g_2^{'}(0) = 1$ and $g_2^{'}(\infty) = 0$. Note that this implies that $g_2(s) = s$ near the origin and is asymptotically flat–two reasonable assumptions.

With these assumptions in place, consider a transformation,

$$z_1 = 1 - g_2^{'}(s). \tag{25}$$

Since $g_2^{'}(s)$ varies from one to zero, $z_1(s)$ varies from zero to one. The monotonicity of the transformation requires $\frac{dz_1(s)}{ds} > 0$, which translates to the condition $g_2^{''}(s) < 0$. Thus, the concavity condition has been exploited twice; to ensure the existence of a minimum at $s = f_p^2$ and the existence of an analog line process. We have satisfied our requirement of a minimum of extra constraints on $g_2(s)$.

Now consider a $g_2(s)$, which, in addition to being concave, is restricted by a different set of criteria: $g_2(0) = 0$, $g_2(\infty) = 1$, and $g_2'(s)$ goes to zero faster than s goes to infinity. The first two criteria are met trivially in most cases. The third criterion demands that $g_2'(s)$ go to zero faster than $\frac{1}{s}$ as s tends to infinity. This is a stronger criterion than the earlier one, which demanded that $g_2'(s)$ merely go to zero as s tended to infinity. Now the *approach* to zero has to be fast.

With the second set of criteria, consider a transformation,

$$z_2 = g_2(s) - s g_2'(s). \tag{26}$$

The monotonicity of the transformation requires $\frac{dz_2(s)}{ds} > 0$, which translates to the condition, $g_2''(s) < 0$, which is the same as required for $z_1(s)$. Once again the concavity of $g_2(s)$ has been exploited twice.

The two transformations can be combined into one transformation that needs no further assumptions. Consider

$$\begin{aligned} z &= \delta(1 - g_2'(s)) + (1 - \delta)(g_2(s) - s g_2'(s)), \\ &= \delta\ z_1(s) + (1 - \delta)\ z_2(s),\ \delta\ \in\ [0, 1], \end{aligned} \tag{27}$$

where z is a convex combination of z_1 and z_2. The external parameter δ can be adjusted to satisfy other criteria. When z is interpolated between z_1 and z_2, the constraints from z_1 and z_2 both are active. For each value of δ in $[0, 1]$, we get a transformation. We refer to this as a *spectrum* of line processes.

Until now, we have not made a firm distinction between using a sequence of $g_2^{(k)}$ functions and using just one g_2 function. A sequence of g_2 functions typically is used in deterministic continuation schemes, whereas solitary g_2 functions are used in stochastic relaxation algorithms [17].

4 General Solutions in the Presence of Interactions

Once we have an equivalent energy function defined over the intensities and any attribute process (gradient magnitude squared, line process), it is natural to ask if the weak membrane[‡] model can be improved with the addition of extra constraints to the energy function. When suitable constraints are added, the next question is how the new energy function with constraints relates to the weak membrane and in what sense does it improve upon it. From a physical point of view, the additional constraints

[‡]In general, the $g_2(s)$ function need not correspond to the weak membrane.

are added to enforce prior beliefs held by us. However, the interaction between the additional constraints and the previous model is non-trivial. These are common problems of complex systems, since it usually is difficult to understand the extent of interaction among simple subsystems [9].

We examine the effect of adding interaction terms that benefit the organization of the line process. We are interested in keeping the interactions sufficiently general and yet deriving closed-form solutions for the line process in terms of its neighbors.[§]

The natural domain for adding interaction energy terms is the analog line process. The spectrum line process sees the following effective energy,

$$E(s(z)) = g_s(s(z)) + \sum_{i \in N(i)} V_c(z, \mathcal{N}(z)), \tag{28}$$

where i is the summation index over the neighbors $\mathcal{N}(z)$ of z. We now restrict the interactions to products of the line processes with weighting factors dependent on the topology.[¶]. This means we have terms like $zz_a z_b$, but not $z^2 z_b^2$, etc. One consequence is that the derivative of V_c with respect to z is just a function of the neighbors,

$$\frac{\partial V_c}{\partial z} \stackrel{\text{def}}{=} F_{\text{int}} \tag{29}$$

where F_{int} can be treated as a number when we are just interested in finding the minimum with respect to z, keeping the neighbors fixed.

When we attempt to solve for the line process in the presence of interactions, we get

$$\frac{dg_s}{ds} / \frac{ds}{dz} + F_{\text{int}} = 0. \tag{30}$$

The first term is different for different line processes and this is why the choice of the line process is crucial for success of the algorithm. Now, $\frac{dg_s}{ds} = g_2''(s)(f_p^2 - s)$. When the spectrum line process is used, we get

$$\frac{dz}{ds} = -g_2''(s)(s(1-\delta)+\delta)$$

and

$$\frac{dg_s}{dz} = -\frac{f_p^2 - s}{s(1-\delta)+\delta}. \tag{31}$$

We now can solve for s in Eq. (30), giving

$$s = \frac{f_p^2 - \delta F_{\text{int}}}{1 + (1-\delta) F_{\text{int}}} \tag{32}$$

[§]The neighbors need not have converged to their respective minima.

[¶]The weights have to be the same for similar cliques.

We have solved for s instead of z, since it emerged naturally from the equations. Solving directly for z in the general case is quite difficult without going through an intermediate route. Once s is known, z can be recovered through the spectrum line process transformation. We now specialize to the two ends of the spectrum (z_1 and z_2). For $z = z_1$,

$$s = f_p^2 - F_{\text{int}}, \tag{33}$$

and for $z = z_2$,

$$s = \frac{f_p^2}{1 + F_{\text{int}}}. \tag{34}$$

When F_{int} is negative, s exceeds f_p^2, indicating a *hysteresis* like role of the interactions. When F_{int} is positive, a *non-maximum suppression*-like behavior is observed.

5 Continuation Method Using GNC Sequence

In this section, we specialize to the GNC sequence and develop the specific algorithms resulting from application of the results of the previous section. The GNC sequence is indexed by the control parameter c and is constructed by piecewise polynomials [3]. The first stage is convex and c varies from low to high values:

$$g_2(s) = \begin{cases} s & 0 \leq s \leq \frac{c}{1+c} \\ 2\sqrt{sc(1+c)} - c(1+s) & \frac{c}{1+c} \leq s \leq \frac{(1+c)}{c} \\ 1 & s \geq \frac{(1+c)}{c} \end{cases}, \tag{35}$$

$$g_2'(s) = \begin{cases} 1 & 0 \leq s \leq \frac{c}{1+c} \\ c\left(\sqrt{\frac{1+c}{sc}} - 1\right) & \frac{c}{1+c} \leq s \leq \frac{(1+c)}{c} \\ 0 & s \geq \frac{(1+c)}{c} \end{cases}, \tag{36}$$

$$g_2''(s) = \begin{cases} 0 & 0 \leq s \leq \frac{c}{1+c}. \\ -\frac{1}{2}\sqrt{\frac{c(1+c)}{s^3}} & \frac{c}{1+c} \leq s \leq \frac{(1+c)}{c} \\ 0 & s \geq \frac{(1+c)}{c} \end{cases}. \tag{37}$$

We see that the GNC sequence satisfies all the criteria for obtaining the spectrum line process: $g_2(s) = 0$ for $s = 0$, $g_2(s) = 1$ for $s \to \infty$, $g_2'(0) = 1$ for $s = 0$, $g_2'(s) = 0$ for $s \to \infty$, $g_2(s)$ exhibits fast convergence to zero as $s \to \infty$, and $g_2''(s)$ is negative. A crucial point is that these criteria are met because of the construction of the GNC sequence using piecewise polynomials that translate into thresholds for the line process.

Now we add the interaction terms that are a special case of the interaction terms of the previous section:

$$\begin{aligned}\mathcal{H}(\mathbf{f},\mathbf{v},\mathbf{h}) &= \mathcal{H}_b(\mathbf{f},\mathbf{v},\mathbf{h})\\ &-\alpha\epsilon_1\sum_{\{i,j\}}\Big(v(i,j)v(i,j+1)+h(i,j)h(i+1,j)\Big)\\ &+\alpha\epsilon_2\sum_{\{i,j\}}\Big(v(i,j)v(i+1,j)+h(i,j)h(i,j+1)\Big),\end{aligned} \tag{38}$$

where the parameters ϵ_1 and ϵ_2 are greater than zero. Physically, we are trying to decrease the penalty on a vertical (horizontal) line if its vertical (horizontal) neighbors are *on* and increase the penalty on a vertical (horizontal) line if its adjacent vertical (horizontal) neighbors are *on*. This corresponds to encouraging hysteresis [4], that leads to the formation of unbroken contours and non-maximum suppression [5] which prevents the formation of multiple edges and adjacent parallel lines. Hysteresis and non-maximum suppression are used routinely in early vision tasks [4]. The penalty is controlled by the parameter ϵ (which subsumes ϵ_1 and ϵ_2). The hysteresis term induces mutual cooperation between line processes that are part of a line segment. The non-maximum suppression term results in competition and a winner-take-all situation [18] between adjacent line processes.

Proceeding along the lines suggested in the previous section, we get for the vertical line process (the method being identical for the horizontal line process),

$$s_v(i,j) = \frac{f_x^2(i,j) + \alpha/\lambda^2\ \delta\left(\epsilon_1\overline{v^v(i,j)} - \epsilon_2\overline{v^h(i,j)}\right)}{1-(1-\delta)\left(\epsilon_1\overline{v^v(i,j)} - \epsilon_2\overline{v^h(i,j)}\right)}, \tag{39}$$

where $\overline{v^v(i,j)} = \frac{v(i,j+1)+v(i,j-1)}{2}$ and $\overline{v^h(i,j)} = \frac{v(i+1,j)+v(i-1,j)}{2}$. Corresponding to the spectrum line process, we have a free parameter δ. When δ is chosen close to zero, the interactions are not that significant in the final result. This corresponds to assuming that there are far more *zeros* than *ones* in the image– a valid assumption. When δ is chosen close to one, the effect of the interactions is much higher. A peculiar thing happens for very low values of c with δ set to one. The line processes self-organize into a set of alternating lines creating a *checkerboard* effect. To see this, consider the following. As a first approximation, ignore the data term. This can be justified when λ and α are large. The energy function can be minimized by setting all alternate vertical and horizontal lines *on*. Only alternate lines are turned on because the non-maximum suppression term acts as a winner-take-all and the support for the line process is the first-order neighborhood. The simple pattern of self-organization illustrates an important principle: the line interactions case is no mere extension of the weak membrane. We have shown that the influence is global [7, 19]. It is likely that more complex and interesting patterns arise when the interactions are extended beyond this initial scheme [20].

When δ is zero, the energy function can be minimized for low values of c by setting the intensities equal to the noisy image and the lines to zero. In fact, the energy itself is equal to zero. As c is increased, more lines become active and the interaction terms slowly come to play. When $\delta = 1$, we get the checkerboard. Since most images contain more empty spaces than edges, the conservative scenario is to keep δ close to zero, thereby introducing interactions gradually.

The adiabatic approximation does not seem to hold for $\delta = 1$. ICM takes much longer to converge in this case. This is to be expected, since the interactions are most significant now. We can no longer expect a close relationship between the GNC sequence and the new cost function.

For us to obtain *good* suboptimal solutions, we require the new cost function to be closely related to the GNC sequence. The new sequence is closely related to the GNC sequence if the adiabatic approximation holds. The adiabatic approximation holds if the line processes still are fast after the addition of interactions. The line processes still are fast if the effect of the interactions is localized. The effect of the interactions is localized when δ is closer to zero. Localized interactions imply that ICM will be fast and that the algorithm will deviate from the GNC algorithm only in areas where the extra constraints are strictly necessary. The interactions systematically enforce hysteresis and non-maximum suppression and do not have any long-range effects.

- Description of the Algorithm:

  1. Set $c = c_*$ (usually, $c_* = 0.25$).
  2. Run the CG algorithm on the intensities.
  3. Update the s processes at the end of the CG iterations using ICM till convergence.
  4. Return to Step 2 til convergence.
  5. Increase c, usually $c = 2^k c_*, k = 0, 1, ...$
  6. Return to Step 2 until the final $c = c^*$ is reached.

The CG algorithm that we have used is the *Polak–Ribiere* line search extension method [21].

6 Experimental Results and Discussion

We executed the algorithm on a noisy, aerial image.

Figures 2a and 2b contain the aerial view of an airport and the corresponding degraded image (5dB, variance=147). Figures 3 and 4 are obtained from the GNC and our algorithm with the same parameters. The line process image using our algorithm is vastly improved. The variation of the parameter δ is the focus of Figs. 5 and 6. When δ is set to zero and ϵ set to 0.5, we obtain results similar to Figs. 3 and 4. However, when δ is set to one, the effect of the interactions and their limitations can be seen clearly. The line process images show a marked preference for vertical and horizontal lines. This is due to our neglect of the contribution from diagonal line processes and the deviation from the properties of the GNC sequence. Values of δ closer to one represent a departure from our major objective to achieve *good* suboptimal solutions using the properties of the GNC sequence. When the interactions dominate, the adiabatic approximation no longer holds and as can be seen from Figs. 5 and 6, there no longer is any relationship between the GNC sequence and our algorithm.

7 Conclusions

We highlight here some of the new ideas and results in this chapter. We have presented new energy functions for MAP estimation and shown the importance of the adiabatic approximation in deriving this new energy function from other continuation schemes like the GNC and MFT methods [7, 22]. We have suggested an algorithm to minimize the new energy functions using a combination of CG and ICM.

The same approach is valid for general Markov random field (MRF) models of the underlying image. Also, further research is needed to move the validity of the adiabatic approximation to higher levels of sophistication. The more fundamental problem of the unstable versus the stable modes remains unsolved. Ideally, we would like to reparametrize the cost function in terms of the unstable modes. The unstable modes for general cost functions would be a linear combination of the intensities and the discontinuities [9]. These problems present interesting areas for future research.

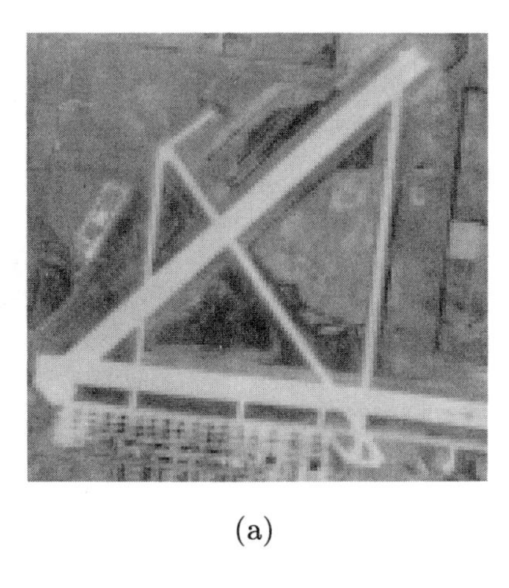

(a)

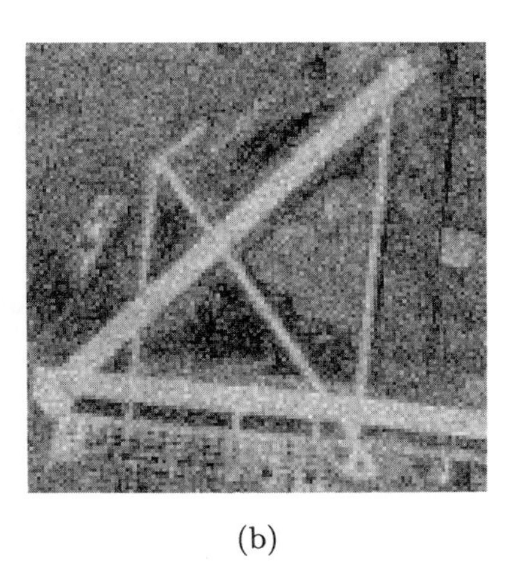

(b)

Figure 2: (a) Original airport, (b) noisy airport.

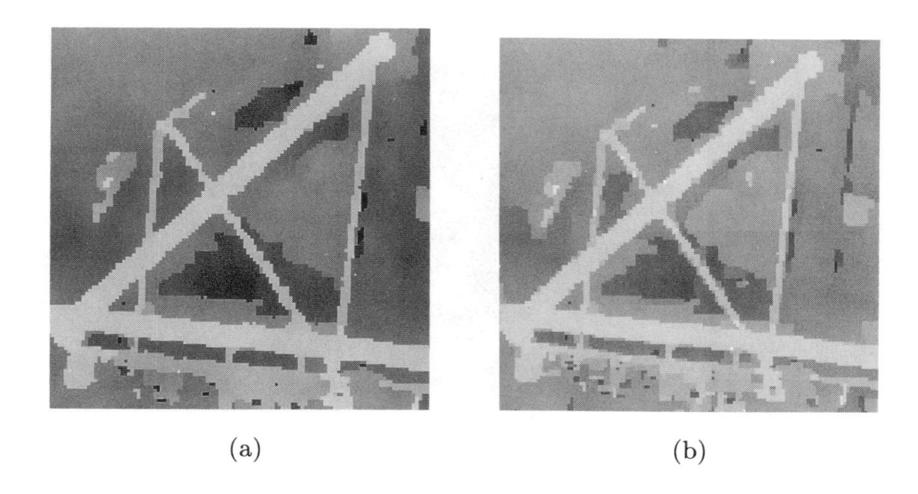

Figure 3: (a) Restored image using the GNC algorithm, (b) restored image using our algorithm.

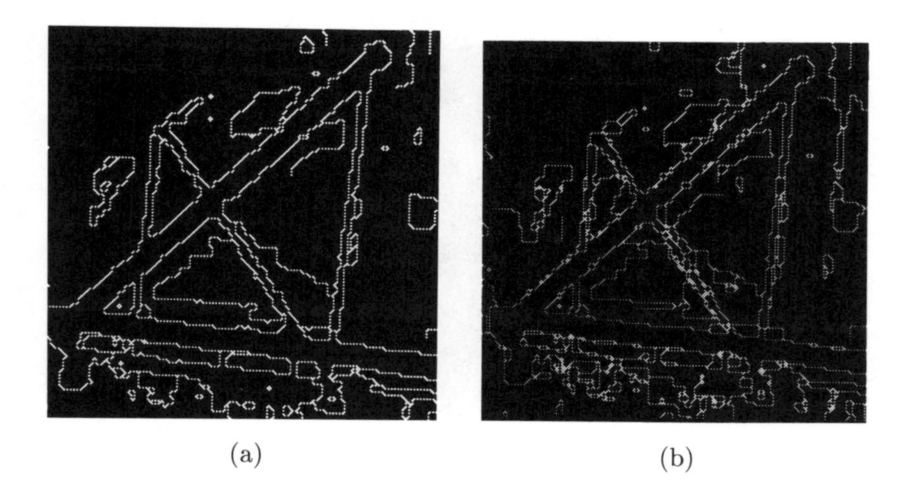

Figure 4: (a) Line process image using the GNC algorithm, (b) line process image using our algorithm.

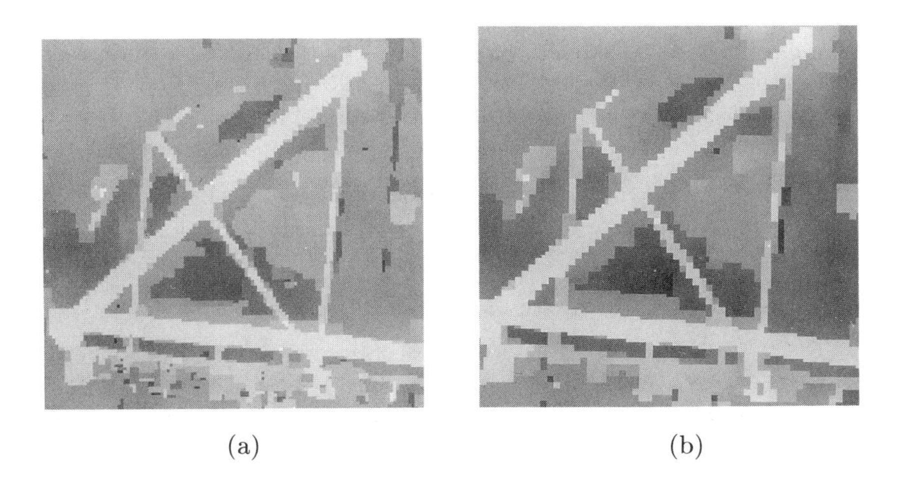

(a) (b)

Figure 5: (a) Restored airport with $\epsilon = 0.5$ and $\delta = 0$, (b) restored airport with $\epsilon = 0.5$ and $\delta = 1$.

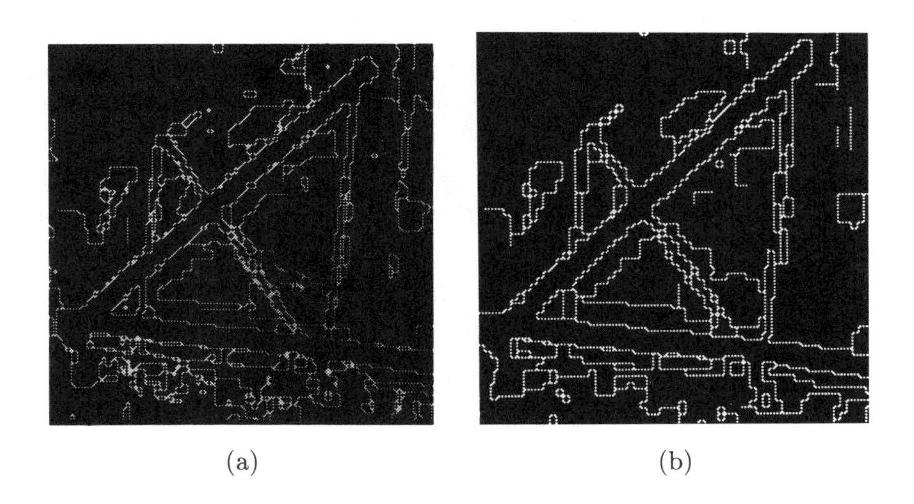

(a) (b)

Figure 6: (a) Line process airport with $\epsilon = 0.5$ and $\delta = 0$, (b) line process airport with $\epsilon = 0.5$ and $\delta = 1$.

Bibliography

[1] S. Geman and D. Geman, "Stochastic Relaxation, Gibbs Distributions and the Bayesian Restoration of Images," *IEEE Trans. Pattern Analysis and Machine Intelligence*, vol. PAMI-6, pp. 721–741, November 1984.

[2] A. Morgan, *Solving Polynomial Systems Using Continuation for Engineering and Scientific Problems,* Prentice-Hall, Englewood Cliffs, NJ, 1987.

[3] A. Blake and A. Zisserman, *Visual Reconstruction*, MIT Press, Cambridge, MA, 1987.

[4] J. Canny, "A Computational Approach to Edge Detection," *IEEE Trans. Patt. Anal. and Mach. Intell.*, vol. PAMI-8, pp. 679–698, Nov. 1986.

[5] R. Nevatia and K. R. Babu, "Linear Feature Extraction and Description," *Computer Graph. and Image Processing*, vol. 13, pp. 257–269, July 1980.

[6] T. Poggio, E. B. Gamble, and J. J. Little, "Parallel Integration of Vision Modules," *Science*, vol. 242, pp. 436–440, October 1988.

[7] D. Geiger and F. Girosi, "Parallel and Deterministic Algorithms for MRFs: Surface Reconstruction and Integration", Technical Report A. I. Memo, No. 1114, Artificial Intelligence Lab, MIT, June 1989.

[8] Y. G. Leclerc, "Constructing Simple Stable Descriptions for Image Partitioning," *International Journal of Computer Vision*, vol. 3, pp. 73–102, 1989.

[9] H. Haken, *Synergetics–An Introduction*, *Springer Series in Synergetics*, Springer-Verlag, vol. 1, 3rd edition, New York, 1983.

[10] J. Besag, "On the Statistical Analysis of Dirty Pictures," *Journal of the Royal Statistical Society B*, vol. 48(3), pp. 259–302, 1986.

[11] S. Geman and D. E. McClure, "Statistical Methods for Tomographic Image Reconstruction," *Proceedings of the 46th Session of the ISI, Bulletin of the ISI*, 1987.

[12] D. Shulman and J. Y. Hervé, "Regularization of Discontinuous Flow Fields," *Proceedings of the IEEE Compt. Soc. Workshop on Visual Motion '89*, pp. 81–86, March 1989.

[13] H. Haken, *Advanced Synergetics"*, *Springer Series in Synergetics*, vol. 20, 2nd edition, Springer-Verlag, New York, 1987.

[14] J. J. Hopfield, "Neurons with Graded Response Have Collective Computational Properties like Those of Two-State Neurons," *Proc. Natl. Acad. Sci., U.S.A.*, vol. 81, pp. 3088–3092, 1984.

[15] A. L. Yuille, "Energy Functions for Early Vision and Analog Networks," *Biological Cybernetics*, vol. 61, pp. 115–123, 1989.

[16] A. Rangarajan and R. Chellappa, "Generalized Graduated Non-Convexity Algorithm for Maximum A Posteriori Image Estimation," *Proceedings of the International Conf. on Pattern Recognition, ICPR-90*, Atlantic City, NJ, June 1990.

[17] D. Geman and G. Reynolds, "Constrained Restoration and the Recovery of Discontinuities," *IEEE Trans. Patt. Anal. and Mach. Intell.*, vol. PAMI-14, pp. 367–383, March 1992.

[18] D. Geiger and A. Yuille, "A Common Framework for Image Segmentation," *Proceedings of the International Conf. on Pattern Recognition, ICPR-90*, Atlantic City, NJ, June 1990.

[19] S. Geman, Remarks at the MRF Workshop in San Diego, Stuart Geman emphasized that a fully connected graph is obtained after the line process is integrated out of the Gibbs distribution".

[20] A. C. Newell, "The Dynamics and Analysis of Patterns", In D. L. Stein, editor, *Lectures in the Sciences of Complexity*, pp. 107–173, Addison-Wesley, Menlo Park, CA, 1989.

[21] D. Luenberger, *Linear and Nonlinear Programming*, Addison-Wesley, Menlo Park, CA, 1984.

[22] J. Zerubia and R. Chellappa, "Mean Field Approximation Using Compound Gauss–Markov Random Field for Edge Detection and Image Estimation", *Proc. ICASSP '90, IEEE Conf. on Acoust., Speech, and Signal Processing*, pp. 2193–2196, April 1990.

A Comprehensive Statistical Model for Single-Photon Emission Tomography

Stuart Geman, Kevin M. Manbeck, and Donald E. McClure
Division of Applied Mathematics
Brown University
Providence, Rhode Island

1 Introduction

The field of tomography is broadly concerned with the application of non-invasive imaging techniques to determine the internal structure of three-dimensional objects. Tomographic images must be reconstructed, typically with the aid of computers and computational algorithms, from data that are collected or observed external to the object under study. In the industrial environment, tomographic techniques have found successful application to nondestructive internal inspection; but the primary applications of tomography have been in medicine, where it has been widely used over the past two decades. In particular, computed tomography (CT) scanning, ultrasound, and, more recently, magnetic resonance imaging (MRI) have revolutionized diagnostic medicine.

Somewhat less successful has been the introduction of emission tomography, positron emission tomography (PET), and single-photon emission computed tomography (SPECT) to the clinical setting. In these modalities, one seeks to learn the internal distribution of a pharmaceutical. This is accomplished by introducing a radiopharmaceutical—a pharmaceutical that has been chemically combined with a radioactive isotope—into a patient, and then measuring, externally, a distribution of radioactive events. These modalities produce generally lower-quality images, with lower spatial resolution, than either CT or MRI.

Despite this relative lack of resolving power, PET and SPECT hold much promise because they can measure *metabolic activity*, which is impossible to measure with CT or, thus far, with MRI. Furthermore, the development of highly discriminating monoclonal antibodies for use in SPECT promises to allow clinicians to accurately locate specific tissue types, both

ISBN 0-12-170608-7

normal and pathological.

The relatively low cost of SPECT systems, the ability of SPECT to measure metabolic activity, and the introduction of new monoclonal antibody imaging agents combine to make SPECT a potentially important clinical tool. Unfortunately, the application of SPECT is currently limited by its poor image quality.

It is popular lore—but incorrect—that a major technological or scientific advance in one tomographic modality (recently MRI) will antiquate and supplant the other modalities. In fact, each modality has a distinct diagnostic or therapeutic application, and it is therefore imperative to develop each imaging method to its fullest potential.

In this chapter, we will describe mathematical methods that may significantly improve SPECT reconstruction techniques. Previous experience with these methods [5], [16], [4], [29], [25], [9], [10], [13], has demonstrated a potential utility of Markov random field (MRF) image models, in a Bayesian framework, for various image restoration and reconstruction tasks. There have been, in particular, several studies of MRF-based Bayesian methods ([11], [12], [21], [22], [27], [14], [7], [6]), and related regularization methods ([28], [34]) for emission tomography. In this chapter, we demonstrate that these methods are well-suited to take advantage of the rather detailed available information on the physics of SPECT imaging, and we derive, analytically, some isotropic properties of the particular random fields employed.

In Section 2 we will briefly review SPECT imaging, and identify the primary physical factors that influence the collection of data in a SPECT imaging session. The Bayesian approach is reviewed in Section 3, emphasizing the dual requirements for both a likelihood model, describing the distribution of the externally observable data for any fixed internal configuration of isotope intensity, and a prior model, describing *a priori* likely and unlikely configurations. The likelihood (or data) model is studied in detail in Section 4. The primary physical effects that govern the SPECT imaging modality have been modeled up to machine-specific parameters. Experiments were designed to measure these parameters, and the results were used to build a machine-specific model. Reconstruction experiments were performed with phantoms of known structure and with real patient data. Results are reported in Section 5. A detailed discussion of the prior probability model is postponed to Section 6, where the model is presented together with an analytic study of its isotropic properties.

2 Physics of SPECT

In a SPECT imaging session, a patient is injected with a pharmaceutical that is tagged with a radioactive isotope. The pharmaceutical is application-dependent and is chosen to concentrate in a region of interest. As this so-called *radiopharmaceutical* undergoes radioactive decay, photons are released from the patient's body and counted by a gamma camera that rotates around the patient and collects data at numerous angles. Once these data are recorded, one is faced with the difficult task of reconstructing a map of the isotope, and thereby the pharmaceutical, concentration.

The physics that govern the release of a photon and the path that a photon travels are well-understood:

1. Photon release follows a Poisson distribution with mean proportional to isotope concentration. In a patient study, the concentration of isotope must be kept low to avoid the effects of high dose radiation. This adversely affects the signal-to-noise ratio and contributes to low-quality reconstructions.

2. After being released, a photon may—at random—experience absorption or Compton scatter. These two effects, together, are known as *attenuation.* A patient's body contains essentially three types of attenuating material: bone, soft tissue, and lung. Hence, the attenuation of a photon is intrinsically nonuniform. The amount of scatter is depth-dependent: photons that are released deep in an attenuating substance have a higher probability of scattering than those that are released near the surface. The effects of absorption and scatter further contribute to degraded reconstructions and low resolution. See Fig. 1 a and b.

3. The geometry of the gamma camera introduces a *collimator effect,* which amounts to a depth-dependent blurring of the reconstructed image. See Fig. 1 c.

The collective effect of these influences on the observed photon counts can be accurately modeled and thereby used to improve reconstruction quality. Since the release of photons, absorption, and Compton scatter are all inherently random events, it is natural to model the SPECT problem in a statistical framework. In this regard, we follow the lead of Shepp and Vardi [33], who were the first to derive reconstruction algorithms for emission tomography from statistical models and basic principles of inference. (Earlier, Rockmore and Makovski [32] recognized the natural role of statistical models in formulating the emission tomography reconstruction problem.)

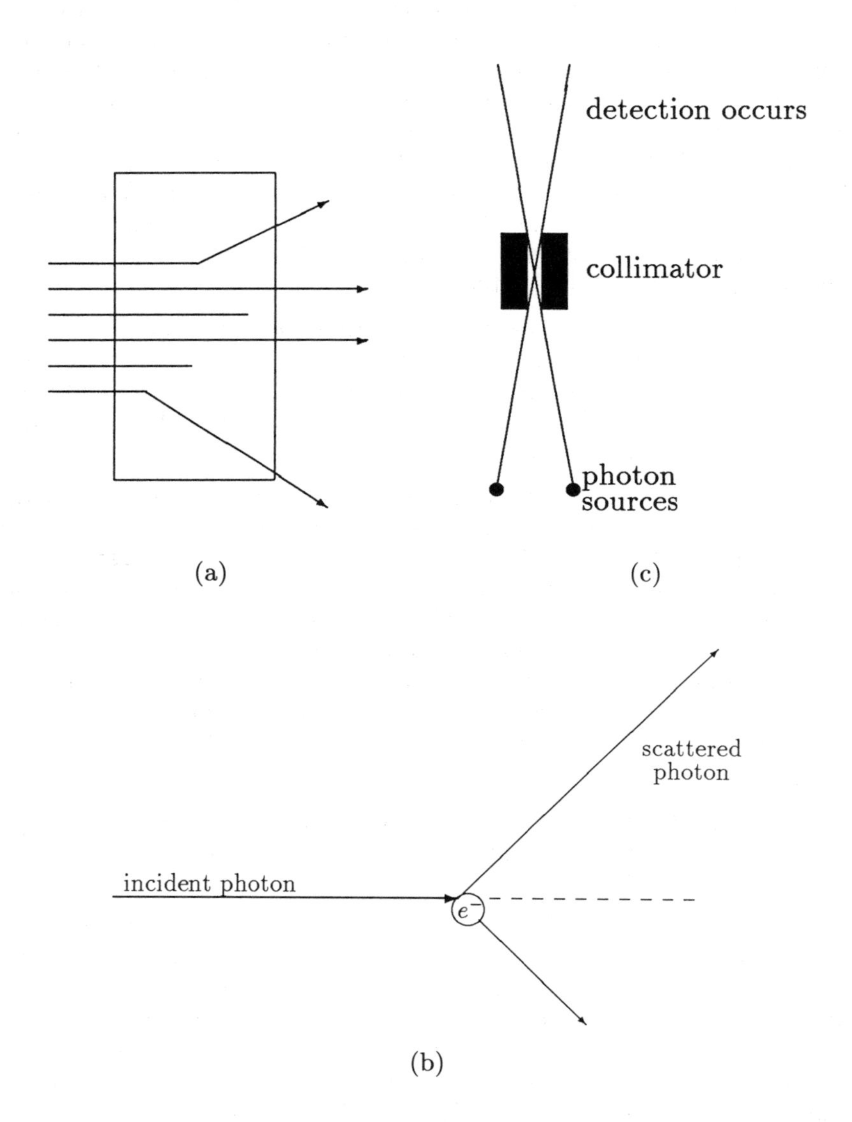

Figure 1: Relevant physical effects. (a) A photon may be attenuated by either being deflected or absorbed. (b) A photon may change direction and energy via Compton scatter with an electron, e^-. (c) Imperfect collimation introduces the solid angle effect: in addition to accepting true projections, a collimator will accept *near* projections.

3 Bayesian Formulation

Commercially available SPECT machines reconstruct isotope concentrations using a version of the filtered backprojection (FB) algorithm. This approach is limited in its ability to accommodate the degrading effects discussed earlier. In particular, FB techniques ignore the Poisson nature of decay and typically allow for only uniform absorption and depth-*independent* scatter and blur. As a result, the commercially available reconstruction methods are characterized by a lack of resolution and other blurring artifacts. Oftentimes, it is safe to ignore the random elements in an image restoration task. For example, backprojection is used successfully in CT where the signal-to-noise ratio is favorable, and scattering and collimator effects are negligible. However, the signal-to-noise ratio in SPECT is low, and absorption, scattering, and collimator effects are highly significant.

The model developed here will explicitly account for quantum noise, nonuniform absorption, and depth-dependent scatter and blur. The approach is based on realistic models of the stochastic and deterministic components of a SPECT data acquisition system.

The model involves the unknown isotope concentration map, which we denote by X, and the observable photon counts, which we denote by Y. The reconstruction problem is to estimate X from Y. The physics of SPECT imaging determine that Y is distributed according to a Poisson distribution with mean,

$$E[Y] = \mathcal{A}X, \tag{1}$$

where the discrete *modified Radon transform* $\mathcal{A}$ describes the process by which an image X is transformed into data Y. $\mathcal{A}$ incorporates the effects of nonuniform absorption and depth-dependent scatter and blur. Later (in Section 4), we will discuss how $\mathcal{A}$ can be *measured* with a simple experimental apparatus. We thus specify a Poisson *likelihood* model $P(Y|X)$ (with mean $\mathcal{A}X$) to evaluate the probability that a given isotope map gives rise to the observed data.

At this stage, one could, in principle, solve the inverse problem by finding that X that maximizes $P(Y|X)$. This is, in fact, the *maximum likelihood* approach proposed by Shepp and Vardi [33] for PET imaging. (See also Rockmore and Makovski [32].) Unfortunately, maximum likelihood reconstructions are critically dependent on the chosen pixel resolution, and in general are badly degraded at clinically interesting pixel resolutions. This difficulty is not entirely unexpected: the reconstruction problem is inherently a nonparametric estimation problem. One seeks to reconstruct a completely general function describing the internal radiopharmaceutical concentration. As is well-known in the context of nonparametric estimation (See e.g., Grenander [15]), as the pixel resolution becomes finer, the variance of the maximum likelihood estimator increases. Maximum likelihood *per se* is generally not consistent for nonparametric estimation. Some sort of regularization is needed.

We regularize using a Bayesian framework in which a *smoothing prior* $P(X)$ is placed on the isotope concentration map X. More specifically, we propose to model, via a prior probability distribution, the expectation that isotope concentration maps are likely to consist of locally smooth regions separated by discontinuities (*boundaries*). (See Section 6 for details.) Applying Bayes's formula, we determine the *posterior distribution,*

$$P(X|Y) = \frac{P(Y|X)P(X)}{\sum_X P(Y|X)P(X)}. \tag{2}$$

It is at this stage that we are ready to solve the inverse problem by estimating X from the data Y. In particular, the estimate we will seek is the posterior mean ($\sum_X XP(X|Y)$), or, in practice, an approximation thereof.

Obviously, a central role is played by the particular expectations modeled in the prior. One is tempted to design a highly *informative* prior, perhaps anticipating, for example, known anatomical shapes and their expected locations. While restorations based on problem-specific prior distributions can be effective, we have chosen instead a more conservative and universal approach by describing a rather general model (the phi model Section 6), which has been applied in such diverse areas as tomography [12], movie restoration [13], infrared image enhancement [8], and astronomical image restoration [24]. The phi model is minimal in the sense that it encourages only the most basic regularity properties: local smoothness and the existence of discontinuities associated with boundaries. The idea, of course, is to lessen the chances of introducing artifacts, via the prior probability distribution, into the reconstructions. As we shall see (Section 6), the model also has, at least approximately, a desirable invariance property with respect to rotation.

4 SPECT Model

Accurately defining the modified Radon transform $\mathcal{A}$, introduced in the previous section, is essential to the proper functioning of the reconstruction process. To describe $\mathcal{A}$, it is necessary to be more precise in the formulation of the SPECT model.

In a typical SPECT imaging session, a patient is injected with a radiopharmaceutical and lies face up on a horizontal table. The imaging session begins with the gamma camera directly above and facing the patient. In this position, the camera detects and records photons leaving the patient's body in a certain time period, often 20 seconds. The camera then rotates about an axis parallel to the table, stops after arcing $360/A$ degrees (A is often 64), and collects counts at this new angle. This process is repeated until the gamma camera returns to its original position directly above the patient, at which time the imaging session is over and data have been collected for projections through A different angles.

The gamma camera is discretized into an array of detection bins with R rows and C columns (64×64 is typical). The scintillation crystal is circular, but the array of bins is rectangular, so many of the corner bins do not have any counts. See Fig. 2 for an example of four gamma pictures taken at different angles around a patient's head. Notice that the counts were only collected in a circular region. Also notice the lack of resolution present in the pictures; this is due to the various sources of noise inherent in SPECT. (See Section 2.)

Once the data have been collected, one is faced with the task of reconstructing a map of isotope intensity within the patient. To save on computations, we have followed convention in that the reconstruction of a single transverse slice, r, was accomplished by using only data from row r in each gamma picture, despite the fact that, due to scatter and collimator effects, data in other camera rows are relevant. Evidently, we have not yet made full use of the available information for reconstructions. From the ensuing discussion, it should be clear to the reader that, computational issues aside, there is no difficulty in accommodating a complete three-qdimensional SPECT model within the proposed framework.

The transverse slice under consideration is discretized into an $N \times N$ array of pixels. A common value for N is 64. See Fig. 3 for a diagram of the camera and image in one transverse slice. In a given slice, label the image pixels, or image *sites*, S_i for $i = 1$ to N^2, and label the gamma camera bins B_j for $j = 1, C{*}A$. (Recall that A is the number of discrete angles through which the camera is rotated.) Let X_i be the isotope intensity at site i and let Y_j be the number of photons counted at bin B_j.

There is a standard mathematical/physical argument (which we will not go into) that leads to the conclusion that the observable photon counts, $Y = \{Y_j\}_{j=1}^{C*A}$, are approximately independent Poisson random variables, with means related linearly to the isotope intensity distribution, $X = \{X_i\}_{i=1}^{N^2}$, as in Eq. (1). It is thus the modified Radon transform (MRT), $\mathcal{A}$, that determines the relation between the unknown isotope intensities and the observable photon counts. The j, i component of the MRT, $\mathcal{A}_{ji}$, is proportional to the probability that a photon emitted at S_i is recorded at B_j.

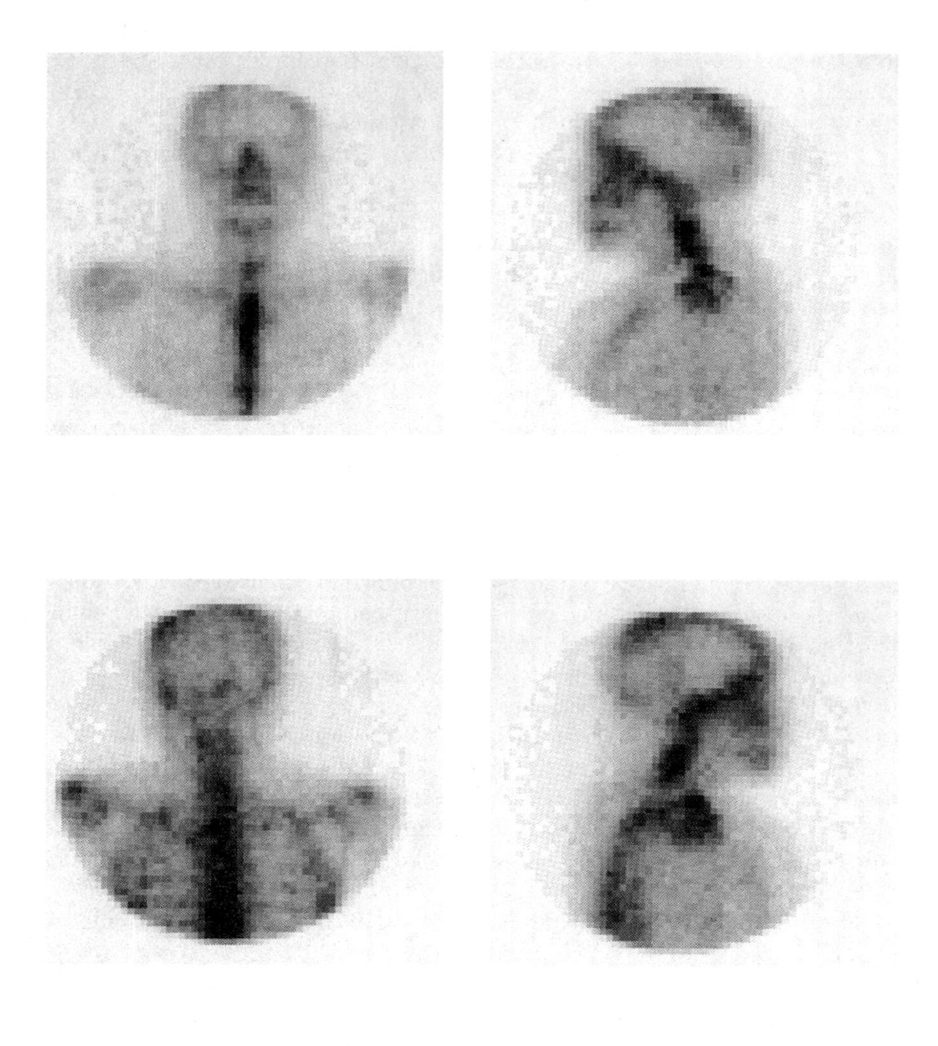

Figure 2: Four gamma pictures captured as the camera rotates around a patient's head. Dark areas represent bins with a large number of photon counts. The lack of resolution present in the images is typical of SPECT.

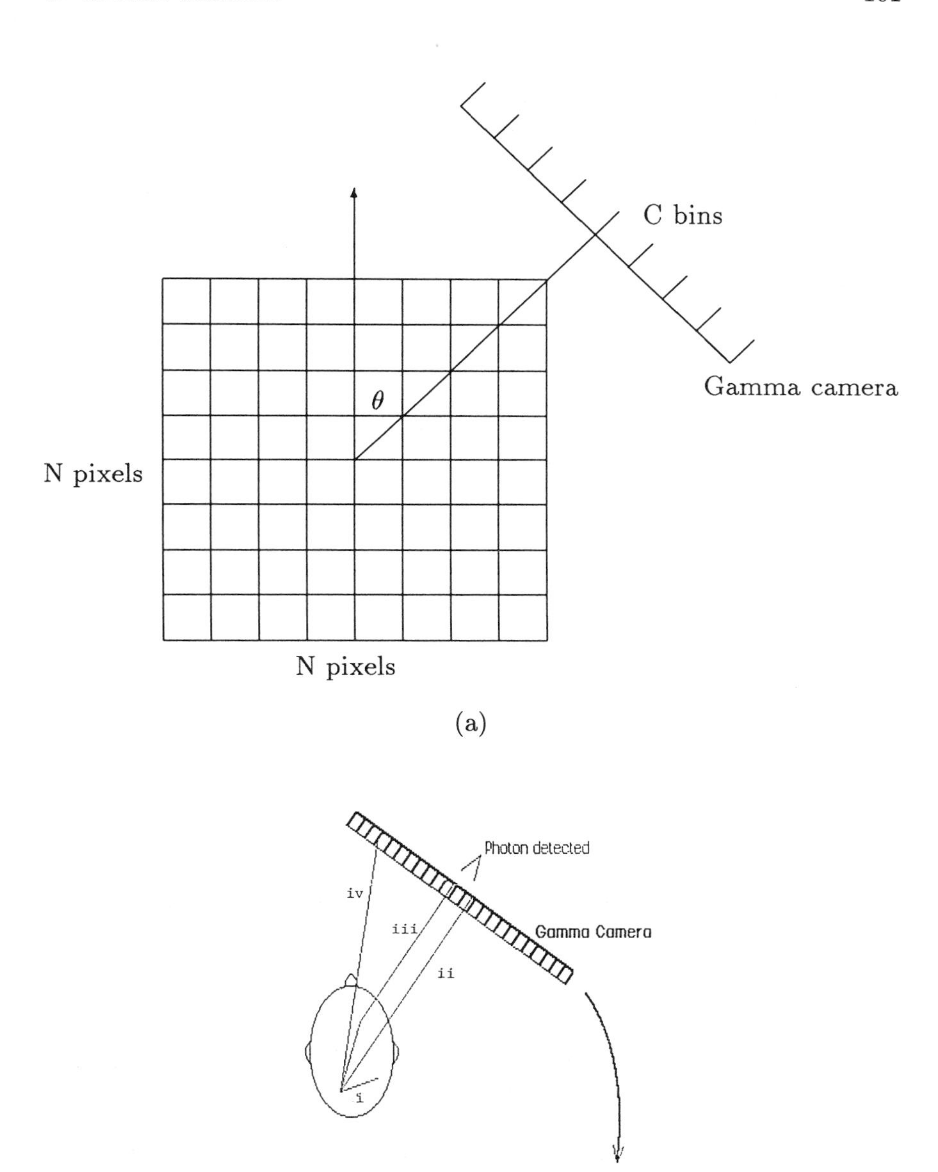

Figure 3: Diagram of a transverse slice. (a) The slice is discretized into an $N \times N$ array. In this figure, the camera has rotated through angle θ. (b) A transverse slice of a patient's head and some of the physical effects modeled in this chapter: i) an absorbed photon, ii) a direct count, iii) a scattered count, iv) a photon rejected by a collimator.

(As we shall see, the proportionality constant is somewhat arbitrary, and will be used to adjust the scale of the estimated isotope intensity values.) The details of absorption, scattering, and collimator effects are modeled through the MRT. An additional effect that could be accounted for in the MRT is the loss and redistribution of an isotope during the imaging session, due to metabolism, diffusion, and radioactive decay. This effect is generally minimal, given the isotopes and pharmaceuticals actually used; we will make the assumption that the distribution and intensity of a radiopharmaceutical is unchanged over the imaging session.

$\mathcal{A}_{ji}$ is modeled through a *detection density*, whose integral over the width of the jth detection bin, B_j, is proportional to $\mathcal{A}_{ji}$. Referring to Fig. 4, the detection density, α, is modeled as a function of: 1) the (perpendicular) distance, d, from location S_i to the camera containing bin B_j; 2) the distance, D, along this same perpendicular, from location S_i to the *boundary* of the object being imaged; and 3) the distance, x, along the camera surface, measured from the perpendicular projection of S_i. Thus, $\alpha = \alpha(D, d, x)$, and

$$\mathcal{A}_{ji} = k \int_{B_j} \alpha(D, d, x) dx. \tag{3}$$

The proportionality constant k is independent of i and j. It is a scale parameter, and its value depends on how we choose physical units such as the time unit and the scale for the observed projection counts Y. For convenience, we adjust k in our experiments so that the reconstructions approximately fill the dynamic range [0,255].

A set of experiments was devised to measure the dependence of the detection density on the three parameters, D, d, and x. In these experiments, a narrow catheter filled with a radioisotope, called a *line source*, is placed in a cylindrical tank. The line source is inclined at an angle θ relative to the imaging table. See Fig. 4. A gamma camera positioned directly above the line source records the photons released as the isotope decays. The observed photon counts form a *line spread function* (LSF). Using data gathered in this manner, it is possible to determine the effects of distance on attenuation, scatter, and imperfect collimation, and, from this, to deduce the dependence of α on D, d, and x. By filling the cylindrical tank with different materials, it is possible to measure these physical effects in relation to the surrounding environment.

The angled line source experiment was performed in two environments: air and water. Air was chosen as a medium for the photons to traverse because it has such a small density that a photon will essentially never be scattered or absorbed. Hence, the degradation process for a photon in

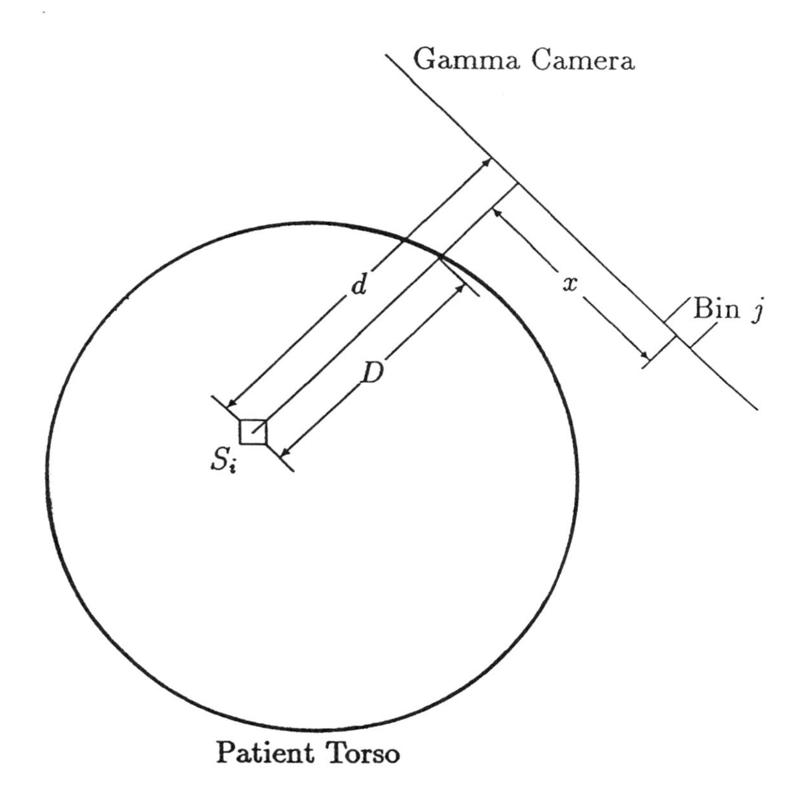

Figure 4: Detection density, due to a source at site S_i, is modeled as a function of: 1) distance d from the source to the camera; 2) distance D from the source to the boundary of the object; 3) distance x from the projection of S_i.

air is relatively simple, and it is possible to isolate the effects of camera geometry. A photon traveling through denser media, on the other hand, will experience a significant amount of absorption and scatter. Water was chosen as an experimental environment because its attenuating properties approximate those of most human body tissue reasonably well. (See, for example, Johns and Cunningham [20].) The two major exceptions to this approximation are bone and lung tissues—the absorption and scatter coefficients for bone are higher than those of water, and the air present in lung tissue dictates that the absorption and scatter coefficients of lung are

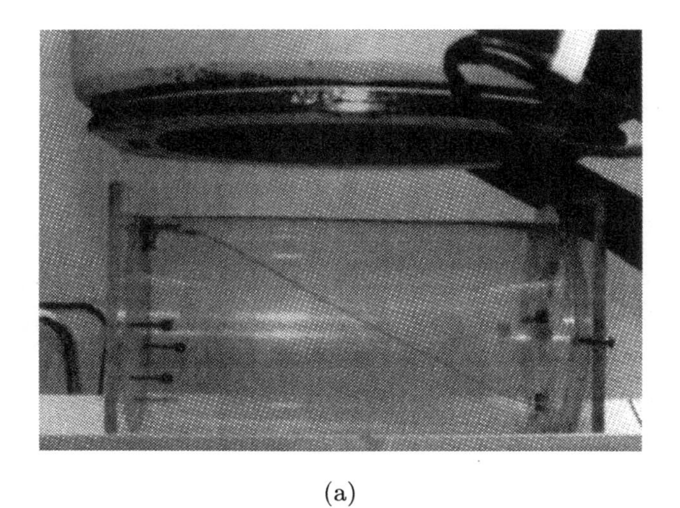

(a)

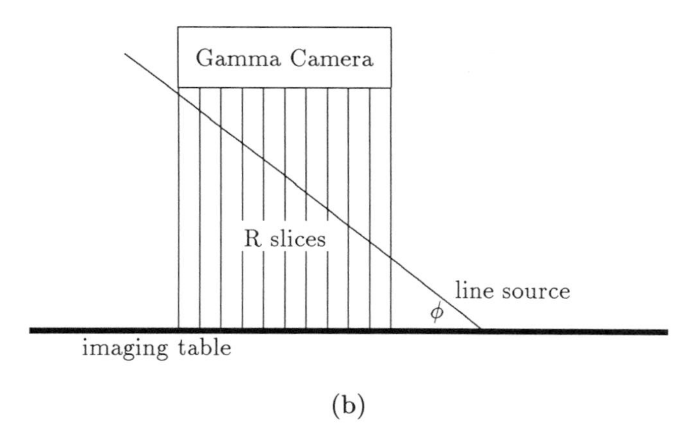

(b)

Figure 5: Experimental apparatus. (a) Photograph of equipment used in line source imaging experiment. The cylinder is resting on the imaging table, and the gamma camera is directly above the cylinder. (b) Diagram of a side view of the line source imaging experiment. Note that different positions on the gamma camera correspond to photons that originated at different distances from the camera.

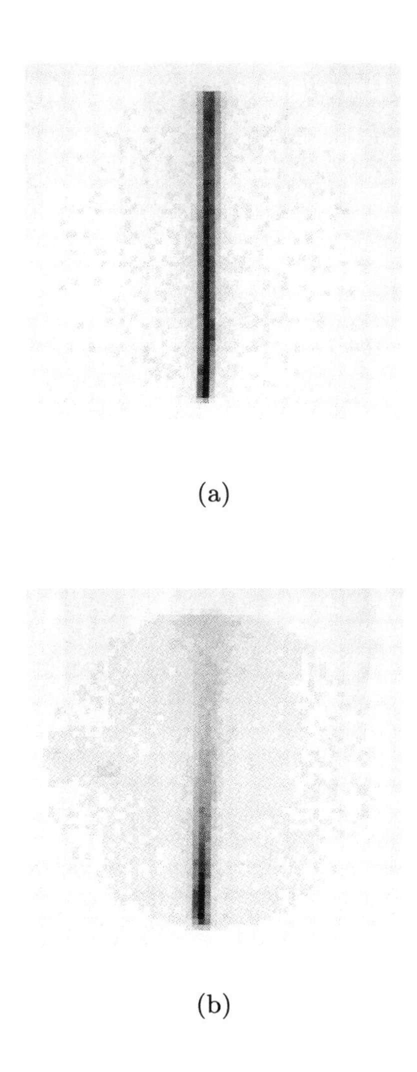

(a)

(b)

Figure 6: Gamma pictures of (a) an angled line source in air and (b) an angled line source in water. The distance between the camera and the line source is greatest at the top of the picture.

lower than those of water. In the experiments reported here (Section 5), we have used the attenuating characteristics of water to approximate all body tissues. We thus assume uniform attenuation. This approximation is not necessary; the reconstruction method is unchanged by the inclusion of nonuniform attenuation. In this case, however, the MRT must be constructed to reflect the inhomogeneous attenuation function. In any case, even with uniform attenuation, there is an overall depth-dependent scatter and absorption, as we shall now see.

Results of the angled line source experiments, for air and water, are shown in Fig. 6. The model used here to fit these results is essentially the one presented in Penny *et al.* [31]. (Beck [2] proposed the same form of decomposition of the line spread function, and strong empirical evidence in support of the model was presented in Floyd *et al.* [7].) There are two contributions to the detection density: *direct* contribution from detected photons that have not been scattered, and a *scatter* contribution from photons that have undergone Compton scatter prior to being detected. Thus,

$$\alpha(D,d,x) = \alpha_{\text{direct}}(D,d,x) + \alpha_{\text{scatter}}(D,d,x). \tag{4}$$

It is convenient to factor each of the three detection densities into the product of a depth-dependent attenuation term and a residual line spread function, where, by definition, the former depends only on D and d, and the latter integrates to one with respect to x:

$$\begin{aligned}
\alpha(D,d,x) &= A(D,d)\zeta(x;D,d), \\
\alpha_{\text{direct}}(D,d,x) &= A_{\text{direct}}(D,d)\zeta_{\text{direct}}(x;D,d), \\
\alpha_{\text{scatter}}(D,d,x) &= A_{\text{scatter}}(D,d)\zeta_{\text{scatter}}(x;D,d), \\
\int_x \zeta(x;D,d) &= \int_x \zeta_{\text{direct}}(x;D,d) = \int_x \zeta_{\text{scatter}}(x;D,d) = 1 \ .
\end{aligned}$$

Modeling the MRT amounts to estimating the four functions, A_{direct}, A_{scatter}, ζ_{direct}, and ζ_{scatter}, from the experimental data.

The contribution from α_{direct} can be deduced from basic physical models for photon attenuation and from the empirical results of the line source experiment in air. Examination of the data (depicted in Fig. 6a) indicates that there is no apparent loss in *total* counts as a function of distance to the camera; the number of counts in each row is essentially the same. On the other hand, the *shape* of observed counts within a row of detectors is well fit by a normal (Gaussian) probability distribution with standard error increasing linearly as a function of distance to the detector. Since essentially all counts depicted in Fig. 6a (line source in air) are direct counts:

$$\zeta_{\text{direct}}(x; D, d) \approx \eta_{\text{direct}}(x; \sigma_{\text{direct}}(D, d)),$$

where

$$\begin{aligned}
\eta_{\text{direct}}(x; \sigma) &= \frac{1}{\sqrt{2\pi\sigma^2}} e^{-x^2/2\sigma^2}, \\
\sigma_{\text{direct}}(D, d) &= \sigma_{\text{direct}}(d) = m_{\text{direct}} d + b_{\text{direct}}.
\end{aligned}$$

m_{direct} and b_{direct} were estimated, by simple regression, to be .0315 and .525, respectively.

As for the attenuation term, $A_{\text{direct}}(D, d)$, the effect is well-known to be exponential in the distance traveled through the uniform attenuating medium. We have already observed that there is essentially no attenuation in air. Therefore, $A_{\text{direct}}(D, d) \approx A_{\text{direct}}(D) = \exp(-\mu D)$. (Because of the proportionality constant k in Eq. (3), A_{direct} is defined only up to an arbitrary multiplicative constant, which we take to be 1.) Coefficients of attenuation μ can be found in the physics and radiology literature for a wide variety of radiation sources and propagation media; see, for example, Johns and Cunningham [20]. The two sources used in our experiments were isotopes of technetium (Tc) and thallium (Tl). The coefficients of attenuation for these isotopes in water are similar, and approximately equal to that of a 150 keV photon source, for which $\mu = 0.15\text{cm}^{-1}$. This means that approximately 15% of the remaining photons are absorbed, or deflected from their original path, for each centimeter traveled. For comparison, values of μ for a 150 keV source and other propagating media are: Muscle—0.155cm^{-1}; Bone—0.246cm^{-1}; Fat—0.137cm^{-1}; Air—$0.16 \times 10^{-3}\text{cm}^{-1}$; and Lucite—$0.17\text{cm}^{-1}$.

The direct model, α_{direct}, together with the angled line source data in water (Fig. 6b) can now be used to develop a model for the contribution from scattered photons, α_{scatter}. One must bear in mind that the counts received from the line source placed in water are the *sum* of direct and scattered contributions. Recalling again that attenuation in air is negligible, the attenuation terms A and A_{scatter} can be assumed to depend on D alone. The overall attenuation, $A = A(D)$, can be inferred from the Fig. 6, panel b, data by examining the total number of counts received as a function of distance through the attenuating medium. The rows of the camera represented at the top of the figure are further from the line source, and these received fewer photons. The actual data indicates that the total number of counts closely follows an exponential decrease with distance, with the coefficient of attenuation approximately equal to $.12\text{cm}^{-1}$. Hence, $A(D)$ is proportional to $\exp(-.12D)$: $A(D) \approx c\exp(-.12D)$. Since $A_{\text{direct}}(D) = \exp(-.15D)$, c governs the ratio of direct to total counts. At $D = 0$, there is no scatter (all counts are direct), which gives the *boundary condition*, $A_{\text{direct}}(0)/A(0) = 1 \Rightarrow c = 1$. We can now integrate Eq. (3) with respect to x to obtain A_{scatter} : $A_{\text{scatter}}(D) \approx \exp(-.12D) - \exp(-.15D)$.

The remaining term is $\zeta_{\text{scatter}}(x; D, d)$, and this, too, can be estimated from the data in Fig. 6b by first accounting for the known contribution from direct counts, via the model $\alpha_{\text{direct}}(D, d, x)$. When the estimated direct count contribution has been subtracted, and when correction has been made for the attenuation term, $A_{\text{scatter}}(D)$, the remaining counts are well fit by a double exponential whose standard error increases linearly with row number, and, therefore, also with distance from the line source. Since there is essentially no scatter in air, the standard error was assumed, *a priori*, to depend solely on the depth of the line source within water, i.e., only on D. Therefore:

$$\zeta_{\text{scatter}}(x; D, d) \approx \eta_{\text{scatter}}(x; \sigma_{\text{scatter}}(D, d)),$$

where

$$\begin{aligned}
\eta_{\text{scatter}}(x; \sigma) &= \frac{1}{\sqrt{2\sigma^2}} e^{-\sqrt{2}|x|/\sigma}, \\
\sigma_{\text{scatter}}(D, d) &= \sigma_{\text{scatter}}(D) = m_{\text{scatter}} D + b_{\text{scatter}}.
\end{aligned}$$

The regression estimates for m_{scatter} and b_{scatter} were .212 and 1.615, respectively.

In summary, the following MRT model was estimated from line source data and was used in the experiments reported in Section 5:

$$\mathcal{A}_{ji} = k \int_{B_j} \alpha(D, d, x) dx,$$

where

$$\begin{aligned}
\alpha(D, d, x) &= e^{-.15D} \frac{1}{\sqrt{2\pi\sigma^2_{\text{direct}}(d)}} e^{-x^2/2\sigma^2_{\text{direct}}(d)} \\
&\quad + \left(e^{-.12D} - e^{-.15D}\right) \frac{1}{\sqrt{2\sigma^2_{\text{scatter}}(D)}} e^{-\sqrt{2}|x|/\sigma_{\text{scatter}}(D)}, \\
\sigma_{\text{direct}}(d) &= .0315d + .525, \\
\sigma_{\text{scatter}}(D) &= .212D + 1.615.
\end{aligned}$$

Given $\mathcal{A}$, and given the isotope intensity distribution $X = \{X_i\}_{i=1}^{N^2}$, the Poisson model for Y is then

$$P(Y|X) = \prod_{j=1}^{C*A} \frac{\left(\sum_{i=1}^{N^2} \mathcal{A}_{ji} X_i\right)^{Y_j}}{Y_j!} \exp\Big(-\sum_{i=1}^{N^2} \mathcal{A}_{ji} X_i\Big).$$

5 Reconstruction Experiments

5.1 Computational Algorithm

Having fixed a transverse slice of interest, we denote by $X = \{X_i\}_{i=1}^{N^2}$ the discretized isotope intensity distribution. The components, X_i, represent concentrations at the internal sites, or pixels, $\{S_i\}_{i=1}^{N^2}$. In all of our experiments, we used the more or less standard discretization, $N = 64$. The reconstruction problem is to estimate these isotope concentrations, and thereby the pharmaceutical concentrations, from the observed photon counts, $Y = \{Y_j\}_{j=1}^{C*A}$, where, as discussed already in Section 4, we have restricted ourselves to using only that row of counts, at each angle of rotation, that corresponds to the transverse slice of interest.

Our reconstructions are based on the posterior distribution of X given Y, which is proportional to $P(Y|X)P(X)$:

$$P(X|Y) = \gamma P(Y|X)P(X).$$

The data, or likelihood, term $P(Y|X)$ was developed in Section 4. The prior term $P(X)$ will be laid out, and discussed in detail, Section 6. The actual reconstructions are an approximation of the posterior mean, $E[X|Y] = \sum_X XP(X|Y)$, derived by an iterative algorithm known as ICE (for *Iterated Conditional Expectations*, see Owen [30]). The posterior mean itself is intractable, since X has $64 \times 64 = 4,096$ components, each of which could attain, in our experiments, any of 32 (equally spaced) values.

The ICE algorithm begins with some initial estimate of isotope concentration X^0. We used $X^0 \equiv 0$. A domain of interest is defined (an ellipse known to contain the patient's body or the phantom under study), and X^1 is then computed from X^0 by sequentially modifying each component of X^0 whose corresponding site is within the domain of interest. We followed a *raster scan* ordering of these components, although there is reason to believe that there are better *site visitation* schedules (see Amit and Grenander [1].) Upon visiting a site, S_i, the value X_i^0 is replaced by its conditional mean, given Y and given the current values of the components associated with the remaining sites S_k, $k \neq i$. The conditional mean is easy to compute from the posterior distribution, despite the fact that the posterior distribution involves an unknown multiplicative (normalizing) constant. We refer the reader to Owen [30] and Manbeck [23] for details. The process is continued, recursively, producing a sequence of images, $X^0, X^1, \ldots$, until there are essentially no further changes. This generally requires about 10 or 15 iterations.

5.2 The Backprojection Algorithm

Most clinical reconstructions are computed by some variant of the backprojection algorithm (*cf.* [19]). In backprojection, the reconstructed isotope concentration at a given pixel is taken to be proportional to the sum of observed counts over *relevant* bins. A particular bin is relevant to a given pixel if a perpendicular line from the pixel to the camera face falls into the bin. In short, photon counts are backprojected to form the reconstructed image. Attenuation can be approximately corrected for by a depth-dependent weighting of the backprojected counts.

An important modification, which can partially accommodate the effects of scatter and collimator geometry, and, to a degree, mitigate the effects of (Poisson) statistical fluctuations, is the *filtered* backprojection. A suitable filter is first applied to the raw data (observed photon counts), and the result is then backprojected to form the reconstructed image. In general terms, scatter and collimator effects contribute to blurring in the reconstructed image, whereas statistical variation in observed photon counts is manifested in the reconstruction by high frequency noise. It is difficult to design filters that *simultaneously* address both of these undesirable effects; noise removal tends to blur boundaries, whereas deblurring tends to accentuate noise. We believe that an important advantage of the statistical approach advocated here is its ability to smooth the reconstructed image, while preserving sharp boundaries and, consequently, maintaining good resolution.

5.3 Results

The results of three experiments are presented. Two are with real patient data—a liver scan and a head scan—and one is with a phantom of known structure. The phantom and liver studies used a technetium isotope, and the head scan used a thallium isotope. In each experiment, a reconstruction by filtered backprojection is presented for comparison. The backprojection algorithm used was the one provided with the imaging machine, and, because it is proprietary, we do not know details about the method of attenuation correction or the particular filter used. We should point out that research continues on filtered backprojection for SPECT and other imaging modalities. It is likely that the method can produce better reconstructions than those that we obtained from the package provided with this particular machine.

In addition to the two reconstructions (by filtered backprojection and by Bayesian estimation via ICE), we present for each experiment a gamma camera picture from a single angle, and the *sinogram* associated with the particular transverse slice being reconstructed. The gamma picture simply depicts the 64×64 array of recorded counts at one of the 64 angles of rotation. The actual data used in reconstructing a transverse slice is depicted in the sinogram. The first row of the sinogram is a row of data from the gamma picture at the first angle of rotation. The second row of the sinogram is from the second angle of rotation, and so on. The data come from the same level in each gamma picture—corresponding to the particular transverse slice being reconstructed. Notice that the first and last rows of the sinogram are similar, since they come from gamma camera pictures at neighboring angles.

5.3.1 Two Lines in Water

A test for the resolution of the reconstruction was devised. Two catheters were filled with isotope and placed parallel and in close proximity to each other within a tank of water. The tank was situated within the SPECT machine so that the catheters were parallel to the axis of rotation of the gamma camera. Ideally, a transverse slice would contain just two small regions of high intensity. Under the (64×64) digitization used, a single pixel in the image array was larger than the bore of the catheters. Ideally, there would be only one pixel in each intensity peak.

Figure 7 shows the result of one such experiment in which the catheters were separated by 25 mm. Severe degradation from collimator effect, scatter, and Poisson noise is evident in the gamma picture (Fig. 7a) and the sinogram (Fig. 7b). The filtered backprojection and Bayesian reconstructions are displayed in Figs. 7c and d, respectively.* Obviously, there is a considerable amount of artifact in the filtered backprojection reconstruction.

*Even in the absence of an isotope source, there is a nonzero level of counts recorded by the gamma camera. This is seen as a more or less uniform background intensity in Fig. 7a. These counts produce a systematic artifact in our reconstructions: the boundary of the elliptical domain of reconstruction invariably contains an inappropriately high level of intensity, reflecting the best possible choice for source locations, given the counts in the more extreme columns of the camera. In all of our experiments, this artifact was removed from the reconstructions by setting the intensity values to zero along the bounding ellipse.

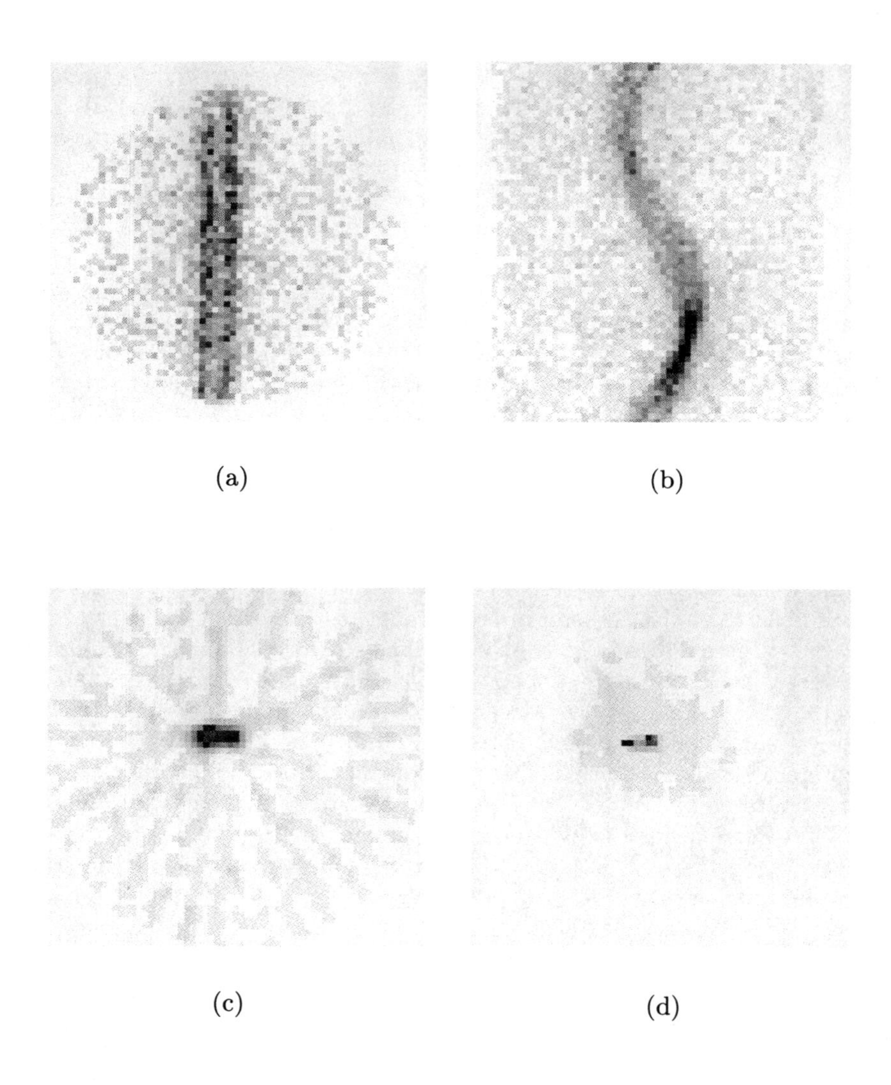

Figure 7: Transverse reconstruction of two line sources in water. (a) Gamma picture. (b) Sinogram. (c) Reconstruction by filtered backprojection. (d) Bayesian reconstruction using ICE.

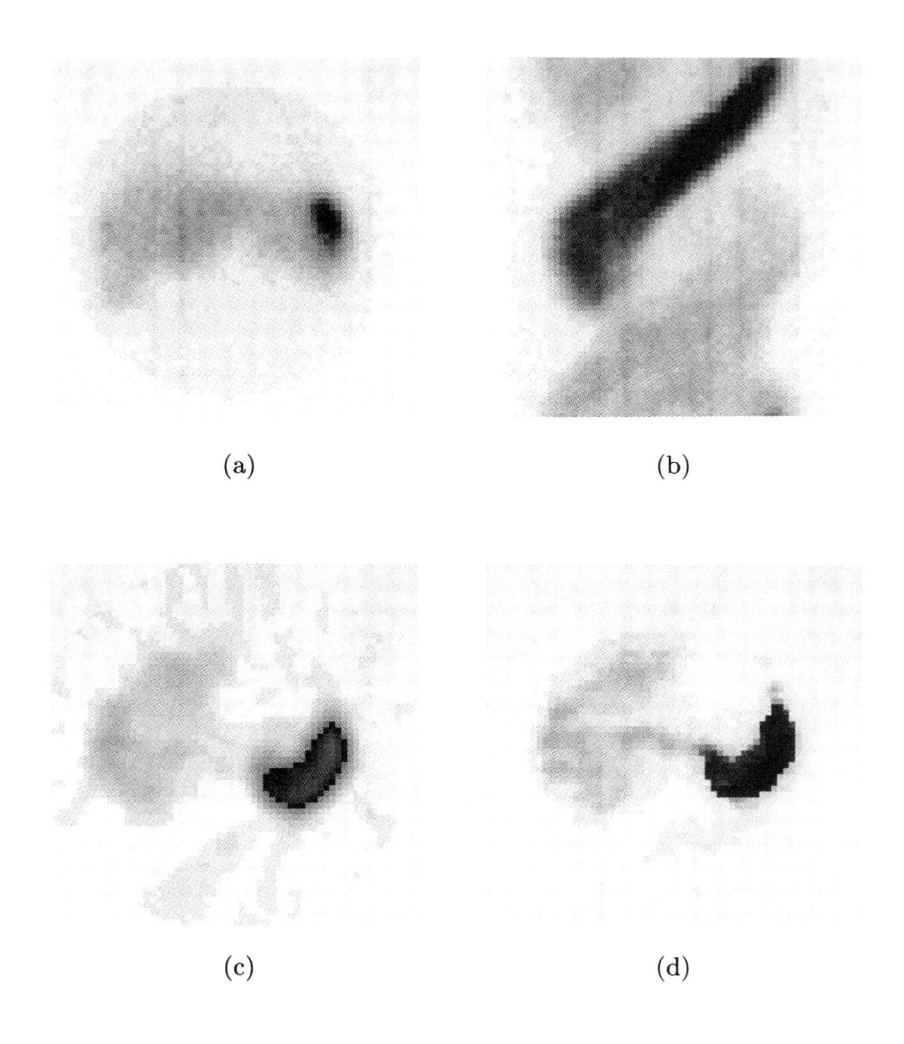

Figure 8: Liver scan. (a) Gamma picture. (b) Sinogram. (c) Reconstruction by filtered backprojection. (d) Bayesian reconstruction using ICE. The darker region is the spleen; the lighter and larger region is the liver. The low concentration of radiopharmaceutical in the liver indicates disease.

5.3.2 Liver Scan

The results are shown in Fig. 8. The actual distribution of isotope intensity is, of course, unknown. The larger and lighter structure is the liver; the smaller and darker structure is the spleen. Poor uptake of radiopharmaceutical by the liver indicates disease. The intense ring of radiopharmaceutical seen on the spleen in the filtered backprojection reconstruction (Fig. 8c) is probably artifact.

5.3.3 Head Scan

The particular pharmaceutical used in this scan is retained mostly by bone, which was the object of interest. Reconstructions are shown in Fig. 9. Again, there is no *true image* to go by, and, in fact, the two reconstructions are rather different. Given the apparent tendency of this filtered backprojection algorithm to blur the isotope distribution (See Fig. 7), we are inclined to believe that the true distribution is less homogeneous than indicated in Fig. 9c, and perhaps more along the lines of the reconstruction in Fig. 9d. In any case, this experiment points out the importance of using phantoms, where structure is known *a priori*, in comparing and assessing reconstruction strategies.

In an experiment such as this one, involving both bone and soft tissue, it would be highly desirable to use a binary-valued attenuation map, thereby taking full advantage of the known gross anatomy of the two tissue types. The distribution of bone and soft tissue can be conveniently encoded into the MRT ($\mathcal{A}$), and there is little additional computational cost for using the resulting inhomogeneous attenuation.

6 A Smoothing Prior and Its Isotropic Properties

One can adopt two points of view towards the role of the prior probability distribution $P(X)$ on possible reconstructions X. On one hand, we can regard $P(X)$ from a decidedly Bayesian perspective as an *a priori* assignment of probabilities to possible states of nature for X. This perspective is useful when we design P, that is, when we prescribe its exact functional form. On the other hand, we can simply regard $P(X)$ as a methodological tool introduced for the purpose of regularizing the otherwise highly variable reconstructions obtained, for example, by the method of maximum likelihood. From this perspective, $P(X)$ is analogous to a *penalty function* introduced as part of the method of penalized maximum likelihood. The

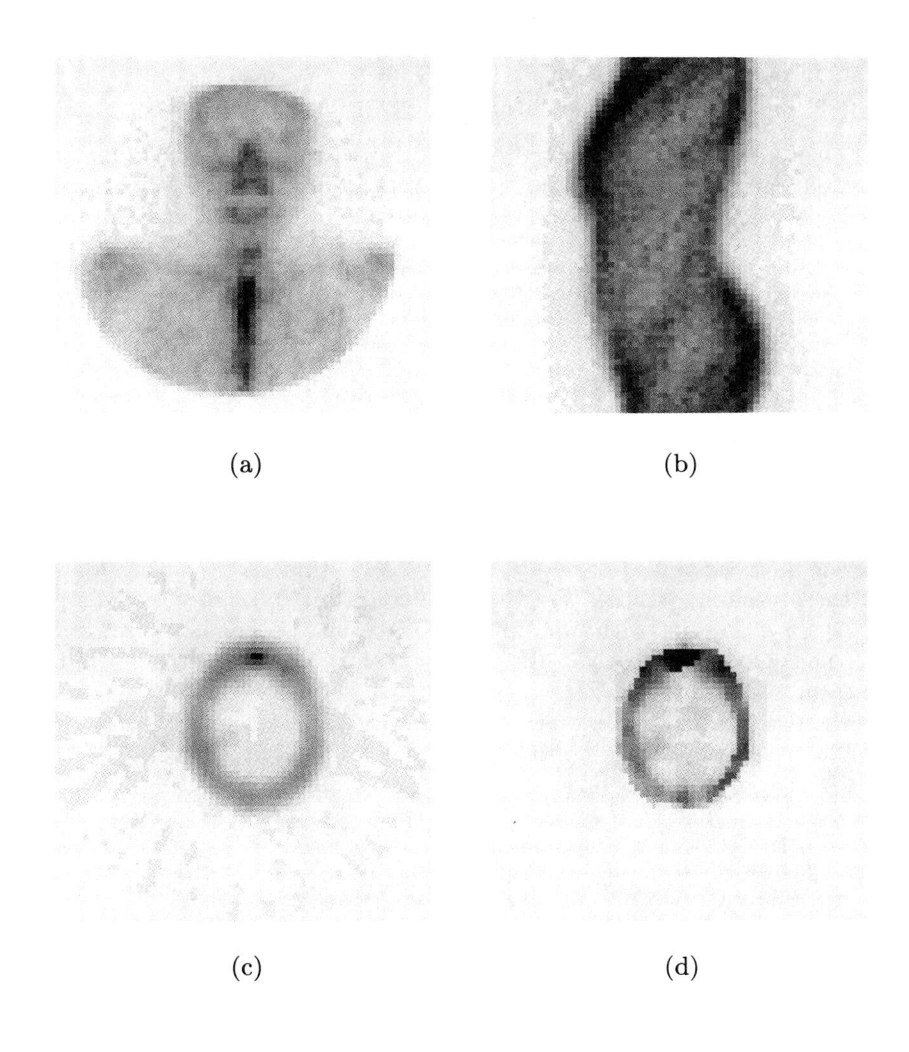

Figure 9: Bone scan of the head. (a) Gamma picture. (b) Sinogram. (c) Reconstruction by filtered backprojection. (d) Bayesian reconstruction using ICE.

two viewpoints complement each other and both are worthwhile to keep in mind in the implementation of the Bayesian approach.

The prior distribution quantifies likelihoods of characteristics of the images X. We adopt a rather general approach that seeks to quantify likelihoods of local characteristics of X, rather than to model highly detailed structural information, such as the likely shapes or locations of structures in X. In particular, we design $P(X)$ to embody two local regularity properties of radiopharmaceutical concentrations:

1. Isotope concentrations tend to be fairly constant within small regions of common tissue type and common metabolic activity. Neighboring pixels are more likely than not to have similar isotope concentrations.

2. Still, sharp boundaries will occur. For example, at the interface between two tissue types, or between two organs, there may be a sharp gradient in the intensity values of neighboring pixels.

Local characteristics such as these are conveniently modeled in terms of the Gibbs representation of P. Let

$$P(X) = \frac{1}{Z} e^{-U(X)},$$

where U is an *energy function* and Z is a normalizing constant; X will have discrete finite range and $Z = \sum_X \exp\{-U(X)\}$. Under P, the likely states are the low-energy states. We therefore construct U so that low-energy states are consistent with our expectations about isotope concentrations X. One could, equivalently, work directly with P, but this turns out to be much more difficult.

The two characteristics we wish to quantify suggest defining U in terms of differences $X_s - X_t$ between neighboring pixel values. Let Λ_N be the $N \times N$ square lattice of pixel sites S_i; s and t will refer to generic points in Λ_N. Then define

$$U(X) = \beta \left[\sum_{[s,t]} \phi(X_s - X_t) + c \sum_{<s,t>} \phi(X_s - X_t) \right]; \tag{5}$$

here, $[s, t]$ indicates that s and t are nearest horizontal or vertical neighbors in the finite lattice Λ_N, commonly referred to as the *first-order neighbors*, and $< s, t >$ indicates nearest diagonal neighbors, commonly referred to as *second-order neighbors*. The constants β and c are positive. c determines the relative contributions of the first-order and second-order neighbors to U as a whole. We shall return to β's interpretation shortly. The function

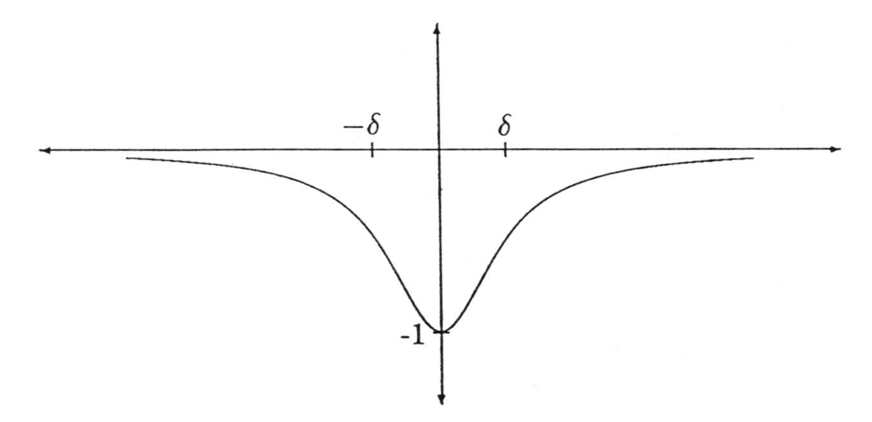

Figure 10: The phi function, which is used to specify an energy function U.

$\phi(\xi)$ is even and minimized at $\xi = 0$. Thus, U is minimized by images X of constant intensity. Images composed of homogeneous subregions are more likely than ones that have a high degree of local variability.

This definition of U induces a graph on Λ_N in which each pixel site s is linked to its eight nearest neighbors in the square lattice. The distribution $P(X)$ then determines a Markov random field with this neighborhood structure.

The qualitative behavior of ϕ is crucial for balancing the competing demands of (i) local smoothness and (ii) permitting sharp boundaries. In a variety of applications, we have used the function ϕ defined by

$$\phi(\xi) = \frac{-1}{1 + (\xi/\delta)^2}, \tag{6}$$

where δ, like β, is a constant. ϕ is depicted in Fig. 10.

The shape of ϕ, especially the fact that it is bounded above, is important for accommodating the second of the two structural properties of isotope concentrations—that sharp gradients may occur across boundaries between regions of different tissue type or metabolic activity. If two pixels s and t are on opposite sides of a boundary, then ϕ, and hence U, do not associate substantially greater penalties with larger values of the difference, $X_s - X_t$. Once a boundary is introduced, the function ϕ is rather indifferent to the size of the jump across the boundary. Relatively high prior probabilities are assigned to images with constant subregions separated by sharp boundaries.

Variations of the energy model of Eqs. (5) and (6) have been explored in the context of tomography, as well as other applications. Green [14] studies the effect of a different phi function, $\log(\cosh(\frac{\xi}{\delta}))$, which is not bounded, but which is asymptotically linear and more amenable to certain optimization algorithms. We have recently reported on variations of the energy in Eq. (5) for which low-energy (high probability) images are piecewise planar or piecewise quadratic, rather than piecewise constant [13].

The energy function U depends on three parameters: δ is easily interpreted as a scale parameter on the range of values of X_s; β controls the ***strength*** of the interactions between a pixel and its neighbors, and in *maximum a posteriori* (MAP) estimation, it corresponds to a weighting factor balancing the contributions of a data term and a penalty term, just as in penalized maximum likelihood; as noted before, c controls the relative contributions of first-order and second-order neighbors to the total energy.

We have always found reconstructions to be relatively insensitive to the choice of δ, provided only that δ is large enough so that the shape of ϕ near the origin is not too singular. As a rule of thumb, we set δ to be approximately 20% of the dynamic range of X. In the experiments reported in Section 5, δ was set to 50.

The reconstructions are more sensitive to the choice of β, but again, as long as β is within a reasonably conservative range of values, the precise value of β is not too critical. If β is too large, the reconstructions will be oversmoothed and if β is too small, they will be undersmoothed. In [12], [27], and [23], experiments are reported that show the effect of different choices of β. The model parameter β can actually be estimated from the observed data Y using the principle of maximum likelihood and the EM algorithm; again we refer to [12], [27], and [23]. In the experiments reported in Section 5, for convenience, we chose the scale parameter k in Eq. (3) so that the reconstructions always nearly fill the dynamic range $[0, 255]$. Then β was fixed once and for all to the *moderate* value $\beta = 1$. The same values of β and δ were used for all three experiments shown in Section 5.

Less obvious is the effect of different choices for c. One invariance property that we would like the energy function U to possess is isotropy—invariance to rotations of the underlying coordinate system. At first, it is natural to expect that c might influence isotropy of U, since c controls how interactions between sites vary with their relative orientations. Individual horizontal and vertical interactions have weight one in Eq. (5), and individual diagonal interactions have weight c. Further, diagonal neighbor pairs are separated by a greater distance than are horizontal/vertical neighbor pairs. Based on (i) the goal of isotropy and (ii) information about the relative orientations and separations of the two types of neighbors, a number of heuristic arguments have been espoused in support of certain specific choices of c. In [12], in fact, we suggested that the *natural* choice for c is $1/\sqrt{2}$; but as we shall show here, a quadratic approximation of U is isotropic in a continuum limit for any choice of c. The value of c does not affect rotational invariance. For the experiments reported in Section 5, c was set to one, largely as a matter of computational convenience and because there is no solid analytical support for any other choice.

6.1 A Two-Dimensional Continuum Limit

The results on isotropy of U presented here are developed in Manbeck [23], and are patterned after similar arguments developed in the context of mathematical physics [35]. See also Besag [3] for a discussion of isotropic properties in the *absence* of a diagonal contribution: $c = 0$. The issue of isotropy is a natural one to question concerning the fidelity between our discrete lattice-based mathematical model on one hand and the continuous physical phenomenon it models (radiopharmaceutical concentration) on the other hand.

The plausibility of rotational invariance, regardless of c, is supported by a simple local asymptotic expansion of the two sums in U. Most neighboring pixel pairs s and t will have similar values, X_s and X_t. Thus, it is reasonable to approximate $\phi(X_s - X_t)$ by low-order terms of the Taylor series for ϕ. In particular,

$$\phi(X_s - X_t) \sim -1 + \frac{(X_s - X_t)^2}{\delta^2}.$$

Assuming that, in the continuous domain, horizontal neighbors have the same separation as vertical neighbors (call the spacing Δ), and imagining the values of X_s to lie on a smooth surface, then

$$\sum_{[s,t]} \phi(X_s - X_t) \sim \text{Constant} + \frac{\Delta^2}{\delta^2} \sum_{\Lambda_N} \nabla X_s{}^2$$

and

$$\sum_{<s,t>} \phi(X_s - X_t) \sim \text{Constant} + 2c\frac{\Delta^2}{\delta^2} \sum_{\Lambda_N} \nabla X_s{}^2.$$

Each of the sums is similar to an approximate integral of the squared gradient of X. The gradient expression by itself is rotationally invariant, and thus each of the sums in U is (approximately) isotropic. Of course, this is not a rigorous argument; there is actually no smooth surface from which we can legitimately imagine the values of X_s to be sampled.

A more formal argument can be developed using a Gaussian approximation of U. From the asymptotic expansion of ϕ above and assuming the differences $X_s - X_t$ are small, it follows that up to an additive constant, the function $\delta^2 U(X)/\beta$ is approximately equal to the quadratic form,

$$V(X) = \sum_{[s,t]} (X_s - X_t)^2 + c \sum_{<s,t>} (X_s - X_t)^2, \tag{7}$$

for s and t in Λ_N. Let $\mathcal{X}$ denote the vector $\mathcal{X} = \{X_s\}_{s \in \Lambda_N}$ and let Q_N be the $N^2 \times N^2$ array of coefficients of the quadratic form $2 \times V(X)$;

$$V(X) = \frac{1}{2} \mathcal{X}' Q_N \mathcal{X}.$$

The factor 1/2 simply makes it easier to relate the expression for $V(X)$ to the standard representation of a multivariate Gaussian distribution. Strictly speaking, since $s \in \Lambda_N$ is a multi-index, $\mathcal{X}$ is not a vector and Q_N is not a matrix in the traditional sense; but there is no inconsistency or harm in the simplified notation.

Q_N is analogous to the inverse of the covariance operator of a Gaussian random vector $\mathcal{X}$. However, Q_N is a singular operator, since the energy functions U and V are defined in terms of first differences of X. Q_N has rank $N^2 - 1$. To remove the singularity and analyze a proper, nondegenerate Gaussian distribution, we approximate Q_N by adding a small homogeneous perturbation. Let $\epsilon > 0$, and define

$$\begin{aligned} V_\epsilon(X) &= \sum_{[s,t]} (X_s - X_t)^2 + c \sum_{<s,t>} (X_s - X_t)^2 + \epsilon \sum_s X_s^2, \qquad (8) \\ &= \frac{1}{2} \mathcal{X}' Q_{N,\epsilon} \mathcal{X}. \end{aligned}$$

The operator $Q_{N,\epsilon}$ is a finite $N^2 \times N^2$ section of an infinite Toeplitz form [17]. Let Q_ϵ denote the doubly infinite Toeplitz form. By simply expanding the squares in Eq. (8), it is easy to see that $(Q_\epsilon)_{ss} = 8 + 8c + 2\epsilon$, $(Q_\epsilon)_{st} = -2$ for first-order neighbors, and $(Q_\epsilon)_{st} = -2c$ for second-order neighbors. As a homogeneous, positive-definite quadratic form, Q_ϵ admits a spectral representation [17], [36]. Let $g(\xi, \eta)$ be the spectral density function,

$$g(\xi, \eta) = 8 + 8c + 2\epsilon - 4\cos\xi - 4\cos\eta - 8c\cos\xi\cos\eta.$$

Then

$$(Q_\epsilon)_{st} = \frac{1}{4\pi^2} \int\int_{[-\pi,\pi]^2} g(\xi, \eta) e^{-i(s-t)\cdot(\xi,\eta)} d\xi d\eta.$$

This particular integral representation can also be verified trivially by substituting the simple expression for g into the integral.

The spectral representation for Q_ϵ is extremely useful for computing and analyzing the inverse of Q_ϵ. Note that $g(\xi, \eta)$ is strictly positive, and bounded below by $2\epsilon > 0$. The reciprocal of g is a bounded, continuous, strictly positive function on $[-\pi, \pi]^2$. Define $\mathcal{R}_\epsilon$, which we shall relate to the inverse of Q_ϵ, by

$$(\mathcal{R}_\epsilon)_{st} = \frac{1}{4\pi^2} \int\int_{[-\pi,\pi]^2} \frac{1}{g(\xi, \eta)} e^{-i(s-t)\cdot(\xi,\eta)} d\xi d\eta. \tag{9}$$

$\mathcal{R}_\epsilon$ is a doubly infinite, symmetric Toeplitz form. All of its finite sections are positive definite, since its spectral density is positive and integrable. Thus, $\mathcal{R}_\epsilon$ is uniquely identified with a zero-mean stationary random field $\tilde{X}_s$, for $s \in \Lambda_\infty$, the doubly infinite integer square lattice. $\mathcal{R}_\epsilon$ is the covariance operator of $\tilde{X}_s$;

$$\text{Cov}(\tilde{X}_s, \tilde{X}_t) = (\mathcal{R}_\epsilon)_{st}.$$

The process $\tilde{X}_s$ and its covariance $\mathcal{R}_\epsilon$ have two important properties. First, when N is large, the section of $\mathcal{R}_\epsilon$ identified with the finite square lattice Λ_N approximates the covariance of our original process X_s with energy function $V(X)$. Second, in the sense of an appropriate continuum limit, $\mathcal{R}_\epsilon$ is rotationally invariant.

The actual covariance matrix of the Gaussian random vector $\mathcal{X}$ is $(Q_{N,\epsilon})^{-1}$. For large N, $(Q_{N,\epsilon})^{-1}$ is approximately equal to the finite $N^2 \times N^2$ section $(\mathcal{R}_\epsilon)_N$, that is, the finite section of $\mathcal{R}_\epsilon$ that restricts the indices s and t to Λ_N. By using the techniques of Grenander and Szegö [17], Section 7.4, it is straightforward to show that the so-called *trace norm* of $(Q_{N,\epsilon})^{-1} - \mathcal{R}_\epsilon$ goes to zero as $N \to \infty$. This is a form of l_2 convergence of the covariance operators:

$$\lim_{N\to\infty} \frac{1}{N^2} \sum_{s\in\Lambda_N} \sum_{t\in\Lambda_N} \left\{ \left[(Q_{N,\epsilon})^{-1} \right]_{st} - (\mathcal{R}_\epsilon)_{st} \right\}^2 = 0.$$

The methods of Grenander and Szegö are used in [26], for example, to compute approximate inverses of covariance operators for random fields indexed on a two-dimensional lattice, exactly the type of approximation needed here for $Q_{N,\epsilon}$.

In this sense, the covariance structure of $\mathcal{X}$ is approximated by $\mathcal{R}_\epsilon$.

We have used the spectral representation of $\mathcal{R}_\epsilon$ to compute values of $(\mathcal{R}_\epsilon)_{st}$ for various choices of the parameters ϵ and c, to examine the dependence of $r(h, v) = (\mathcal{R}_\epsilon)_{(0,0),(h,v)} = \text{Cov}(\tilde{X}_{(0,0)}, \tilde{X}_{(h,v)})$ on the orientation of the ray from the origin to the point (h, v). The numerical evidence of isotropy is compelling. When $\epsilon = 0.01$ and $c = 2$, a comparison of two points five units from the origin shows $r(0, 5) = 0.026 = r(3, 4)$. Also, the values of r at points on the horizontal axis were compared to values of r on the 45° diagonal. First, the values of r on the integer lattice were interpolated to form an extension of r that could be evaluated for noninteger arguments. Simple linear interpolation was done on the diagonal, so a value $r(h/\sqrt{2}, h/\sqrt{2})$ at distance h from the origin is fit by the linear interpolant between the values of r at the two integer-lattice points on the diagonal nearest to $(h/\sqrt{2}, h/\sqrt{2})$. Then the ratios $\rho(h) = r(h,0)/r(h/\sqrt{2}, h/\sqrt{2})$ were calculated for $h = 2, 3, \ldots, 20$. When $\epsilon = 0.01$ and $c = 2$, the ratios decreased from $\rho(2) = 1.06$ to $\rho(20) = 1.001$. Similar results were obtained when $\epsilon = 0.01$ and $c = \sqrt{2}$, $c = 1$, and $c = 0$. For further numeric evidence of isotropy in the latter ($c = 0$) case, see Besag [3].

The continuum limit will involve shrinking the lattice spacing, or, equivalently, examining $r(h, v)$ at progressively larger distances $\sqrt{h^2 + v^2}$. The numerical results, however, strongly indicate that the distribution is already nearly isotropic well before the continuum is approached.

To prove a limiting form of rotation invariance for $\mathcal{R}_\epsilon$, we need to identify lattice points with points in the continuous plane. We shall define a sequence of lattices, indexed by n, that become progressively finer discretizations of the continuum as $n \to \infty$. Assume that the lattice spacing between first-order neighbors is $\Delta = 1/n$, the same in both the horizontal and vertical directions. Further, we shall allow the parameter ϵ to depend on n ($\epsilon \to \epsilon_n$) so that we can obtain a nondegenerate limit for the associated spectral density functions.

Consider a fixed point (x, y) in the plane, and assume for now that nx and ny are integers. The Cartesian coordinates (x, y) correspond to the lattice coordinates (nx, ny), since the spacing of the refined lattice points is assumed to be $1/n$. The covariance of the lattice-based process $\tilde{X}_s$ between the two sites, (nx, ny) and $(0, 0)$, is given by $(\mathcal{R}_{\epsilon_n})_{(0,0),(nx,ny)}$. We will show that there is a rotationally invariant (in (x, y)) limit function (as $n \to \infty$).

Rather than worry about whether nx and ny are integers, we will *define* a function on all of R^2 by extending the formula for $\mathcal{R}_\epsilon$ (Eq. (9)) to the continuum. Formally, $r_n(x, y) = (\mathcal{R}_{\epsilon_n})_{(0,0),(nx,ny)}$, or, explicitly, for any $(x, y) \in R$,

$$r_n(x,y) = \frac{1}{4\pi^2} \int\int_{[-\pi,\pi]^2} \frac{1}{g_n(\xi,\eta)} e^{-i(nx,ny)\cdot(\xi,\eta)} d\xi d\eta, \qquad (10)$$

where

$$g_n(\xi,\eta) = 8 - 4\cos\xi - 4\cos\eta + 8c - 8c\cos\xi\cos\eta + 2\epsilon_n. \qquad (11)$$

The change-of-variable $n\xi \to \xi$ and $n\eta \to \eta$ in Eq. (10) gives

$$r_n(x,y) = \frac{1}{4\pi^2} \int\int_{[-n\pi,n\pi]^2} \frac{1}{n^2 g_n(\xi/n, \eta/n)} e^{-i(x,y)\cdot(\xi,\eta)} d\xi d\eta. \tag{12}$$

Let $F_n(\xi,\eta)$ denote the function whose Fourier transform is performed in Eq. (12):

$$F_n(\xi,\eta) = \chi_{[-n\pi,n\pi]^2} \cdot \frac{1}{n^2 g_n(\xi/n, \eta/n)}.$$

The limiting properties of F_n are readily seen by substituting the first two terms of the Taylor series for the cosine in Eq. (11). If $\epsilon_n = \epsilon/n^2$, then

$$\lim_{n\to\infty} F_n(\xi,\eta) = \frac{1}{(2+4c)(\xi^2+\eta^2)+2\epsilon}; \tag{13}$$

call this limiting function $F_\infty(\xi,\eta)$.

The pointwise convergence of F_n does not imply the pointwise convergence of r_n; this step needs to be analyzed carefully, since F_n is not integrable on R^2.

We can, however, show that F_n converges to F_∞ in $L_2(R^2)$, which guarantees convergence of r_n, also in $L_2(R^2)$. We start by observing the following properties of $\{F_n\}$ and the limit F_∞.

Proposition 1 *Let F_∞ be the pointwise limit of the sequence F_n as defined by Eq. (13).*

1. *$F_\infty \in L_2(R^2)$;*
2. *There exists a function $F \in L_2(R^2)$ such that for all n, $F_n(\xi,\eta) \le F(\xi,\eta)$;*
3. *The sequence $\{F_n\}_{n=1}^\infty$ converges to F_∞ in $L_2(R^2)$.*

Proof. 1. Obvious. In fact, $F_\infty \in L_p$ for all $p > 1$.
2. The inequality $(1-\cos x) \ge \alpha x^2$ for all $x \in [-\pi,\pi]$ and $\alpha \le 2/\pi^2$ is proved in Simon [35]. Fix any $\alpha \le 2/\pi^2$:

$$\begin{aligned}
F_n(\xi,\eta) &= \frac{\chi_{[-n\pi,n\pi]^2}}{4n^2\left[1-\cos\frac{\xi}{n}+1-\cos\frac{\eta}{n}+c(2-2\cos\frac{\xi}{n}\cos\frac{\eta}{n})+\frac{\epsilon_n}{2}\right]} \\
&\le \frac{1}{4n^2\left[1-\cos\frac{\xi}{n}+1-\cos\frac{\eta}{n}\right]+2\epsilon_n n^2}, \\
&\le \frac{1}{4\alpha\left[\xi^2+\eta^2\right]+2\epsilon}.
\end{aligned}$$

The first inequality follows from dropping positive terms from the denominator on the left, and the second inequality follows from Simon's lower bound on $1 - \cos x$. The final upper bound on F_n is in $L_2(R^2)$, as in the first part of the proposition.
3. The second part of the proposition allows us to apply the dominated convergence theorem to $|F_n(\xi, \eta) - F_\infty(\xi, \eta)|^2$ to conclude L_2 convergence of F_n to F_∞.

Since the limit F_∞ is *not* in L_1, but only in L_2, it is not a spectral density function itself. F_∞ is *not* identified with the covariance function of any continuous parameter Gaussian random field. Nonetheless, the covariance functions r_n do converge in an L_2 sense to the L_2 Fourier transform, i.e., the Plancherel transform, of F_∞.

To see this, define, for any function F in L_2,

$$\hat{F}(x, y) = \frac{1}{2\pi} \lim_{A \to \infty} \int_A^A \int_A^A F(\xi, \eta) e^{-i(x,y)\cdot(\xi,\eta)} d\xi d\eta, \tag{14}$$

where the limit is taken in L_2. See, for example, Helmberg [18]. This is the Plancherel transform, and it is unitary on L_2. From Eq. (12), we have

$$r_n(x, y) = \frac{1}{2\pi} \hat{F}_n(x, y).$$

Since $F_n \to F_\infty$ in L_2, and since the Plancherel transform is an isometry on L_2,

$$\lim_{n \to \infty} \hat{F}_n = \hat{F}_\infty$$

in L_2. Even though $\hat{F}_\infty(x, y)$ is not a covariance function, the limiting function is still well-defined as an element of L_2. Thus, r_n has a limit in L_2 as $n \to \infty$.

Finally, we observe that the limit $\hat{F}_\infty$ is rotationally invariant.

Proposition 2 $\hat{F}_\infty(x \cos \theta - y \sin \theta, x \sin \theta + y \cos \theta) = \hat{F}_\infty(x, y)$ *i.e., for all* θ *in* $[0, 2\pi)$.

Proof. This follows from a simple change of variable in the asymptotic analysis before. The sequence $F_n(x \cos \theta + y \sin \theta, -x \sin \theta + y \cos \theta)$ has the same pointwise and L_2 limits as $F_n(x, y)$, and hence the respective Plancherel transforms have the same L_2 limit, $\hat{F}_\infty$.

Consequently, the L_2 limit $\lim_{n \to \infty} r_n$ is isotropic, for any choice of the constant c.

6.2 Extensions

The three-dimensional (3-D) analog of the model for the *a priori* distribution is of interest when analyzing three-dimensional data. In the case of tomography, this 3-D analog is required for performing true 3-D reconstructions, as opposed to the ordinary 2-D slice reconstructions. The 3-D model adds interactions *between* slices to the prior information *within* slices embodied in Eq. (5), and the two-dimensional picture elements (pixels) become three-dimensional volume elements (voxels). Since the thickness of slices depends on camera geometry, the voxels will not, in general, be cubes, and there are again the issues of scaling and isotropy. The mathematical analysis of the 3-D prior follows precisely the pattern used earlier to deduce approximate isotropy for the 2-D phi model.

Consider a prior distribution $P(X)$ on 3-D images $X = \{X_s\}_{s \in \Lambda}$, where Λ is a finite 3-D lattice of size $N_x \times N_y \times N_z$. In analogy with Eq. (7), we consider Gaussian priors with a local neighborhood structure. The lattice neighbors of a site s may include any lattice point in the $3 \times 3 \times 3$ discrete cube centered at s.

When we associate points of the lattice Λ with points in continuous 3-space, we specifically allow for different spacing in the x, y, and z dimensions. In other words, in the context of tomography, we do not require that pixels within a 2-D cross section be square nor that the thickness of a slice be constrained by values used to quantize the plane of the slice. Let a, b, and c denote the Euclidean distance between neighboring lattice sites in the x, y, and z directions, respectively.

The neighborhood of a site s admits a natural decomposition into seven subsets, depending on the difference $(s - t)$. Sites s and t are said to be neighbors of type i — denoted $<s, t>_i$—for $i = 1, 2, \ldots, 7$ according to the following scheme:

$$\begin{aligned}
i &= 1 \Longleftrightarrow \quad (s-t) = (\pm 1, 0, 0), \\
i &= 2 \Longleftrightarrow \quad (s-t) = (0, \pm 1, 0), \\
i &= 3 \Longleftrightarrow \quad (s-t) = (0, 0, \pm 1), \\
i &= 4 \Longleftrightarrow \quad (s-t) = (\pm 1, \pm 1, 0), \\
i &= 5 \Longleftrightarrow \quad (s-t) = (\pm 1, 0, \pm 1), \\
i &= 6 \Longleftrightarrow \quad (s-t) = (0, \pm 1, \pm 1), \\
i &= 7 \Longleftrightarrow \quad (s-t) = (\pm 1, \pm 1, \pm 1).
\end{aligned}$$

Then, in analogy with Eq. (8), we consider

$$V_\epsilon(X) = \sum_{i=1}^{7} c_i \sum_{<s,t>_i} (X_s - X_t)^2 + \epsilon \sum_s X_s^2, \tag{15}$$

with $s, t \in \Lambda$. Our goal is to specify values for c_i, $i = 1, ...7$, so that the associated distribution possesses rotational invariance in a suitable continuum limit. The quadratic form in Eq. (15) has a spectral representation, identified with a strictly positive, continuous spectral density function g on $[-\pi, \pi]^3$. The reciprocal of g is, accordingly, the spectral density of a well-defined Gaussian process $\tilde{X}_s$, for s in the infinite lattice Z^3. Just as in the 2-D case, finite sections of the covariance operator $\mathcal{R}_\epsilon$ of $\tilde{X}_s$ approximate the covariance operator of the original X_s process in the trace norm, as $\Lambda \to Z^3$.

To carry out the asymptotic analysis of rotation invariance of $\mathcal{R}_\epsilon$, we assume that the lattice spacing becomes successively finer in the continuum. In particular, we consider a sequence of lattices, indexed by n, with spacings a/n, b/n, and c/n in the x, y, and z directions, respectively. To obtain a nondegenerate limit for the 3-D analog of Eq. (10), it is necessary to scale the coefficients in Eq. (15) to depend on n. Let

$$c_i = \frac{C_i}{n}$$

and

$$\epsilon = \frac{\mathcal{E}}{n^3}$$

for fixed constant values of C_i and $\mathcal{E}$. Here, we require that $\mathcal{E} > 0$ and that the C_i are nonnegative. The asymptotic analysis then proceeds step-by-step as in the 2-D case. Only minor variations occur. For example, the 3-D analog of F_∞ (Eq. (13)) is in L_p, for $p > 3/2$ (rather than for $p > 1$, as in the 2-D case).

The analysis yields conditions on the constants C_i, which assure rotation invariance of the Plancherel transform of F_∞, and hence, approximate isotropy for the process with energy function V_ϵ. Simply stated, the conditions reduce to a *balance condition* on the constants:

$$\begin{aligned} & (C_1 + 2C_4 + 2C_5 + 4C_7)\frac{a}{bc} \\ = \quad & (C_2 + 2C_4 + 2C_6 + 4C_7)\frac{b}{ac} \\ = \quad & (C_3 + 2C_5 + 2C_6 + 4C_7)\frac{c}{ab}. \end{aligned}$$

In addition, we require that the common value in this equation be positive.

The balance condition describes explicitly how the parameters of the energy function are affected by the different scales a, b, and c in the three dimensions. Of course, there is a 2-D analog of the balance condition for non-square pixels.

Bibliography

[1] Y. Amit and U. Grenander, *Comparing Sweep Strategies for Stochastic Relaxation*, *J. Multivariate Analysis*, Vol. 37, pp. 197–222, 1991.

[2] J. W. Beck. *Analysis of a Camera Based Single Photon Emission Computed Tomography (SPECT) System*, Ph.D. Thesis, Duke University, Durham, North Carolina, 1982.

[3] J. Besag, "On a System of Two-Dimensional Recurrence Equations," *J. Royal Statist. Soc., Series B*, Vol. 43, pp. 302–309, 1981.

[4] J. Besag, "On the Statistical Analysis of Dirty Pictures (with Discussion)," *J. Royal Statist. Soc., Series B*, Vol. 48, pp. 259–302, 1986.

[5] J. Besag, "Spatial Interaction and the Statistical Analysis of Lattice Systems (with Discussion)," *J. Royal Stat. Soc., Series B*, Vol. 36, pp. 192–236, 1974.

[6] J.-M. Dinten, *Tomographic Reconstruction with a Limited Number of Projections: Regularization Using a Markov Model*, Technical Report, Laboratoire de Statistique Appliquée, Université Paris-Sud, Orsay, 1988.

[7] E. C. Floyd, R. J. Jaszczak, C. C. Harris, and R. E. Coleman. "Energy and Spatial Distribution of Multiple Order Compton Scatter in SPECT," *Phys. Med. Biol.*, Vol. 29, pp. 1217–1230, 1984.

[8] D. Geman and S. Geman, "Bayesian Image Analysis," in E. Bienenstock, F. Fogelman, and G. Weisbuch, editors, *Disordered Systems and Biological Organization*, Springer–Verlag, Berlin, 1986.

[9] D. Geman, S. Geman, C. Graffigne, and P. Dong, "Boundary Detection by Constrained Optimization," *IEEE Trans. Pattern Analysis and Machine Intelligence*, Vol. 12, pp. 609–628, 1990.

[10] D. Geman and G. Reynolds, "Constrained Restoration and the Recovery of Discontinuities," *IEEE Trans. Pattern Anal. and Machine Intell.*, vol. 14, pp. 367–383, 1992.

[11] S. Geman and D.E. McClure, "Bayesian Image Analysis: An Application to Single Photon Emission Tomography," in *1985 Proceedings of the Statistical Computing Section, American Statistical Association*, pp. 12–18, 1985.

[12] S. Geman and D.E. McClure, "Statistical Methods for Tomographic Image Reconstruction," *Bulletin of the International Statistical Institute*, Vol. 52, pp. 5–21, 1987.

[13] S. Geman, D.E. McClure, and D. Geman, A Nonlinear Filter for Film Restoration and Other Problems in Image Processing. *CVGIP: Graphical Models and Image Processing*, Vol. 54:4, pp. 281–289, 1992.

[14] P. J. Green, "Bayesian Reconstruction from Emission Tomography Data Using a Modified EM Algorithm," *IEEE Trans. Med. Imaging*, Vol. 9, pp. 84–93, 1990.

[15] U. Grenander, *Abstract Inference*, John Wiley and Sons, New York, 1978.

[16] U. Grenander, *Tutorial in Pattern Theory*, Technical Report, Division of Applied Mathematics, Brown University, Providence, Rhode Island, 1983.

[17] U. Grenander and G. Szegö, *Toeplitz Forms and Their Applications*, Chelsea Publishing Company, New York, Second Edition, 1984.

[18] G. Helmberg, *Introduction to Spectral Theory in Hilbert Space*, North-Holland Publishing Company, Amsterdam, 1969.

[19] G. T. Herman, *Image Reconstruction from Projections, the Fundamentals of Computerized Tomography, Computer Science and Applied Mathematics*, Academic Press, New York, 1980.

[20] H. E. Johns and J. R. Cunningham, *The Physics of Radiology*, Charles C. Thomas, Springfield, Illinois, Fourth Edition, 1983.

[21] V. E. Johnson, W. H. Wong, X. Hu, and C.-T. Chen, Statistical Aspects of Image Restoration, *IEEE Trans. Pattern Analysis and Machine Intelligence*, vol. 13, pp. 413–425, 1991.

[22] E. Levitan and G. T. Herman, "A Maximum *a posteriori* Probability Expectation Maximization Algorithm for Image Reconstruction in Emission Tomography," *IEEE Trans. on Medical Imaging*, Vol. 6, pp. 185–192, 1987.

[23] K. M. Manbeck, *Bayesian Statistical Methods Applied to Physical Phantom and Patient Data*, Ph.D. Thesis, Division of Applied Mathematics, Brown University, Providence, Rhode Island, 1990.

[24] K. M. Manbeck, *Hubble Telescope Image Restoration By Statistical Methods*, Technical Report, Division of Applied Mathematics, Brown University, Providence, Rhode Island, 1990.

[25] J. L. Marroquin, S. Mitter, and T. Poggio, Probabilistic Solution of Ill-Posed Problems in Computational Vision, *J. Amer. Stat. Assoc.*, Vol. 82, pp. 76–89, 1987.

[26] D. E. McClure and S. C. Shwartz, *A Method of Image Representation Based on Bivariate Splines*, Technical Report, Division of Applied Mathematics, Brown University, Providence, Rhode Island, 1989.

[27] J. A. Mertus, *Self Calibrating Methods for Image Reconstruction in Emission Computed Tomography*, Ph.D. Thesis, Division of Applied Mathematics, Brown University, Providence, Rhode Island, 1988.

[28] M. I. Miller, D. L. Snyder, and T. R. Miller, "Maximum-Likelihood Reconstruction for Single-Photon Emission Tomography," *IEEE Trans. Nuclear Science*, Vol. 32, pp. 769–778, 1985.

[29] D. W. Murray and B. B. Buxton. "Scene Segmentation from Visual Motion Using Global Optimization" *IEEE Trans. Pattern Analysis and Machine Intelligence*, Vol. 9, pp. 220–228, 1987.

[30] A. B. Owen, "Discussion of: Statistics, Images, and Pattern Recognition", by B. D. Ripley, *Can. J. Statist.*, Vol. 14, pp. 106–110, 1986.

[31] B. C. Penny, M. A. King, and K. Knesaurek, "A Projector, Back-Projector Pair which Accounts for the Two-Dimensional Depth and Distance Dependent Blurring in SPECT, *IEEE Trans. Nuclear Science*, Vol. 37, pp. 681–686, 1990.

[32] A. Rockmore and A. Makovski, "A Maximum Likelihood Approach to Emission Image Reconstruction from Projections, *IEEE Trans. Nuclear Science*, Vol. 23, pp. 1428–1432, 1976.

[33] L. A. Shepp and Y. Vardi. "Maximum-Likelihood Reconstruction for Emission Tomography," *IEEE Trans. Medical Imaging*, Vol. 1, pp. 113–121, 1982.

[34] B. W. Silverman, M. C. Jones, J. D. Wilson, and D. W. Nychka, "A Smoothed EM Approach to Indirect Estimation Problems, with Particular Reference to Stereology and Emission Tomography, *J. Royal Stat. Soc.*, Series B, Vol. 52, pp. 271–324, 1990.

[35] B. Simon, *The* $P(\phi)_2$ *Euclidean (Quantum) Field Theory*, Princeton University Press, Princeton, New Jersey, 1974.

[36] A. M. Yaglom, *An Introduction to the Theory of Stationary Random Functions*, Prentice-Hall, Englewood Cliffs, New Jersey, 1962.

Gaussian Markov Random Fields at Multiple Resolutions

Sridhar Lakshmanan† **and Haluk Derin**‡

†Electrical and Computer Engineering, University of Michigan at Dearborn, Dearborn, Michigan

‡Department of Electrical and Computer Engineering
University of Massachusetts
Amherst, Massachusetts

1 Introduction

Representation of images at multiple resolutions have come about due to three primary considerations: Such representations support highly efficient algorithms based on the *divide and conquer* principle. The hierarchy of resolutions provides a (smooth) transition between pixel-level (local) features and region-level (global) features, an illustration of the *action at a distance* principle. The multi-resolution framework provides a model for certain types of early processing in natural (human) vision. Motivated by such considerations, a number of multi-resolution algorithms have been proposed and used for a variety of applications [21].

In recent years, Markov random fields (MRFs) have been, and are increasingly being, used to model *a priori* beliefs about the continuity of image features such as region labels, textures, edges, and so on. Introduction of these Markov models in a Bayesian framework has resulted in a coherent framework for processing images. Specifically, such a framework has allowed researchers to pose image processing problems as well-defined statistical inference problems, and develop a variety of relaxation-type algorithms for processing images, based on a Monte Carlo computational theory. The chapters in this volume, and the references therein, attest to this statement.

In spite of these desirable properties of the MRF framework, there have been some fundamental drawbacks. The computational algorithms, though based on a well-defined theory, are usually slow; primarily because, due to the inherent assumption of Markovianity, the global features are allowed

ISBN 0-12-170608-7

to exist and propagate only through local interactions. A multi-resolution MRF framework, on the other hand, would support efficient algorithms that start processing images at a coarse resolution, and then progressively refine them to finer resolutions. Such processing of images at multiple resolutions would allow global features to propagate relatively quickly, and to change in bigger increments, which are the primary reasons for their efficiency [4, 12, 13, 18]. Another drawback arises from a modeling point of view. In the MRF framework, *a priori* beliefs about the image features are modelled at exactly one scale (or resolution). A multi-resolution MRF framework would alleviate this shortcoming by providing a vehicle (or framework) for expressing apriori beliefs about image features over a range of scales (or resolutions).

The fusion of multi-resolution and MRF ideas has been investigated before. In [12], a combination of the multi-resolution and the MRF ideas were explored, and an important connection between these ideas and analogous ones in the renormalization group studies in statistical mechanics [5, 9] was established. The result was a multi-resolution MRF framework that was used to process binary images efficiently. The key problem in developing a multi-resolution MRF framework was to provide consistent model descriptions for MRFs at multiple resolutions. It turned out that even for binary images (modeled as MRFs), simple resolution transformations (such as sampling) resulted in the loss of Markovianity; so, to process images (modeled as MRFs) at multiple resolutions, a methodology for approximating (possibly) non-Markov fields by MRFs was needed. The cumulant and bond-moving approximations, which are used for analogous purposes in renormalization group studies, were explored as candidates, and certain variations of those approximations turned out to be useful. In [11], the multi-resolution MRF ideas were extended to images modeled as Gaussian Markov random fields (GMRFs). For GMRFs under the influence of an inhomogeneous external field, the power spectra at multiple resolutions were obtained analytically. This analysis was extended to include the cases when the lattice size and the number of resolution levels go to infinity. The results therein provided a generalization of analogous results that were obtained for GMRFs under the influence of a homogeneous external field in [3]. Neither of the two studies, however, considered the approximation schemes needed to process images (modeled as GMRFs) at multiple resolutions.

In [7] also, the fusion of the multi-resolution and the MRF ideas were explored, and an important connection with similar studies in economics [1, 22] was established. For 1-D time series modeled as auto-regressive (AR) processes, the problem of consistent model descriptions under resolution

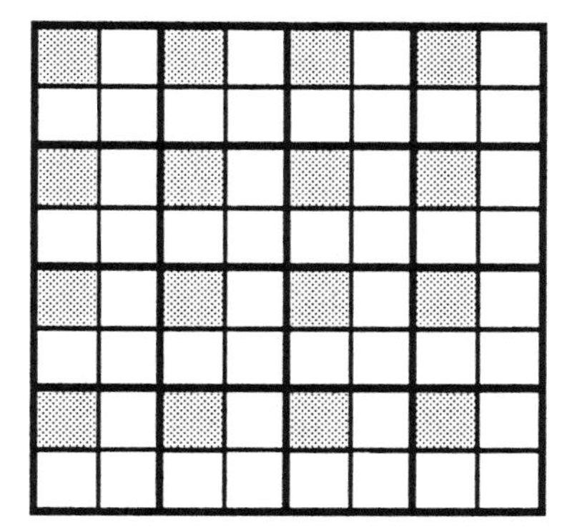

Figure 1: Sampling type resolution transformation

transformations was considered. The fact that, under resolution transformations, AR processes become auto-regressive moving-average (ARMA) processes, and lose the Markovianity property, was established. A methodology was developed for estimating the AR parameters for 1-D processes, given their lower resolution versions. This study also indicated possible applications of such ideas to image processing tasks, such as spectral feature extraction and image synthesis.

In [4] and [13], multi-resolution MRF-based algorithms were designed to process images hierarchically. These studies resulted in computationally efficient multi-resolution MRF-based image processing algorithms. However, the problem of consistent model descriptions was not addressed in these studies. Finally, in [18], a multi-resolution MRF-based algorithm was designed based on the renormalization group ideas in [24].

A continuation of these previous contributions (and understanding), leads us to the topic of this chapter: development of a multi-resolution framework for MRFs, with a view towards processing images. We specialize further and concentrate on a multi-resolution Gaussian Markov random field (GMRF) framework. The key problem in developing such a framework would be to provide consistent model descriptions for GMRFs at multiple resolutions. Following [12], we consider two types of resolution transformation models:

1. *Sampling* (Fig. 1)

2. *Block-to-Point* (Fig. 2)

Starting with either a finite- or an infinite-lattice GMRF $X^{(0)}$ and obtaining coarser resolution versions of it, $X^{(1)}, X^{(2)}, \ldots$, by successive ap-

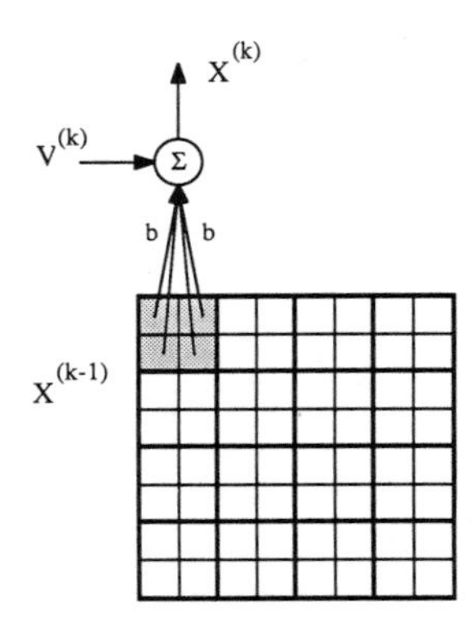

Figure 2: Block-to-point type resolution transformation.

plications of the two resolution transformations, we show that the random fields, $X^{(1)}, X^{(2)}, \ldots$, are in general non-Markov (analogous to such observations in [1, 3, 5, 7, 9, 12, 22, 23, 24]), and obtain exact descriptions for $X^{(1)}, X^{(2)}, \ldots$, along with their covariances and power spectra. As an illustration of this result, we consider a special case of it. We show that if $X^{(0)}$ is a second-order infinite-lattice GMRF with a separable autocorrelation, then $X^{(1)}, X^{(2)}, \ldots$, are ARMA, hence non-Markov (analogous to results in [1, 7, 22]). We carry the analysis of this special case further, and obtain certain characteristics that are pertinent to the covariances and power spectra of these multi-resolution fields. As a by-product of considering this example, we identify a new class of GMRFs that retain Markovianity under sampling. This is one of a small collection of models having this invariance property [9], the 1-D Ising model [23] being the most commonly referred example.

The fact that, in general, GMRFs lose Markovianity upon resolution transformations implies that modeling problems arise in processing images modeled as GMRFs at multiple resolutions. To remedy this problem, we need to be able to *approximate* non-Markov Gaussian fields as GMRFs. There exist a number of possible (*free energy*-based) approximations [5, 9, 23, 24], the cumulant and bond-moving approximations [5] being the most commonly cited. For reasons that are explained, we introduce a new approximation that is based on approximating the covariance (as opposed to the *free energy*), and call it the *Covariance Invariance Approximation* (CIA). We investigate the theoretical and computational aspects of the CIA, and observe that these considerations lead to some fundamental issues regarding GMRFs. We show that the CIA is optimal according to a number of different information theoretic notions such as *Maximum Entropy*, *Minimum Kullback–Leibler Distance*, and *Maximum Likelihood.* Because it approximates the covariance, the CIA is better suited to image processing than the cumulant and the bond-moving approximations. There are some basic problems associated with using the cumulant and bond-moving approximations, as also stated in [12]. Both of these approximations deal with approximating a quantity called the *free energy* [23], which is not a quantity of any significance or interest in image modeling. An important argument in favor of the CIA is that the principles of Maximum Likelihood and Maximum Entropy, which the CIA stems from, are exactly the same principles invoked when modeling the original image $X^{(0)}$ as a GMRF. With our present understanding, it appears the CIA scheme is distinct from, and has advantages over, the free energy-based schemes used in renormalization group studies. It also appears that the CIA scheme could be potentially useful as one of the alternatives in renormalization group studies [17]. An exploration of these ideas would take us too far afield from the objectives of this communication, so we do not pursue it any further.

The rest of the chapter is organized as follows: In Section 2, we briefly describe the GMRF model, and introduce the two types of resolution transformation models. In Section 3, we obtain analytical expressions for the joint densities, the covariances, and the power specta of $X^{(1)}, X^{(2)}, \ldots,$ based on the GMRF model assumption on $X^{(0)}$. In Section 4, we illustrate our results from Section 3, by considering a special class, namely, the separable autocorrelation GMRFs. Finally, in Section 5, we introduce the CIA scheme and investigate its properties, followed by some concluding remarks in Section 6.

2 Gaussian Markov Random Field and Resolution Transformation Models

2.1 The GMRF Model

We briefly describe the finite-lattice GMRF model. Let $\Omega^{(0)} = \{(i,j) : 0 \leq i \leq M-1,\ 0 \leq j \leq N-1\}$ be a rectangular lattice. We denote elements of $\Omega^{(0)}$ by $\underline{s}$, $\underline{t}$, $\underline{u}$ and so on, where $\underline{s}=(s_1, s_2)$. We assume that sums (and differences) of the form $\underline{s}+\underline{t}$ are evaluated in Modulo(M,N) - which is equivalent to assuming that $\Omega^{(0)}$ is a toroidal lattice.

Let $X^{(0)} = \{X^{(0)}_{\underline{s}},\ \underline{s} \in \Omega(0)\}$, at level "0" of the image pyramid (Fig. 3), be a GMRF, with respect to a neighborhood system η; that is, $X^{(0)}$ is a

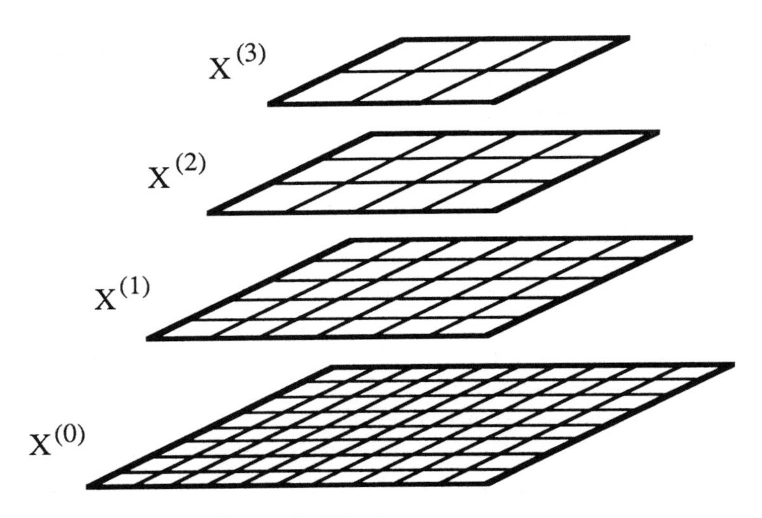

Figure 3: The image pyramid.

set of jointly Gaussian stationary r.v.'s that also has the Markov property. Let $\underline{X}^{(0)}$ be a column vector corresponding to a (row-wise) raster scan ordering of the r.v.'s comprising $X^{(0)}$, and $\Sigma^{(0)}$ represent the $(MN \times MN)$ covariance matrix of $\underline{X}^{(0)}$. Restricting our attention, without any loss of generality, to zero-mean GMRFs, and denoting realizations of r.v.'s and r.f.'s by the corresponding lower-case letters, the joint pdf of $X^{(0)}$ (also of $\underline{X}^{(0)}$) has the form,

$$P(X^{(0)} = x^{(0)}) = \frac{1}{(2\pi)^{\frac{MN}{2}} (\det \Sigma^{(0)})^{\frac{1}{2}}} \exp\{-\frac{1}{2}[\underline{x}^{(0)}]^T [\Sigma^{(0)}]^{-1} [\underline{x}^{(0)}]\}. \tag{1}$$

The Markov property on $X^{(0)}$, that is,

$$P(X_{\underline{s}}^{(0)} | X_{\underline{t}}^{(0)}, \underline{t} \neq \underline{s}) = P(X_{\underline{s}}^{(0)} | X_{\underline{s}+\underline{r}}^{(0)}, \underline{r} \in \eta), \tag{2}$$

implies that the inverse covariance $[\Sigma^{(0)}]^{-1}$ is a sparse matrix, with zero entries corresponding to pairs of sites that are not neighbors of each other. The stationarity assumption, in conjunction with the toroidal lattice assumption, implies that the inverse covariance is block-circulant, and it involves only a few nonzero parameters: $\sigma^{(0)}$ and $\underline{\theta}^{(0)}$. Specifically, all the diagonal entries of $[\Sigma^{(0)}]^{-1}$ are equal to $1/[\sigma^{(0)}]^2$; entries corresponding to sites $\underline{r}$ apart, for $\underline{r} \in \eta$, are $-\theta_{\underline{r}}^{(0)}/[\sigma^{(0)}]^2$, and all other entries are zero. It is important to note that, due to the symmetry of the inverse covariance matrix $[\Sigma^{(0)}]^{-1}$, $\theta_{\underline{r}}^{(0)} = \theta_{-\underline{r}}^{(0)}$, for any $\underline{r} \in \eta$. The GMRF defined by Eq. (1) is valid, if and only if the covariance matrix $\Sigma^{(0)}$, or equivalently, the inverse covariance $[\Sigma^{(0)}]^{-1}$, is positive definite. We assume that the parameters, $\sigma^{(0)}$ and $\underline{\theta}^{(0)}$, are such that the resulting $\Sigma^{(0)}$ is positive definite.

The following theorem gives the autocorrelation $R^{(0)}$ (also the autocovariance, since $X^{(0)}$ is zero-mean), and the power spectrum $S^{(0)}$ of $X^{(0)}$, in terms of the parameters $\sigma^{(0)}$ and $\underline{\theta}^{(0)}$.

Theorem 1 *For $\{X_{\underline{s}}^{(0)}, \underline{s} \in \Omega^{(0)}\}$, a zero-mean GMRF,*

$$S_{\underline{u}}^{(0)} \triangleq [\sigma^{(0)}]^2 \{1 - \sum_{\underline{r} \in \eta} \theta_{\underline{r}}^{(0)} \cos[\frac{2\pi}{M} r_1 u_1 + \frac{2\pi}{N} r_2 u_2]\}^{-1}, \quad \underline{u} \in \Omega^{(0)}, \tag{3}$$

is the power spectrum, and

$$R_{\underline{s}}^{(0)} \triangleq \frac{1}{MN} \sum_{\underline{u} \in \Omega^{(0)}} S_{\underline{u}}^{(0)} \cos[\frac{2\pi}{M} s_1 u_1 + \frac{2\pi}{N} s_2 u_2]\}^{-1}, \quad \underline{s} \in \Omega^{(0)}, \tag{4}$$

is the autocorrelation. It is also true that $\{[S_{\underline{u}}^{(0)}]^{-1}, \underline{u} \in \Omega(0)\}$ are the eigenvalues of the inverse covariance matrix $[\Sigma^{(0)}]^{-1}$.

Proof. Refer to [14]. □

This completes our description of the finite-lattice GMRF model; Eqs. (1), (2), and Theorem 1 specify the random field $X^{(0)}$ completely.

We now briefly describe the infinite-lattice GMRF model. Following [19], we let the lattice size (M and N) go to infinity. This leads to the following result [19]:

Theorem 2 *For $X^{(0)} = \{X_{\underline{s}}^{(0)}, \underline{s} \in \mathcal{Z}^2\}$, a zero-mean infinite-lattice GMRF, its power spectrum is given by:*

$$S_{\underline{\omega}}^{(0)} = [\sigma^{(0)}]^2 \{1 - \sum_{\underline{r} \in \eta} \theta_{\underline{r}}^{(0)} \cos[r_1 \omega_1 + r_2 \omega_2]\}^{-1}, \tag{5}$$

$\underline{\omega} \in \tilde{\Omega}^{(0)} \triangleq [-\pi, \pi] \times [-\pi, \pi]$. □

(Note that we use the same symbols for denoting analogous quantities in finite- and infinite-lattice models; the one intended should be clear from the context.) The infinite-lattice random field $X^{(0)}$ also satisfies the Markov property Eq. (2), with $\underline{t} \neq \underline{s}$ replaced by $\underline{t} = \underline{s} + \underline{r}$, $\underline{r} \in A$, for A any subset of $\mathcal{Z}^2$ that contains η, but not $\underline{0}$. Furthermore, any subset of r.v.'s from the random field $X^{(0)}$ are jointly Gaussian. As in the finite-lattice case, we assume that $\sigma^{(0)}$ and $\underline{\theta}^{(0)}$ are chosen so that $S_{\underline{\omega}}^{(0)} > 0$, for all $\underline{\omega} \in \tilde{\Omega}^{(0)}$. This completes our brief description of the infinite-lattice GMRF model.

Both the finite- and infinite-lattice GMRFs satisfy the Markov property. Specifically, the conditional pdf, in both cases, has the form,

$$P(X_{\underline{s}}^{(0)} = x_{\underline{s}}^{(0)} | X_{\underline{s}+\underline{r}}^{(0)} = x_{\underline{s}+\underline{r}}^{(0)}, \underline{r} \in \eta) = \frac{1}{\sqrt{2\pi[\sigma^{(0)}]^2}} \exp\{-\frac{1}{2[\sigma^{(0)}]^2}[x_{\underline{s}}^{(0)} - \sum_{\underline{r}\in\eta} \theta_{\underline{r}}^{(0)} x_{\underline{s}+\underline{r}}^{(0)}]^2\}. \tag{6}$$

The finite- and infinite-lattice GMRFs described before have an equivalent characterization in terms of the following AR representation,

$$X_{\underline{s}}^{(0)} = \sum_{\underline{r}\in\eta} \theta_{\underline{r}} X_{\underline{s}+\underline{r}}^{(0)} + U_{\underline{s}}^{(0)}, \tag{7}$$

where $U^{(0)}$ is zero-mean, Gaussian, and has an autocorrelation given by:

$$E\{U_{\underline{s}}^{(0)} U_{\underline{s}+\underline{r}}^{(0)}\} = \begin{cases} -\theta_{\underline{r}}^{(0)} \left[\sigma^{(0)}\right]^2 & \text{if } \underline{r} \in \eta \\ \left[\sigma^{(0)}\right]^2 & \text{if } \underline{r} = \underline{0} \\ 0 & \text{otherwise} \end{cases}. \tag{8}$$

For finite-lattice models, Eqs. (7) and (8) are to be interpreted on the toroidal lattice.

Having described the finite- and infinite-lattice GMRF models, we now describe the two types of resolution transformation models. Before we do that, however, a few words about the multi-resolution notation is in order: Elements of the rectangular lattices at any level of resolution are denoted by $\underline{s}$, $\underline{t}$, $\underline{u}$, and so on, and in the quantities of interest that are defined on those lattices, these elements appear as subscripts. The level of the resolution is denoted by 0, 1, ..., and for multi-resolution quantities, their resolution levels appear as superscripts. The neighborhood set η is an index set, and so it will be used to denote lattice site increments at all levels of resolution. M and N, defining the lattice size (M,N), without loss of generality, are assumed to be powers of 2.

2.2 Sampling

For sampling as the resolution transformation, we have:

$$X_{\underline{s}}^{(k)} = X_{2\underline{s}}^{(k-1)}, \tag{9}$$

for all $\underline{s} \in \Omega^{(k)} \triangleq \{(i,j) : 0 \leq i \leq \frac{M}{2^k} - 1, \ 0 \leq j \leq \frac{N}{2^k} - 1\}$ (for finite-lattice fields), or for all $\underline{s} \in \mathcal{Z}^2$ (for infinite-lattice fields). The random field $X^{(k)}$ thus obtained is a coarser resolution version of $X^{(k-1)}$, and, in the finite-lattice case, the lattice $\Omega^{(k)}$, upon which $X^{(k)}$ is defined, is one-half the size of the lattice $\Omega^{(k-1)}$. By successive application of Eq. (9), we have, for every $\underline{s} \in \Omega^{(k)}$ or $\mathcal{Z}^2$:

$$X_{\underline{s}}^{(k)} = X_{\underline{s}'}^{(0)}, \tag{10}$$

where $\underline{s}' = 2^k \underline{s}$. Figure 1 depicts the sampling-type resolution transformation.

2.3 Block-to-Point Transformation

Under the block-to-point resolution transformation, we have:

$$X_{\underline{s}}^{(k)} = b \sum_{\underline{r} \in \mathcal{C}_1} X_{2\underline{s}+\underline{r}}^{(k-1)} + V_{\underline{s}}^{(k)}, \tag{11}$$

for all $\underline{s} \in \Omega^{(k)}$ (for finite-lattice fields), or for all $\underline{s} \in \mathcal{Z}^2$ (for infinite-lattice fields), where $\mathcal{C}_k = \{\underline{r} : 0 \leq r_1 \leq 2^k - 1, \ 0 \leq r_2 \leq 2^k - 1\}$ and the $V_{\underline{s}}^{(k)}$'s are *iid*, independent of $X^{(k-1)}$, Gaussian-distributed with zero mean and variance "a". Reinterpreting Eq. (11), for each $\underline{s} \in \Omega^{(k)}$ or $\mathcal{Z}^2$, $X_{\underline{s}}^{(k)}$ is Gaussian-distributed, about the average of $X^{(k-1)}$ over non-overlapping blocks of sites $\{\underline{t} : \underline{t} = 2\underline{s} + \underline{r}, \ \underline{r} \in \mathcal{C}_1\}$, with variance a. When $a = 0$, $X_{\underline{s}}^{(k)}$ becomes the average of $X^{(k-1)}$ over the block. As in sampling, the random field $X^{(k)}$ is a coarser resolution version of $X^{(k-1)}$, and in the finite-lattice case, the lattice $\Omega^{(k)}$ upon which $X^{(k)}$ is defined, is one-half the size of the lattice $\Omega^{(k-1)}$. Successive application of Eq. (11) yields (also in [3]) for every $\underline{s} \in \Omega(k)$ or $\mathcal{Z}^2$,

$$X_{\underline{s}}^{(k)} = b^k \sum_{\underline{r} \in \mathcal{C}_k} X_{\underline{s}'+\underline{r}}^{(0)} + W_{\underline{s}}^{(k)}, \tag{12}$$

where $\underline{s}' = 2^k \underline{s}$, and $W_{\underline{s}}^{(k)}$'s are again *iid*, Gaussian distributed with mean zero and variance $a_k = a[1 - (4b^2)^k]/[1 - 4b^2]$. Note that the $W_{\underline{s}}^{(k)}$'s are appropriately defined in terms of the $V_{\underline{s}}^{(k)}$'s through successive application of Eq. (11). The interpretations following Eq. (11) are applicable here as well. Figure 2 depicts the block-to-point-type resolution transformation.

For finite-lattice fields, because of the reduction in lattice size under both types of resolution transformations, the number of levels of resolution can be no greater than $\min(\log_2 M, \log_2 N)$. In the following section, our objective is to obtain analytical expressions for the joint pdf's, the power spectra, and the covariances of $X^{(1)}, X^{(2)}, \ldots$, based on the GMRF assumption on $X^{(0)}$.

3 Descriptions at Multiple Resolutions

We consider the descriptions of the r.f.'s at multiple resolutions with the two types of resolution transformations separately.

Theorem 3 *Let $X^{(0)}$ be a finite-lattice GMRF with respect to a neighborhood system η; also let the type of resolution transformation be sampling. Then*

1. *The covariance of $\underline{X}^{(k)}$ is*

$$\Sigma^{(k)} = [D_0^{(k)}]\Sigma^{(0)}[D_0^{(k)}]^T, \tag{13}$$

where, $D_0^{(k)} = D_{k-1}^{(k)} \cdot D_{k-2}^{(k-1)} \cdots D_0^{(1)}$,

$$D_{k-1}^{(k)} \triangleq \begin{bmatrix} \underline{d}_{k-1}^{(k)} & \underline{0} & \underline{0} & \underline{0} & \cdots & \underline{0} & \underline{0} \\ \underline{0} & \underline{0} & \underline{d}_{k-1}^{(k)} & \underline{0} & \cdots & \underline{0} & \underline{0} \\ & \cdots & & \cdots & & \cdots & \\ \underline{0} & \underline{0} & \underline{0} & \underline{0} & \cdots & \underline{d}_{k-1}^{(k)} & \underline{0} \end{bmatrix}, \tag{14}$$

$D_{k-1}^{(k)}$ is a $\left(\frac{M}{2^k} \cdot \frac{N}{2^k}\right) \times \left(\frac{M}{2^{k-1}} \cdot \frac{N}{2^{k-1}}\right)$ matrix, and the $\underline{d}_{k-1}^{(k)}$ and the $\underline{0}$ blocks are submatrices of size $\frac{N}{2^k} \times \frac{N}{2^{k-1}}$ that have the same structure as $D_{k-1}^{(k)}$ with $\underline{d}_{k-1}^{(k)}$ replaced by 1 and $\underline{0}$ replaced by 0.

2. *The power spectrum of $X^{(k)}$ is*

$$S_{\underline{u}}^{(k)} = \frac{1}{2^{2k}} \sum_{\underline{r} \in C_k} S_{\underline{u}+\underline{r}'}^{(0)}, \tag{15}$$

where $\underline{r}' = (\frac{M}{2^k} r_1, \frac{N}{2^k} r_2)$, for all $\underline{u} \in \Omega^{(k)}$.

3. *The joint pdf of $X^{(k)}$ is*

$$P(X^{(k)} = x^{(k)}) = \frac{1}{(2\pi)^{\frac{MN}{2^{2k+1}}}(\det\Sigma^{(k)})^{\frac{1}{2}}} \exp\{-\frac{1}{2}[\underline{x}^{(k)}]^T[\Sigma^{(k)}]^{-1}[\underline{x}^{(k)}]\}, \tag{16}$$

where $\underline{x}^{(k)}$ is a row-wise ordering of $x^{(k)}$. In general, the random field $X^{(k)}$ is not Markov. In fact the entries in $[\Sigma^{(k)}]^{-1}$ corresponding to sites $\underline{r}$ apart, for $\underline{r} \in \Omega^{(k)} (\underline{r} \neq \underline{0})$, are

$$- \theta_{\underline{r}}^{(k)} \; [\sigma^{(k)}]^{-2} = \frac{2^{2k}}{MN} \sum_{\underline{u} \in \Omega^{(k)}} [S_{\underline{u}}^{(k)}]^{-1} \cos[\frac{2\pi}{M} r_1 u_1 + \frac{2\pi}{N} r_2 u_2], \quad (17)$$

and for $\underline{r} = \underline{0}$, the entries are

$$[\sigma^{(k)}]^{-2} = \frac{2^{2k}}{MN} \sum_{\underline{u} \in \Omega^{(k)}} [S_{\underline{u}}^{(k)}]^{-1}. \quad (18)$$

Proof. From Eq. (10), and the definition of $D_0^{(k)}$ in Eq. (14), the relationship between $X^{(k)}$ and $X^{(0)}$ can be rewritten as:

$$\underline{X}^{(k)} = D_0^{(k)} \underline{X}^{(0)}. \quad (19)$$

The covariance of $\underline{X}^{(k)}$ given in Eq. (13) is a direct consequence of Eq. (19).

Specializing Eq. (13) further, and using the fact that $X^{(0)}$ (hence $X^{(k)}$) is zero-mean, the autocorrelation (hence the covariance) of $X^{(k)}$ is

$$R_{\underline{s}}^{(k)} = R_{\underline{s}'}^{(0)}, \underline{s} \in \Omega^{(k)}, \quad (20)$$

where $\underline{s}' = 2^k \underline{s}$. Using Eqs. (4) and (20), the autocorrelation of $X^{(k)}$ can be rewritten as:

$$R_{\underline{s}}^{(k)} = \frac{1}{MN} \sum_{\underline{u} \in \Omega^{(0)}} S_{\underline{u}}^{(0)} \cos[\frac{2\pi}{M} 2^k s_1 u_1 + \frac{2\pi}{N} 2^k s_2 u_2]. \quad (21)$$

Notice that for any $\underline{s}$ and $\underline{u}$,

$$\cos[\frac{2\pi}{M} s_1' u_1 + \frac{2\pi}{N} s_2' u_2] = \cos[\frac{2\pi}{M} s_1'(u_1 + \frac{M}{2^k} r_1) + \frac{2\pi}{N} s_2'(u_2 + \frac{N}{2^k} r_2)], \quad (22)$$

for all $\underline{r} \in \mathcal{C}_k$. Using Eq. (22) and rearranging Eq. (21), we obtain:

$$R_{\underline{s}}^{(k)} = \frac{2^{2k}}{MN} \sum_{\underline{u} \in \Omega^{(k)}} [\frac{1}{2^{2k}} \sum_{\underline{r} \in \mathcal{C}_k} S_{\underline{u}+\underline{r}'}^{(0)}] \cos[\frac{2\pi}{\frac{M}{2^k}} u_1 s_1 + \frac{2\pi}{\frac{N}{2^k}} u_2 s_2], \quad (23)$$

for all $\underline{s} \in \Omega^{(k)}$. It follows from Eq. (23) that the power spectrum $S_{\underline{u}}^{(k)}$ is the expression given by Eq. (15).

By Eq. (19), since $X^{(0)}$ is multivariate Gaussian, $X^{(k)}$ is also multivariate Gaussian. It is zero-mean and has a covariance given by Eq. (13). Hence, the joint pdf of $X^{(k)}$ is given by Eq. (16). The covariance $\Sigma^{(k)}$ is symmetric and block-circulant; hence, its inverse can be calculated by the procedure given in [14]: the entries of $[\Sigma^{(k)}]^{-1}$ corresponding to sites $\underline{r}$ apart are given in Eqs. (17) and (18). Notice that since $S_{\underline{u}}^{(k)}$ cannot be written in the same form as the RHS of Eq. (3), with coefficients $(\theta_{\underline{r}}^{(k)})$ that are zero outside an appropriate neighborhood, $X^{(k)}$ does not have a Markov spectrum. Hence, $X^{(k)}$ is non-Markov, thereby completing the proof. □

Following a line of reasoning similar to the proof of Eq. (15), for infinite-lattice GMRFs, we have:

$$S_{\underline{\omega}}^{(k)} = \frac{1}{2^{2k}} \sum_{\underline{r} \in C_k} S_{\underline{\omega}" + \underline{\gamma}"}^{(0)}, \tag{24}$$

for all $\underline{u} \in \tilde{\Omega}^{(k)}$ where, $\underline{\omega}" = 2^{-k}\underline{\omega}$, $\underline{\gamma}" = \frac{2\pi}{2^k}\underline{r}$, and $\tilde{\Omega}^{(k)} = \tilde{\Omega}^{(0)}$.

The preceding theorem yields the analytical expressions for the multiresolution quantities when the model for resolution transformation is sampling. The next theorem gives the analogous results for the block-to-point resolution transformation.

Theorem 4 *Let $X^{(0)}$ be a finite-lattice GMRF with respect to a neighborhood system η; also let the type of resolution transformation be blok-to-point. Then:*

1. *The covariance of $\underline{X}^{(k)}$ is*

$$\Sigma^{(k)} = a_k \mathcal{I}^{(k)} + [B_0^{(k)}]\Sigma^{(0)}[B_0^{(k)}]^T, \tag{25}$$

where $\mathcal{I}^{(k)}$ is the identity matrix of size $(\frac{M}{2^k} \cdot \frac{N}{2^k}) \times (\frac{M}{2^k} \cdot \frac{N}{2^k})$, $B_0^{(k)} = B_{k-1}^{(k)} \cdot B_{k-2}^{(k-1)} \cdots B_0^{(1)}$,

$$B_{k-1}^{(k)} \triangleq \begin{bmatrix} \underline{b}_{k-1}^{(k)} & \underline{b}_{k-1}^{(k)} & \underline{0} & \underline{0} & \cdots & \underline{0} & \underline{0} \\ \underline{0} & \underline{0} & \underline{b}_{k-1}^{(k)} & \underline{b}_{k-1}^{(k)} & \cdots & \underline{0} & \underline{0} \\ & \cdots & & \cdots & & \cdots & \\ \underline{0} & \underline{0} & \underline{0} & \underline{0} & \cdots & \underline{b}_{k-1}^{(k)} & \underline{b}_{k-1}^{(k)} \end{bmatrix}, \tag{26}$$

$B_{k-1}^{(k)}$ is a $\left(\frac{M}{2^k} \cdot \frac{N}{2^k}\right) \times \left(\frac{M}{2^{k-1}} \cdot \frac{N}{2^{k-1}}\right)$ matrix, and the $\underline{b}_{k-1}^{(k)}$ and the $\underline{0}$ blocks are submatrices of size $\frac{N}{2^k} \times \frac{N}{2^{k-1}}$ that have the same structure as $B_{k-1}^{(k)}$ with $\underline{b}_{k-1}^{(k)}$ replaced by b and $\underline{0}$ replaced by 0.

2. *The power spectrum of* $X^{(k)}$ *is*

$$S_{\underline{u}}^{(k)} = \frac{1}{2^{2k}} \sum_{\underline{r} \in C_k} [\Lambda_{\underline{u}+\underline{r}'}^{(k)} S_{\underline{u}+\underline{r}'}^{(0)} + a_k], \tag{27}$$

where $\underline{r}' = (\frac{M}{2^k} r_1, \frac{N}{2^k} r_2)$, *and*

$$\Lambda_{\underline{u}}^{(k)} = b^{2k} \sum_{\underline{t} \in C_k'} (2^k - |t_1|)(2^k - |t_2|) \cos[\frac{2\pi}{M} u_1 t_1 + \frac{2\pi}{N} u_2 t_2], \tag{28}$$

where, $C_k' = \{\underline{r} : -2^k + 1 \leq r_1 \leq 2^k - 1,\ -2^k + 1 \leq r_2 \leq 2^k - 1\}$.

3. *The joint pdf of* $X^{(k)}$ *is the same as that in Eq. (16), the difference being that the covariance* $\Sigma^{(k)}$ *is now given by Eq. (25). The entries in* $[\Sigma^{(k)}]^{-1}$ *are given in Eqs. (17) and 18, the only difference being that power spectrum* $S_{\underline{u}}^{(k)}$ *is now given by Eq. (27). Hence, the random field* $X^{(k)}$ *is not Markov.*

Proof. From Eq. (12), and the definition of $B_0^{(k)}$ in Eq. (26), the relationship between $X^{(k)}$ and $X^{(0)}$ can be rewritten as:

$$\underline{X}^{(k)} = B_0^{(k)} \underline{X}^{(0)} + \underline{W}^{(k)}, \tag{29}$$

where $\underline{W}^{(k)}$ is a column vector corresponding to a (row-wise) raster scan ordering of the *iid* random variables comprising $W^{(k)}$. The covariance of $\underline{X}^{(k)}$ given in Eq. (25) is a direct consequence of Eq. (29).

Specializing Eq. (25) further, and using the fact that $X^{(0)}$ and $W^{(k)}$ (hence $X^{(k)}$) are zero-mean, the autocorrelation (hence the covariance) of $X^{(k)}$ is

$$R_{\underline{s}}^{(k)} = a_k \delta(\underline{s}, \underline{0}) + b^{2k} \sum_{\underline{t} \in C_k'} (2^k - |t_1|)(2^k - |t_2|) R_{\underline{s}'+\underline{t}}^{(0)}, \tag{30}$$

where $\underline{s}' = 2^k \underline{s}$. Notice that a_k affects only $R_0^{(k)}$; the correlation between two distinct points is independent of a_k. Using Eqs. (4) and (30), and rearranging terms, the autocorrelation of $X^{(k)}$ can be rewritten as:

$$R_{\underline{s}}^{(k)} = \frac{1}{MN} \sum_{\underline{u} \in \Omega^{(0)}} [\Lambda_{\underline{u}}^{(k)} S_{\underline{u}}^{(0)} + a_k] \cos[\frac{2\pi}{M} 2^k s_1 u_1 + \frac{2\pi}{N} 2^k s_2 u_2], \tag{31}$$

where $\Lambda_{\underline{u}}^{(k)}$ is given in Eq. (28). Now using Eq. (22), and rearranging Eq. (31), we obtain:

$$R_{\underline{s}}^{(k)} = \frac{2^{2k}}{MN} \sum_{\underline{u} \in \Omega^{(k)}} \{\frac{1}{2^{2k}} \sum_{\underline{r} \in C_k} [\Lambda_{\underline{u}+\underline{r}'}^{(k)} S_{\underline{u}+\underline{r}'}^{(0)} + a_k]\} \cdot \cos[\frac{2\pi}{\frac{M}{2^k}} u_1 s_1 + \frac{2\pi}{\frac{N}{2^k}} u_2 s_2], \tag{32}$$

for all $\underline{s} \in \Omega^{(k)}$. It follows from Eq. (32) that the power spectrum $S_{\underline{u}}^{(k)}$ is the one given in Eq. (27).

By Eq. (29), since $X^{(0)}$ and $W^{(k)}$ are multivariate Gaussian, $X^{(k)}$ is also multivariate Gaussian. It is zero-mean and has a covariance given by Eq. (25). Hence, the joint density of $X^{(k)}$ is as given by Eq. (16). The covariance $\Sigma^{(k)}$ is symmetric and block-circulant, hence, its inverse can be calculated by the procedure given in [14]: its entries are as given in Eqs. (17) and (18). Notice that, as in the case of sampling, $S_{\underline{u}}^{(k)}$ cannot be expressed as in Eq. (3) with coefficients $(\theta_{\underline{r}}^{(k)})$ that are zero outside an appropriate neighborhood. Hence, $X^{(k)}$ is non-Markov, thereby completing the proof. □

Following a line of reasoning similar to the proof of Eq. (27), for infinite lattice GMRFs, we have (analogous to results in [3, 11]):

$$S_{\underline{\omega}}^{(k)} = 2^{2k} \sum_{\underline{r} \in C_k} [\Lambda_{\underline{\omega}"+\underline{\gamma}"}^{(k)} S_{\underline{\omega}"+\underline{\gamma}"}^{(0)} + a_k], \tag{33}$$

for all $\omega \in \tilde{\Omega}^{(k)}$, where $\underline{\omega}" = 2^{-k}\underline{\omega}$, and $\underline{\gamma}" = \frac{2\pi}{2^k}\underline{r}$.

Theorems 3 and 4 present the joint density, the covariance, and the power spectrum of $X^{(k)}$ in terms of the corresponding quantities of $X^{(0)}$, for both types of resolution transformations. The next section provides a specific illustration of these relationships for a simple non-trivial GMRF.

4 Special Case

We specialize the results of Section 3, by considering a second-order infinite-lattice GMRF with a separable autocorrelation. The analytical tractability of this example enables us to further understand the reasons why this special class of GMRFs lose-Markovianity under resolution transformations. Imposing the separability assumption leads to the following: $\underline{\theta}^{(0)}$ and η are such that

$$S_{\underline{\omega}}^{(0)} = \mathcal{A}_{\omega_1}^{(0)} \cdot \mathcal{B}_{\omega_2}^{(0)}, \text{ for all } \underline{\omega} = (\omega_1, \omega_2) \in \tilde{\Omega}^{(0)}, \tag{34}$$

where

$$\begin{aligned} \mathcal{A}_{\omega_1}^{(0)} &\triangleq [\sigma_v^{(0)}]^2 \{1 - \sum_{r_1 \in \eta_h} \theta_{(r_1,0)}^{(0)} \cos[r_1\omega_1]\}^{-1}, \\ \mathcal{B}_{\omega_2}^{(0)} &\triangleq [\sigma_h^{(0)}]^2 \{1 - \sum_{r_2 \in \eta_v} \theta_{(0,r_2)}^{(0)} \cos[r_2\omega_2]\}^{-1}, \text{and} \\ [\sigma^{(0)}]^2 &= [\sigma_h^{(0)}]^2 [\sigma_v^{(0)}]^2, \end{aligned} \tag{35}$$

with η_h and η_v appropriately defined. Specifically,

$$\eta_v = \{r_1 : (r_1, 0) \in \eta\}, \text{and } \eta_h = \{r_2 : (0, r_2) \in \eta\} \tag{36}$$

Specializing Eqs. (34)–(36) to second-order η, we have:

$$\begin{aligned} \eta_h = \eta_v &= \{r : -1 \leq r \leq 1,\ r \neq 0\}, \\ \mathcal{A}_{\omega_1}^{(0)} &= \mathcal{A}^{(0)}(z = e^{j\omega_1}), \\ \mathcal{B}_{\omega_2}^{(0)} &= \mathcal{B}^{(0)}(z = e^{j\omega_2}), \end{aligned} \tag{37}$$

where

$$\begin{aligned} \mathcal{A}^{(0)}(z) &= [\sigma_v^{(0)}]^2[K_v^{(0)}]^{-1}(1 - \rho_h^{(0)}z)(1 - \rho_h^{(0)}z^{-1}), \text{and} \\ \mathcal{B}^{(0)}(z) &= [\sigma_h^{(0)}]^2[K_h^{(0)}]^{-1}(1 - \rho_v^{(0)}z)(1 - \rho_v^{(0)}z^{-1}), \end{aligned} \tag{38}$$

with $\rho_h^{(0)}$, $\rho_v^{(0)}$, $K_h^{(0)}$, and $K_v^{(0)}$ defined to be:

$$\begin{aligned} \rho_h^{(0)} &= \begin{cases} \frac{1}{2}\{ [\theta_{(0,1)}^{(0)}]^{-1} + \sqrt{[\theta_{(0,1)}^{(0)}]^{-2} - 4} \} & \text{if } -\frac{1}{2} \leq \theta_{(0,1)}^{(0)} \leq 0 \\ \frac{1}{2}\{ [\theta_{(0,1)}^{(0)}]^{-1} - \sqrt{[\theta_{(0,1)}^{(0)}]^{-2} - 4} \} & \text{if } 0 \leq \theta_{(0,1)}^{(0)} \leq \frac{1}{2} \end{cases}, \\ \rho_v^{(0)} &= \begin{cases} \frac{1}{2}\{ [\theta_{(1,0)}^{(0)}]^{-1} + \sqrt{[\theta_{(1,0)}^{(0)}]^{-2} - 4} \} & \text{if } -\frac{1}{2} \leq \theta_{(1,0)}^{(0)} \leq 0 \\ \frac{1}{2}\{ [\theta_{(1,0)}^{(0)}]^{-1} - \sqrt{[\theta_{(1,0)}^{(0)}]^{-2} - 4} \} & \text{if } 0 \leq \theta_{(1,0)}^{(0)} \leq \frac{1}{2} \end{cases}, \\ K_h^{(0)} &= 1 + [\rho_h^{(0)}]^2, \text{and} \\ K_v^{(0)} &= 1 + [\rho_v^{(0)}]^2. \end{aligned} \tag{39}$$

The conditions imposed on $\sigma^{(0)}$ and $\underline{\theta}^{(0)}$ in order for $S_{\underline{\omega}}^{(0)} > 0$ for all $\underline{\omega} \in \tilde{\Omega}^{(0)}$, translate into $|\rho_h^{(0)}| < 1$ and $|\rho_v^{(0)}| < 1$.

We now analyze this special class under the block-to-point type of resolution transformation. We first need the following result, analogous to the ones in [1, 22]:

Theorem 5 *Let $X^{(0)}$ satisfy the following difference equation, for all $\underline{s} \in \mathcal{Z}^2$:*

$$X_{\underline{s}}^{(0)} - \rho_h^{(0)} X_{\underline{s}-(0,1)}^{(0)} - \rho_v^{(0)} X_{\underline{s}-(1,0)}^{(0)} + \rho_h^{(0)} \rho_v^{(0)} X_{\underline{s}-(1,1)}^{(0)} = \mathcal{U}_{\underline{s}}^{(0)}, \tag{40}$$

where $\mathcal{U}_{\underline{s}}^{(0)}$'s are not necessarily independent, although they are identically distributed. Furthermore, let $X^{(k)}$ denote a sampled version of $X^{(0)}$ as in Eq. (10), then

$$X_{\underline{t}}^{(k)} - \rho_h^{(k)} X_{\underline{t}-(0,1)}^{(k)} - \rho_v^{(k)} X_{\underline{t}-(1,0)}^{(k)} + \rho_h^{(k)} \rho_v^{(k)} X_{\underline{t}-(1,1)}^{(k)} = \mathcal{V}_{\underline{t}}^{(k)}, \tag{41}$$

for all $\underline{t} \in \mathcal{Z}^2$, where

$$\begin{aligned} \rho_h^{(k)} &= [\rho_h^{(0)}]^{2^k}, \\ \rho_v^{(k)} &= [\rho_v^{(0)}]^{2^k}, \text{ and} \\ \mathcal{V}_{\underline{t}}^{(k)} &= \sum_{\underline{r} \in \mathcal{C}_k} [\rho_v^{(0)}]^{r_1} [\rho_h^{(0)}]^{r_2} \mathcal{U}_{\underline{t}'-\underline{r}}^{(0)}, \text{ where } \underline{t}' = 2^k \underline{t}. \end{aligned} \tag{42}$$

Proof. The expression in Eq. (40) can be rewritten as:

$$[1 - \rho_h^{(0)} D_h][1 - \rho_v^{(0)} D_v] X_{\underline{s}}^{(0)} = \mathcal{U}_{\underline{s}}^{(0)}, \tag{43}$$

where D_h and D_v denote the horizontal and vertical lag operators, respectively. Specializing Eq. (43) to those sites ($\underline{s}$) such that $\underline{s} = 2^k \underline{t}$, for some $\underline{t} \in \mathcal{Z}^2$, and after multiplying it on both sides by

$$\frac{1 - [\rho_h^{(0)} D_h]^{2^k}}{1 - \rho_h^{(0)} D_h} \; \frac{1 - [\rho_v^{(0)} D_v]^{2^k}}{1 - \rho_v^{(0)} D_v}, \tag{44}$$

we obtain the following equation; refer to [1, 22]:

$$\begin{aligned} &\{1 - [\rho_h^{(0)} D_h]^{2^k}\}\{1 - [\rho_v^{(0)} D_v]^{2^k}\} X_{\underline{s}}^{(0)} \\ &\qquad = \frac{1 - [\rho_h^{(0)} D_h]^{2^k}}{1 - \rho_h^{(0)} D_h} \; \frac{1 - [\rho_v^{(0)} D_v]^{2^k}}{1 - \rho_v^{(0)} D_v} \mathcal{U}_{\underline{s}}^{(0)}. \end{aligned} \tag{45}$$

Using the definition of $X_{\underline{s}}^{(k)}$ in Eq. (10), and the fact that Eq. (45) is specialized to those $\underline{s} = 2^k \underline{t}$, for some $\underline{t} \in \mathcal{Z}^2$, the LHS of Eq. (45) can be rewritten as:

$$\{1 - [\rho_h^{(0)}]^{2^k} D_h\} \; \{1 - [\rho_v^{(0)}]^{2^k} D_v\} \; X_{\underline{s}}^{(k)}. \tag{46}$$

The expression in Eq. (44), on the other hand, can be rewritten as:

$$\sum_{\underline{r}\in\mathcal{C}_k} [\rho_v^{(0)} D_v]^{r_1}\ [\rho_h^{(0)} D_h]^{r_2}, \tag{47}$$

and hence the RHS of Eq. (45) becomes:

$$\sum_{\underline{r}\in\mathcal{C}_k} [\rho_v^{(0)}]^{r_1}\ [\rho_h^{(0)}]^{r_2}\ \mathcal{U}^{(0)}_{\underline{s}-\underline{r}}, \text{ where } \underline{s} = 2^k \underline{t}, \text{ for some } \underline{t} \in \mathcal{Z}^2. \tag{48}$$

Eqs. (41), (42) are direct consequences of Eqs. (45), (46), (48), which completes the proof of the theorem. □

The following interpretation of Theorem 5 is of importance: The r.f. $X^{(0)}$, a second-order GMRF with a separable autocorrelation, admits an AR representation given in Eq. (40), where the $\mathcal{U}^{(0)}_{\underline{s}}$'s are *iid* zero-mean Gaussian; refer to [15]. Now, by applying Theorem 5, we conclude that the r.f. $X^{(k)}$ obtained by sampling $X^{(0)}$ (as in Eq. (10)) is also a second-order GMRF with a separable autocorrelation. In other words, the random fields $X^{(k)}$ for $k \geq 1$ have an AR representation of the same order (and type) as that of $X^{(0)}$; hence, they are Markov with respect to the same neighborhood η. So, the second-order GMRFs with separable autocorrelations are a class of Markov fields that retain Markovianity under the sampling-type resolution transformation. The class of models having this invariance property is rather small [9], the most commonly referred example being the 1-D Ising model [23]. We note, however, that analogous results do not necessarily hold for higher-order GMRFs with separable autocorrelations.

We further analyze the special class under consideration:

Theorem 6 *Let $X^{(0)}$ be a second-order infinite lattice GMRF with a separable autocorrelation. Also, let the type of resolution transformation be block-to- point (for simplicity, we assume $a_k = 0$). Then, for any $k \geq 1$, $X^{(k)}$ is an ARMA process (as also noted in [1, 7, 22]). Specifically, the random field $X^{(k)}$, for any $k \geq 1$, obeys the following difference equation:*

$$\begin{aligned} X^{(k)}_{\underline{s}} - \rho_h^{(k)} X^{(k)}_{\underline{s}-(0,1)} - \rho_v^{(k)} X^{(k)}_{\underline{s}-(1,0)} + \rho_h^{(k)} \rho_v^{(k)} X^{(k)}_{\underline{s}-(1,1)} = \\ \mathcal{W}^{(k)}_{\underline{s}} - \beta_h^{(k)} \mathcal{W}^{(k)}_{\underline{s}-(0,1)} - \beta_v^{(k)} \mathcal{W}^{(k)}_{\underline{s}-(1,0)} + \beta_h^{(k)} \beta_v^{(k)} \mathcal{W}^{(k)}_{\underline{s}-(1,1)}, \end{aligned} \tag{49}$$

where $\mathcal{W}^{(k)}_{\underline{s}}$'s are iid zero-mean Gaussian-distributed with variance ν^k, and ν^k, $\beta_h^{(k)}$, and $\beta_v^{(k)}$ are appropriately defined in terms of $\sigma^{(0)}$, $\rho_h^{(0)}$, and $\rho_v^{(0)}$ (as given in the proof).

Proof. A second-order GMRF with a separable autocorrelation admits an AR representation as in Eq. (40) (refer to [19]), - with $\mathcal{U}_{\underline{s}}^{(0)}$ being zero-mean *iid* Gaussian with variance, $\nu^{(0)} = \frac{[\sigma^{(0)}]^2}{K_h^{(0)} K_v^{(0)}}$. Let $Y^{(k)}$ denote the random field obtained from $X^{(0)}$ in the following fashion:

$$Y_{\underline{s}}^{(k)} = b^k \sum_{\underline{r} \in \mathcal{C}_k} X_{\underline{s}+\underline{r}}^{(0)}. \tag{50}$$

Now, $X^{(k)}$, the block-to-point transformed version of $X^{(0)}$ (as given in Eq. (12), with $a_k = 0$) is obtained from $Y^{(k)}$ by sampling, that is,

$$X_{\underline{s}}^{(k)} = Y_{\underline{s}'}^{(k)}, \tag{51}$$

where $\underline{s}' = 2^k \underline{s}$. If $X^{(0)}$ admits an AR representation as in Eq. (40), then by Eq. (50), $Y^{(k)}$ admits the following representation:

$$\begin{aligned} Y_{\underline{s}}^{(k)} - \rho_h^{(0)} Y_{\underline{s}-(0,1)}^{(k)} - \rho_v^{(0)} Y_{\underline{s}-(1,0)}^{(k)} + \rho_h^{(0)} \rho_v^{(0)} Y_{\underline{s}-(1,1)}^{(k)} \\ = b^k \sum_{\underline{r} \in \mathcal{C}_k} \mathcal{U}_{\underline{s}+\underline{r}}^{(0)} \triangleq \mathcal{Q}_{\underline{s}}^{(0)}, \end{aligned} \tag{52}$$

for all $\underline{s} \in \mathcal{Z}^2$. Now using Eqs. (51), (52), and Theorem 5, we obtain:

$$X_{\underline{t}}^{(k)} - \rho_h^{(k)} X_{\underline{t}-(0,1)}^{(k)} - \rho_v^{(k)} X_{\underline{t}-(1,0)}^{(k)} + \rho_h^{(k)} \rho_v^{(k)} X_{\underline{t}-(1,1)}^{(k)} = \mathcal{H}_{\underline{t}}^{(k)}, \tag{53}$$

for all $\underline{t} \in \mathcal{Z}^2$, where

$$\mathcal{H}_{\underline{t}}^{(k)} = \sum_{\underline{r} \in \mathcal{C}_k} [\rho_v^{(0)}]^{r_1} [\rho_h^{(0)}]^{r_2} \mathcal{Q}_{\underline{t}'-\underline{r}}^{(0)}, \text{ where } \underline{t}' = 2^k \underline{t}. \tag{54}$$

Since $\mathcal{U}_{\underline{s}}^{(0)}$'s are *iid*, the RHS of Eq. (54) is, for each $\underline{s} \in \mathcal{Z}^2$, the MA of $(2^{k+1}-1)^2$ terms. The $\mathcal{H}_{\underline{s}}^{(k)}$'s are stationary r.v.s, and their autocorrelation $B_{\underline{t}}^{(k)} \triangleq E\{\mathcal{H}_{\underline{s}}^{(k)} \mathcal{H}_{\underline{s}+\underline{t}}^{(k)}\} = 0$, for any $\underline{t} \notin \eta \cup \underline{0}$ (where η is the second-order neighborhood). Furthermore, for $\underline{t} \in \eta \cup \underline{0}$, $B_{\underline{t}}^{(k)}$ is the coefficient corresponding to $z_v^{t_1'} z_h^{t_2'}$ in the following rationale:

$$\begin{aligned} \nu^{(0)} b^{2k} \{ \sum_{\underline{r} \in \mathcal{C}_k} z_v^{-r_1} z_h^{-r_2} \} \{ \sum_{\underline{r} \in \mathcal{C}_k} z_v^{r_1} z_h^{r_2} \} \{ \sum_{\underline{t} \in \mathcal{C}_k} [\rho_v^{(0)}]^{t_1} [\rho_h^{(0)}]^{t_2} z_v^{-t_1} z_h^{-t_2} \} \\ \cdot \{ \sum_{\underline{t} \in \mathcal{C}_k} [\rho_v^{(0)}]^{-t_1} [\rho_h^{(0)}]^{-t_2} z_v^{t_1} z_h^{t_2} \}, \end{aligned} \tag{55}$$

where $t'_1 = 2^k t_1$, and $t'_2 = 2^k t_2$. That being the case, the MA process $\mathcal{H}^{(k)}$ admits an MA representation given by the RHS Eq. (49), where $\nu^{(k)}$, $\beta_h^{(k)}$, and $\beta_v^{(k)}$ are given by the following set of equations:

$$\begin{aligned}
\beta_h^{(k)} &= \frac{B_{(0,0)}^{(k)} + \sqrt{[B_{(0,0)}^{(k)}]^2 - 4[B_{(0,1)}^{(k)}]^2}}{2B_{(0,1)}^{(k)}}, \\
\beta_v^{(k)} &= \frac{B_{(0,0)}^{(k)} + \sqrt{[B_{(0,0)}^{(k)}]^2 - 4[B_{(1,0)}^{(k)}]^2}}{2B_{(1,0)}^{(k)}}, \text{ and} \\
\nu^{(k)} &= \frac{B_{(0,0)}^{(k)}}{\{1 + [\beta_h^{(k)}]^2\}\{1 + [\beta_v^{(k)}]^2\}}.
\end{aligned} \tag{56}$$

At this point, all that remains to be shown is that $\beta_h^{(k)}$ and $\beta_v^{(k)}$ are real, and do not have an absolute value of 1. $B^{(k)}$ is a real valued, even sequence, hence $\beta_h^{(k)}$ and $\beta_h^{(k)}$ are real. Because it is a linear transformation of $X^{(0)}$, $X^{(k)}$ has a positive power spectrum, and from Eq. (55), this positivity implies $|\beta_h^{(k)}| \neq 1$ and $|\beta_v^{(k)}| \neq 1$, thereby completing the proof. □

The fact that the infinite-lattice second-order GMRF with a separable autocorrelation becomes an ARMA process under the block-to-point-type resolution transformation illustrates the result of Theorem 4, namely that block-to-point transformed multi-resolution GMRFs in general are not Markov.

We continue this investigation of this special class a little further. Since $X^{(0)}$ is a second-order GMRF with a separable autocorrelation, it has an AR representation of the form in Eq. (40), and its autocorrelation is given by:

$$R_{\underline{t}}^{(0)} = \nu^{(0)}\{1 - [\rho_v^{(0)}]^2\}^{-1}\ \{1 - [\rho_h^{(0)}]^2\}^{-1}\ [\rho_v^{(0)}]^{|t_1|}\ [\rho_h^{(0)}]^{|t_2|}. \tag{57}$$

By Theorem 6, for such an $X^{(0)}$, the random field $X^{(k)}$ admits an ARMA representation given by Eq. (49). The autocorrelation of $X^{(k)}$ is given by the following equation:

$$\begin{aligned}
R_{\underline{t}}^{(k)} &= \nu^{(k)} T_h^{(k)} T_v^{(k)}\ \{G_v^{(k)}\delta(t_1) + [\rho_v^{(0)}]^{|t'_1|}\} \\
&\quad \cdot \{G_h^{(k)}\delta(t_2) + [\rho_h^{(0)}]^{|t'_2|}\},
\end{aligned} \tag{58}$$

where $\underline{t}' = 2^k \underline{t}$, and $T_h^{(k)}$, $T_v^{(k)}$, $G_h^{(k)}$, and $G_v^{(k)}$ are related to $\rho_h^{(0)}$, $\rho_v^{(0)}$, $\beta_h^{(k)}$, and $\beta_v^{(k)}$ (but for brevity, we leave their definitions out).

The impulses in $R_{\underline{t}}^{(k)}$, at $\underline{t} = \underline{0}$, cause the r.f. $X^{(k)}$ to lose Markovianity. On the other hand, since $|\rho_h^{(0)}| < 1$ and $|\rho_v^{(0)}| < 1$, Eq. (58) implies that the correlation $(R_{\underline{t}}^{(k)})$ at the resolution level k decays 2^k times faster than at level 0. In other words, although non-Markov, $X^{(k)}$ is less correlated than $X^{(0)}$. This, on the other hand, supports the argument that it would be reasonable to approximate the coarser-resolution r.f.'s with GMRFs. The next section investigates such approximation schemes.

5 Approximations

The fact that GMRFs lose Markovianity upon resolution transformations imply that such random fields do not possess a parsimonious description at those resolutions. Hence, we encounter difficulties in processing images modeled as GMRFs at multiple resolutions. To remedy this situation, we need to be able to *approximate* non-Markov Gaussian random fields by appropriate GMRFs. Familiar candidates for such approximation are the (free energy-based) cumulant and bond-moving approximations that are used for analogous purposes in renormalization group studies [5, 9]. As we already mentioned, and also recognized in [12], the free energy is not a quantity of any particular interest in image modeling. The quantities of interest in image modeling are the covariance and the power spectrum. Motivated by such considerations, we devise a new approximation methodology, the *Covariance Invariance Approximation* (CIA), that seeks to approximate the covariance (as opposed to the free energy).

In the following sequel, we present the CIA methodology. As defined earlier, let $\Sigma^{(k)}$ denote the covariance matrix of the actual (possibly) non-Markov Gaussian random field $X^{(k)}$. Also, let ν denote an arbitrary neighborhood system ($\nu = \eta \cup \{\underline{0}\}$, for example). We propose to approximate the statistical behavior of $X^{(k)}$ by a random field $\tilde{X}^{(k)}$, with an associated covariance matrix $\tilde{\Sigma}^{(k)}$ such that:

1. The entries of $\tilde{\Sigma}^{(k)}$ corresponding to pairs of sites that are neighbors of each other (with respect to ν) are equal to those of $\Sigma^{(k)}$.

2. The entries of $[\tilde{\Sigma}^{(k)}]^{-1}$ corresponding to pairs of sites that are not neighbors are zero.

3. $\tilde{\Sigma}^{(k)}$ is positive definite.

The resulting random field $\tilde{X}^{(k)}$, with $\tilde{\Sigma}^{(k)}$ as its associated covariance matrix, is a GMRF with respect to the neighborhood system ν. We call this approximation criterion the *Covariance Invariance Approximation* (CIA). Although we stated this criterion only for finite-lattice r.f.'s, the same methodology applies for infinite-lattice r.f.'s as well.

The relevant issues that arise when we consider such an approximation are the rationale, the existence, the uniqueness, and the computation. The following set of results addresses and settles these relevant issues :

Theorem 7 (Rationale) .

1. *Among all covariance matrices $\hat{\Sigma}^{(k)}$ whose entries, corresponding to pairs of sites that are neighbors of each other, agree with those of $\Sigma^{(k)}$, the choice of $\tilde{\Sigma}^{(k)}$ yields the maximum entropy.*

2. *Among all covariance matrices $\hat{\Sigma}^{(k)}$, such that the entries of $[\hat{\Sigma}^{(k)}]^{-1}$ corresponding to pairs of sites that are not neighbors are zero, the choice of $\tilde{\Sigma}^{(k)}$ is the one that maximizes the likelihood of an observation having $\Sigma^{(k)}$ as its sample covariance.*

3. *For ν fixed, among all GMRFs, $\tilde{X}^{(k)}$ with the covariance matrix $\tilde{\Sigma}^{(k)}$ has the minimum Kullback–Leibler distance from the Gaussian field $X^{(k)}$ with covariance matrix $\Sigma^{(k)}$.*

Proof. (The proof presented here is based on Dempster's work on Covariance Selection [8]). Let $\hat{\mathcal{P}}$ denote any element in the class of all multivariate Gaussian densities at level k, whose inverse covariance matrix has zero entries corresponding to sites that are not neighbors (with respect to ν). Let $\mathcal{P}^*$ denote any element in the class of all multivariate Gaussian densities at level k, whose covariance matrix agrees with $\Sigma^{(k)}$ on entries corresponding to sites that are neighbors. Let $\tilde{\mathcal{P}}$ denote any element in the class of multivariate Gaussian densities at level "k" that possesses both of these properties. We now derive a set of inequalities that yield the assertions (i), (ii), and (iii) in the theorem. To do that, we need the following general inequality from [20]:

$$-\int g(X^{(k)})\ \log g(X^{(k)})\ dX^{(k)} < -\int g(X^{(k)})\ \log f(X^{(k)})\ dX^{(k)}, \quad (59)$$

for any two distinct densities, $g(\cdot)$ and $f(\cdot)$.

A direct application of Eq. (59) yields:

$$-\int \mathcal{P}^*(X^{(k)}) \log \mathcal{P}^*(X^{(k)}) dX^{(k)} < -\int \mathcal{P}^*(X^{(k)}) \log \tilde{\mathcal{P}}(X^{(k)}) dX^{(k)}. \quad (60)$$

Now, using the facts that the associated covariance matrices of $\mathcal{P}^*$ and $\tilde{\mathcal{P}}$ agree at entries corresponding to sites that are neighbors, and that $\tilde{\mathcal{P}}$ is a Gaussian Markovian density (GMD), we obtain the following equality:

$$- \int \mathcal{P}^*(X^{(k)}) \log \tilde{\mathcal{P}}(X^{(k)})\, dX^{(k)} = - \int \tilde{\mathcal{P}}(X^{(k)}) \log \tilde{\mathcal{P}}(X^{(k)})\, dX^{(k)}. \tag{61}$$

The LHS of Eq. (60) and the RHS of Eq. (61) are the entropies associated with the pdf's, $\mathcal{P}^*$ and $\tilde{\mathcal{P}}$; therefore, the assertion (i) of the theorem is a direct implication of Eqs. (60) and (61).

Another application of Eq. (59) gives the following inequality:

$$\int \tilde{\mathcal{P}}(X^{(k)}) \log \hat{\mathcal{P}}(X^{(k)})\, dX^{(k)} < \int \tilde{\mathcal{P}}(X^{(k)}) \log \tilde{\mathcal{P}}(X^{(k)})\, dX^{(k)}. \tag{62}$$

Again, using the covariance matching properties between $\mathcal{P}^*$ and $\tilde{\mathcal{P}}$, and the fact that $\tilde{\mathcal{P}}$ and $\hat{\mathcal{P}}$ are GMDs, we obtain the following equality:

$$\int \mathcal{P}^*(X^{(k)}) \log \hat{\mathcal{P}}(X^{(k)})\, dX^{(k)} = \int \tilde{\mathcal{P}}(X^{(k)}) \log \hat{\mathcal{P}}(X^{(k)})\, dX^{(k)}. \tag{63}$$

The LHS of Eq. (63) is, under $\hat{\mathcal{P}}$, the log-likelihood of an observation with $\Sigma^{(k)}$ as its sample covariance, and the LHS of Eq. (61) is, under $\tilde{\mathcal{P}}$, the log-likelihood of a similar observation. Hence, assertion (ii) of the theorem is a direct implication of Eqs. (62) and (63).

The Kullback–Leibler distance between $\hat{\mathcal{P}}$ and $\mathcal{P}$ (the actual, possibly non-Markov, Gaussian pdf at level k) is (refer to [16]):

$$I(\mathcal{P}, \hat{\mathcal{P}}) = \int \mathcal{P}(X^{(k)}) \log \frac{\mathcal{P}(X^{(k)})}{\hat{\mathcal{P}}(X^{(k)})}\, dX^{(k)}. \tag{64}$$

Therefore, the $\hat{\mathcal{P}}$ that minimizes the Kullback–Leibler distance is the same as the one that maximizes:

$$\int \mathcal{P}(X^{(k)}) \log \hat{\mathcal{P}}(X^{(k)})\, dX^{(k)}. \tag{65}$$

Using the correlation matching property of $\tilde{\mathcal{P}}$ and $\mathcal{P}$, and the fact that $\hat{\mathcal{P}}$ is a GMD, we obtain the following equality:

$$\int \mathcal{P}(X^{(k)}) \log \hat{\mathcal{P}}(X^{(k)})\, dX^{(k)} = \int \tilde{\mathcal{P}}(X^{(k)}) \log \hat{\mathcal{P}}(X^{(k)})\, dX^{(k)}. \tag{66}$$

Eqs. (62), (65), and (66), imply assertion (iii) of the theorem. This completes the proof. □

Theorem 7 provides the rationale for CIA. We now present a result that addresses the existence, the uniqueness, and the computational issues regarding the CIA:

Theorem 8 *If* $\sigma^{(0)}$ and $\underline{\theta}^{(0)}$ *are chosen so that* $\Sigma^{(0)}$ *is positive definite, then* $\tilde{\Sigma}^{(k)}$ *exists for all k, and it is unique. Moreover, the nonzero elements of* $[\tilde{\Sigma}^{(k)}]^{-1}$ *can be obtained by a simple gradient-ascent procedure.*

Proof. If $\Sigma^{(0)}$ is positive definite, then the $\Sigma^{(k)}$ given in Eqs. (13) and (25) are also positive definite. By the results in [8, 25], the existence and uniqueness of $\tilde{\Sigma}^{(k)}$ is a direct consequence of the positive definiteness of $\Sigma^{(k)}$. The primary reason for the existence and uniqueness of $\tilde{\Sigma}^{(k)}$ is, starting from $\Sigma^{(k)}$, we can specify a sequence of positive definite matrices such that, for each of the matrices in that sequence, the entries corresponding to sites that are neighbors with respect to ν agrees with that of $\Sigma^{(k)}$, with $\tilde{\Sigma}^{(k)}$ as the limit of that sequence. (See [8] for proof.)

Due to the maximum-likelihood interpretation of $\tilde{\Sigma}^{(k)}$ (Theorem 7), and the fact that the likelihood is a concave function of the nonzero entries of $[\tilde{\Sigma}^{(k)}]^{-1}$, $[\tilde{\Sigma}^{(k)}]^{-1}$ can be obtained by a gradient ascent procedure [10]. For Gaussian fields (unlike general exponential families [10]), the gradient is actually computable, and so the gradient ascent procedure can be actually implemented. This completes the proof. □

The CIA is closely related to the ideas of maximum entropy spectrum estimation [25] and of covariance selection [8]. In both of those approaches, the objective is to select a model covariance that agrees with an observed sample covariance over an appropriate region, and where, in addition, the model covariance has some desirable information–theoretic properties. Such a model covariance can be selected if and only if the sample covariance over that region has a positive definite extension, i.e., if there exists a positive definite covariance that agrees with the observed sample covariance over that region. In general, the extendability condition is not easily verifiable. In the case of CIA, this extendability condition is automatically satisfied (Theorem 8), thereby enabling us to make use of some of the results that have been established in those related areas.

In principle, the CIA methodology can be directly extended to the case of non-Gaussian fields, such as Ising fields, and Multi-level Logistic fields. An exact analog of Theorems 7 and 8 can be proven. In practice, however, due to the intractability of the partition function for such non-Gaussian fields, the calculations necessary to obtain the CIA-based approximations cannot be carried out either analytically or by any deterministic (gradient-based) algorithm. To compute the CIA-based approximations, stochastic algorithms such as the ones in [6] and [26] would have to be used.

As we mentioned earlier, the (free energy-based) cumulant and bond-moving approximations are familiar candidates for analogous purposes in renormalization group studies [5, 9]. For Gaussian fields, the two quantities, *free energy* and *entropy*, are equivalent. This implies that, due its maximum entropy interpretation, the CIA belong to the class of *upper bound* approximations of the free energy. It is of interest to determine whether or not an upper-bounding approximation such as the cumulant method, used with appropriate modifications (such as a specific splitting of the quadratic potential between inter- and intra-cell energies), could give an approximating covariance having the same properties as the CIA. The cumulant and bond-moving approximations are popular in statistical mechanics because of their ability to accurately predict certain quantities of interest called *critical exponents* for a multitude of models [5]. When the quantities of interest are not the critical exponents, but those that are obtained by differentiating the free energy (such as the covariance), the performances of these approximations are not so satisfactory [2]. The CIA, on the other hand, is designed to approximate the covariance, and it has a number of nice theoretical and computational properties. Notwithstanding the unsettled issues, the CIA appears to be fundamentally different from the known approximations such as the cumulant and bond-moving methods [5]. A detailed exploration of such ideas warrants further investigation.

6 Conclusions

We have presented a multi-resolution treatment for GMRFs. As noted in such analogous studies, we reaffirmed that, in general, images modeled as GMRFs lose Markovianity upon resolution transformations. We derived exact expressions for the joint density, covariance, and power spectrum of these random fields at multiple resolutions, under two types of resolution transformations, namely, the sampling and the block-to-point transformations.

We illustrated these results by specializing them to the class of second-order GMRFs with a separable autocorrelation. The analytical tractability of this example enabled us to further understand the reasons why is special class of GMRFs in general loses Markovianity upon resolution transformation. In the process, we identified a class of Markov fields (second-order GMRFs with a separable autocorrelation) that retain their Markovianity under sampling-type resolution transformation.

To process images modeled as GMRFs at multi-resolutions, we need to approximate (possibly) non-Markov Gaussian fields at coarser resolutions, by GMRFs. This leads us to proposing a new approximation methodology we called the Covariance Invariance Approximation (CIA). We argued that the CIA was more suitable in the context of image modeling than the renormalization group based approximation methodologies. We showed that the CIA is based on the same information–theoretic rationale that is used to justify the GMRF model assumptions for the original image $X^{(0)}$. We brought forth a number of theoretical and computational issues that are relevant to the CIA, and conclusively settled them. We think the CIA methodology may be potentially useful in renormalization group studies as well.

Acknowledgments

The authors would like to thank Professor Johnathan L. Machta for very helpful discussions regarding analogous studies in renormalization group theory.

Bibliography

[1] T. Amemiya and R. Y. Wu, *The Effect of Aggregation on Prediction in the Autoregressive Model,* J. Am. Stat. Assoc., Vol. 67, pp. 628–632, 1972.

[2] M. N. Barber, *Optimal Variational Approximations to Renormalization Group Transformations I. Theory,* J. Phys. A: Math. Gen., Vol. 10, pp. 1721–1736, 1977.

[3] T. L. Bell and K. G. Wilson, *Finite-Lattice Approximations to Renormalization Groups,* Phys. Rev. B, Vol. 11, No. 9, pp. 3431–3444, 1975.

[4] C. Bouman and B. Liu, *Segmentation of Textured Images Using a Multi-resolution Approach,* Proc. IEEE Int. Conf. Acoust. Speech Signal Process., pp. 1124–1127, 1988.

[5] T. W. Burkhardt and J. M. J. van Leeuwen (Eds.), *Real-Space Renormalization,* Springer-Verlag, New York, 1982.

[6] B. Chalmond, *An iterative Gibbsian Technique for Reconstruction of m-ary Images, Pattern Recognition*, Vol. 22, pp. 747–761, 1989.

[7] R. Chellappa, *Time Series Models for the Representation of Multi-Resolution Images,* in *Multiresolution Image Processing and Analysis*, A. Rosenfeld (Ed.), Springer-Verlag, New York, 1984.

[8] A. P. Dempster, *Covariance Selection, Biometrics*, Vol. 28, pp. 157–175, 1972.

[9] C. Domb and M. S. Green (Eds.), *Phase Transition and Critical Phenomena,* Vol. 6, Academic Press, Boston, 1976.

[10] S. Geman and C. Graffigne, *Markov Random Field Image Models and Their Application to Computer Vision,* in *Proc. Internat. Congr. Math.*, A. M. Gleason (Ed.), Amer. Math. Soc., Providence. Rhode Island, 1987.

[11] B. Gidas, *A Multilevel–Multiresolution Technique for Computer Vision via Renormalization Group Ideas,* Proc. SPIE — High Speed Computing, Vol. 880, pp. 214–218, 1988.

[12] B. Gidas, *A Renormalization Group Approach to Image Processing,* IEEE Trans. Pattern Anal. Mach. Intell., Vol. 11, No. 2, pp. 164–180, 1989.

[13] M. Jubb and C. Jennison, *Aggregation and Refinement in Binary Image Restoration,* presented in the AMS-IMS-SIAM Jt. Summer Conf. Math. Sci. — Spat. Stat. Img., Maine, 1988.

[14] R. L. Kashyap, *Random Field Models on Torous Lattices for Finite Images,* Proc. Ann. Conf. Inf. Sci. Syst., pp. 215–220, 1981.

[15] R. L. Kashyap, *Analysis and Synthesis of Image Patterns by Spatial Interaction Models,* in *Progress in Pattern Recognition*, Vol. 1, L. N. Kanal and A. Rosenfeld (Eds.), pp. 149–186, North-Holland, 1981.

[16] S. Kullback and R. A. Leibler, *On information and sufficiency,* Ann. Math. Stat., Vol. 22, pp. 79–86, 1951.

[17] J. L. Machta, personal communication, 1989.

[18] I. Matsuba, *Renormalization Group Approach to Hierarchical Image Analysis,* Proc. IEEE Int. Conf. Acoust. Speech Signal Process., pp. 1044–47, 1988.

[19] P. A. P. Moran, *A Gaussian Markovian Process on a Square Lattice,* J. Appl. Probab., Vol. 10, pp. 54–62, 1973.

[20] C. R. Rao, *Linear Statistical Inference and its Applications,* Wiley, 1965.

[21] A. Rosenfeld (Ed.), *Multiresolution Image Processing and Analysis,* Springer-Verlag, New York, 1984.

[22] L. G. Telser, *Discrete Samples and Moving Sums in Stationary Stochastic Processes,* J. Am. Stat. Assoc., Vol. 62, pp. 484–499, 1967.

[23] C. J. Thompson, *The Statistical Mechanics of Phase Transitions,* Contemp. Phys. (GB), Vol. 19, No. 3, pp. 203–224, 1978.

[24] K. G. Wilson and J. Kogut, *The Renormalization Group and ε Expansion,* Phys. Rep., Vol. 12, pp. 75–200, 1974.

[25] J. W. Woods, *Two-Dimensional Markov Spectral Estimation,* IEEE Trans. Inform. Theory, Vol. 22, pp. 552–559, 1976.

[26] L. Younes, *Estimation and Annealing for Gibbsian Fields,* Ann. Inst. Henri Poincaré, Vol. 24, pp. 269–294, 1988.

Classification of Natural Textures Using Gaussian Markov Random Field Models

Shankar Chatterjee
Department of Electrical Engineering
University of California at San Diego
La Jolla, California

1 Introduction

Textural and contextual cues are two important pattern elements in human interpretation of visual data. There are many applications—like identification of large-scale geological formations, land use patterns, and interpretation of aerial and medical images—where texture classification and identifying homogeneous regions are important. A good review of the current state-of-the-art can be found in [2] and [3]. The texture (or any other) classification problem can be stated as follows: There are a finite number of classes C_i, $i = 1, 2, \ldots, r$. A number of training textures belonging to each class are available. Based on the information extracted from these sets, a rule is designed to classify the given test texture of unknown class to one of the known r classes. The important step in a classification procedure is the choice of features, which reduces the dimension of data to a computationally reasonable amount so that one could get by with a simple classifier, such as a nearest neighbor rule.

Extracting features (cues) for textured images has been a subject of research for a long time. For micro-textures, which we will be mostly concerned with here, currently used methods utilize:

1. Features derived from second-order gray-level statistics, including the gray-level co-occurence matrices [4], [2].

2. Features derived from gray-level run length statistics [4].

3. Features derived from gray-level difference statistics [4].

ISBN 0-12-170608-7

4. Features, e.g., moments derived from decorrelation methods [5].

5. Features, e.g., rings and wedges, derived from the Fourier power spectrum [4].

Classification using features obtained from the gray-level co-occurence matrices is set on the premise that human texture perception relies on the second-order probabilities of the underlying stochastic field. The second-order statistics for a random field can be empirically computed by counting the number of occurences (matches) of a pair (*dipole*) of values (gray levels) at a specified distance apart and orientation. The counts are arranged in a matrix where the rows and columns refer to the two gray levels. Thus, for a given length and orientation, a $G \times G$ co-occurence matrix is formed where $[0, G-1]$ is the range of gray-level values. Due to the large amount of data storage, several simplifications are usually made. First, the number of gray levels is reduced by equal probability quantization (EPQ) to keep the matrix size small. The separating distance and orientation are quantized, as the observations are on a discrete grid. Most importantly, these matrices are not used directly. Instead, features, e.g., energy, entropy, local homogeneity, etc., computed from these matrices are used to discriminate textures. One could generalize second-order co-occurence statistics to third-order by using a triangle of varying size, to fourth-order by using a rectangle of varying size, etc.

The following theorem [27] establishes the relationship between MRF parameters and co-occurence probability: *For a given set of random variables y taking integer values in a bounded set G on a plane $\mathcal{R}^2$, which may be characterized by a Gibbs random field (GRF), the elements of the co-occurence matrix* (dipole statistics) *are sufficient statistics for GRF parameters.*

In the gray-level run length method, a matrix, whose elements denote the number of occurences of a particular gray level, I_g, of a specified length d is constructed. Depending on the orientation, θ of the runs and the length, d of the runs, a set of matrices is formed. From each of these matrices, features, e.g., short run emphasis, long run emphasis, etc., are computed to be used in classifying textures.

In the gray-level difference method, one computes the occurence of the difference of two gray levels, I_{g1} and I_{g2}, separated by a specified distance d and orientation θ. It thus generates a set of single-variable probability distributions. One computes features, e.g., contrast, angular second moment, etc., from these distributions. This is a modified version of the co-occurence model in the sense that here the dipole has a gray value of zero at one of the two ends.

The decorrelation method of feature extraction [5] using a Laplacian window has this underlying modeling interpretation: suppose the texture is modeled as a 2-D separable causal Markov field, where the observation $y(s)$ at a location s is written as:

$$\begin{aligned} y(s) &= \theta_1 y(s+(-1,0)) + \theta_2 y(s+(0,-1)) \\ &\quad - \theta_1\theta_2 y(s+(-1,-1)) + \sqrt{\beta}\omega(s). \end{aligned} \quad (1)$$

Then as $\theta_1, \theta_2 \to 1$, the decorrelated residuals generated by the Laplacian operators are similar to those obtained by:

$$\begin{aligned} \hat{\omega}(s) &= \frac{1}{\sqrt{\beta^*}}[y(s) - \theta_1^* y(s+(-1,0)) - \theta_2^* y(s+(0,-1)) \\ &\quad + \theta_1^*\theta_2^* y(s+(-1,-1))], \end{aligned}$$

where θ_1^*, θ_2^*, and β^* are the least-square (LS) estimates. Thus, from the point of view of image modeling, in the limit, the decorrelation features are extracted from the moments of the residuals obtained by fitting Eq. (1) to the given texture. The loss of information during this process of dimensionality reduction of the data may be significant, since the residuals themselves contain only partial information. It is interesting to note that no classification experiments have been reported using these features. The performances of the feature vectors are justified using the Bhattacharyya distance measure [8]. The relation between the Bhattacharyya distance used and the bounds on the error in classification is true only when the feature vectors are Gaussian, with known mean vector and covariance matrices [22]. However, in practice, one or both of the assumptions are not true.

The directionality (or lack thereof) feature of repetitive textures influences some of the dominant spatial frequency components. For a two-dimensional field (homogeneous), an obvious choice is to compute a spatial frequency transform, e.g., Fourier, Walsh, etc., and the corresponding power spectrum. The features extracted from the power spectrum can be used to discriminate textures. In [4], the power spectrum was computed from the periodogram by computing a two-dimensional Fourier transform (by taking its magnitude and squaring it) and features, e.g., energy contained in angular frequencies (rings and wedges), were extracted from it. Well-documented results in 1-D spectral estimation [6] show that spectral estimates obtained by fitting a model possess better statistical properties compared to the periodogram estimates. Using the maximum entropy method, an improved estimate of the power spectrum and the ring and wedge features were obtained and used in classifying textures [9].

A major disadvantage of the aforementioned methods is that the features are not rotation-invariant; i.e., given a rotated version of a known texture, the values of the extracted features might lead us to believe that it belongs to a different texture class. In frequency domain approaches, rotation invariance in texture recognition can be obtained by filtering the image with a set of Gabor filters for all possible orientations and considering the responses separately [29]. Modifying a two-dimensional autoregressive model (e.g., a non-causal autoregressive (NCAR) model), which is usually defined with respect to a rectangular (square) grid to a circular grid, a set of rotation-invariant features is proposed in [30]. The model used here is a modified version of a second-order NCAR model, where the nearest eight neighbors are interpolated on a circle according to the following formula,

$$y(s) = \alpha \sum_{r \in N_{2c}} g_t y(s \oplus r) + \sqrt{\beta} w(s), \tag{2}$$

where the neighborhood N_{2c} contains the lattice points,

$$\begin{aligned} N_{2c} = \; & \{(-1,-1),(-1,0),(-1,1),(0,-1),(0,0),(0,1),(1,-1), \\ & (1,0),(1,1)\} \end{aligned}$$

and the interpolation kernel g_t is given by,

$$g_t = \begin{bmatrix} 0.005 & 1.4336 & 0.4005 \\ 1.4336 & 0.6636 & 1.4336 \\ 0.4005 & 1.4336 & 0.4005 \end{bmatrix}.$$

The parameters α and β can be estimated by LS technique and can be used as discriminating features.

Here, we assume that the image textures are Gaussian and that one could fit Gaussian MRF models to them. We also look at binarized images later and model them with appropriate MRFs [26]. Binarization of the natural textures (gray-level) was done with equal probability quantization. The assumption of Gaussianity for the lattice variables (pixels) for gray-level images has been verified empirically by several researchers in image restoration [34]. The justification for using Markov fields stems from the fact that textures are usually homogeneous and local characteristics are consistent with global properties. However, this may not be true for macro-textures, e.g., straw, raffia, etc., which have lots of structures in them. This model-based assumption leads to two methods of feature extraction. The first method consists of fitting a GMRF model to the texture and constructing the feature vector with LS estimates of the parameters [12]. Since textures close to the original one can be constructed using LS estimates in most situations, the LS estimates are suggested as information-preserving features. Using a simple minimum distance classifier and leave-one-out strategy, an accuracy of over 99 percent is achieved for textures of size 64×64.

If the texture under study is truly generated by a Gaussian MRF model, then the sample correlations over a specific window defining the model are sufficient statistics [16] for the parameters of the model. Using sample correlations, one can identify an underlying Gaussian MRF model as described in [24],[23] and synthesize a texture close to the original. The vector of sample correlations is suggested as a lossless optimal feature vector.

In the following, we will look at the performance of these feature vectors on some natural texture images from the Brodatz album.

2 Information-Preserving Features

A motivation in using this method of feature extraction is its ease in going from the observed data to the feature space and vice versa. The other conventional methods of textural feature extraction do not permit this. If we assume that the zero mean observations from the given texture $[y(s),\ s \in \Omega,\ \Omega = \{s = (i,j) : 0 \leq i,j \leq M-1\}]$ are Gaussian and obey the following difference equation [10] [11]:

$$y(s) = \sum_{r \in N_s} \theta_r (y(s+r) + y(s-r)) + e(s), \tag{3}$$

where the zero mean stationary Gaussian sequence $e(s)$ is a correlated innovations sequence,

$$E[e(s)e(r)] = -\theta_(s-r)\nu I_{N^*}(s-r), \tag{4}$$

and $I_{N^*}(r)$ is an indicator function, which is equal to 1 when $r \in N^* \equiv N \cup (0,0)$, zero otherwise. N is the set of neighborhood pixels, $N = \{\{r\} \cup \{-r\} | r \in N_s\}$. The unknown parameters, $\mathbf{\Theta} = (\theta_r, r \in N_s)$ and ν, can be estimated [1] using the LS method as given here:

$$\begin{aligned} \mathbf{\Theta}^* &= \left[\sum_{\Omega_I} \mathbf{y}_s \mathbf{y}_s^T\right]^{-1} \left[\sum_{\Omega_I} \mathbf{y}_s y(s)\right], \\ \nu^* &= \frac{1}{M^2} \sum_{\Omega_I} \left[y(s) - \mathbf{\Theta}^{*T} \mathbf{y}_s\right]^2, \end{aligned} \tag{5}$$

where $\mathbf{y}_s =$ col. $[y(s+r) + y(s-r),\ r \in N_s]$, and Ω_I, the interior region, is defined as [12], $\Omega_I = \Omega - \Omega_B$, and the boundary set $\Omega_B = \{s = (i,j) : s \in \Omega \text{ and } (s+r) \notin \Omega \text{ for at least one } r \in N\}$.

Parameter	Grass	Calf Leather	Pigskin
$\theta^*_{1,0}$	0.5928	0.5831	0.3855
$\theta^*_{0,1}$	0.4054	0.2164	0.4498
$\theta^*_{-1,1}$	-0.2351	-0.1242	-0.1100
$\theta^*_{1,1}$	-0.2161	-0.0514	-0.1481
$\theta^*_{2,0}$	-0.1831	-0.1592	-0.0592
$\theta^*_{0,2}$	-0.0033	-0.0060	-0.0491
$\theta^*_{-1,2}$	-0.0078	-0.0055	-0.0027
$\theta^*_{1,2}$	-0.0184	-0.0164	0.0089
$\theta^*_{-2,1}$	0.0807	0.0419	0.0043
$\theta^*_{2,1}$	-0.0807	0.0114	0.0148
ν^*	464.56	209.31	78.394

Table 1: Least-squares Estimates of the Parameters Corresponding to a Fourth-Order Model

One can synthesize real textures close to the original one using $\mathbf{\Theta}^*, \nu^*$ [13], [15]. Since the textures synthesized using a fourth-order model retain most of the statistical characteristics of the original textures, we have used a feature vector using the estimates of model parameters corresponding to a fourth-order MRF model. Model order dictates the set of neighbors chosen [13]. To make the feature vector robust against illumination changes, we have used $\eta = \{\theta^*, \nu^*/\hat{\rho}^2\}$, where $\hat{\rho}^2$ is the sample variance of the texture as a feature vector. Typical values of θ^* and ν^* corresponding to a fourth-order Gaussian MRF fitted to various textures are given in Table 1.

A similar set of features was extracted for texture classification using non-causal autoregressive (NCAR) models in [14]. The NCAR models form a class that is a subset of Gaussian MRF models in the sense that, given a Gaussian NCAR model, an equivalent Gaussian MRF model always exists, but the converse is not always true. For non-Gaussian variates, the probability structures of these two classes of models are different. Since LS estimates are inconsistent [12] for NCAR models, approximate maximum likelihood (ML) estimates obtained using gradient techniques have been used for parameter estimation. The texture classification problem using Gaussian MRF models has also been considered in [19]. Assuming pixel-wise independence for non-neighbor sites, a set of coding estimates [20] has been used for parameter estimation. As pointed out in [20], even for the first-order MRF model, the coding estimate is statistically very inefficient compared to the LS or ML estimates.

3 Lossless Feature Set

The transformation from the given texture to the feature vector is many-to-one and usually involves some loss of information. One can, however, derive a lossless feature vector for texture classification using Gaussian MRF models by using the notion of sufficient statistics [16]. Suppose that the $M^2 \times 1$ lexicographically ordered vector $\mathbf{y}$ from the texture is represented by a Gaussian MRF model Eq. (3) characterized by θ and ν. Then the probability density of $\mathbf{y}$ has the exponential family form [21],

$$p(\mathbf{y}|\boldsymbol{\Theta}, \nu) = \frac{|H(\theta)|^{1/2}}{(2\pi\nu)^{M^2/2}} \exp\left(-\frac{\mathbf{y}^T H(\theta)\mathbf{y}}{2\nu}\right), \tag{6}$$

where $H(\theta)$ is the transformation matrix from $\{e(s)\}$ to $\{y(s)\}$ ($H(\theta)\mathbf{y} = \mathbf{e}$). For an exponential family, the parameterized probability density can be written as [7]:

$$p(y|\Theta) = exp\{g(T(y), \Theta)) + c(\Theta) + h(y)\}, \tag{7}$$

where $T(y)$ is the sufficient statistic for the parameter set Θ. Now we define the sample correlation function on the lattice,

$$C_d(r) = \frac{1}{M^2} \sum_{s \in \Omega_I} y(s)y(s+r). \tag{8}$$

The quadratic form $\mathbf{y}^T H(\theta)\mathbf{y}$ in (6) can be simplified as:

$$\mathbf{y}^T H(\theta)\mathbf{y} = M^2[C_d(0) - \boldsymbol{\Theta}^T \mathbf{C}_d], \tag{9}$$

where $\mathbf{C}_d =$ col. $[C_d(r), r \in N]$.

Thus Eq. (6) can be written as:

$$p(\mathbf{y}|\theta, \nu) = \frac{|H(\theta)|^{1/2}}{(2\pi\nu)^{M^2/2}} \exp\left(-\frac{M^2}{2\nu}\{C_d(0) - \theta^T \mathbf{C}_d\}\right). \tag{10}$$

Using the Neyman–Fisher factorization theorem [16], [7],

$$\xi = \{C_d(0), C_d(r) | r \in N\}$$

is a sufficient statistic for $(\boldsymbol{\Theta}, \nu)$, and hence is a lossless feature vector. It is termed as such since one can generate a random field close to the original one given these correlations. One can now design a Bayes rule or other standard classification rules like the nearest neighbor [17] using ξ as the feature vector. Note that the dimensionality of the feature vector is $(m+1)$, where m is the number of independent θ_r's.

4 Experimental Results

Using ξ, ζ, and η as the feature vectors, experiments were carried out to test their effectiveness in classifying textures. A set of seven textures were chosen from a Brodatz album, each available as a 256×256 digitized picture from the USC Signal and Image Processing Institute database [18]. Each image was divided into 16 non-overlapping subimages. Thus, each texture class had 16 samples. Two sets of subimages were used — 64×64 and 32×32. No histogram equalization was done on the subimages. The mean of each subimage was subtracted and the data was assumed to be a realization of a fourth-order Gaussian MRF model (10 dimensional feature vector), since it was adequate enough to synthesize the original 256×256 images.

A minimum distance classifier using the class mean and class covariance was designed. A leave-one-out strategy was adopted: to classify a particular sample, it was taken out and the classifier was trained on the remaining 15 samples of that class and 16 samples from each of the classes. This was repeated for all 112 samples.

Let each of the K classes contain n samples, whose feature dimension is $(r+1)$ corresponding to the set $\xi = \{C_d(0), \mathbf{C}_d(r) | r \in N_s\}$ or to the set $\eta = \{\theta^*, \nu^*/\hat{\rho}^2\}$, where $\hat{\rho}^2$ is the sample variance. First, the feature mean $x^{(i^*)}$ the feature variance $\sigma^{2(i^*)}$ for the ith class, $i = 1, \ldots, K$, were computed. Then the test sample was taken out and the feature mean $x^{(i)}$ and and the feature variance $\sigma^{2(i)}$ for that class were recomputed using other samples from it. The test sample $x^{(t)}$ is then assigned to the class K^* for which the weighted distance,

$$d(x^{(t)}, i^*) = \sum_f [x^{(t)}(f) - x^{(i^*)}(f)]^2 / \sigma^{2(i^*)}(f), \tag{11}$$

is minimum.

For 64×64 sized texture samples, using η as the feature vector, only one out of the 112 samples was misclassified, yielding an accuracy of over 99 percent. Using ξ as the feature vector, eight samples were misclassified, the classification accuracy being 93 percent. If we use normalized correlation values $\zeta = \{1, C_d(r)/C_d(0)\}$ as the feature vector, the classification rate

	Wood	Grass	Tree Bark	Pigskin	Leather	Raffia	Wool
Wood	16	0	0	0	0	0	0
Grass	0	16	0	0	0	0	0
Tree Bark	0	1	15	0	0	0	0
Pigskin	0	0	0	16	0	0	0
Leather	0	0	0	0	16	0	0
Raffia	0	0	0	0	0	16	0
Wool	0	0	0	0	0	0	16

Table 2: Results of Classification Using $\{\theta^*, \nu^*/\hat{\rho}^2\}$ as Feature Vector for 64×64 texture samples. Misclassification: 1 out of 112.

is about 94 percent. Similar features were extracted for 32×32 samples. Using normalized correlation values and the LS estimates, the classification rates are 82 and 93.75 percent, respectively. The results of classification experiments are given in Tables 2–7.

5 Binary Textures

So far, we have considered gray-level textures in which the random variates are assumed to be Gaussian. In some applications the data is often obtained in a binary form and the task is to distinguish between two non-similar textures, based on features extracted from them.

If we model the binarized texture as an MRF, the conditional probability distribution of $\{y\}$ is given by [26]:

$$P(y(\mathbf{s})|T) = \frac{e^{y(\mathbf{s})T}}{1+e^{T}} = \frac{e^{y(\mathbf{s})T}}{Z}, \tag{12}$$

where $\mathbf{s}$ refers to any pixel coordinate and T is a linear function of the neighborhood pixel intensities. Z is called the partition function. For a first-order model, T can be written as:

$$\begin{aligned} T = \alpha \quad &+ \quad \beta_1(y(\mathbf{s}+(0,1)) + y(\mathbf{s}+(0,-1))) \\ &+ \quad \beta_2(y(\mathbf{s}+(1,0)) + y(\mathbf{s}+(-1,0))). \end{aligned} \tag{13}$$

α, β_1, β_2 are the parameters of the model; α relates to the probability of occurrence of the center pixel; $\beta_{(.)}$'s are parameters to capture clustering of

	Wood	Grass	Tree Bark	Pigskin	Leather	Raffia	Wool
Wood	16	0	0	0	0	0	0
Grass	0	12	4	0	0	0	0
Tree Bark	0	0	16	0	0	0	0
Pigskin	0	0	0	15	0	1	0
Leather	0	0	0	0	16	0	0
Raffia	0	0	0	0	0	16	0
Wool	0	0	0	3	0	0	13

Table 3: Results of Classification Using $\{C_d(0), C_d(r)\}$ as Feature Vector for 64×64 Texture Samples. Misclassification: 8 out of 112

	Wood	Grass	Tree Bark	Pigskin	Leather	Raffia	Wool
Wood	16	0	0	0	0	0	0
Grass	0	14	0	0	0	0	2
Tree Bark	0	0	16	0	0	0	0
Pigskin	0	1	0	15	0	0	0
Leather	0	1	0	0	15	0	0
Raffia	0	0	0	1	0	15	0
Wool	0	2	0	3	0	0	14

Table 4: Results of Classification Using $\{1, C_d(r)/C_d(0)\}$ as Feature Vector for 64×64 Texture Samples. Misclassification: 7 out of 112

	Wood	Grass	Tree Bark	Pigskin	Leather	Raffia	Wool
Wood	16	0	0	0	0	0	0
Grass	0	16	0	0	0	0	0
Tree Bark	0	0	13	0	0	3	0
Pigskin	0	0	0	16	0	0	0
Leather	0	1	0	0	15	0	0
Raffia	1	0	2	0	0	13	0
Wool	0	0	0	0	0	0	16

Table 5: Results of Classification Using $\{\theta^*, \nu^*/\hat{\rho}^2\}$ as Feature Vector for 32×32 Texture Samples. Misclassification: 7 out of 112

	Wood	Grass	Tree Bark	Pigskin	Leather	Raffia	Wool
Wood	14	0	1	0	0	0	1
Grass	0	11	5	0	0	0	0
Tree Bark	0	0	16	0	0	0	0
Pigskin	0	0	0	6	0	4	6
Leather	0	0	3	0	13	0	0
Raffia	0	0	2	1	0	13	0
Wool	1	0	0	4	0	0	11

Table 6: Results of Classification Using $\{C_d(0), C_d(r)\}$ as Feature Vector for 32×32 Texture Samples. Misclassification: 28 out of 112

	Wood	Grass	Tree Bark	Pigskin	Leather	Raffia	Wool
Wood	15	0	0	0	1	0	0
Grass	0	13	0	0	2	0	1
Tree Bark	0	0	14	1	0	0	1
Pigskin	0	1	2	11	0	1	1
Leather	0	2	0	0	14	0	0
Raffia	0	0	0	3	0	13	0
Wool	0	3	0	1	0	0	12

Table 7: Results of Classification Using $\{1, C_d(r)/C_d(0)\}$ as Feature Vector for 32×32 Texture Samples. Misclassification: 20 out of 112

black or white pixels in the vertical or horizontal directions. When $\beta_1 = \beta_2$, we have an *isotropic* binary MRF (first-order).

For a second-order model, the diagonal neighbors are also included and T is given by (Fig. 1):

$$\begin{aligned} T = \alpha \quad &+ \quad \beta_1(y(s + (0,1)) + y(s + (0,-1))) \\ &+ \quad \beta_2(y(s + (1,0)) + y(s + (-1,0))) \\ &+ \quad \gamma_1(y(s + (1,1)) + y(s + (-1,-1))) \\ &+ \quad \gamma_2(y(s + (1,-1)) + y(s + (-1,1))) \end{aligned} \tag{14}$$

In our study, we have assumed that the parameters for the diagonal neighbors are the same, i.e., $\gamma_1 = \gamma_2$.

The problem of fitting a model to data involves formulating an estimation scheme for the model. This is the scope of both the descriptive and generative model. In the descriptive modeling approach, the goodness of fit is validated by classification using a large sample set.

The parameters for a binary MRF can be evaluated using the *coding* technique in practice, as was suggested in [28] and later verified in [26]. One way to obtain a good estimate of the parameters, $\alpha, \beta_1, \beta_2, \gamma_1, \gamma_2$, is the maximum likelihood estimation. If $p(y|.)$ denote the conditional probability $p(y = Y|\,\text{neighbors of Y})$, where y is a lattice variable, the likelihood function is given by:

$$\mathcal{L} = \prod_y p(y|.). \tag{15}$$

Even though this estimate is *optimum*, it is difficult to compute, as the joint probability function requires the value of Z, the partition function, which in turn contains all the possible realizations of the field. In the coding method, the lattice is partitioned into different disjoint sets. Each coding is chosen so that its site variables are conditionally independent, given the variables at other coding sites. Thus, in Eq. (13) or (14), y's could be assumed to belong to one particular coding site, and the likelihood maximized to obtain a set of parameter values. It was observed in [28] that the corresponding parameter values for the different coding sites are very close to each other. In practice, we take the average of these values. Thus, if there are $1, 2, \ldots, C$ codings, the maximum likelihood coding estimate $\hat{\vartheta}_{ML}$ for the whole field is given by:

$$\hat{\vartheta}_{ML} = \frac{1}{C} \sum_{i=1}^{C} \hat{\vartheta}_i, \tag{16}$$

where $\hat{\vartheta}_i$ is the parameter vector for ith coding.

6 Classification of Binary Textures

The same set of seven textures used in the earlier experiment with a GMRF model was considered here. Each texture class (e.g., wood, grass, etc.) had 16 samples. The texture images were binarized by using a histogram equalization procedure, a practice commonly used to reduce the number of gray levels in an image. The textures were assumed to be generated by a second-order binary Markov random field. We use both isotropic and anisotropic versions of the second-order model. As was mentioned earlier, in the isotropic version, $\beta_1 = \beta_2$ and $\gamma_1 = \gamma_2$ whereas in the anisotropic counterpart $\beta_1 \neq \beta_2$. The reason for such choices is derived from the definition of two pixel cliques in a 3×3 neighborhood.

The same minimum distance classifier was used here. The *leave-one-out* strategy was adopted for all 112 samples. Two sets of experiments were performed on both 64×64 and 32×32 samples. The results are tabulated in Tables 8–11.

7 Discussions

The classification accuracy decreases with the size of the subimages. This is expected, since for smaller images, the estimates are based on a smaller number of samples. Although correlation functions have been used earlier

	Wood	Grass	Tree Bark	Pigskin	Leather	Raffia	Wool
Wood	13	0	0	0	0	3	0
Grass	0	9	4	3	0	0	0
Tree Bark	0	1	11	2	0	2	0
Pigskin	0	5	4	7	0	0	0
Leather	0	0	0	0	15	1	0
Raffia	0	0	0	1	0	15	0
Wool	0	0	1	0	0	0	15

Table 8: Results of Classification Using $\hat{\vartheta}_{\mathrm{ML}}$ (with Isotropic Second-Order MRF) Feature Vector for 64×64 Texture Samples. Misclassification: 27 out of 112

	Wood	Grass	Tree Bark	Pigskin	Leather	Raffia	Wool
Wood	14	0	1	0	0	1	0
Grass	0	10	5	3	0	1	0
Tree Bark	0	2	12	0	0	2	0
Pigskin	0	0	0	16	0	0	0
Leather	0	0	0	0	16	0	0
Raffia	0	0	0	1	0	15	0
Wool	0	0	0	0	0	0	16

Table 9: Results of Classification Using $\hat{\vartheta}_{\mathrm{ML}}$ (with Anisotropic Second-Order MRF) Feature Vector for 64×64 Texture Samples. Misclassification: 13 out of 112

	Wood	Grass	Tree Bark	Pigskin	Leather	Raffia	Wool
Wood	12	0	0	1	2	1	0
Grass	3	6	4	3	0	0	0
Tree Bark	0	2	13	1	0	0	0
Pigskin	1	4	0	10	0	0	1
Leather	6	0	0	0	8	2	0
Raffia	2	0	0	0	2	12	0
Wool	0	0	0	1	0	0	15

Table 10: Results of Classification Using $\hat{\vartheta}_{\mathrm{ML}}$ (with isotropic second-order MRF) Feature Vector for 32×32 Texture Samples. Misclassification: 36 out of 112

	Wood	Grass	Tree Bark	Pigskin	Leather	Raffia	Wool
Wood	13	0	0	1	2	0	0
Grass	3	8	4	1	0	0	0
Tree Bark	0	2	13	1	0	0	0
Pigskin	1	3	0	11	0	0	1
Leather	4	0	0	0	10	2	0
Raffia	2	0	0	0	2	12	0
Wool	0	0	0	1	0	0	15

Table 11: Results of Classification Using $\hat{\vartheta}_{\mathrm{ML}}$ (with Anisotropic Second-Order MRF) Feature Vector for 32×32 Texture Samples. Misclassification: 30 out of 112

in the context of texture segmentation [31], no quantitative justification is available. Using the notion of sufficient statistics with a GMRF model, it is shown that by using sample correlations, one can achieve reasonable success in texture classification.

An important aspect of the modeling textures by random fields is that one could extract features that are information-preserving. The LS estimates for GMRFs are information-preserving, as is evidenced in their capability to synthesize textures that are perceptually close to the original ones [15]. The sample correlations are also information-preserving, as one can regenerate the underlying GMRF that have similar statistical properties as the original, from the sample correlation array [25]. An attractive application of modeling textures by random fields can be found in the problem of segmenting an image based on the homogeneity (e.g., textural) property of groups of pixels [33], [32].

Bibliography

[1] J. Besag, "On the Statistical Analysis of Dirty Pictures" (with discussion), *J. Royal Statist. Soc.*, Vol. 48, Series B, pp. 259–302, 1986.

[2] R. M. Haralick, "Statistical and Structural Approaches to Textures," *Proc. of IEEE*, Vol. 67, no. 5, pp. 786–804, May 1979.

[3] R. M. Haralick, K. Shanmugam, and I. Dinstein, "Textural Feaures for Classification," *IEEE Trans. on Systems, Man and Cybernetics*, Vol. 3, pp.610–621, November 1973.

[4] J. Weszka, C. R. Dyer, and A. Rosenfeld, "A Comparative Study of Texture Measures for Terrain Classification," *IEEE Trans. on Systems, Man and Cybernetics*, Vol. 6, pp. 269–285, April 1976.

[5] O. D. Faugeras and W. K. Pratt, "Decorrelation Methods of Texture Feature Extraction", *IEEE Trans. on Pattern Analysis and Machine Intelligence*, Vol. 2, pp. 323–332, July 1980

[6] D. G. Childers, *Modern Spectral Estimation*, IEEE Press, New York, 1975.

[7] P.J. Bickel and K.J. Doksum, *Mathematical Statistics*, Holden-Day, Oakland, California, 1977.

[8] S. Kullback, *Information Theory and Statistics*, John Wiley & Sons Inc., New York, 1959.

[9] C. H. Chen, "A Study of Texture Classification Using Spectral Features," *Proc. of Intl. Conf. on Pattern Recognition, Munich, W. Germany*, 1982.

[10] Y. A. Rosanov, "On Gaussian Fields with Given Conditional Distributions," *Theory of Probability and its Applications*, Vol. II, pp. 381–391, 1967.

[11] J. W. Woods, "Two-Dimensional Discrete Markovian Fields", *IEEE Trans. on Information Theory*, Vol. 18, pp. 232–240, March 1972.

[12] R. L. Kashyap and R. Chellappa, "Estimation and Choice of Neighbours in Spatial Interaction Models of Images" *IEEE Trans. on Information Theory*, Vol. 29, pp. 60–72, January 1983.

[13] R. Chellappa, "Two-Dimensional Discrete Gaussian Markov Random Field Models for Image Processing", *Progress in Pattern Recognition*, L. N. Kanal and A. Rosenfeld (editors), Vol. II, North-Holland, Amsterdam, 1983.

[14] R. L. Kashyap, R. Chellappa, and A. Khotanzad, "Texture Classification Using Features Derived from Random Field Models," *Pattern Recognition Letters*, Vol. 1, no. 1, pp. 43–50, October 1982.

[15] R. Chellappa, S. Chatterjee, and R. Bagdazian, "Texture Synthesis and Compression Using Gaussian Markov Random Field Models", *IEEE Trans. on Systems, Man and Cybernetics*, Vol. 15, no. 2, pp. 298–303, Mar./Apr. 1985.

[16] M. H. DeGroot, *Optimal Statistical Decisions*, McGraw-Hill, New York, 1970.

[17] R. H. Duda and P. E. Hart, *Pattern Classification and Scene Analysis*, John Wiley, New York, 1973.

[18] A. Weber, "Image Data Base," *USCIPI Report No. 1070*, Image Processing Institute, University of Southern California, Los Angeles, March 1983.

[19] H. Kaneko and E. Yodogawa, "A Markov Random Field Application to Texture Classification," *Proc. of Conf. on Pattern Recognition and Image Processing*, IEEE Computer Society, Las Vegas, Nevada, 1982, pp. 221–225.

[20] P. A. P. Moran and J. E. Besag, "On the Estimation and Testing of Spatial Interaction in Gaussian Lattice Processes," *Biometrika*, Vol. 62, pp. 555–572, 1975..

[21] H. Kunsch, "Thermodynamics and Statistical Analysis of Gaussian Random Fields," *Zeitschrift für Wahrscheinlickeitstheorie und Verwandte Gebiete*, Vol. 58, pp. 407-421, November 1981.

[22] K. Fukunaga, *Introduction to Statistical Pattern Recognition*, Academic Press, New York, 1972.

[23] G. Sharma, "Modern Two-Dimensional Spectral Estimation Using Noncausal Spatial Models," University of Southern California, Los Angeles, 1984.

[24] R. Chellappa and G. Sharma, "Realizing Gaussian Markov Random Field Models from True Correlations," *Conf. on Information Sciences and Systems*, Princeton, March 1984.

[25] R. Chellappa, Y.H. Hu, and S.Y. Kung, "On Two-Dimensional Markov Spectral Estimation," *IEEE Trans. on Acoust., Speech and Signal Proc.*, Vol. 31, pp. 836–841, August 1983.

[26] G. R. Cross and A. K. Jain, "Markov Random Field Texture Models," *IEEE Trans. on Pattern Analysis and Machine Intelligence*, Vol. 5, pp. 25–39, January 1983.

[27] M. Hassner and J. Sklansky, "The Use of Markov Random Fields as Models of Textures," *Computer Graphics and Image Processing*, Vol. 12, pp. 376–406, April 1980.

[28] J. E. Besag, "Spatial Interaction and Statistical Analysis of Lattice Systems," *Jour. Royal Stat. Soc.*, Vol. B-36, Series B, pp. 192–236, 1974.

[29] A. Bovik, M. Clark, and W. Geisler, "Multichannel Texture Analysis Using Localized Spatial Filters," *IEEE Trans. on Pattern Analysis and Machine Intelligence*, Vol. 12, no. 1, pp. 55–73, January 1990.

[30] R.L Kashyap and A. Khotanzad, "A Model-Based Method for Rotation Invariant Texture Classification," *IEEE Tran. on Pattern Analysis and Machine Intelligence*, Vol. 18, no. 4, pp. 472–481, July 1986.

[31] P.C. Chen and T. Pavlidis, "Segmentation by Texture Using Correlation," *IEEE Trans. on Pattern Analysis and Machine Intelligence*, Vol. 15, no. 1, pp. 464–469, Jan. 1983.

[32] H. Derin and H. Elliott, "Modeling and Segmentation of Noisy and Textured Images Using Gibbs Random Fields," *IEEE Trans. on Pattern Analysis and Machine Intelligence*, Vol. 9, pp. 39–55, January 1987.

[33] C. W. Therrien, "An Estimation-Theoretic Approach to Terrain Image Segmentation," *Comp. Vis. Graph. Im. Proc.*, Vol. 22, pp. 313–326, 1983.

[34] D. T. Kuan, "Nonstationary 2-D Recursive Restoration of Images with Signal-Dependent Noise with Application to Speckle Reduction," University of Southern California, Los Angeles, 1982.

Spectral Estimation for Random Fields with Applications to Markov Modelling and Texture Classification

J. Yuan and T. Subba Rao
Department of Mathematics
University of Manchester Institute of Science and Technology
Manchester, United Kingdom

1 Introduction

We are concerned with the statistical analysis of a real-valued stationary random field $\{x(t),\ t \in \mathbf{Z}^d\}$, where $\mathbf{Z}$ is the space of integers and the dimension $d \geq 1$. When $d = 1$ and t designates time, $\{x(t),\ t \in \mathbf{Z}^d\}$ is known as a *time series.* Statistical analysis of linear stationary time series has been well-established theoretically, and has found many applications in various fields of applied sciences.

It is useful to generalize time series modelling techniques to the random field case, but there exist many inherent analytical problems that are not present for $d = 1$. For example, the fundamental theorem of algebra is important for ARMA modelling of time series, but it no longer holds for $d > 1$. The situation is similar to the transition from ordinary differential equations to partial differential equations, although the purpose of statistical modelling is to fit an equation to the solution instead of solving an equation.

We can avoid the analytical problems with polynomials by adopting the *frequency domain* approach; that is, to make statistical inferences from the cumulant spectra estimated by smoothing the periodograms. The estimated spectra are relatively simple by construction, and their asymptotic cumulant properties can be derived under fairly general conditions on the

ISBN 0-12-170608-7

random field. From the spectral estimates and their cumulant properties, we can construct statistics useful for the analysis of the random field, and study the asymptotic properties of the statistics.

While the spectrum of a random field can be estimated alternatively by fitting a parametric model with a spectrum closest in certain measure to the periodogram (smoothed and/or tapered when necessary), we also can obtain estimated parameters of a model from the nonparametric estimate of the spectrum. This spectral approach is particularly effective in Markov modelling, where the inverse autocovariances estimated using the smoothed periodogram not only provide sensible estimates of the parameters, but also help in identifying the model.

The estimated spectra also can be used in the classification of textures as realizations of stationary random fields. Each class of textures is characterized by a spectrum that we estimate from a training sample, and a texture is assigned to a class in such a way that its periodogram (smoothed when necessary) is closest in certain measure to the estimated spectrum of the class. From the asymptotic cumulant properties of the spectral estimates, we can study the limiting behavior of the probability of misclassification as the sample size tends to infinity.

The following notation is used throughout this chapter.

For any $a = (a_1, \ldots, a_d)$, $b = (b_1, \ldots, b_d) \in \mathbf{R}^d$, $a{\cdot}b = a_1b_1 + \ldots + a_db_d \in \mathbf{R}^d$, and $a/b = (a_1/b_1, \ldots, a_d/b_d) \in \mathbf{R}^d$, if $b_1, \ldots, b_d \neq 0$.

For any $j, j^{(1)}, \ldots, j^{(r)}, N \in \mathbf{Z}^d$ with $N_s > 0$, $s = 1, \ldots, d$, we write $2\pi j/N$ as ω_j, $2\pi j^{(s)}/N$ as $\omega_j^{(s)}$, $s = 1, \ldots, r$, the product $N_1 \cdots N_d$ as $|N|$, and $j_1 = 1, \ldots, N_1, \ldots, j_d = 1, \ldots, N_d$ as $j = 1, \ldots, N$, or $1 \leq j \leq N$.

2 Spectral Estimation

Let $\{x(t),\ t \in \mathbf{Z}^d\}$ be a real-valued random field on $\mathbf{Z}^d$, where $\mathbf{Z}$ is the space of integers and the dimension $d \geq 1$.

We say that $\{x(t),\ t \in \mathbf{Z}^d\}$ is (weakly) stationary, if for any $t, s \in \mathbf{Z}^d$,

$$Ex(t) = Ex(s), \operatorname{cov}\{x(t+s), x(t)\} = \operatorname{cov}\{x(s), x(0)\}. \tag{1}$$

In this case, the autocovariance function,

$$R(s) = \operatorname{cov}\{x(s), x(0)\},\ s \in \mathbf{Z}^d, \tag{2}$$

is positive definite in the sense that

$$\sum_{j=1}^{n} \sum_{k=1}^{n} a(j) R\left(s^{(j)} - s^{(k)}\right) a(k)^* \geq 0 \tag{3}$$

for any $a(1), \ldots, a(n) \in \mathbf{C}$, $s^{(1)}, \ldots, s^{(n)} \in \mathbf{Z}^d$. It follows [37] that there exists a nonnegative measure F on $\mathcal{B}(\Pi^d)$, the class of Borel sets on $\Pi^d = (0, 2\pi]^d$, such that

$$R(s) = \int_{\Pi^d} \exp\{is \cdot \omega\} F(d\omega), s \in \mathbf{Z}^d. \tag{4}$$

The measure F is known as the spectral distribution of $\{x(t),\ t \in \mathbf{Z}^d\}$. It has a unique Lebesgue decomposition,

$$F(d\omega) = f(\omega)d\omega + F^{(s)}(d\omega), \tag{5}$$

where f is nonnegative and Lebesgue integrable, and $F^{(s)}$ is singular with respect to the Lebesgue measure. We say that f is the spectrum of $\{x(t)\}$, if F is absolutely continuous with respect to the Lebesgue measure. It is easy to verify that F is absolutely continuous with the spectrum,

$$f(\omega) = (2\pi)^{-d} \sum_{s \in \mathbf{Z}^d} R(s) \exp\{-is \cdot \omega\},\ \omega \in \Pi^d, \tag{6}$$

if the autocovariances $\{R(s),\ s \in \mathbf{Z}^d\}$ are absolutely summable.

More generally, we say that $\{x(t),\ t \in \mathbf{Z}^d\}$ is stationary up to order r, if for any $t,\ t^{(1)}, \ldots, t^{(p)} \in \mathbf{Z}^d,\ p \leq r$,

$$\operatorname{cum}\left\{x(t^{(1)}), \ldots, x(t^{(p)})\right\} = \operatorname{cum}\left\{x(t^{(1)} + t), \ldots, x(t^{(p)} + t)\right\}. \tag{7}$$

In this case, the rth-order cumulant spectrum is given by

$$\begin{aligned} f^{(r)}&\left(\omega^{(1)}, \ldots, \omega^{(r-1)}\right) \\ &= (2\pi)^{-(r-1)d} \sum_{u^{(1)}} \cdots \sum_{u^{(r-1)}} c\left(u^{(1)}, \ldots, u^{(r-1)}\right) \exp\left\{-i \sum_{p=1}^{r-1} u^{(p)} \cdot \omega^{(p)}\right\}, \\ &\omega^{(1)}, \ldots, \omega^{(r-1)} \in \Pi^d, \end{aligned} \tag{8}$$

provided that the rth-order cumulant function,

$$c(u^{(1)}, \ldots, u^{(r-1)}) = \operatorname{cum}\{x(t + u^{(1)}), \ldots, x(t + u^{(r-1)}), x(t)\}, \tag{9}$$

is absolutely summable.

Given a finite realization $\{x(t),\ t = 1, \ldots, N\}$ of a stationary random field $\{x(t),\ t \in \mathbf{Z}^d\}$ with an absolutely continuous spectral distribution, we wish to estimate its spectrum f and spectral integrals of the form,

$$\eta(\phi, \psi) = \int_{\Pi^d} \phi(\omega)\psi(f(\omega))d\omega, \tag{10}$$

where ϕ and ψ are functions on Π^d and $(0, +\infty)$, respectively.

There exists a comprehensive literature on spectral estimation for time series—*cf.* [7], [28], [30], and [31]—but this is not the case with random fields. Grenander and Rosenblatt [15] proved the asymptotic consistency of the smoothed periodogram of a Gaussian stationary random field on $\mathbf{Z}^2$. Priestley [29] gave asymptotic expressions for the covariances of weighted integrals of the periodogram of a random field for testing the presence of a mixed spectrum. Brillinger [6] considered the estimation of the cumulant spectra of a stationary random field on $\mathbf{R}^d$ from the observations $\{x(t); h(t/T) \neq 0\}$,where $T > 0$ and h is a function on $\mathbf{R}^d$ with $\int h(t)^2 dt > 0$, and derived the asymptotic properties of the spectral estimates as $T \to \infty$.

As a simple estimate of the spectrum, the periodogram

$$I_N(\omega) = (2\pi)^{-d}|N|^{-1}\left|\sum_{t=1}^{N} x(t)\exp\{-it\cdot\omega\}\right|^2, \quad \omega \in \Pi^d, \tag{11}$$

is not consistent although asymptotically unbiased [15]; thus, it is smoothed to get a consistent estimate,

$$\hat{f}_N(\omega) = (2\pi)^d|N|^{-1}\sum_{j=1}^{N} W_N(\omega-\omega_j)\, I_N(\omega_j), \quad \omega \in \Pi^d, \tag{12}$$

where W_N is an appropriate function on $\mathbf{R}^d$ periodic in each dimension with period 2π.

Theorem 1 *If* $\{x(t),\ t \in \mathbf{Z}^d\}$ *is stationary with a continuous spectrum* f, W_N *satisfies*

$$\sup_{\omega}(2\pi)^d|N|^{-1}\sum_{j=1}^{N}|W_N(\omega-\omega_j)| = O(1), \text{ as } N \to \infty, \tag{13}$$

and for any fixed $t \in \mathbf{Z}^d$,

$$(2\pi)^d|N|^{-1}\sum_{j=1}^{N} W_N(\omega-\omega_j)\exp\{it\cdot(\omega-\omega_j)\} = 1+o(1), \text{ as } N \to \infty, \tag{14}$$

uniformly in $\omega \in \Pi^d$, *then we have*

$$\max_{\omega}\left|E\hat{f}_N(\omega) - f(\omega)\right| \to 0, \text{ as } N \to \infty. \tag{15}$$

The conditions on W_N ensure that it behaves like a delta function as N becomes large, in the sense that

$$(2\pi)^d |N|^{-1} \sum_{j=1}^{N} W_N(\omega - \omega_j)\, \phi(\omega_j) = \phi(\omega) + o(1), \text{ as } N \to \infty, \tag{16}$$

uniformly in ω, for any continuous function ϕ on $\mathbf{R}^d$ that is periodic in each dimension with period 2π. This can be proved easily using the Weierstrass approximation theorem, *cf.* [38].

Analogous to the approach of [8] for multiple time series, we can derive the asymptotic properties of the spectral estimates from those of the discrete Fourier transform of the observations,

$$d_N(\omega_j) = \sum_{t=1}^{N} x(t) \exp\{-it \cdot \omega_j\},\ j = 1, \ldots, N, \tag{17}$$

using the product cumulant theorem due to [21].

Lemma 1 *If $\{x(t),\ t \in \mathbf{Z}^d\}$ is stationary up to order $r \geq 2$, with mean zero and cumulants satisfying*

$$\sum_{p=1}^{r-1} \sum_{u^{(1)}} \cdots \sum_{u^{(p)}} \left(\sum_{s=1}^{p} \|u^{(s)}\|^d \right) \left| c\left(u^{(1)}, \ldots, u^{(p)}\right) \right| < \infty. \tag{18}$$

then as $N \to \infty$ we have,

$$cum\left\{ d_N\left(\omega_j^{(1)}\right), \ldots, d_N\left(\omega_j^{(r)}\right) \right\}$$

$$= (2\pi)^{d(r-1)} |N| \left\{ f^{(r)}\left(\omega_j^{(1)}, \ldots, \omega_j^{(r-1)}\right) + O(1) \sum_{s=1}^{d} N_s^{-1} \right\} \delta_N(j) \tag{19}$$

$$+ O(1)\left\{1 - \delta_N(j)\right\} \prod_{s=1}^{d} \left\{ N_s \delta_N^{(s)}(j) + 1 \right\}$$

uniformly in $j^{(1)}, \ldots,\ j^{(r)} \in \mathbf{Z}^d$, where $f^{(r)}\left(\omega^{(1)}, \ldots, \omega^{(r-1)}\right)$ is the rth-order cumulant spectrum as given in Eq. (8),

$$j = j^{(1)} + \cdots + j^{(r)}, \tag{20}$$

$$\delta_N^{(s)}(t) = \begin{cases} 1, & \text{if } t_s = 0 \pmod{N_s}, \\ 0, & \text{otherwise}, \end{cases} \quad t \in \mathbf{Z}^d,\ s = 1, \ldots, d, \tag{21}$$

$$\delta_N(t) = \prod_{s=1}^{d} \delta_N^{(s)}(t),\ t \in \mathbf{Z}^d. \tag{22}$$

This is an extension of the result of [8] for multiple time series. It can be proved easily using the orthogonality of

$$(\exp\{-it \cdot \omega_j\}\,;\; t = 1, \dots, N)\,, j = 1, \dots, N,$$

and the Lebesgue bounded convergence theorem; *cf.* [38].

Lemma 2 *Let $\{v_1, \dots, v_p\}$, $p \geq 1$, be an indecomposable partition of the set $\{(i,j) \mid i = 1, \dots, r,\; j = 1, 2\}$, and define a matrix $A = (a_{ik})_{p \times r}$ with elements,*

$$a_{ik} = \begin{cases} 1, & \text{if}(k,1) \in v_i,\; (k,2) \notin v_i, \\ -1, & \text{if}(k,2) \in v_i,\; (k,1) \notin v_i,\; i = 1, \dots, p,\; k = 1, \dots, r. \\ 0, & \text{otherwise}, \end{cases} \tag{23}$$

Then

(a) A is of rank $p-1$, and

(b) for any submatrix A_n composed of $n \leq p-1$ rows of A, we can permutate the rows and columns of A_n so that it becomes

$$\begin{bmatrix} \pm 1 & * & \dots & \dots & * & * & \dots & * \\ 0 & \pm 1 & \ddots & & & & & \vdots \\ \vdots & \ddots & \ddots & \ddots & & & & \vdots \\ \vdots & & \ddots & \pm 1 & * & & & \vdots \\ 0 & \dots & \dots & 0 & \pm 1 & * & \dots & * \end{bmatrix}. \tag{24}$$

Proof. (a) Since $(1, \cdots, 1)A = (0, \cdots, 0)$, it suffices to show that the first $p-1$ rows of A are linearly independent.

Suppose that for some constants $c_1, \dots, c_{p-1}$,

$$(c_1, \cdots, c_{p-1})A_{p-1} = (0, \cdots, 0), \tag{25}$$

where A_{p-1} is the submatrix composed of the first $p-1$ rows of A, then

$$(c_1, \cdots, c_{p-1}, c_p)A = (0, \cdots, 0), \tag{26}$$

where $c_p = 0$.

If for some $i \neq j$, v_i and v_j hook, then by definition there exist $(s_i, t_i) \in v_i, (s_j, t_j) \in v_j$, such that $s_i = s_j$. It follows from Eq. (26) that

$$c_i - c_j = \pm(c_1, \cdots, c_p)(a_{1,s_i}, \cdots, a_{p,s_i})^T = 0. \tag{27}$$

For any $i, j = 1, \ldots, p$, we have by definition of indecomposability that there exist $k_1, \ldots, k_s = 1, \ldots, p$, such that v_i and v_{k_1} hook, v_{k_1} and v_{k_2} hook,$\ldots$,v_{k_s} and v_j hook. It follows from Eq. (27) that

$$c_i = c_{k_1} = \cdots = c_{k_s} = c_j. \tag{28}$$

Therefore,

$$c_1 = \cdots = c_{p-1} = c_p = 0, \tag{29}$$

so that the first $p-1$ rows of A are linearly independent.

(b) Since $n \leq \text{rank}(A)$, A_n is of full rank. Thus,

$$(1, \cdots, 1)A_n \neq (0, \cdots, 0). \tag{30}$$

Note that the only possible nonzero elements of A_n are 1 and -1, each appearing at most once in each column. Thus, from Eq. (30), there exists a column of A_n with exactly one nonzero element. Rearrange the rows of A_n so that this nonzero element sits in row 1, and then rearrange the columns so that it sits in column 1. Now A_n becomes of the form,

$$\begin{bmatrix} \pm 1 & * & \ldots & * \\ 0 & & & \\ \vdots & & A_{n-1} & \\ 0 & & & \end{bmatrix}. \tag{31}$$

Next, consider the submatrix A_{n-1}. By similar arguments to the preceding, we can rearrange the rows and then the columns of A_{n-1} and correspondingly those of A_n, so that A_n becomes of the form,

$$\begin{bmatrix} \pm 1 & * & * & \ldots & * \\ 0 & \pm 1 & * & \ldots & * \\ \vdots & 0 & & & \\ \vdots & \vdots & & A_{n-2} & \\ 0 & 0 & & & \end{bmatrix}. \tag{32}$$

We can carry on with this procedure until the final form is achieved.

The asymptotic cumulant property of the spectral estimates will be given in terms of the magnitude of the weight function W_N as defined in the following. A function W_N on $\mathbf{R}^d$ is said to be of magnitude M, where $M \in \mathbf{R}^d$ depends on N with

$$1 \leq M_s \leq N_s, s = 1, \ldots d, \tag{33}$$

if as $N \to \infty$,

$$\sup_{\omega} |N|^{-1} \sum_{j=1}^{N} |W_N(\omega - \omega_j)| \prod_{s=1}^{d} \left\{1 + \delta_N^{(s)}(j) N_s M_s^{-1}\right\} = O(1), \tag{34}$$

where $\delta_N^{(s)}(j)$ is as given by Eq. (21).

Theorem 2 *If $\{x(t),\ t \in \mathbf{Z}^d\}$ is stationary up to order $2r$ with mean zero and cumulants satisfying*

$$\sum_{p=1}^{2r-1} \sum_{u^{(1)}} \cdots \sum_{u^{(p)}} \left(\sum_{s=1}^{p} \|u^{(s)}\|^d \right) \left| c\left(u^{(1)}, \ldots, u^{(p)}\right) \right| < \infty, \tag{35}$$

and W_N is of magnitude M, then as $N \to \infty$,

$$cum\left\{ \hat{f}_N(\omega^{(1)}), \ldots, \hat{f}_N(\omega^{(r)}) \right\} = O(1) \left(\frac{|M|}{|N|} \right)^{r-1} \tag{36}$$

uniformly in $\omega^{(1)}, \ldots, \omega^{(r)} \in \Pi^d$.

Proof. For any $\omega^{(1)}, \cdots, \omega^{(r)} \in \Pi^d$, we have

$$\begin{aligned} &\mathrm{cum}\left\{ \hat{f}_N\left(\omega^{(1)}\right), \ldots, \hat{f}_N\left(\omega^{(r)}\right) \right\} \\ &= |N|^{-2r} \sum_{j^{(1)}=1}^{N} \cdots \sum_{j^{(r)}=1}^{N} \left\{ \prod_{s=1}^{r} W_N\left(\omega^{(s)} - \omega_j^{(s)}\right) \right\} \\ &\quad \mathrm{cum}\left\{ d_N\left(\omega_j^{(s)}\right) d_N\left(-\omega_j^{(s)}\right);\ s = 1, \ldots, r \right\}. \end{aligned} \tag{37}$$

From the product cumulant theorem (*cf.*, [7], p. 21), we have

$$\begin{aligned} &\mathrm{cum}\left\{ d_N\left(\omega_j^{(s)}\right) d_N\left(-\omega_j^{(s)}\right);\ s = 1, \ldots, r \right\} \\ &= \sum_{p=1}^{r} \sum_{\{v_1, \ldots, v_p\}} \prod_{i=1}^{p} \mathrm{cum}\left\{ d_N\left((-1)^{t-1} \omega_j^{(s)}\right); (s,t) \in v_i \right\}, \end{aligned} \tag{38}$$

where the second summation is over all indecomposable partitions $\{v_1, \cdots, v_p\}$ of the set $\{(s,t) \mid s = 1, \cdots, r,\ t = 1, 2\}$. Note that we need only consider $p \leq r$ in the first summation, since $\{x(t)\}$ has zero mean, so that its first-order cumulants vanish.

It then follows from Lemma 1 that

$$\operatorname{cum}\left\{\hat{f}_N\left(\omega^{(1)}\right),\ldots,\hat{f}_N\left(\omega^{(r)}\right)\right\}$$

$$= O(1)\,|N|^{-2r}\sum_{p=1}^{r}\sum_{\{v_1,\ldots,v_p\}}\sum_{q_1=0}^{p}\cdots\sum_{q_d=0}^{p} N_1^{q_1}\cdots N_d^{q_d}$$

$$\sum\nolimits^{(q_1\ldots q_d)}\prod_{s=1}^{r}\left|W_N\left(\omega^{(s)}-\omega_j^{(s)}\right)\right|, \tag{39}$$

where the last summation is over all $j^{(1)},\ldots,j^{(r)}=1,\ldots,N$, such that

$$\sum_{i=1}^{p}\delta_N^{(s)}\left(\sum_{k=1}^{r}a_{ik}j^{(k)}\right)=q_s,\, s=1,\ldots,d, \tag{40}$$

where $\{a_{ik}\}$ are as defined in Eq. (23).

When $p>1$, let A_n be a matrix composed of the first $n=p-1$ rows of $A=(a_{ik})_{p\times r}$, and consider

$$g^{(i)}=\sum_{k=1}^{r}a_{ik}j^{(k)},\, i=1,\ldots,n, \tag{41}$$

or as written in matrix form,

$$\begin{bmatrix} g^{(1)} \\ \vdots \\ g^{(n)} \end{bmatrix} = A_n \begin{bmatrix} j^{(1)} \\ \vdots \\ j^{(r)} \end{bmatrix}. \tag{42}$$

From Lemma 2, we can find a permutation $(i_1,\ldots,i_r)$ of $(1,\ldots,r)$ and $(s_1,\ldots,s_n)$ of $(1,\ldots,n)$, so that Eq. (42) can be rewritten as

$$\begin{bmatrix} \pm1 & * & \cdots & \cdots & * & * & \cdots & * \\ 0 & \pm1 & \ddots & & & & & \vdots \\ \vdots & \ddots & \ddots & \ddots & & & & \vdots \\ \vdots & & \ddots & \pm1 & * & & & \vdots \\ 0 & \cdots & \cdots & 0 & \pm1 & * & \cdots & * \end{bmatrix} \begin{bmatrix} j^{(i_1)} \\ \vdots \\ j^{(i_r)} \end{bmatrix} = \begin{bmatrix} g^{(s_1)} \\ \vdots \\ g^{(s_n)} \end{bmatrix}. \tag{43}$$

This suggests that we write

$$\sum\nolimits^{(q_1\ldots q_d)}\prod_{s=1}^{r}\left|W_N\left(\omega^{(s)}-\omega_j^{(s)}\right)\right|$$

$$= \sum^{(i_r)} \left| W_N \left(\omega^{(i_r)} - \omega_j^{(i_r)} \right) \right| \cdots \sum^{(i_1)} \left| W_N \left(\omega^{(i_1)} - \omega_j^{(i_1)} \right) \right|, \tag{44}$$

where for each $t = 1, \ldots, r$, $\sum^{(i_t)}$ is over all $j^{(i_t)} = 1, \ldots, N$, such that $j^{(1)}, \ldots, j^{(r)}$ satisfy Eq. (40) for some $j^{(i_1)}, \ldots, j^{(i_{t-1})} = 1, \ldots, N$, since we have from Eqs. (34) and (43) that

$$\sum^{(i_t)} \left| W_N \left(\omega^{(i_t)} - \omega_j^{(i_t)} \right) \right| = O(1) \prod_{u=1}^{d} \left\{ (M_u/N_u)^{\delta_N^{(u)}(g^{(s_t)})} N_u \right\} \tag{45}$$

uniformly in $\omega^{(i_t)}, j^{(i_{t+1})}, \ldots, j^{(i_r)}$, for $t = 1, \ldots, n$, and

$$\sum^{(i_r)} \left| W_N \left(\omega^{(i_r)} - \omega_j^{(i_r)} \right) \right| \cdots \sum^{(i_{n+1})} \left| W_N \left(\omega^{(i_{n+1})} - \omega_j^{(i_{n+1})} \right) \right|$$
$$= O(|N|^{r-n}) \tag{46}$$

uniformly in $\omega^{(i_{n+1})}, \ldots, \omega^{(i_r)}$. It follows that

$$\sum^{(q_1 \ldots q_d)} \prod_{s=1}^{r} \left| W_N \left(\omega^{(s)} - \omega_j^{(s)} \right) \right|$$
$$= O\left(|N|^r\right) \prod_{u=1}^{d} (M_u/N_u)^{q_u'} \tag{47}$$

uniformly in $\omega^{(1)}, \ldots, \omega^{(r)}$, where

$$q_u' = \sum_{t=1}^{n} \delta_N^{(u)} \left(g^{(s_t)} \right) \geq q_u - 1, \; u = 1, \ldots, d, \tag{48}$$

from Eq. (40). Thus, we have

$$\sum^{(q_1 \ldots q_d)} \prod_{s=1}^{r} \left| W_N \left(\omega^{(s)} - \omega_j^{(s)} \right) \right| = O\left(|N|^r\right) \prod_{s=1}^{d} (M_s/N_s)^{q_s - 1} \tag{49}$$

uniformly in $\omega^{(1)}, \ldots, \omega^{(r)}$. Note that Eq. (49) obviously holds for $p = 1$.

It thus follows from Eqs. (39) and (49) that

$$\operatorname{cum}\left\{ \hat{f}_N \left(\omega^{(1)} \right), \ldots, \hat{f}_N \left(\omega^{(r)} \right) \right\}$$
$$= O(1) |N|^{1-r} \max_{1 \leq p \leq r} \max_{0 \leq q_1, \ldots, q_d \leq p} \prod_{s=1}^{d} M_s^{q_s - 1} \tag{50}$$
$$= O(1) \left(\frac{|M|}{|N|} \right)^{r-1} \tag{51}$$

uniformly in $\omega^{(1)}, \ldots, \omega^{(r)}$.

Corollary 1 *If* $\{x(t),\ t \in \mathbf{Z}^d\}$ *is stationary up to order 4 with mean zero and cumulants satisfying Eq. (35) for* $r = 2$, *and* W_N *is of such magnitude* M *that*

$$1 \leq M_s = \circ (N_s), \ s = 1, \ldots, d, \ as\ N \to \infty, \tag{52}$$

then we have

$$E\left(\left|\hat{f}_N(\omega) - E\hat{f}_N(\omega)\right|^2\right) \to 0, \ as\ N \to \infty \tag{53}$$

uniformly in $\omega \in \Pi^d$.

This follows easily from Eqs. (36) and (52).

Corollary 2 *If* $\{x(t),\ t \in \mathbf{Z}^d\}$ *is stationary up to order* $2r$ *with mean zero and cumulants satisfying Eq. (35), and* W_N *is of magnitude* M, *then for any positive integer* $p \leq r/2$, *we have*

$$E\left(\left|\hat{f}_N(\omega) - E\hat{f}_N(\omega)\right|^{2p}\right) = O(1)\left(\frac{|M|}{|N|}\right)^p, \ as\ N \to \infty \tag{54}$$

uniformly in $\omega \in \Pi^d$.

Proof. For any $p \leq r/2$, we have from the product cumulant theorem and Eq. (36) that

$$\begin{aligned} & E\left(\left|\hat{f}_N(\omega) - E\hat{f}_N(\omega)\right|^{2p}\right) \\ &= O(1)\sum_{s=1}^{p} \sum_{\substack{n_1, \ldots, n_s > 1 \\ n_1 + \cdots + n_s = 2p}} \sum_{u=1}^{s} \left(\frac{|M|}{|N|}\right)^{n_u - 1} \\ &= O(1)(|M|\,/\,|N|)^p. \end{aligned}$$

Corollary 3 *If* $\{x(t),\ t \in \mathbf{Z}^d\}$ *is stationary up to order* $2r$ *for some even integer* $r > 4$ *with mean zero and cumulants satisfying Eq. (35), and* $W_N(\cdot)$ *is of such magnitude* M *that*

$$1 \leq M_s \leq N_s^{1-a}, s = 1, \ldots, d \tag{55}$$

for some constant $a \in (4r^{-1}, 1)$. *Then we have as* $N \to \infty$,

$$\max_j \left|\hat{f}_N(\omega_j) - E\hat{f}_N(\omega_j)\right| \to 0, \ a.s.. \tag{56}$$

Proof. For any $\epsilon > 0$, and any $T \in \mathbf{Z}^d$, we have from Eqs. (54) and (55) that

$$\begin{aligned}
&P\left\{\bigcup_{N=T}^{\infty}\left\{\max_{1\leq j\leq N}\left|\hat{f}_N(\omega_j) - E\hat{f}_N(\omega_j)\right| > \epsilon\right\}\right\} \\
&\leq \sum_{N=T}^{\infty}\epsilon^{-r}E\left(\max_{1\leq j\leq N}\left|\hat{f}_N(\omega_j) - E\hat{f}_N(\omega_j)\right|^r\right) \\
&\leq \epsilon^{-r}\sum_{N=T}^{\infty}|N|\max_{1\leq j\leq N}E\left(\left|\hat{f}_N(\omega_j) - E\hat{f}_N(\omega_j)\right|^r\right), \\
&= O(1)\sum_{N=T}^{\infty}|N|^{1-ra/2}, \\
&= o(1), \text{ as } N \to \infty.
\end{aligned}$$

Thus, the strong convergence follows.

Note that $\hat{f}_N$ essentially is a weighted average of the periodogram ordinates $I_N(\omega_j)$, and in Theorem 2 and its corollaries, we only assume that the weight function is of certain magnitude M; thus, the results are applicable not only to the estimator of the spectrum, but also to the estimation of the spectral integral,

$$\eta(\phi, 1) = \int_{\Pi^d}\phi(\omega)f(\omega)d\omega, \tag{57}$$

by

$$\hat{\eta}_N(\phi, 1) = (2\pi)^d|N|^{-1}\sum_{j=1}^{N}\phi(\omega_j)\hat{f}_N(\omega_j), \tag{58}$$

where ϕ is a continuous function on $\mathbf{R}^d$ periodic in each dimension with period 2π, 1 stands for the identity function $\psi(x) = x$, and $\hat{f}_N$ is the periodogram smoothed by a weight function W_N. If the conditions of Theorem 1 on f and W_N are satisfied, then

$$\phi_N(\omega) = (2\pi)^d|N|^{-1}\sum_{j=1}^{N}W_N(\omega - \omega_j)\phi(\omega_j) = \phi(\omega) + o(1) \tag{59}$$

uniformly in $\omega \in \Pi^d$, and we can write:

$$\hat{\eta}_N(\phi, 1) = (2\pi)^d|N|^{-1}\sum_{j=1}^{N}\phi_N(\omega_j)I_N(\omega_j) \tag{60}$$

as a weighted average of the periodogram ordinates $I_N(\omega_j)$ with weight function ϕ_N of magnitude $(1, \ldots, 1)$. Thus, we have:

Theorem 3 *If $\{x(t),\ t \in \mathbf{Z}^d\}$ is stationary up to order $2r$ with mean zero and cumulants satisfying Eq. (35) for any r, and W_n satisfies the conditions of Theorem 1, then for any functions $\phi^{(1)}, \ldots, \phi^{(r)}$ continuous on $\mathbf{R}^d$ and periodic in each dimension with period 2π, we have*

$$E\hat{\eta}_N\left(\phi^{(i)}, 1\right) = \eta\left(\phi^{(i)}, 1\right) + o(1), \tag{61}$$

$$E\left\{\left(\hat{\eta}_N\left(\phi^{(i)}, 1\right) - E\hat{\eta}_N\left(\phi^{(i)}, 1\right)\right)^{2[r/2]}\right\} = O\left(|N|^{-[r/2]}\right), \tag{62}$$

$$\hat{\eta}_N\left(\phi^{(i)}, 1\right) \to \eta\left(\phi^{(i)}, 1\right), \ \ a.s., \tag{63}$$

$$cum\left\{\hat{\eta}_N\left(\phi^{(1)}, 1\right), \ldots, \hat{\eta}_N\left(\phi^{(r)}, 1\right)\right\} = O\left(|N|^{-r}\right), \tag{64}$$

and

$$|N|^{1/2}\left\{\hat{\eta}_N\left(\phi^{(s)}, 1\right) - E\hat{\eta}_N\left(\phi^{(s)}, 1\right)\right\}, s = 1, \ldots, r, \tag{65}$$

are asymptotically normal with mean zero and covariances,

$$(2\pi)^d \int_{\Pi^d} \phi^{(i)}(\omega)\left\{\phi^{(j)}(\omega) + \phi^{(j)}(-\omega)\right\} f(\omega)^2 d\omega$$

$$+ (2\pi)^d \int_{\Pi^d}\int_{\Pi^d} f^{(4)}(\omega, -\omega, \nu)\phi^{(i)}(\omega)\phi^{(j)}(\nu) d\omega d\nu. \tag{66}$$

The asymptotic normality is established by showing that the joint cumulants of (65) converge to those of a normal distribution, analogous to the approach of [8] for time series. Note that the results hold if we replace each $\phi^{(i)}$ by some $\phi_N^{(i)}$ with

$$\sup_{\omega}\left|\phi_N^{(i)}(\omega) - \phi^{(i)}(\omega)\right| \to 0, \text{ as } N \to \infty. \tag{67}$$

For the general case of estimating $\eta(\phi, \psi)$ by

$$\hat{\eta}_N(\phi, \psi) = (2\pi)^d |N|^{-1} \sum_{j=1}^{N} \phi(\omega_j)\, \psi\left(\hat{f}_N(\omega_j)\right), \tag{68}$$

where ψ is not necessarily the identity function, but has continuous second-order derivatives, we can write:

$$\psi\left(\hat{f}_N(\omega)\right) = \psi\left(\bar{f}_N(\omega)\right) + \psi'\left(\bar{f}_N(\omega)\right)\Delta\hat{f}_N(\omega)$$

$$+\tfrac{1}{2}\psi''\left(\tilde{f}_N(\omega)\right)\left(\Delta\hat{f}_N(\omega)\right)^2, \tag{69}$$

where $\bar{f}_N(\omega) = E\hat{f}_N(\omega), \Delta\hat{f}_N(\omega) = \hat{f}_N(\omega) - \bar{f}_N(\omega)$, and $\tilde{f}_N(\omega)$ is between $\hat{f}_N(\omega)$ and $\bar{f}_N(\omega)$. In this case, we can write:

$$\hat{\eta}_N(\phi, \psi) = \bar{\eta}_N(\phi, \psi) + \hat{\eta}_N\left(\phi\psi'(\bar{f}_N), 1\right) - E\hat{\eta}_N\left(\phi\psi'(\bar{f}_N), 1\right) + \gamma_N, \quad (70)$$

where

$$\bar{\eta}_N(\phi, \psi) = (2\pi)^d |N|^{-1} \sum_{j=1}^{N} \phi(\omega_j)\, \psi\left(\bar{f}_N(\omega_j)\right), \quad (71)$$

$$\gamma_N = \frac{1}{2}(2\pi)^d |N|^{-1} \sum_{j=1}^{N} \phi(\omega_j)\, \psi''\left(\tilde{f}_N(\omega_j)\right)\left(\Delta\hat{f}_N(\omega_j)\right)^2. \quad (72)$$

If the conditions in Theorem 1 are satisfied and W_N is of such magnitude M that

$$1 \leq M_s = \circ\left(N_s^{1/2}\right), \; s = 1, \ldots, d, \; as N \to \infty, \quad (73)$$

then we can show that

$$|N|^{1/2}\gamma_N \to 0, \text{ in prob., as } N \to \infty, \quad (74)$$

so that

$$|N|^{1/2}\left\{\eta_N(\phi\psi) - \bar{\eta}_N(\phi\psi)\right\}$$

has the same asymptotic normality as

$$|N|^{1/2}\left\{\hat{\eta}_N(\phi\psi'\left(\bar{f}_N\right), 1) - E\hat{\eta}_N(\phi\psi'\left(\bar{f}_N\right), 1)\right\},$$

whose asymptotic properties are available from Theorem 3.

We note that Taniguchi [33] considered the estimation of spectral integrals of the form (10) for time series, but his condition on $\psi(\cdot)$ rules out the inverse autocovariances. Bhansali [2] derived asymptotic properties of estimated inverse autocovariances for time series using both parametric and nonparametric approaches.

For the estimation of the spectrum, we can use a scale parameter window,

$$W_N(\omega) = (2\pi)^{-d} \sum_{t \in \mathbf{Z}^d} k(t/M) \exp\{-it\omega\}, \; \omega \in \mathbf{R}^d, \quad (75)$$

as in the time series case—*cf.* [5], [27], [30]—where the scale parameter $M \in \mathbf{R}^d$ depends on N in such a way that

$$M, \; \frac{N}{M} \to \infty, \text{ as } N \to \infty, \quad (76)$$

and the lag window generator $k(\cdot)$ is a function on $\mathbf{R}^d$ symmetric and square integrable with $|k(x)| \leq k(0) = 1, x \in \mathbf{R}^d$. The Fourier transform (in L^2 sense),

$$K(\theta) = (2\pi)^{-d} \int_{\mathbf{R}^d} k(x) \exp\{-ix \cdot \theta\} dx, \tag{77}$$

is known as a *spectral window generator.*

Under certain regularity conditions, we can show that W_N as given by Eq. (75) is of magnitude M and satisfies the conditions in Theorem 1, and obtain asymptotic expressions for the bias and covariances of the spectral estimates; details are given in [38].

Theorem 4 *Let $\{x(t),\ t \in \mathbf{Z}^d\}$ be stationary up to order $2r$ with mean zero and cumulants satisfying Eq. (35) for any r, and W_N be a scale parameter window with $k(\cdot)$ satisfying certain regularity conditions; then for any fixed $\omega^{(1)}, \ldots, \omega^{(r)} \in \Pi^d$,*

$$\left(\frac{|N|}{|M|}\right)^{1/2} \left\{\hat{f}_N\left(\omega^{(i)}\right) - E\hat{f}_N\left(\omega^{(i)}\right)\right\},\ i = 1, \ldots, r, \tag{78}$$

are asymptotically normal with mean zero and covariances,

$$f\left(\omega^{(i)}\right)^2 \left\{\eta\left(\omega^{(i)} + \omega^{(j)}\right) + \eta\left(\omega^{(i)} - \omega^{(j)}\right)\right\} \int_{\mathbf{R}^d} k(x)^2 dx, \tag{79}$$

where $\eta(\omega) = 1$ if $\omega_s = 0 \pmod{2\pi}$, $s = 1, \ldots, d$, and $\eta(\omega) = 0$ otherwise.

Theorem 5 *Let $\{x(t),\ t \in \mathbf{Z}^d\}$ be stationary up to order 2 with mean zero and autocovariances satisfying*

$$\sum_{j \in \mathbf{Z}^d} \|j\|^2 \, |R(j)| < \infty, \tag{80}$$

and W_N be a scale parameter window with $k(\cdot)$ satisfying certain regularity conditions. Then we have

$$E\hat{f}_N(\omega) - f(\omega)$$

$$= (2d)^{-1} \int_{\mathbf{R}^d} \|\theta\|^2 K(\theta) d\theta \sum_{s=1}^{d} M_s^{-2} f_{ss}^{(2)}(\omega) + \sum_{s=1}^{d} \left\{O\left(N_s^{-1}\right) + o\left(M_s^{-2}\right)\right\} \tag{81}$$

uniformly in $\omega \in \Pi^d$, *where*

$$f_{ss}^{(2)}(\omega) = -(2\pi)^{-d} \sum_{j \in \mathbf{Z}^d} j_s^2 R(j) \exp\{-ij \cdot \omega\}, \; s = 1, \ldots, d \tag{82}$$

are second-order partial derivatives of the spectrum.

This is an extension of the time series result of [27].

Following the idea of Priestley [30] for time series, we can obtain a mean square error criterion,

$$C(K) = \int_{\mathbf{R}^d} K(\theta)^2 d\theta \left(\int_{\mathbf{R}^d} \|\theta\|^2 K(\theta) d\theta \right)^{d/2}, \tag{83}$$

for the choice of the spectral window generator K, and prove that the Bartlett–Priestley spectral window generator,

$$K_{\mathrm{BP}}(\theta) = \begin{cases} \frac{d(d+2)\Gamma(d/2)}{4\Gamma(1/2)^d \pi^d} \left\{1 - \pi^{-2}\|\theta\|^2\right\}, & \text{if } \|\theta\| \leq \pi, \\ 0, & \text{otherwise}, \end{cases} \tag{84}$$

minimizes the criterion $C(K)$ among all nonnegative spectral window generators K, where $\Gamma(\cdot)$ is the Gamma function; *cf.* [38].

The Bartlett–Priestley lag window generator is given by

$$k_{\mathrm{BP}}(x) = \frac{d(d+2)\Gamma(d/2) J_{d/2+1}(\pi\|x\|)}{2^{1-d/2}(\pi\|x\|)^{d/2+1}}, \; x \in \mathbf{R}^d, \tag{85}$$

where $\Gamma(\cdot)$ is the Gamma function and $J_r(\cdot)$ is the r-th order Bessel function of the first kind; *cf.* ([6] p. 46) and the references cited there for the Fourier transform of $\phi(\theta) = \left(1 - \|\theta\|^2\right)^a, \|\theta\| \leq 1$.

Alternatively, we can use a multiplicative lag window generator,

$$k(x) = \prod_{s=1}^{d} k_1(x_s), x = (x_1, \ldots, x_d) \in \mathbf{R}^d, \tag{86}$$

as suggested in [29], where k_1 is a lag window generator for time series, *e.g.*, the Parzen lag window generator [27].

Lag window generators with bounded support offer computational advantages; otherwise, we can use a modified scale parameter window,

$$W_N(\omega) = (2\pi)^{-d} \sum_{j=-Q}^{T} \left\{ \prod_{s=1}^{d} \left(1 - j_s T_s^{-1}\right) \right\} k(\frac{j}{M}) \exp\{-ij \cdot \omega\}, \; \omega \in \mathbf{R}^d, \tag{87}$$

where $T, \ Q \in \mathbf{Z}^d$ with

$$T_s = \left[\frac{N_s}{2}\right], \ Q_s = \left[\frac{(N_s - 1)}{2}\right], \ s = 1, \ldots, d. \tag{88}$$

The modification is to ensure that W_N is nonnegative when K is, and that its discrete Fourier transform is easily obtainable. The previous results for scale parameter windows remain true; *cf.* [38].

The spectral estimates $\hat{f}_N(\omega_j)$ at the discrete frequencies ω_j are the convolution of $I_N(\omega_j)$ with $W_N(\omega_j)$, and thus can be calculated using fast Hartley transforms, which have certain advantages over fast Fourier transforms; *cf.*, [17] [19]. A FORTRAN program for $2-d$ FHT is available in [38]. The original description of Hartley transform may be found in [17].

The preceding results can be extended to the estimation of higher-order cumulant spectra of real-valued random fields on $\mathbf{Z}^d$; details are available in a Ph.D. thesis of [38].

3 Markov Modelling of Random Fields

The Markov property for stationary random fields has been studied by many authors; *cf.* [1], [10], [18] [20], [24], [25], [26], [32], [36]. It usually is assumed that the random field is Gaussian, and the Markov property often is defined in terms of conditional probabilities. The following definition (*cf.* [18]) is concerned only with the second order property of the random field.

A stationary random field $\{x(t), \ t \in \mathbf{Z}^d\}$ is said to be Markov with neighbor set S, if

$$\text{Proj}\,\{x(t) \mid x(t+s), \ s \neq 0\} = \text{Proj}\,\{x(t) \mid x(t+s), \ s \in S\}, \ t \in \mathbf{Z}^d, \tag{89}$$

where S is a finite and symmetric subset of $\mathbf{Z}^d \backslash \{0\}$, and *Proj* denotes Hilbert space linear projection.

The Markov property also can be characterized equivalently by a parametric model, or in terms of the spectral distribution of the random field; *cf.* [18] and [32].

Theorem 6 *For any stationary random field $\{x(t), \ t \in \mathbf{Z}^d\}$ and any finite and symmetric subset S of $\mathbf{Z}^d \backslash \{0\}$, the following statements are equivalent:*

(a) $\{x(t), \ t \in \mathbf{Z}^d\}$ is Markov with neighbor set S;

(b) $\{x(t), \ t \in \mathbf{Z}^d\}$ satisfies

$$x(t) + \sum_{s \in S} a(s) x(t+s) = \epsilon(t), \ t \in \mathbf{Z}^d \tag{90}$$

for certain constants $\{a(s), s \in S\}$ *and some stationary random field* $\{\epsilon(t), t \in \mathbf{Z}^d\}$ *with autocovariances,*

$$R_\epsilon(u) = \begin{cases} \sigma^2, & \text{if } u = 0, \\ \sigma^2 a(u), & \text{if } u \in S, \\ 0, & \text{otherwise,} \end{cases} \quad u \in \mathbf{Z}^d, \tag{91}$$

where $\sigma^2 = R_\epsilon(0)$*;*

(c) the Radon–Nikodym derivative of the absolutely continuous part of the spectral distribution of $\{x(t),\ t \in \mathbf{Z}^d\}$ *is of the form,*

$$f(\omega) = (2\pi)^{-d}\sigma^2 \left\{1 + \sum_{s \in S} a(s)\exp\{is \cdot \omega\}\right\}^{-1}, \quad \omega \in \Pi^d, \tag{92}$$

for some $\sigma \geq 0$ *and certain constants* $\{a(s), s \in S\}$*, and the singular part concentrates on*

$$\left\{\omega \in \Pi^d; 1 + \sum_{s \in S} a(s)\exp\{is \cdot \omega\} = 0\right\}. \tag{93}$$

The difference equation, Eq. (90), together with the autocovariance specification Eq. (91), is called a Markov model with neighbor set S.

Let $\{x(t),\ t \in \mathbf{Z}^d\}$ be a stationary random field, and $f(\omega)$ the Radon–Nikodym derivative of the absolutely continuous part of its spectral distribution F. If $f(\omega)^{-1} \in L^2(\Pi^d, F)$, then

$$Ri(s) = (2\pi)^{-2d} \int_{\Pi^d} f(\omega)^{-1} \exp\{is \cdot \omega\} d\omega, \quad s \in \mathbf{Z}^d \tag{94}$$

are known as the inverse autocovariances, and

$$ri(s) = \frac{Ri(s)}{Ri(0)}, s \in \mathbf{Z}^d, \tag{95}$$

the inverse autocorrelations.

It follows from Theorem 6 that a stationary random field $\{x(t),\ t \in \mathbf{Z}^d\}$ with $f(\omega)^{-1} \in L^2(\Pi^d, F)$ is Markov with neighbor set S iff its inverse autocovariances vanish outside $S \cup \{0\}$, and the parameters $\{a(s), s \in S\}$ of the corresponding Markov model are identical to the inverse autocorrelations. This fact enables effective analysis of Markov models using the frequency domain approach.

The parameters of a Markov model usually are estimated by maximizing a Whittle-type criterion (*cf.*, [10], [13], [16], [20], and [35]).

$$C_W(\theta) = -(2\pi)^{-d} \int_{\Pi^d} \left\{\log\left(f(\omega, \theta)\right) + \hat{f}(\omega)/f(\omega, \theta)\right\} d\omega, \tag{96}$$

where $\hat{f}(\omega)$ is a nonparametric estimator of the spectrum, e.g., the tapered periodogram ([13]). It originates from an approximation to the Gaussian log-likelihood, but without the Gaussian assumption it always can be interpreted as a measure of the closeness of $\hat{f}(\omega)$ to $f(\omega, \theta)$. This criterion involves θ in a nonlinear fashion, and it thus is computationally intensive to maximize unless there are few parameters. Lim and Malik [22] developed an iterative algorithm, but its convergence property has not been established.

Alternatively, we can obtain estimates of the parameters of a Markov model from the estimated inverse autocovariances,

$$\hat{Ri}(s) = (2\pi)^{-d}|N|^{-1}\sum_{j=1}^{N} \hat{f}_N(\omega_j)^{-1}\cos(s\cdot\omega_j),\ s\in S\cup\{0\}, \tag{97}$$

where $\hat{f}_N$ is the periodogram smoothed with a window W_N, and as special cases of the estimation of spectral integrals, we have:

Theorem 7 *If $\{x(t),\ t\in\mathbf{Z}^d\}$ is stationary up to order $2r$ with mean zero and cumulants satisfying Eq. (35) for any $r > 0, f(\omega)$ is strictly positive, W_N satisfies Eq. (14) for any fixed $t\in\mathbf{Z}^d$, and is of magnitude M satisfying Eq. (73), then for any $r>0$, and any fixed $s^{(1)},\ldots,s^{(r)}\in\mathbf{Z}^d$,*

$$|N|^{1/2}\left\{\hat{Ri}\left(s^{(i)}\right)-\bar{Ri}\left(s^{(i)}\right)\right\},\ i=1,\ldots,r \tag{98}$$

are asymptotically normal with mean zero and covariances,

$$2(2\pi)^{-3d}\int_{\Pi^d}\cos\left(s^{(i)}\cdot\omega\right)\cos\left(s^{(j)}\cdot\omega\right)f(\omega)^{-2}d\omega$$

$$+(2\pi)^{-3d}\int_{\Pi^d}\int_{\Pi^d}\frac{f^{(4)}(\omega,-\omega,\nu)}{f(\omega)^2f(\nu)^2}\cos\left(s^{(i)}\cdot\omega\right)\cos\left(s^{(j)}\cdot\nu\right)d\omega d\nu, \tag{99}$$

and for each $i=1,\ldots,r$,

$$\hat{Ri}(s^{(i)})\to Ri(s^{(i)}),\ a.s.,\ as\ N\to\infty, \tag{100}$$

where

$$\bar{Ri}(s) = (2\pi)^{-d}|N|^{-1}\sum_{j=1}^{N}\left(E\hat{f}_N(\omega_j)\right)^{-1}\cos\left(s\cdot\omega_j\right),\ s\in\mathbf{Z}^d. \tag{101}$$

s	a(s)	$\hat{a}(s)$	$\tilde{a}(s)$
(-1, 0)	0.120	0.120	0.120
(1, 1)	-0.080	-0.079	-0.078
(0, 1)	0.140	0.142	0.141
(-1, 1)	0.100	0.097	0.096

Table 1: True and estimated parameters of a Markov model

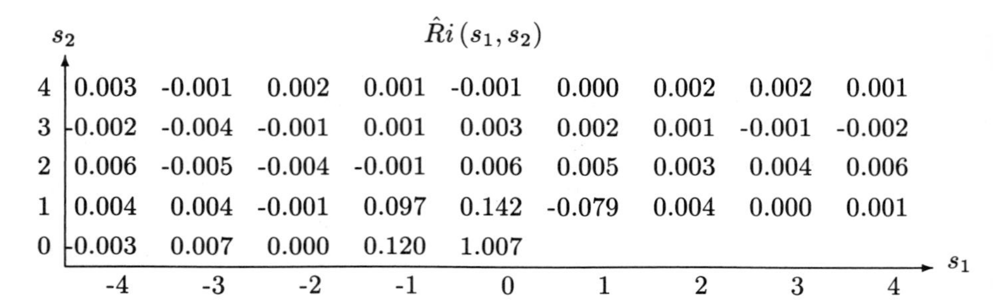

$\hat{Ri}\,(s_1, s_2)$

s_2 \ s_1	-4	-3	-2	-1	0	1	2	3	4
4	0.003	-0.001	0.002	0.001	-0.001	0.000	0.002	0.002	0.001
3	-0.002	-0.004	-0.001	0.001	0.003	0.002	0.001	-0.001	-0.002
2	0.006	-0.005	-0.004	-0.001	0.006	0.005	0.003	0.004	0.006
1	0.004	0.004	-0.001	0.097	0.142	-0.079	0.004	0.000	0.001
0	-0.003	0.007	0.000	0.120	1.007				

Figure 1: Estimates of $Ri\,(s_1, s_2)$ using Bartlett–Priestley window.

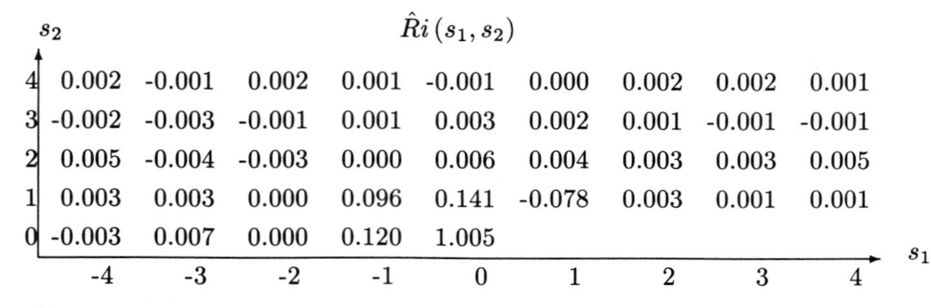

$\hat{Ri}\,(s_1, s_2)$

s_2 \ s_1	-4	-3	-2	-1	0	1	2	3	4
4	0.002	-0.001	0.002	0.001	-0.001	0.000	0.002	0.002	0.001
3	-0.002	-0.003	-0.001	0.001	0.003	0.002	0.001	-0.001	-0.001
2	0.005	-0.004	-0.003	0.000	0.006	0.004	0.003	0.003	0.005
1	0.003	0.003	0.000	0.096	0.141	-0.078	0.003	0.001	0.001
0	-0.003	0.007	0.000	0.120	1.005				

Figure 2: Estimates of $Ri\,(s_1, s_2)$ using Parzen window.

s	a(s)	$\hat{a}(s)$	$\tilde{a}(s)$
(-1, 0)	0.040	0.042	0.041
(-2, 0)	0.030	0.028	0.027
(2, 1)	-0.030	-0.026	-0.025
(1, 1)	-0.020	-0.021	-0.019
(0, 1)	0.120	0.123	0.122
(-1, 1)	-0.080	-0.080	-0.079
(-2, 1)	0.020	0.019	0.018
(0, 2)	0.060	0.064	0.061
(-1, 2)	0.050	0.047	0.043

Table 2: True and estimated parameters of another Markov model.

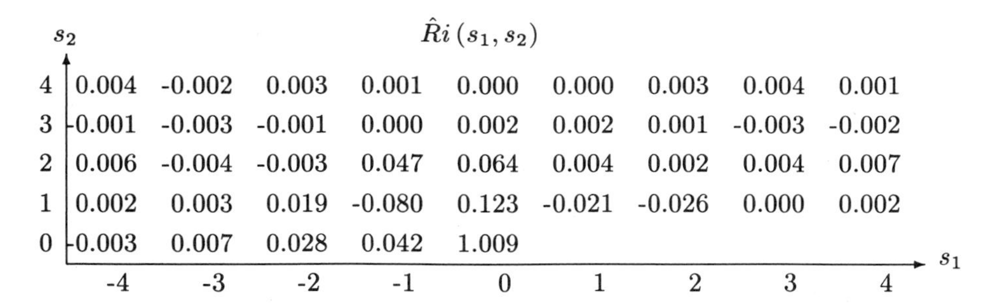

$\hat{Ri}\,(s_1, s_2)$

s_2 \ s_1	-4	-3	-2	-1	0	1	2	3	4
4	0.004	-0.002	0.003	0.001	0.000	0.000	0.003	0.004	0.001
3	-0.001	-0.003	-0.001	0.000	0.002	0.002	0.001	-0.003	-0.002
2	0.006	-0.004	-0.003	0.047	0.064	0.004	0.002	0.004	0.007
1	0.002	0.003	0.019	-0.080	0.123	-0.021	-0.026	0.000	0.002
0	-0.003	0.007	0.028	0.042	1.009				

Figure 3: Estimates of $Ri\,(s_1, s_2)$ using Bartlett–Priestley window.

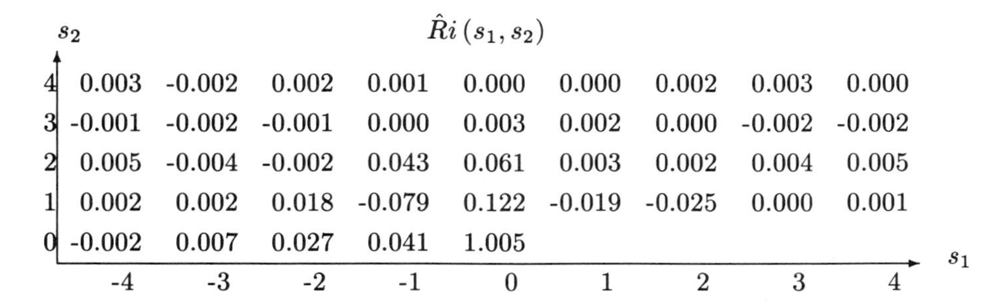

$\hat{Ri}\,(s_1, s_2)$

s_2 \ s_1	-4	-3	-2	-1	0	1	2	3	4
4	0.003	-0.002	0.002	0.001	0.000	0.000	0.002	0.003	0.000
3	-0.001	-0.002	-0.001	0.000	0.003	0.002	0.000	-0.002	-0.002
2	0.005	-0.004	-0.002	0.043	0.061	0.003	0.002	0.004	0.005
1	0.002	0.002	0.018	-0.079	0.122	-0.019	-0.025	0.000	0.001
0	-0.002	0.007	0.027	0.041	1.005				

Figure 4: Estimates of $Ri\,(s_1, s_2)$ using Parzen window.

The estimation scheme (97) is nonparametric in the sense that the estimates are not obtained by numerically maximizing a criterion involving them as parameters, and they provide sensible estimates of the inverse autocovariances even without the assumption of a Markov model. If the random field is, in fact, Markov, they provide consistent and asymptotically normal estimates of the model parameters.

For a Markov random field with $f(\omega)^{-1} \in L^2(\Pi^d, F)$, its neighbor set S simply is $\{s; Ri(s) \neq 0, s \neq 0\}$. Thus, the estimated inverse autocovariances not only provide sensible estimates of the model parameters, but also can be expected to indicate the neighbor set S. In this case, the $\hat{Ri}(s)$'s are more effective than in the time series case, since we are looking only for the sites at which they differ significantly from zero instead of what pattern they exhibit, exponential, sinusoidal or a mixture of both (*cf.* [3]). Moreover, if the true model is not Markov, then the inverse autocovariances should not decay rapidly at sites away from the origin, and this should be reflected in their estimates.

It would be desirable to have the asymptotic normality of

$$|N|^{1/2} \left\{\hat{Ri}\left(s^{(i)}\right) - Ri\left(s^{(i)}\right)\right\}, \; i = 1, \ldots, r, \tag{102}$$

so that we formally could test hypotheses of the form

$$Ri\left(s^{(i)}\right) = 0, \; i = 1, \ldots, r, \tag{103}$$

for the identification of the neighbor set of a Markov model. This can be achieved for the case $d = 1$, since we always can choose a scale parameter M for the window in such a way that

$$N^{1/4} \ll M \ll N^{1/2}, \text{ as } N \to \infty, \tag{104}$$

so that

$$N^{1/2}\left(\bar{Ri}\left(s^{(i)}\right) - Ri\left(s^{(i)}\right)\right) = O(1)N^{1/2}\sup_{\omega}\left|E\hat{f}_N(\omega) - f(\omega)\right| \tag{105}$$

$$= O(1)N^{1/2}\left(M^{-2} + N^{-1}\right) \to 0, \; as \; N \to \infty, \tag{106}$$

but when $d \geq 2$, it is impossible for the bias of $\hat{f}_N(\omega)$ to be of order $o\left(|N|^{-1/2}\right)$ due to the effect of the increased dimensionality. The smoothing of the periodogram improves its higher-order cumulant properties, but the bias is increased by $O\left(M_1^{-2} + \ldots + M_d^{-2}\right)$, and this leaves room for improvement.

We simulated several Markov random fields on $\mathbf{Z}^2$ using the algorithm of Chellappa [10], and estimated the inverse autocovariances. Two examples are given here. The sample size is (256, 256), and the variance of $\{\epsilon(t)\}$ is known to be 1. In Tables 1 and 2, $\{a(s)\}$ denote the true parameters, $\{\hat{a}(s)\}$ are parameters estimated using the Bartlett–Priestley window with $M = (10, 10)$ for the first model and $M = (12, 12)$ for the second, and $\tilde{a}(s)$ estimated using the Parzen window with $M = (16, 16)$ for both cases. In Figures 1–4, we give at each site the estimated inverse autocovariance using each of the windows. It seems that the estimates are reasonably accurate, and they provide a clear indication of the neighbor set of the corresponding Markov model.

4 Classification of Textures

Texture classification is an important subject in image processing. The problem is to classify a texture image to one of several classes, provided that a number of realizations is available for each class as training samples. Due to the size of digitized texture data, it is usual practice to extract features from the texture, and classify on the basis of a feature vector of manageable size using standard techniques from multivariate analysis; *cf.* [11], [12], [34].

Alternatively, we can classify the textures directly as realizations of stationary random fields using their estimated spectra. This avoids the subjective choice of feature vectors, and the convergence property of the probability of classification can be studied using the asymptotic properties of the spectral estimates.

Suppose that there are n classes of stationary random fields on $\mathbf{Z}^d$ characterized by their second-order spectra $\{f^{(i)}, i = 1, \ldots, n\}$, in the sense that a stationary random field on $\mathbf{Z}^d$ belongs to some class, $i = 1, \ldots, n$, iff it has spectrum $f^{(i)}$. Given a finite realization $\{x(t),\ t = 1, \ldots, N\}$ of a stationary random field $\{x(t),\ t \in \mathbf{Z}^d\}$ from one of the classes, we wish to decide to which class it belongs.

We use the criterion,

$$c(X_N, i) = -|N|^{-1} \sum_{t=1}^{N} \left\{ \log\left(f^{(i)}(\omega_t)\right) + \frac{I_N(\omega_t)}{f^{(i)}(\omega_t)} \right\}, \ i = 1, \ldots, n, \quad (107)$$

for the classification. In other words, we assign X_N to class i, iff

$$c(X_N, i) > c(X_N, j), \ j \neq i. \quad (108)$$

The preceding classification rule is equivalent to maximizing an approximate log-likelihood function if the random fields are Gaussian, as proved by Yuan and Subba Rao in [39]. It was used by Dargahi-Noubary and Laycock [14] for the discrimination analysis of seismic records modelled as Gaussian time series.

In the general case where $\{x(t),\ t \in \mathbf{Z}^d\}$ is not necessarily Gaussian, we still can give it a reasonable interpretation. For any spectra f and g, we can define the discrepancy of f from g as

$$d^{(0)}(f,g) = |N|^{-1} \sum_{j=1}^{N} \left\{ \frac{f(\omega_j)}{g(\omega_j)} - \log\left(\frac{f(\omega_j)}{g(\omega_j)}\right) - 1 \right\}, \tag{109}$$

and the preceding classification rule is to minimize the discrepancy of the periodogram I_N from the spectrum $f^{(i)}$.

The discrepancy also can be defined in one of the following ways,

$$d^{(1)}(f,g) = |N|^{-1} \sum_{j=1}^{N} \left(f(\omega_j) - g(\omega_j)\right)^2, \tag{110}$$

$$d^{(2)}(f,g) = |N|^{-1} \sum_{j=1}^{N} \left(\log\left(f(\omega_j)\right) - \log\left(g(\omega_j)\right)\right)^2, \tag{111}$$

$$d^{(3)}(f,g) = |N|^{-1} \sum_{j=1}^{N} \left(\frac{f(\omega_j) - g(\omega_j)}{g(\omega_j)}\right)^2, \tag{112}$$

$$d^{(4)}(f,g) = |N|^{-1} \sum_{j=1}^{N} \left(\frac{f(\omega_j)}{g(\omega_j)} + \frac{g(\omega_j)}{f(\omega_j)} - 2\right), \tag{113}$$

resulting in alternative classification rules, but the periodogram has to be smoothed with a window for improved higher-order cumulant properties.

The effectiveness of a classification rule is characterized by its probability of misclassification; but even in the simple case of two classes of Gaussian random variables, this is hardly obtainable for the sample maximum likelihood criterion; *cf.*, [23]. It is even more difficult when we classify random fields using finite samples. However, we can obtain an upper bound for the probability of misclassification, which converges to zero as the sample size tends to infinity.

First, we consider the theoretical case where the spectra $\{f^{(i)}\}$ are known exactly, and the criterion $c(X_N, i)$ is used for the classification.

Theorem 8 *If $\{x(t),\ t \in \mathbf{Z}^d\}$ is stationary up to some order $2r$ with mean zero and cumulants satisfying Eq. (35), then the probability of misclassification using $c(X_N, i)$ is $O\left(|N|^{-[r/2]}\right)$ as $N \to \infty$.*

Proof. If X_N comes from class i, then for any $j \neq i$, we have

$$E\left(c\left(X_N, i\right) - c\left(X_N, j\right)\right) = g\left(f^{(i)}, f^{(j)}\right) + \circ(1), \tag{114}$$

where

$$g\left(f^{(i)}, f^{(j)}\right) = (2\pi)^{-d} \int_{\mathbf{Z}^d} \left\{ \frac{f^{(i)}\left(\omega_j\right)}{f^{(j)}\left(\omega_j\right)} - \log\left(\frac{f^{(i)}\left(\omega_j\right)}{f^{(j)}\left(\omega_j\right)} \right) - 1 \right\} d\omega > 0. \tag{115}$$

Thus, from Corollary 2 and the Markov inequality, we have for any $p \leq r/2$,

$$\begin{aligned} & pr\left\{c\left(X_N, i\right) \leq c\left(X_N, j\right)\right\} \\ & \leq \frac{E\left(\left|c\left(X_N, j\right) - c\left(X_N, i\right) - Ec\left(X_N, j\right) + Ec\left(X_N, i\right)\right|^{2p}\right)}{\left(Ec\left(X_N, i\right) - Ec\left(X_N, j\right)\right)^{2p}} \\ & = O(1) g\left(f^{(i)}, f^{(j)}\right)^{-2p} |N|^{-p}. \end{aligned} \tag{116}$$

It follows that the probability of misclassification is of order $O\left(|N|^{-[r/2]}\right)$ as $N \to \infty$.

For the practical case where the spectra $\{f^{(i)}\}$ are unknown, we estimate each $f^{(i)}$ from a sample $\{x^{(i)}(t), t = 1, \ldots, N\}$ belonging to class i by smoothing its periodogram $I_N^{(i)}$ with a window W_N, and use the sample criterion,

$$\hat{c}\left(X_N, i\right) = -|N|^{-1} \sum_{t=1}^{N} \left\{ \log\left(\hat{f}_N^{(i)}\left(\omega_t\right)\right) + \frac{I_N\left(\omega_t\right)}{\hat{f}_N^{(i)}\left(\omega_t\right)} \right\}, \quad i = 1, \ldots, n, \tag{117}$$

for the classification.

Theorem 9 *If $\{x(t)\}$ and $\left\{x^{(i)}(t)\right\}, i = 1, \ldots, n$, are stationary up to some order $2r$ with mean zero and cumulants satisfying Eq. (35), $\{x(t)\}$ is independent of $\left\{x^{(i)}(t)\right\}, i = 1, \ldots, n$, W_N is of magnitude M and satisfies Eq. (14) for any fixed $t \in \mathbf{Z}^d$, then the probability of misclassification using the sample criterion $\hat{c}(X_N, i)$ is of order $O(1)\,|N|\left(|M| / |N|\right)^{[r/2]}$ as $N \to \infty$.*

c1	c2	c3	c4	c5	c6	c7
44444444	22222222	33333333	44444444	55555555	66666666	77777777
44444414	22222222	33333333	44444444	55555555	66666666	77777777
44441111	22222222	33333333	44444444	55555555	66666666	77777777
11111111	22222222	33333333	44444444	55555555	66666666	77777777
1111*111	2222*222	3333*333	4444*444	5555*555	6666*666	7777*777
11111111	22222222	33333333	44444444	55555555	66666666	77777777
11112221	22222222	33333333	44444444	55555555	66666666	77777777
11112111	12112212	62222226	44144411	66666666	66666666	22222222
m/c:23	m/c: 4	m/c: 8	m/c: 3	m/c: 8	m/c: 0	m/c: 8

Table 3: Classification results using $\hat{c}\left(X_N, i\right)$.

This is proved by showing that for any $\epsilon > 0$ and any $p \leq r/2$,

$$pr\left\{\left|\hat{c}\left(X_N, i\right) - c\left(X_N, i\right)\right| \geq \epsilon\right\} = O\left(\frac{|N|^{1-p}}{|M|^p}\right), i = 1, \ldots, n \tag{118}$$

from the cumulant properties of the spectral estimates; details are given in [39].

Similar results can be obtained for the alternative classification rules based on the definitions from Eqs. (110)–(113), of the discrepancy of one spectrum from another.

We have considered the limiting behavior of the probability of misclassification as the sample size tends to infinity, for both the theoretical case where the spectra of the classes are known, and the practical case where they are estimated from independent samples. The basic assumption on the random fields is that they have such short memory that their cumulant functions up to certain order satisfy the summability condition, as seen in Eq. (35). If we have the additional knowledge that the random fields are Gaussian, then the cumulant condition reduces to

$$\sum_{s \in \mathbf{Z}^d} \|s\|^d \, |R(s)| < \infty, \tag{119}$$

where $R(\cdot)$ is the autocovariance function, and we have the stronger result that the probability of misclassification tends to zero faster than any negative power of $|N|$ as $N \to \infty$, provided that the magnitude M of W_N satisfies Eq. (55) for the sample classification rule.

We applied the classification criterion $\hat{c}\left(X_N, i\right)$ to seven sets of digitized texture data from [9], each of size 512×512 with light intensity values in the range 0–255. Each data set is divided into 8×8 subsamples each of size 64×64, and a subsample from each class is used to estimate its spectrum,

c1	c2	c3	c4	c5	c6	c7
11111111	21112212	77333333	44444444	55555555	66666363	77777777
11111111	11112111	33733373	44444444	55555555	66666636	77777777
11111111	11112112	33333333	44444444	55555555	66666666	33777777
11111111	11122122	33733337	44444444	55555545	66666666	77777777
1111*111	2111*221	3773*337	4444*444	5555*555	6666*666	77777*77
11111111	11112122	37333337	44444444	55555555	66666666	77777777
11111111	11212222	33333333	44444444	55545545	66666666	77777777
11111111	11111111	66266666	24446444	22626626	66626666	11111111
m/c: 0	m/c:41	m/c:19	m/c: 2	m/c:11	m/c: 4	m/c:10

Table 4: Classification results using $d^{(1)}(f,g)$.

the estimated spectra are shown in Fig. 5. The rest 63×7 subsamples were classified one by one, and the results were as shown in Table 3. The classes are labelled $c1, \ldots, c7$, the numbers in the table indicate both the subsamples and the classes to which they were classified, the asterisks represent the subsamples from which the spectra were estimated, and the last row gives the number of misclassifications (m/c) for each class.

The classification criterion worked very well for the last six data sets with all the subsamples above the last row classified correctly, while for the first data set, about one-third of the subsamples were misclassified. Upon close examination of the texture data, we found that the last two lines of each data set are all zeros, and the last data set has 26 lines of zeros. This results in the distortion of the spectra estimated from the last row of subsamples, and accounts for the classification errors with subsamples in the bottom row.

We also tried the alternative classification rules based on the definitions from Eqs. (110)–(113) for the discrepancy of one spectrum from another, and the results were similar to the preceding except that the square difference rule using $d^{(1)}(f,g)$ did better with the first class but worse with the second, as shown in Table 4.

Acknowledgements

This work is supported partially by an SERC grant under the Complex Stochastic Systems Initiative.

We would like to thank Dr. R. Chellappa for providing the texture image data digitized by the Signal and Image Processing Unit, University of Southern California.

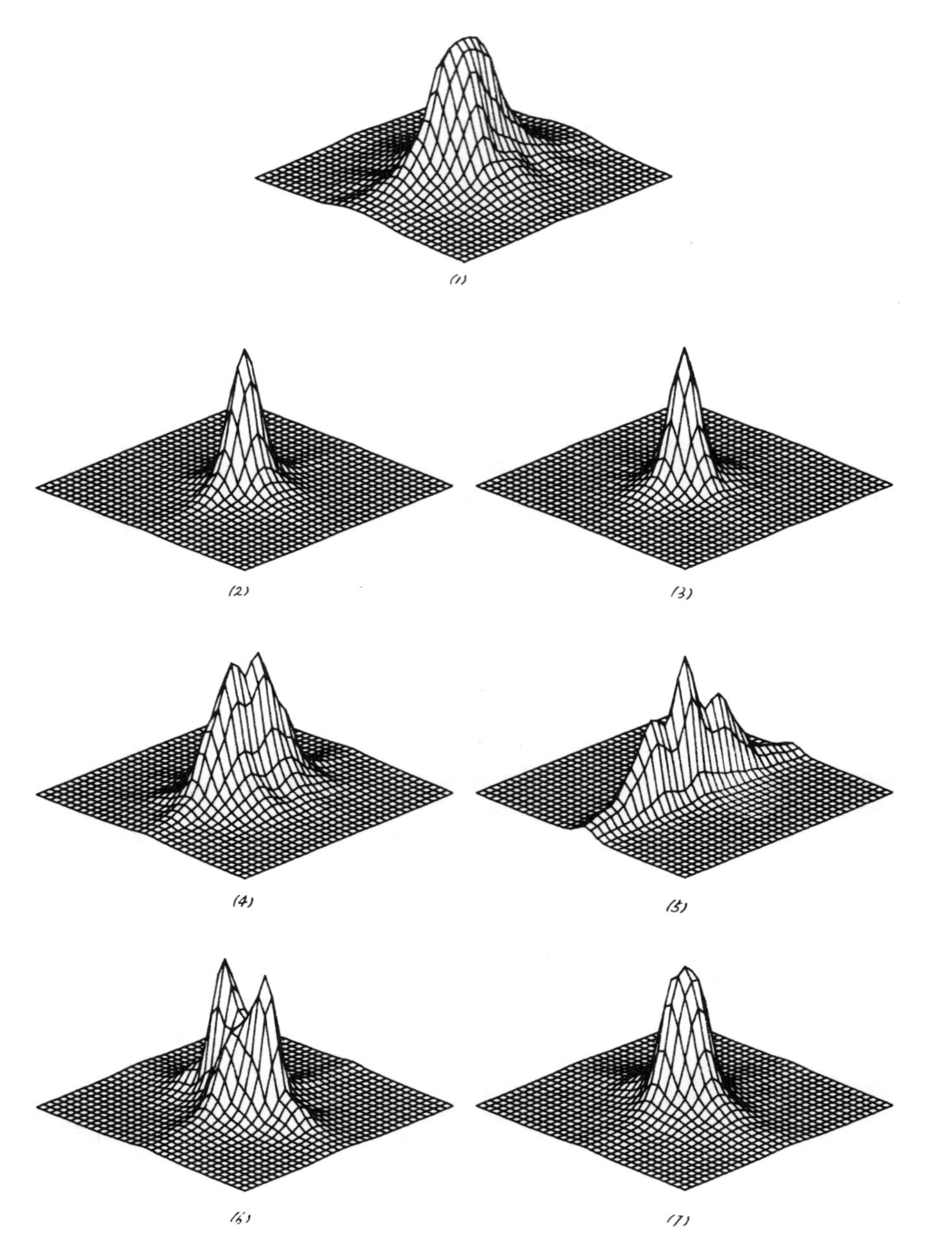

Figure 5: Estimated spectra for seven classes of textures. $N = (64, 64)$, $M = (8, 8)$.

Bibliography

[1] J. Besag, Spatial Interaction and the Statistical Analysis of Lattice Systems, *J. R. Statist. Soc. B*, Vol. 36, pp. 192-236, 1974.

[2] R. J. Bhansali, Autoregressive and Window Estimates of the Inverse Correlation Function, *Biometrika*, Vol. 67, No. 3, pp. 551-66, 1980.

[3] G. E. P. Box, and G. M. Jenkins, "Time Series Analysis, Forecasting and Control," Holden-Day, San Francisco, 1970.

[4] R. N. Bracewell, The Fast Hartley Transform, *Proc. IEEE.*, Vol. 72, No. 8, pp. 1010-1018, 1984.

[5] D. R. Brillinger, Asymptotic Properties of Spectral Estimates of the Second Order, *Biometrika*, Vol. 56, pp. 375-390, 1969.

[6] D. R. Brillinger, The Frequency Analysis of Relations Between Stationary Spatial Series, *Proc. Twelfth Bien. Sem. Canadian Math. Congr*, R. Pyke (ed.), pp. 39-81, Montreal: Can. Math. Congr., 1970.

[7] D. R. Brillinger, "Time Series: Data Analysis and Theory", Holt, Rinehart and Winston, New York, 1975.

[8] D. R. Brillinger, and M. Rosenblatt, Asymptotic Theory of kth Order Spectra, In "Spectral Analysis of Time Series," B. Harris (ed.), pp. 153-188, Wiley, New York, 1967.

[9] P. Brodatz, "Textures: A Photographic Album for Artists and Designers," Dover Publications, New York, 1966.

[10] R. Chellappa, Two-Dimensional Discrete Gaussian Markov Random Models For Image Processing, in Progress in Pattern Recognition 2, L.N. Kanal and A. Rosenfeld (eds.), pp. 79-112, 1985.

[11] R. Chellappa, and S. Chatterjee, Classification of Textures Using Gaussian Markov Random Fields. *IEEE Trans. on Acoustics, Speech and Image Processing*, ASSP-33, pp. 959-963, 1985.

[12] C. H. Chen, A Study of Texture Classification Using Spectral Features, In *Proc. Intl. Conf. Pattern Recognition*, Munich, West Germany, 1982.

[13] R. Dahlhaus, and H. Künsch, Edge Effects and Efficient Parameter Estimation for Stationary Random Fields, *Biometrika*, Vol. 74, No. 4, pp. 877-82, 1987.

[14] G. R. Dargahi-Noubary, and P. J. Laycock, Spectral Ratio Discriminants and Information Theory, *J. Time Ser. Anal.*, Vol. 2, No. 2, pp. 71-86, 1981.

[15] U. Grenander, and M. Rosenblatt, Some Problems in Estimating the Spectrum of a Time Series, *Proc. 3rd Berkeley Symp. on Statist. and Prob.*, Univ. of California, Berkeley, 1957.

[16] X. Guyon, Parameter Estimation for a Stationary Process on a *d*-dimensional Lattice, *Biometrika*, Vol. 69, pp. 95-105, 1982.

[17] R. V. L. Hartley, A More Symmetrical Fourier Analysis Applied to Transmission Problems, *Proceedings of the IRE*, Vol. 30, p. 144, 1942.

[18] Zepei Jiang (Chiang, Tse-pei), On Markov Models of Random Fields, *Acta Mathematicae Applicatae Sinica*, Vol. 3, No. 4, pp. 328-341, 1987.

[19] R. Kumaresan, and P. K. Gupta, Vector-Radix Algorithm for a 2-D Discrete Hartley Transform, *Proc. IEEE*, Vol. 74, No. 5, pp. 755-757, 1986.

[20] H. Künsch, Thermodynamics and Statistical Analysis of Gaussian Random Fields, Z. Wahr. verw. Geb., Vol. 58, pp. 407-21, 1981.

[21] V. P. Leonov, and A. N. Shiryaev, On A Method of Calculation of Semi-Invariants, Theor. Prob. Appl. 4, pp. 319-329, 1959.

[22] J. S. Lim, and N. A. Malik, A New Algorithm for Two Dimensional Maximum Entropy Power Spectrum Estimation, *IEEE Trans. Acoustics, Speech and Signal Processing*, ASSP-29, No. 3, pp. 401-13, 1981.

[23] K. V. Mardia, J. T. Kent, and J. Bibby, Multivariate Analysis, Academic Press, London, 1979.

[24] K. V. Mardia, Multi-Dimensional Multivariate Gaussian Markov Random Fields with Applications to Image Processing, *J. Multivariate Anal.*, Vol. 24, pp. 265-284, 1988.

[25] P. A. P. Moran, A Gaussian Markov Process On a Square Lattice, *J. Appl. Prob.*, Vol. 10, pp. 54-62, 1973.

[26] P. A. P. Moran, Necessary Conditions for Markovian Processes On a Lattice, *J. Appl. Prob.*, Vol. 10, pp. 605-612, 1973.

[27] E. Parzen, On Consistent Estimates of the Spectrum of a Stationary Time Series, *Ann. Math. Statist.*, Vol. 28, pp. 329-348, 1957.

[28] E. Parzen, Time Series Papers, Holden-Day, San Francisco, California, 1967.

[29] M. B. Priestley, Analysis of Two-Dimensional Stationary Processes with Discontinuous Spectra, *Biometrika*, Vol. 51, pp. 195–217, 1964.

[30] M. B. Priestley, Spectral Analysis and Time Series, Academic Press, London, 1981.

[31] M. Rosenblatt, Stationary Sequences and Random Fields, Birkhäuser, Boston, 1985.

[32] Yu. A. Rozanov, On Gaussian Fields with Given Conditional Distributions, *Theo. Prob. Appl.*, Vol. 12, No. 3, pp. 381–391, 1967.

[33] M. Taniguchi, On Estimation of the Integrals of Certain Functions of Spectral Density, *J. Appl. Prob.*, Vol. 17, pp. 73–83, 1980.

[34] J. Weszka, C. R. Dyer, and A. Rosenfeld, A Comparative Study of Texture Measures for Terrain Classification,*IEEE Trans. on Syst., Man., and Cybern.*, Vol. SMC-6, pp. 269–285, 1976.

[35] P. Whittle, On Stationary Processes in the Plane, *Biometrika*, Vol. 41, pp. 450–462, 1954.

[36] J. W. Woods, Two-Dimensional Discrete Markovian Fields, *IEEE Trans. Information Theory*, Vol. IT-18, No. 2, pp. 232-240, 1972.

[37] A. M. Yaglom, An Introduction to the Theory of Stationary Random Functions, Prentice-Hall, Englewood Cliffs, 1962.

[38] J. Yuan, Spectral Analysis of Multidimensional Stationary Processes with Applications in Image Processing, Ph.D. thesis, University of Manchester Institute of Science and Technology, England, 1989.

[39] J. Yuan, and T. Subba Rao, Classification of Textures Using Second Order Spectra, To appear in J. Time Series Analysis, 1992.

Probabilistic Network Inference for Cooperative High and Low Level Vision

P. B. Chou[†], P. R. Cooper[‡], M. J. Swain[§], C. M. Brown[*], and L. E. Wixson[*]

[†]IBM Research Division
T. J. Watson Research Center
Yorktown Heights, New York

[‡]Institute for the Learning Sciences and
Department of Electrical Engineering and Computer Science
Northwestern University
Ann Arbor, Michigan

[§]Department of Computer Science
University of Chicago
Chicago, Illinois

[*]Department of Computer Science
University of Rochester
Rochester, New York

1 Introduction

We present a framework, based on Bayesian-probability theory, for approaching computer vision problems that can be posed as labeling. The labeling problem is: for each image location (site), find the appropriate attribute or label (such as depth or semantic interpretation) describing the corresponding portion of the scene. With various definitions of site and label, this large class of problems includes image processing, early vision, intrinsic property reconstruction, and sensor fusion right through the processes of segmentation and recognition or matching. It has long been recognized that interpretive decisions about the *physical structure* of objects

ISBN 0-12-170608-7

and their *identity* should not be independent, and our framework supports a principled integration of these processes. Although our application domain is computer vision, the results can be taken in a larger context of knowledge representation, reasoning with several distinct but interacting bodies of knowledge, and uncertain inference. We demonstrate the ideas on three instances of the labeling problem—boundary detection, surface reconstruction using sparse range data and irradiance inputs, and object recognition in scenes with obscuration.

The chapter presents the following results:

1. Coupled Markov random fields are presented as a mechanism for combining several sources of *a priori* and observational knowledge in a Bayesian framework. The MRFs encode the assignment of (discrete or continuous) labels to image sites.

2. Knowledge is encoded by the neighborhood structure of the MRF and by the assignment of goodness *potentials* to local structures (cliques) in the MRF. The potentials then determine the probability distribution of the labels in the MRF, whose *a posteriori* probability distribution is derived by combining the pooled external observations and the *a priori* distribution.

3. The Highest Confidence First (HCF) algorithm and its parallel implementation, Local HCF, are robust and efficient estimation methods for finding the solution of the labeling problem given by the *a posteriori* probability distribution. HCF and Local HCF meet the principles of graceful degradation and least commitment. They are deterministic calculations whose running time is predictable and small. Since its introduction in [4], the HCF method has been used in recent work in model-based segmentation from multiple-intensity images and in texture discrimination [12, 20].

4. As a sample application, coupled MRFs are used for a unified treatment of reconstruction and segmentation of three-dimensional surfaces with sparse depth observations and intensity discontinuity information. HCF is extended to handle both symbolic and numerical labels simultaneously using coupled MRFs.

5. As a high-level application of coupled MRFs with non-isotropic field topology, scene structure determination and symbolic matching for object recognition are performed simultaneously.

6. Domain-dependent knowledge may be specified explicitly, but also can be learned from noisy observations.

2 Spatial Priors and Markov Random Fields

Markov Random Fields (MRFs) have been used as the basis of an evidential approach to many computer vision tasks in recent years.* Most of this work has addressed very low-level representations and processes, and has used MRFs that are essentially rectangular arrays. As we shall see, coupled MRFs can unify the segmentation and reconstruction processes.

Fortunately, the theory of Markov random fields extends beyond simple arrays, and can be applied to arbitrarily structured graphs like the one used later to build a high-level structure representation. Some of the relevant aspects of MRF theory and its application to labeling problems are now very briefly reviewed [13].

Consider a set $\mathbf{X}$ of discrete-valued random variables. Associate with the random variables an undirected graph G defined as a set S of sites (or vertices) and a neighborhood system (or set of edges) E. The random variables of the field are indexed by the graph vertices as X_s. Variables are neighbors in the MRF when the associated vertices are adjacent in the graph. In the formulation of a labeling problem as an MRF, the variables in the labeling problem are the random variables of the MRF.

The value ω_s of a random variable may be any member l_i of the state space set L. Because of the application of the field to the labeling problem, the event elements of the set L will be called *labels*. An assignment of values to all the variables in the field is called a *configuration*, and is denoted ω.

We are interested in the probability distributions P over the random field $\mathbf{X}$. Markov random fields have a locality property:

$$P(X_s = \omega_s | X_r = \omega_r, r \in S, r \neq s) = P(X_s = \omega_s | X_r = \omega_r, r \in N_s), \quad (1)$$

that says roughly that the state of site is dependent only upon the state of its neighbors (N_s). MRFs can also be characterized in terms of an energy function U with a Gibbs distribution:

$$P(\omega) = \frac{e^{-U(\omega)/T}}{Z}, \quad (2)$$

where T is the temperature, and Z is a normalizing constant.

If we are interested only in the prior distribution $P(\omega)$, the energy function U is defined as:

$$U(\omega) = \sum_{c \in C} V_c(\omega), \quad (3)$$

*See other chapters in this volume. With regret, we have deleted many relevant references to save space.

where C is the set of cliques defined by the neighborhood graph G, and the V_c are the clique potentials.

Specifying the clique potentials V_c provides a convenient way to specify the global joint prior probability distribution P. The clique potentials can be conveniently viewed as weights in a connectionist network. They provide a mechanism to express soft constraints between labels at related variables. Unary clique potentials in effect express first-order priors, while binary clique potentials express the constraints between pairs of variables in the field.

Suppose we are instead interested in the distribution $P(\omega|O)$ on the field after an observation O. An observation constitutes a combination of spatially distinct observations at each local site. The evidence from an observation at a site is denoted $P(O_s|\omega_s)$ and is called a *likelihood.* Assuming likelihoods are local and spatially distinct, it is reasonable to assume that they are conditionally independent. Then, with Bayes's rule, we can derive:

$$U(\omega|O) = \sum_{c \in C} V_c(\omega) - \sum_{s \in S} \log P(O_s|\omega_s) \tag{4}$$

Inference on the MRF network can be framed in terms of the energy function. For example, the maximum *a posteriori* probability can be computed by finding the minimum of the non-convex energy function U, which corresponds to a particular selection of labels for each variable. Evidence about the hypotheses is expressed as label likelihoods, and prior knowledge is expressed in terms of the clique potentials, generalized weights that express soft constraints between spatially related variables. Only simple local operations are involved in updating the energy measure and local characteristics as new opinions from the early visual modules become available.

This formulation leads to a difficult minimization problem to find the lowest energy. Many techniques are applicable [1], and inference methods depending only upon these local measures, such as stochastic MAP [9, 10] and MPM estimations [14, 16, 17], are common. Swain's local technique (Section 8) can easily be implemented in this framework. We shall first describe a new deterministic approximation algorithm called Highest Confidence First (HCF [3]).

3 Estimation with Highest Confidence First

An estimation method should be: *efficient* (deterministic, and preferably should make the maximum improvement at each step), *predictable* (results depending on the inputs and the *a priori* distribution but not other performance-affecting parameters), and *robust.* The Highest Confidence First (HCF) method meets these requirements. It is not a stochastic method like simulated annealing, but a deterministic one like ICM estimation [1, 2]. Both start in some initial configuration. At each iteration through the image sites, the state of each site is either changed to the state that yields maximal decrease of the energy, or is left unchanged if no energy reduction is possible. The process always stops at a local minimum when no more changes can be made. In a parallel implementation, convergence is assured if the neighboring entities are not updated simultaneously.

In the HCF algorithm all sites initially are specially labeled as *uncommitted*, instead of starting with some specific labeling as with previous optimization methods. Uncommitted sites do not participate in the computation of the energy of the field, nor, therefore, do they actively influence the commitments of their neighbors. However, an uncommitted site always takes into account the states of the active neighbors when making a commitment. For each site, a stability measure is computed. The more negative the stability, the more confidence we have in changing its labeling. On each iteration, the site with minimum stability is selected, and its label is changed to the one that creates the lowest energy. This in turn causes the stabilities of the site's neighbors to change. A commitment to assume a label is not a commitment to a particular label—the label of a committed site will be altered if it is too much at variance with its neighbors. The process is repeated until all changes in the labeling would result in an increase in the energy, at which point the energy is at local minimum in the energy function and the algorithm terminates. The algorithm may be implemented with a priority queue (Fig. 1).

The stability of a site is defined in terms of a quantity known as the *augmented a posteriori local energy* E, which is:

$$E_s(l) = \sum_{c:s\in c} V_c'(\omega') - \sum_{s\in S} \log P(O_s|\omega_s),$$

where ω' is the configuration that agrees with ω everywhere, except that $\omega_s' = l$. Also, V_c' is 0 if $\omega_r = l_0$, the uncommitted state, for any r in c; otherwise, it is equal to V_c.

The stability G of an uncommitted site s is the negative difference between the two lowest energy states that can be reached by changing its label:

$$G_s(\omega) = -min_{k\in L, k\neq\omega_{min}}(E_s(k) - E_s(\omega_{min}))$$

In this expression, $\omega_{\text{min}} = \{k|E_k \text{ is a minimum}\}$. The stability of a committed site is the difference between it and the lowest energy state different

begin
    $\omega = (l_0, \ldots, l_0)$;
    top = Create_Heap(ω);
    while stability$_{top} < 0$ **do**
        s = top;
        Change_State(ω_s);
        Update_Stability(stability$_s$);
        Adjust_Heap(s);
        for $r \in N_s$ **do**
            Update_Stability(stability$_r$);
            Adjust_Heap(r)
        end
    end
end

Figure 1: The HCF algorithm.

from the current state ω_s:

$$G_s(\omega) = \min_{k \in L, k \neq \omega_s} (E_s(k) - E_s(\omega_s)).$$

Thus, stability is a combined measure of the observable evidence and the *a priori* knowledge about the preferences of the current state over the other alternatives.

The HCF algorithm always returns in finite time, with a feasible solution with locally minimal energy. This implementation takes $O(N)$ comparisons to create the heap and $O(\log(N))$ to maintain the heap invariance for every visit to a site, provided the neighborhood size is small relative to N — the number of sites. The procedure makes progress for every visit, in contrast to the iterative relaxation procedure that may make only few changes per iteration (N visits). Our edge detection experiments (Section 4) show that on the average, less than one percent of the sites are visited more than once with HCF, while usual deterministic relaxation procedures take around 10 iterations to reach a local minimum. This advantage becomes more evident as the number of sites gets larger.

4 Probabilistic Boundary Detection

In edge detection, the labeling problem is to assign to each site a label from the set {EDGE, NON-EDGE }, based on discrete intensity measures on the pixels of a square lattice-structured image. Our probabilistic framework uses the outputs of a set of local operators that relate the intensity observations to edge labels, and ignores the intensity values afterwards. Thus, the global optimization is performed over a space much smaller than the ones for full reconstruction.

The edge sites are considered to be situated on the boundary between two pixels. We adopt a basic step-edge model to compute the local likelihoods of a site s being EDGE or NON-EDGE. We use a Bayesian approach that takes into account the pixel values in a 1×4 or 4×1 mask, knowledge about the gray-level distribution (taken from the image histogram or known *a priori*), and a white Gaussian noise model [19]. The *edge-finder* returns a likelihood ratio for the presence of an edge. A binary configuration may be constructed by declaring an edge present where the log of its likelihood ratio is greater than log $P(\text{NON-EDGE})/P(\text{EDGE})$, the logarithm of prior (local) odds. This binary configuration is the *thresholded log likelihood ratio* (TLR), and it can be considered as a MAP estimate obtained without using contextual information [3]. In our experiments, we use TLRs as the initial estimates whenever possible. Each edge site is modeled as a random variable of the binary field. The intensity values are used only to calculate the local likelihoods for the edge sites, and the likelihoods constitute the input of the stand-alone line process.

Here and in Section 5, clique energies were chosen to encourage the growth of continuous line segments and to discourage abrupt breaks in line segments, close parallel lines (competitions), and sharp turns in line segments. *Encouragement* or *discouragement* is associated with a clique by assigning it a negative or positive energy, respectively. To encode these relationships, a second-order neighborhood, in which each site is adjacent to eight others, is used. This neighborhood system is shown in Fig. 2 and typical clique values are shown in Fig. 3. Another interesting method of encoding neighborhood information about edges may be found in [8].

Changing the potential function associated with the 1-clique has the greatest effect on the final result, followed by the 2-clique and 4-clique potential functions, in that order. (The singleton clique controls first-order statistics and the larger cliques higher-order statistics.) More insight on the function of clique values is given in Section 8.

We show the results of one set of experiments (Fig. 4).[†] The anneal-

[†]See [3] and Section 5 for more.

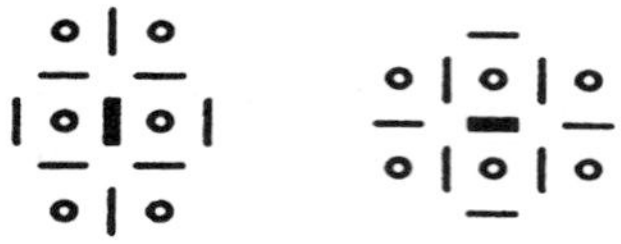

Figure 2: Neighborhoods for vertical and horizontal edge sites. Circles represent pixels, the thick line represents the site, and thin lines represent the neighbors of the site.

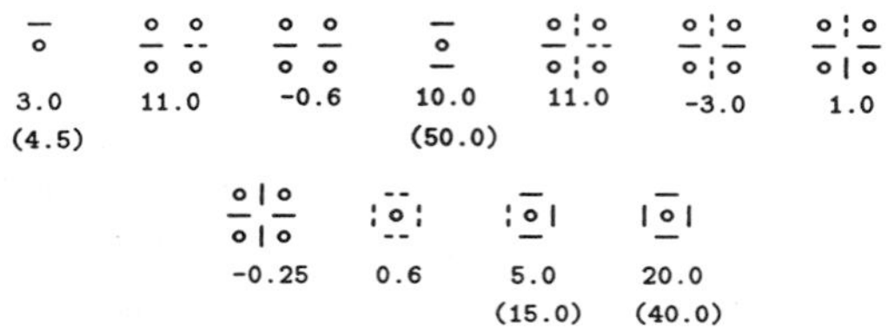

Figure 3: Clique energies. The parenthesized values were used for corrupted edge data. (See text.)

ing schedule for the stochastic MAP follows the one suggested in [4]: The stochastic MAP was run for 1000 iterations and the stochastic MPM for 500 (300 to reach equilibrium, 200 to collect statistics). Except in the case of the HCF algorithm, where the MRF is initialized to all null (uncommitted) states, the MRF is initialized to the TLR configuration. The MRF specification is the same throughout.

Some timing studies were done on stochastic minimization algorithms compared with HCF and Local HCF (Section 5). They indicate that simulated annealing performs quite slowly and is sensitive to its parameters (annealing schedule). Stochastic MPM converges quickly but is sensitive to initial conditions, so would be a good candidate to follow HCF for an improved labelling [22].

The deterministic algorithms have well-defined convergence times, and (HCF and ICM (scanline)) were timed on images of various sizes using a Sun 3/260 with floating point acceleration. HCF exhibits surprisingly linear time increase with size of MRF field, and is approximately as fast as other deterministic methods. HCF's results are consistently better than the other algorithms, giving superior results both with synthetic and real image data. HCF is more robust than other methods to inaccurate edge

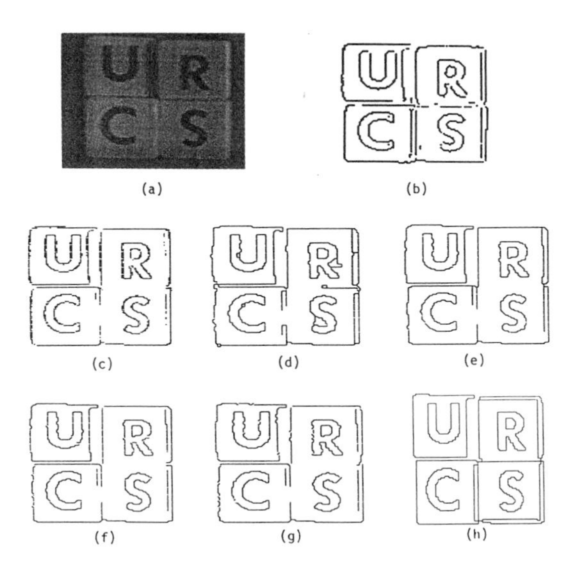

Figure 4: Comparison of different energy minimization schemes. (See text.) (a) Natural 100 x 124 image of four plastic blocks. (b) Thinned and thresholded output of Kirsch operators. (c) TLR configuration. (d) Stochastic MAP estimate. (e) Stochastic MPM estimate. (f) ICM estimate (scanline visiting order). (g) ICM estimate (random visiting order). (h) HCF result.

models. (We compared methods using an edge model that did not object to multiple close parallel edge elements.) The HCF results from this algorithm conform with our *a priori* model of smooth, continuous boundaries, and also are consistent with the observations.

The ICM algorithm performs inconsistently and its results depend to a large extent upon the initialization of the MRF and the visiting order. HCF does not rely on any predefined order, and thus is not biased for any boundary shape. The stochastic MAP algorithm with simulated annealing gets stuck in undesirable local minima. However, our experiments suggest that there are problems with both slow and fast annealing [3]. We evaluated the energy measures obtained by the algorithms (Table 1, Section 5).

The convergence times of the ICM algorithms are unpredictable—they vary with visiting order, MRF initialization and even upon the particular image given as input. It has been established that steepest descent methods in MRF and other symmetric connection nets can, in fact, take exponential time in the worst case [11]. However, HCF seems very well-behaved in practice. HCF must visit every site at least once; in fact, for this application, it visits each site on the *average* less than 1.01 times before converging. The first decision made by a site is nearly always the best one. In theory, the time taken by the HCF algorithm should be given by $c_1 N + c_2 V \log_2 N$, where c_1 and c_2 are positive constants, N the number of sites to be labeled, and V the number of visits. V here is at least N, and we conjecture that on the average it is cN for some small $(1 < c < 2)$ constant c. Since the latter term should dominate, one would expect to see a nonlinear curve in a plot of run time vs. number of sites. However, the curve is very nearly a straight line, indicating either that the constant c_2 is very small, or that the changed stability values do not propagate very far up the heap on the average. The former does not appear to be true: our experiences suggest that the initial heap construction takes far less time than the rest of the algorithm.

5 Local HCF

The HCF algorithm is serial, deterministic, and guaranteed to terminate. We have developed a parallel version of HCF, called Local HCF, suitable for a SIMD architecture in which each processor must only communicate with a small number of its neighbors. Such an architecture would be capable of labeling an image in realtime. Experiments have shown that Local HCF almost always performs better than HCF and thus much better than other previous techniques.

The Local HCF algorithm is a simple extension of HCF: *on each iteration, change the state of each site whose stability is negative and less than the stabilities of its neighbors.* In a preprocessing phase, the sites are each given a distinct rank, and, if two stabilities are equal in value, the site with lower rank is considered to have lower stability. These state changes are done in parallel, as is the recalculation of the stabilities for each site. The algorithm terminates when no states are changed. Pseudo-code for Local HCF is given in Fig. 5, for which you should assume a processor is assigned to every element of the *site* data structure. The algorithm is written in a notation similar to C* [18], a programming language developed for the Connection Machine. In the algorithm, the operator &all returns the result of a global *and* operation.

For the low-level vision tasks that we have studied, the MRFs have uniform spatial connectivity and uniform clique potential functions. Thus, Local HCF applied to these tasks is well suited for a massively parallel SIMD approach that assigns a simple processor to each site. Each processor need only be able to examine the states and stabilities of its neighbors.

```
site: parallel array[1..N_SITES] of record
  stability;
  i; /* rank */
  change;
end

begin
  with site do in parallel
    do
      begin
        change := false;
        Update_Stability(stability);
        (nbhd_stability,k) := min_{n∈N[i]} (site[n].stability,site[n].i);
        if stability < 0 and (stability,i) < (nbhd_stability,k) then
          begin
            Change_State(state);
            change := true;
          end
        any_change := (&all change);
      end
    until any_change = false;
end
```

Figure 5: The Local HCF algorithm.

The testing and updating of the labels of each site can then be executed in parallel. Such a neighborhood interconnection scheme is simple, cheap, and efficient.

Like HCF, Local HCF is deterministic and guaranteed to terminate. It will terminate because the energy of the system decreases on each iteration. We know this because (roughly) (a) at least one site changes state per iteration—there is always a site whose stability is a minimum—and (b) the energy change per iteration is equal to the sum of the stabilities of the sites that are changed. These stabilities are negative and the state changes will not interact with each other because none of the changed sites are neighbors. Therefore, the energy of the system always decreases. A rigorous proof of convergence is given in [22].

Determinism and guaranteed termination are valuable features. Analysis of results is much easier; for each set of parameters, only one run is

Method	Chbd	P	URCS	Bad URCS
TLR	−3952	−572	4785	59719
Anneal	−4282	−680	−349	−5303
MPM	−4392	−723	−503	−5296
ICM(s)	−4364	−693	−503	−4954
ICM(r)	−4334	−715	−513	−3728
HCF	−4392	−750	−380	−9635
Local HCF	−4392	−720	−625	−9648

Table 1: Energy Values for Checkerboard, P Block, URCS Blocks, and Corrupted URCS Block Images. The Smaller the Energy, the Closer the Labeling is to the MAP Estimate.

needed to evaluate the performance, as opposed to a sampling of runs, as with simulated annealing.

5.1 Test Results

Local HCF has been directly compared to HCF on several edge-finding tasks. For both algorithms, the input is produced by Sher's probabilistic edge detector [19] and consists of the log likelihood ratio for an edge at each site. The algorithms were tested on likelihood ratios from a synthetic checkerboard image, a "P" block image, and a "URCS" block image (Fig. 6). The algorithms were also presented with noisy (corrupted) likelihood ratios obtained by using an incomplete edge model to find edges in the URCS image. This test of robustness against inaccurate assumptions is one of the strong points of HCF and it is important that it not be lost.

Cliques and their energies were chosen as in Section 4 (Figs. 2, 3). The correctness of the labeling may be assessed qualitatively by visual inspection, and the efficacy of the energy minimization may be measured quantitatively by energy of the final configuration. Figure 7 shows the labelings produced by Local HCF on the four test cases. Analysis of the time course of the computation reveals that 90% of the sites commit in the first 10 iterations, and 98% in the first 15 iterations. In these trials, Local HCF produces seemingly better labelings than traditional techniques at a much lower computational cost.

Table 1 shows energies of the final configurations yielded by thresholding the TLR, simulated annealing MAP estimation, Monte Carlo MPM estimation, ICM estimation (scan-line order), ICM (random order), HCF, and our results from Local HCF. The only real surprise is the relatively inferior energy minimization performance of HCF on the URCS image (Fig.

Figure 6: Test images (8 bits/pixel). Checkerboard and P images are 50 × 50. URCS image is 100 × 124.

Figure 7: Edge labelings produced by Local HCF for preceding images. Rightmost labeling demonstrates Local HCF on noisy edge data from the URCS image.

4), compared to its clearly superior qualitative results over all schemes except Local HCF. The values of MAP, MPM, and the ICMs are the averages of the results from several runs. In almost every case, Local HCF found the labeling with the least energy. Each Local HCF run took 20–30 iterations (parallel state changes); we expect that the Connection Machine will carry out these labelings almost instantaneously. Local HCF is ideal for problems with a large number of labels, such as recognition problems (Section 7).

5.2 Comparing Local HCF and HCF

When Local HCF produces better labelings, it is usually because HCF has propagated strong local information too far. In HCF, one site s often commits to a certain label, immediately followed by one of its neighbors $s + 1$ committing to a compatible label, and so on for more neighbors. In this manner, the effects of locally high confidence can propagate too far. HCF exhibits a horizon effect, that is, puts off making *unpleasant* choices in a similar way to game playing programs with limited look-ahead. A simple one-dimensional example exhibits the problem with HCF and shows how Local HCF avoids it.

Consider a linear array of variables representing edge (e) or non-edge (n), each neighbors with the two adjacent variables. The neighborhood graph is then a chain containing unary and binary cliques. Assign values to the cliques as follows:

> Unary cliques edge, non-edge: 0 0 (edge, non-edge equally likely)
>
> Binary cliques (e,e) (n,n) (e,n) : −0.5 1 −0.5 (line breaks discouraged)

Suppose an edge detector reports the following log likelihood ratios, log (P(observation | edge) / P(observation | non-edge)):

> 4 -0.2 -0.4 -0.5 -0.3 0.1 -0.3 -0.4

By a dynamic programming method [2] the optimal labeling can be found to be :

> e n n n n n n n

HCF produces the labeling:

> e e e e e e e e

because it is always locally favorable to extend the edge labeling that was initiated by the strong evidence at the left-hand variable, rather than introduce a line break. HCF would propagate the line indefinitely, given continued weak evidence against an edge. Local HCF produces the optimal labeling, because sites of locally minimum stability commit to non-edge before the evidence from the left-hand variable propagates across the entire field.

6 Coupled Segmentation and Reconstruction

HCF can be applied to simultaneous reconstruction and segmentation of three-dimensional scene parameters (intrinsic images) from visual information.[‡] Represent the pixel and discontinuity images as in Section 4. Add a set of *depth process* random variables indexed by sites representing the depth value at pixel locations. These processes will be used to form a coupled Markov random field, in which a configuration of labels corresponds to an admissible solution to our problem.

The depth measurements are considered sparse and independently measured, with noise modeled by unbiased Gaussian distributions. Intensity edges are partial evidence supporting or refuting the hypotheses about depth discontinuities. This computation is effected by a fast hierarchical evidence combination algorithm [3]. Again, we consider the spatially distinct intensity observations to be conditionally independent, and we assume that the depth and intensity observations are only related through the geometry of the surfaces in view.

Surfaces and boundaries tend to be continuous and smooth. MRFs can be separately defined to model these properties, but the depth and line processes are not independent: a line at an edge site breaks the connection between the two variables at the adjacent pixel sites; a small change in the values of two adjacent depth variables suggests the absence of a discontinuity in between. This interdependence is the basis for the concept of coupled MRFs—a unified treatment of reconstruction and segmentation.

Continuous surfaces can be modeled by setting the potential energy V for the cliques consisting of two adjacent depth sites, say, i and j, and the line site in between them , say, ij, proportional to $(1-l_{ij})(f_i - f_j)^2$, where f_i is the depth at site i [14]. Using this potential function, minimizing the energy measure has the effect of fitting membrane patches to the lattice. We use this potential function throughout our experiments, along with the line site potentials described earlier.

[‡]See [3] for details.

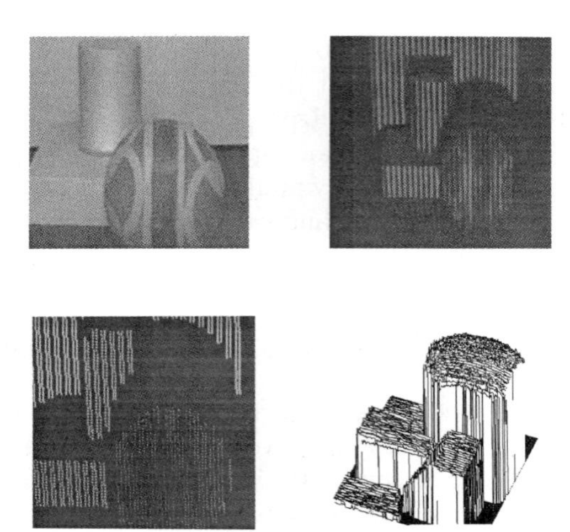

Figure 8: Data for segmentation and reconstruction using intensity and depth. (a) Intensity. (b) Light stripes. (c) One of three sparse depth images. (d) Combined sparse depth images plotted as depth surface.

Bayes's rule combines the *a priori* knowledge, the noise assumption, and the early visual observations to derive the *a posteriori* belief [3].

Coping with the continuous depth labels under the HCF regime calls for defining the augmented local energy functions with respect to augmented configurations (which in this work are quadratic—their shape is controlled by a parameter β), defining stability measures (the stability of uncommitted sites is determined by a parameter α), with large *alpha* encouraging quicker commitment.

We performed depth segmentation and reconstruction on a variety of synthetic and natural scenes. In the latter, the Cooper stereo algorithm [5] yields sparse disparity data for tabletop scenes.

Figure 8(a) shows intensity edges from a scene with a beach ball and a rectangular box sitting in front of a cylindrical object, with a flat background. Vertical (with respect to the epipolar line) light strips were projected onto the scene to create artificial texture needed by the stereo system (b). Disparity observations are shown in (c) and (d). Three depth images were overlaid, but still only 35% of the pixel locations have at least one disparity observation.

A map of those locations overlaid with the TLR estimate of the intensity discontinuities is shown in Fig. 9(a). Using $\alpha = 30$, $\beta = 0.003$, and $P(\text{NON-EDGE}|\text{NON-DEPTH-EDGE}) = 0.9$, the reconstructed dis-

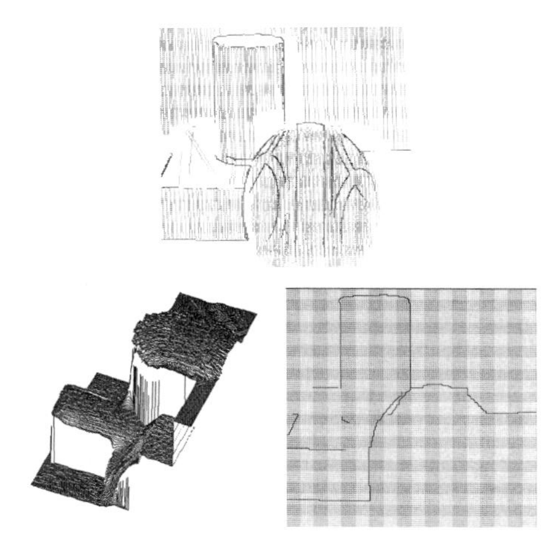

Figure 9: Results of segmentation and reconstruction using intensity and depth. (a) TLR intensity and sparse depth. (b) reconstructed depth using intensity data. (c) disparity discontinuities.

parity map (in perspective) and discontinuities are shown in (b) and (c). The surfaces corresponding to the sphere, cylinder, and the background planes are smoothly reconstructed, and significant disparity discontinuities are detected. When regions have inadequate disparity observations, (e.g., the tabletop or the top of the box), the result is a *leaking* effect: disparity values of neighboring regions leak through the holes of weak intensity gradients, resulting in under-segmentation and erroneous disparity estimates. A possible fix is to identify those regions prior to the reconstruction, possibly by building convex hulls of the *adjacent* pixels with disparity measures, and limit the reconstruction process outside of those regions.

7 Coupled Recognition and Reconstruction

It has long been recognized that interpretive decisions about the *physical structure* of objects and their *identity* should not be independent. Previous sections have described how MRFs can perform interacting low-level vision tasks simultaneously. This section similarly applies MRFs to coupled high-level vision problems. Structure and identity determination are posed as labeling problems and solved with a coupled MRF. HCF optimization, the best available method of optimization, is used in the experiments.

We use a massively parallel network for coupled recognition and structure reconstruction. The domain is Tinkertoy objects, and again the problems are posed as labeling, allowing smooth incorporation of *a priori* knowledge, and scene evidence. Because their identity is defined primarily by the spatial relationships between simple parts, Tinkertoys provide a convenient domain task for examining recognition from structure. Structure recognition with imperfect data requires inexact matching, which requires a *best match* definition and heuristics to circumvent the combinatorial search.

7.1 Network Description

The coupled MRF network is described by its definitions of variables and labels, the connection pattern between the variables (graph structure of the net), the prior probabilities, and the evidence.[§]

Variables: We formulate the structure-matching problem as a labeling problem, using the unit/value principle, which is to consider all possible object–model *part correspondences* simultaneously. The variables are defined to be the object parts, and the possible labels are the potentially corresponding model parts. The part-variables must encode the relevant aspects of the domain. We choose the *slots* on the disk (discrete junction connection points) as logical *parts*. Variables have two types: *slot variables* and *virtual rod variables*, or *vrods*. The vrods represent the set of hypothetical rods that might connect any two slots. This set serves as the variables for *both* recognition and structure inference, allowing considerable elegance.

Labels: The labels in the matching problem are the possibly corresponding parts from the model. For structure inference, they correspond to the hypothetical physical parameters the parts and their composition can achieve in the world. Thus, the labels for the vrod variables are the possible rod lengths, and the labels for the *slot* variables are *empty*, *filled*, and *not_exists* (i.e., the slot has a rod plugged into it or not, or it is a false positive from the lower vision levels). The definitions of the variables and labels for both halves of the field are summarized in Table 2.

Network Structure: The connections define the MRF graph, and represent constraints between related variables in the field; they encode the essence of the computation. The topology of the overall network consists of the same network replicated twice, once for each labeling problem, and a set of coupling connections (Fig. 10).

The connections *within* the replicated subnetwork can be seen in Fig. 11. The coupling connections, mostly connections between the corresponding

[§]A more complete presentation of the network is available in [6].

Structure Inference Subnet		Matching Subnet	
Vars	Labels	Vars	Labels
slots	¬∃	slots′	no matches
	∃, empty		matches mod. slot X
	∃, full		matches mod. slot Y, etc.
vrods	¬∃	vrods′	no matches
	∃, has len. L1		matches mod. rod A
	∃ has len. L2		matches mod. rod B, etc.
	∃ has len. L3		

Table 2: Definition of Variables and Labels

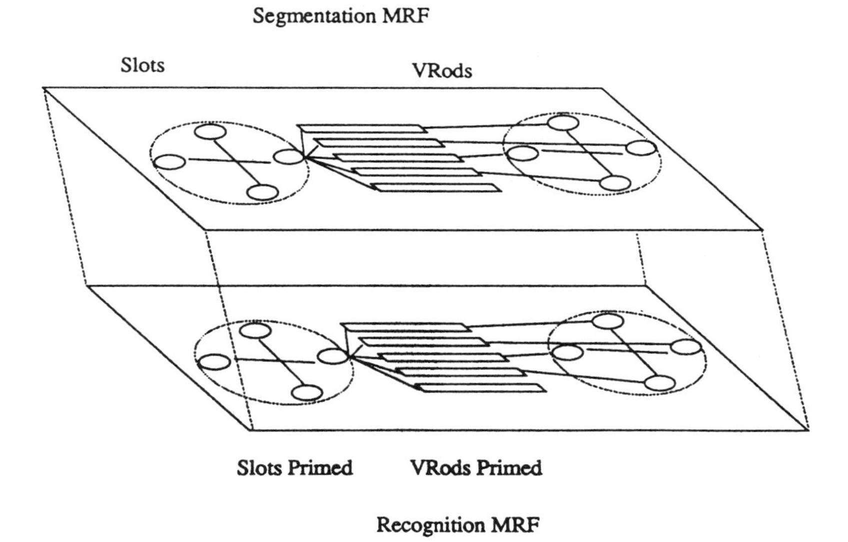

Figure 10: Recognition and segmentation MRFs are structurally identical.

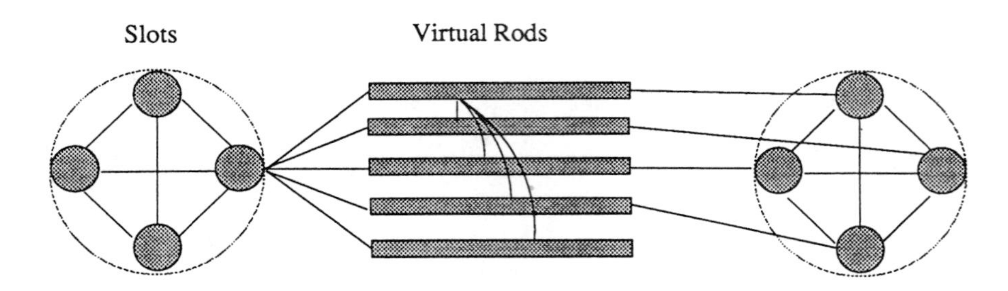

Figure 11: Fragment of MRF graph. The shaded objects are the MRF sites (slots and rods). The solid lines represent edges in the MRF graph. The virtual rods show all the connections from one slot to slots on one other disk, as well as the dangling rod possibility. The set of virtual rods is a clique; only some of the connections are shown.

variables in the two subfields, are not shown. The non-homogeneous, non-isotropic nature of the graph structure reflects the application of MRF theory to the representation of high-level structure and recognition, and differs greatly from traditional image-based array MRF applications.

Prior Knowledge and Likelihoods: We have two sources of information: sensor evidence for this problem instance, and *prior* knowledge. Domain-dependent knowledge, both qualitative and quantitative, is represented by clique potentials in the MRF.

The qualitative facts and evidence that follow are implemented as numerical weights [6]. The network performance was surprisingly unaffected by significant variations in these quantifying choices.

Most of the constraints reflect qualitative facts about Tinkertoys. In the segmentation subnet, the constraints primarily ensure *consistency.* For example, it is inconsistent that two rods be connected to the same slot. The constraints in the recognition subnet are derived from the model. For example, all the slots at an object disk must be matched against all the slots on a single model disk so that they can align.

The *coupling constraints* between subnets are crucial to the entire coupled approach. They reinforce interpretations of the physical world that are consistent with the model, and are based on certain knowledge about the world. Say that object rod "A" matches model rod "2." The length of model rod 2 is known. If the model rod "2" were length $L1$, there would be a potential weight encouraging the physical interpretation of rod A as length $L1$.

Quantitative prior knowledge may be expressed. Clique potentials can represent the frequency with which local properties occurred in past problem instances (but there are some difficulties [15].) For example, clique potentials on the length labels of the rod variables can encode first-order statistics about rod lengths in previous problem instances. Second- and higher-order features (such as junction geometry at a disk) can also be represented in the network. In this way, statistics based on a domain of previous problem instances can influence perceptual inferences in the current problem instance. One might think of the domain-dependent prior knowledge as *smoothing* the current evidence to a solution during the inference process on the network.

As in the coupled segmentation and reconstruction before, the *image evidence* yields likelihoods for labels in the segmentation subnet. For example, the likelihood that a particular rod has length $L1$ might be 0.7, and the likelihood of length $L2$ might be 0.1. These experiments use a synthesized likelihood generator. The evidence was constructed from qualitative judgments, such as *very certain*, *almost no evidence for this*, etc. The experiments deliberately probed a wide range of input conditions, including worst cases, reflecting possible data that could arise from real images.

A schematic overview of the entire system, showing the sources of prior information and evidence and the relationship between the subnets, is given in Fig. 12.

7.2 Experiment: A Man and His Dog

The problem is to segment and identify one of the components (the dog) of a scene in Fig. 13 with two objects (a man and his dog), accidentally aligned so as to be possibly mistaken for one object.¶ The figure shows input evidence for the problem instance. The label likelihoods are given by bar graphs near the spatial hypothesis they describe. The lines are scaled to represent likelihoods between zero and one. The identification problem is to identify the dog.

The problem demonstrates typical segmentation ambiguities. Local labeling ambiguity is represented by the different likelihoods (e.g., for different rod lengths). There is structural uncertainty—is it one object or two? The problem also shows how priors and evidence from segmentation and recognition combine to yield a decision. The evidence about the major segmentation decision is (by design) inconclusive. In fact, it is slightly weighted in favor of the wrong interpretation, especially if considered locally. Thresholds would yield the wrong interpretation. Coupling information from the recognition process to segmentation achieves the correct

¶This is an abbreviated account of an experiment in [6], which also has more.

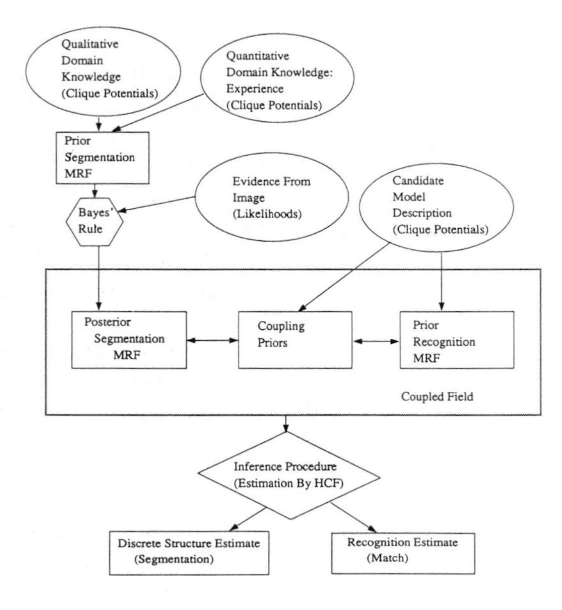

Figure 12: System overview.

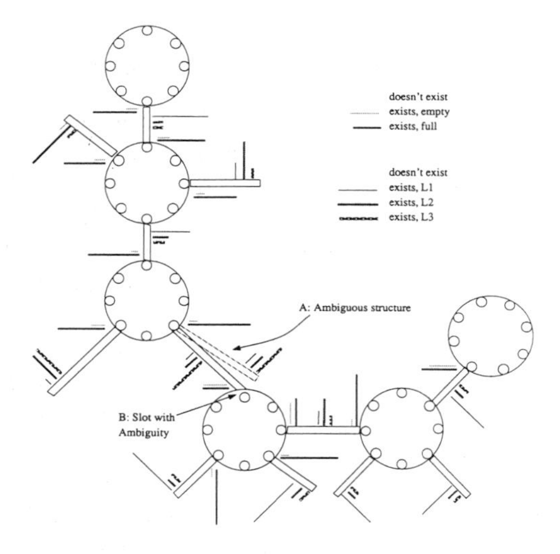

Figure 13: A Man and His Dog: schematic of image showing accidental alignment, with bar graphs of label likelihoods from simulated low-level interpretation procedure.

interpretation. The MRF inference procedure is a powerful way to resolve such an ambiguous decision problem. The sequential trace of the inference process is particularly impressive in this regard, because it involves a local decision-reversal. In this case, the global energy of the later decision is better than the first decision.

The evidence surrounding the connection of the two objects (at point A in Fig. 13) is very ambiguous, with *connected* and *disconnected* hypotheses having very similar evidence. The hypothetical likelihood generator is fairly confident about the length of the rod, but is noncommital on whether it connects the two slots or not. (Both hypotheses have similar likelihoods at each of their labels.) This is an example of how true structural ambiguity is represented in the net. The evidence at the slot hypothesis (B in the figure) is completely ambiguous. The accidental alignment yields evidence for the *full* (plugged in) label. (A more sophisticated likelihood generator might have knowledge about slot-rod junctions, and would thus know that lack of perpendicularity at the junction is evidence for the *empty* label.)

Fig. 14 shows the final stages of the inference process. The segmentation labeling decisions are shown by the shading of the parts in the scene. The recognition or matching decisions are shown by the lines connecting the model parts to object parts. In the model, there is only one rod of length $L2$. As a result, unique labelings can be easily found that map each rod in the image of length $L2$ to the dog's trunk axis.

In Fig. 14(a), the inference procedure has incorrectly labeled the scene as one object. Once the slot representing the man's *hip* joint is labeled *full*, excitation energy exists in favor of some rod being attached. Then the *connecting* virtual rod is chosen because its local evidence is slightly stronger than that of the competing *dangling* virtual rod; but local part-wise matching decisions are being made in parallel as well. The correspondence labeling has matched the dog's shoulders correctly. Eventually (Fig. 14(b)), the correspondence of the dog's trunk axis propagates to match the trunk–hip slot (A in the figure). As a result, all the slots on the dog's hip disk become matched. At this point, because of the correct recognition, a conflict exists. *In the model*, the crucial slot (B) is empty; but the segmentation labeling of the connecting man's leg (C) is reinforcing the *full* label at B. The coupling from the recognition field labels slot B as *empty*, and an inconsistency between the labeling of B and C exists. Ultimately, the segmentation labeling is reversed and the objects are correctly segmented.

This work demonstrates simultaneous, truly coupled segmentation and recognition from structure. The network behavior is extremely robust to variations in the values for the synthesized evidential input, showing that an appropriate selection of variables and labels defines an architecture with

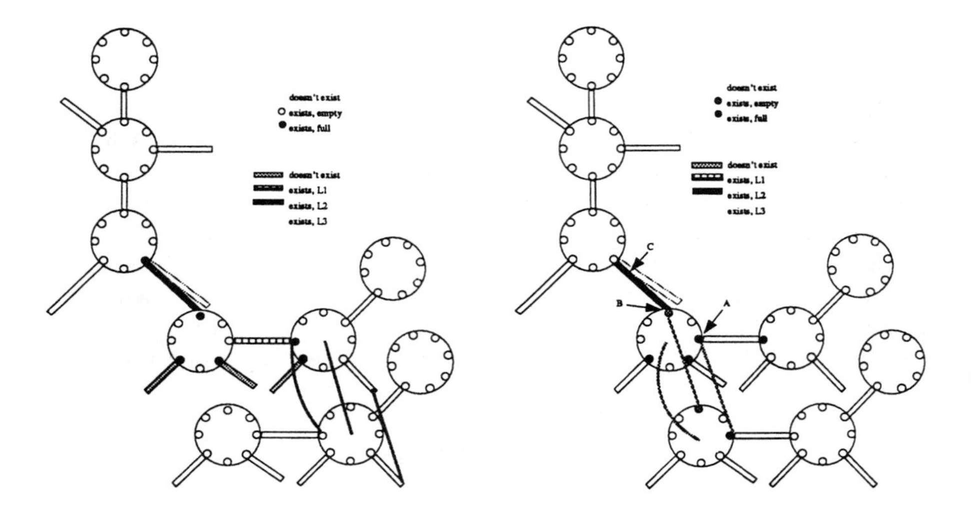

Figure 14: (a) Coupled Experiment: 2—incorrect segmentation decision, and some further correspondence inferences. (b) Expt. 3—match has propagated to slot A, causing a match of all slots at that disk, including B. The labeling at vrod C suggests slot B should be *full*, but the correspondence with the model dictates that B is labeled *empty*.

a well-behaved energy function. The basic parameter of the network is the size of the scene that is representable, in terms of the number of available slots n. $O(n^2)$ virtual rods are required to represent all possible slot/slot connections. Connecting the virtual rods together requires $O(n^3)$ 2-cliques, and other cubic terms arise from the 3-cliques, so the asymptotic space complexity is $O(n^3)$. The simulation on the Rochester Connectionist Simulator requires approximately 30 megabytes of memory.

8 Learning Clique Potentials

The probabilistic semantics of MRFs allow us to tie the parameters of the MRF to probabilities of events in the world. We are particularly interested in estimating (learning) parameters of the MRF from noisy observations. We propose two new approaches and compare them with traditional approaches to parameter estimation for MRFs. Both approaches involve estimating local joint probabilities from experience. In one approach, the joint probabilities are converted to clique parameters of the Gibbs distribution so that the traditional HCF algorithm can be used. In the other approach, the HCF algorithm is modified to run directly with the local probabilities of the MRF instead of the Gibbs distribution.

8.1 Previous Work

Most of the work in parameter estimation for Gibbs distributions has been done for texture models. The most common method for estimating clique parameters in this domain is Besag's pseudolikelihood method [10]. This method applies to homogeneous random fields, as in the edge detection work of Section 4. The problem with using the pseudolikelihood approach in our domain is that the pseudolikelihood function is a product of the conditional probabilities of a site given its neighborhood. What is required for the method is to follow the uphill gradient of this function to find its maximum. The gradient of the function is badly behaved when the conditional probabilities are close to zero and one, as is true in our domain, leading to poor results with this method.

Another approach to parameter estimation is the linear regression approach introduced by Derin and Elliot [8]. It relies on ratios of the conditional probabilities mentioned earlier, and so it, too, leads to poor results in domains such as ours. It is also possible to use stochastic relaxation to search for the best parameters, but this is extremely computationally intensive.

Instead of doing Maximum Likelihood Estimation, we propose to do the simpler task of estimating the joint probabilities of cliques of sites in the MRF, and converting from values to clique potentials. The conversion can be done exactly when the graph associated with the MRF is *chordal.* When the graph is not chordal (as in our case), the conversion is not exact, but a good approximation in practice.

The Mobius inversion theorem [13] provides an exact formula for the conversion of local conditional probabilities to clique parameters. The formula is not appropriate in our case because it relies on ratios of noisy estimates of values very near zero or one—a problem similar to the ones described earlier. The complex formula also provides little insight about the first-order relationships between local probabilities and clique parameters.

8.2 Markov Random Fields on Chordal Graphs

Associated with a Markov random field is a graph G, which defines the neighbors of each element of the field. If G is restricted to be a certain type of graph, called a *chordal* graph, it turns out that it is straightforward to determine the clique parameters given the joint distribution of the random variables.

Definition: *An undirected graph is said to be* **chordal** *if every cycle of length four or more has a chord, i.e., an edge joining two nonconsecutive vertices.*

Theorem 1 ([15], Ch. 3, Theorem 8) *The probability distribution P with a chordal Markov graph G can be written as a product of the distributions of the maximal cliques of G divided by a product of the distributions of their intersections.*

By dividing the values of the intersections of each clique among the cliques, it is possible to represent the joint probability distribution as a product of values associated with the maximal cliques of G. So, the clique parameters of MRFs on chordal graphs can be determined in a straightforward manner from the low-order joint probabilities associated with maximal cliques in the Markov graph. We mainly have good insights about the low-order cliques, however, so it is useful to turn this product into one that involves all the cliques of G. It can be shown that the the joint distribution can be represented as a product of values $g(C)$ associated with *all* the cliques of G [21]. The $g(C)$ are defined to be:

$$g(C) = P(C) \text{ if } c \text{ is a unary clique}$$

$$g(C) = \frac{P(C)}{\prod_{C_j \subset C} g(C_j)} \text{ otherwise.}$$

Thus, for binary cliques,

$$g(C) = \frac{P(l_1, l_2)}{P(l_1)P(l_2)}, \tag{5}$$

where l_1 and l_2 are the labels at the two sites in the clique. For a clique of order 2 or greater, if $g(C)$ has a value of 1, then that means that all the interactions within the clique are determined by the lower-order cliques.

8.3 Clique Parameters for General Graphs

Consider a simple non-chordal graph such as the ring, with every site adjacent to a clockwise and counterclockwise neighbor. Suppose we apply the formula of Eq. 5 to an MRF defined on this graph. Although the graph is not chordal and the theorem may not hold, the formula is still well-defined. How good an approximation to the correct probability distribution can we obtain? One can show that when *any* neighboring pair of variables is, in fact, independent, the formula is exact [21]. Experiments show that when any pair of variables is close to being independent, the formula provides clique values that derive good approximations to the true joint probabilities.

A similar result holds for more general graphs: if enough cliques in the MRF graph G are independent so that deleting their edges from G results in a chordal graph, then the distribution is described exactly by the formula of Eq. 5 applied to G.

8.4 Experimental Results

The experiments applied the clique estimation method introduced in the previous section to the edge detection problem introduced in Section 4. Although there are violations to the conditions under which the parameter estimation technique works best, the resulting clique values give segmentations almost as good as can be obtained by careful hand-tuning, as can be seen in Figure 15. Experiments show that significant deviations from the correct clique parameters severely disrupt the segmentation. (See [21].)

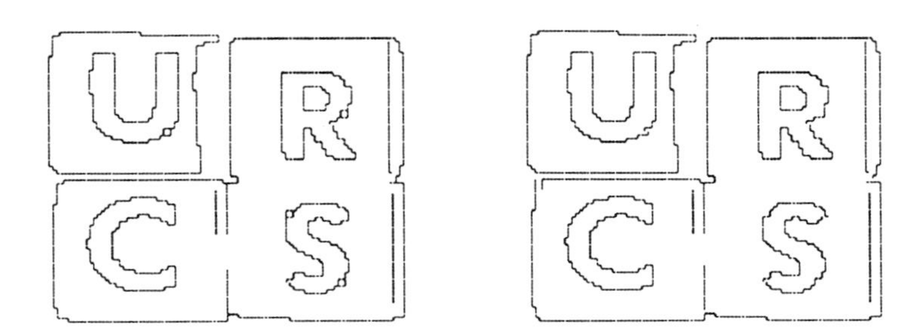

Figure 15: URCS image segmentations obtained with Local HCF using clique potentials derived from the statistics of the (left–URCS, right–P) image.

8.5 HCF Using Local Probabilities

Instead of being defined in terms of the Gibbs distribution, HCF may be defined in terms of the local probabilities of the MRF. This saves the inaccuracies involved in converting from the local probabilities to the Gibbs clique values.

The probability-based HCF is the same algorithm as given in Figure 1. The only change is the definition of *stability.* Now we define the conditional probability,

$$P'_s(\omega_s) = P(X_s = \omega_s | X_r = \omega_r, r \in S, r \neq s, r \text{ has committed}),$$

and *stability* to be:

$$G_s(\omega) = - \min_{k \in L, k \neq \omega_{min}} \log \frac{P'_s(k)}{P'_s(\omega_{\min})}$$

for an uncommitted site, and

$$G_s(\omega) = \min_{k \in L, k \neq \omega_s} \log \frac{P'_s(k)}{P'_s(\omega_s)}$$

for a committed site.

Although the new algorithm is not exactly the same as the traditional HCF algorithm, when all the sites have committed, it is, and it is in the spirit of HCF when they are not.

For HCF, when all sites are committed, the energy difference minimized is:

$$E_s(k) - E_s(\omega_s) = \log P(X_s = k \wedge X - X_s = \omega) \\ - \log P(X_s = \omega_s \wedge X - X_s = \omega)$$

$$= \log \frac{P(X_s = k \wedge X - X_s = \omega)}{P(X_s = \omega_s \wedge X - X_s = \omega)},$$

$$= \log \frac{P(X_s = k | X - X_s = \omega)}{P(X_s = \omega_s | X - X_s = \omega)},$$

$$= \log \frac{P(X_s = k | X_r = \omega_r, r \in S, r \neq s)}{P(X_s = \omega_s | X_r = \omega_r, r \in S, r \neq s)},$$

which is the same as for probability-based HCF. When some sites are uncommitted, HCF attempts to consider information from only the neighboring sites that are committed, exactly as is done in probability-based HCF. Some of the time, the stabilities used by the two algorithms are equivalent; other times, the stability used by the clique-based HCF makes unwarranted independence assumptions that may make its stability value differ from the one using the true conditional probabilities. (See [21].)

A potential problem with using the neighborhood probability information directly is that there may be a large number of neighborhood configurations. For a neighborhood of size n, and k different states at each site, there are $k^{(n+1)}$ different configurations of the variable and its neighborhood. If we are storing information for every partial neighborhood configuration there, are $(k+1)^{(n+1)}$ different values to store. These numbers can sometimes be reduced by symmetries (as is the case for the segmentation examples).

9 Summary and Discussion

We have given a framework, based on Bayesian-probability theory, for posing (as labeling) and performing a range of image analysis tasks from early vision through segmentation, reconstruction, and finally recognition. The central issues addressed by this chapter are the representation of knowledge, reasoning procedures for combining distinct bodies of knowledge, and inference methods for using available knowledge to infer scene properties. The central idea of our approach to knowledge representation and reasoning is the decoupling of external evidence and *a priori* knowledge. Using joint likelihood ratios rather than probability distributions to accumulate evidence enables the integration of the *a priori* knowledge, encoded in terms of a joint probability distribution of all sites (MRFs), with the pooled external evidence in a Bayesian formalism. Knowledge is encoded as clique potentials expressing *tension* or *energy* in local configurations of labels. Cliques and their potentials represent constraints, including qualitative (*hard a priori*) domain-dependent truths, and *soft* probabilistic constraints that reflect statistical domain properties.

MRFs are convenient for the representation of labeling problems, and are particularly convenient for the expression of the arbitrary spatial relationships that arise in the representation of spatially complex objects. The coupling of two MRFs, each one addressing a different inference problem, is a powerful way to let two labeling problems interact in a uniform way. MRFs thus provide a means of attacking the *chicken and egg* problem of computer vision in which reconstruction depends on segmentation and vice versa, and there is a similar circular dependency between structure-finding and recognition.

We introduce the Highest Confidence First (HCF) algorithm and its parallel cousin, Local HCF. The basic computation performed in the labeling context is one of statistical estimation. HCF provides neither the traditional Maximum A Posteriori estimation (which is very difficult to compute) nor the traditional Maximum Likelihood Estimation, but rather a new type of estimation that treats individual variables differently in accordance with the relative significance of the variable observations. The underlying intuition of the Highest Confidence First estimation is simple: in deciding the identities of the variables, the use of contextual information becomes more important as the external observations become less informative. Traditional estimation treats all of the variables equally, thus their results are more likely to be affected by noise and the inaccuracy of prior models.

Although it does not guarantee global optimality, HCF effectively avoids minor local optima by a *greedy search* in an augmented space. The augmented space consists of 3^N elements, including the 2^N hypercube corners in the admissible solution space. The elements in the augmented space form $N + 1$ layers. Layer i consists of the elements with exactly i dimensions of values 0 or 1 (*committed*). The elements are conceptually connected in the sense that the computation starts at the 0-th layer (all *uncommitted*), and terminates at some element in the N-th layer—the admissible solution space. The connections between layers are uni-directional; the computation goes through the layers one by one, from lower-numbered ones to higher-numbered ones. There are also some intra-layer connections that allow the computation to fine tune its directions.

HCF is a serial, global algorithm that takes into account the goodness (compatibility with evidence and neighboring labels) of all sites. We also introduce Local HCF, which is a parallel version that only takes into account a local version of compatibility. It can run much faster than HCF, and yet often produces better results. Experimental comparisons between Local HCF and several other energy minimizing algorithms are provided.

MRFs are usually structured as regular grids, reflecting the lowest level of image structure. However MRF theory does not depend on the regular structure. As an application of MRFs with nonisotropic, non-homogeneous connectivity, we have turned to object recognition. We have described a coupled network that solves the recognition problem from uncertain information by inferring the solution to both the segmentation problem and the matching problem simultaneously. The network is applied to intermediate-level data in a Tinkertoy structure understanding domain (line segments, blobs adjacency relations of the sort that arise from low-level vision.) Within a probabilistic network framework, the evidence and relevant prior constraints interact to yield good global answers to segmentation and recognition, even when either problem on its own is underdetermined, and even when the local evidence is ambiguous or favors the incorrect interpretation. Visual inference decisions can be computed that would be very difficult to achieve with traditional vision system architectures.

Last, the determination of clique potentials (apriori knowledge about label configurations) is usually a time-intensive, ad-hoc task. However, the process can be automated so that the system learns clique potentials through time.

Bibliography

[1] J. E. Besag, "On the Statistical Analysis of Dirty Pictures, *Journal of the Royal Statistical Society*, Vol. 48, pp. 259–302, 1986.

[2] A. Blake, A. Zisserman, and A. V. Papoulias, "Weak Continuity Constraints Generate Uniform Scale Space Descriptions of Plane Curves," *Proceedings of the European Conference on Artificial Intelligence*, pp. 518–528, 1986.

[3] P. B. Chou and C. M. Brown, "The Theory and Practice of Bayesian Image Labeling," *International Journal of Computer Vision*, Vol. 4(3), pp. 185–210, June 1990.

[4] P. B. Chou, "The Theory and Practice of Bayesian Image Labeling," Technical Report TR 258, Department of Computer Science, University of Rochester, Rochester, New York, August 1988.

[5] P. R. Cooper, "Order and Structure in Stereo Correspondence," Technical Report 216, Department of Computer Science, University of Rochester, New York, June 1987.

[6] P. R. Cooper, "Parallel Object Recognition from Structure (the Tinkertoy Project", Technical Report 301, Deptartment of Computer Science, University of Rochester, New York, July 1989.

[7] C. David and S. W. Zucker, Potentials, Valleys, and Dynamic Global Coverings," Technical Report CIM 89-1, McGill Research Centre for Intelligent Machines, McGill University, Montreal, March 1989.

[8] H. Derin and H. Elliot, "Modeling and Segmentation of Noisy and Textured Images Using Gibbs Random Fields," *IEEE Transactions on Pattern Analysis and Machine Intelligence, PAMI*, Vol. 9, pp. 39–55, 1987.

[9] S. Geman and D. Geman, "Stochastic Relaxation, Gibbs Distributions, and the Bayesian Restoration of Images," *PAMI*, Vol. 6(6), pp. 721–741, November 1984.

[10] S. Geman and C. Graffigne, "Markov Random Field Image Models and Their Applications To Computer Vision," *Proceedings of the International Congress of Mathematicians*, pp. 1496–1515, 1986.

[11] A. Haken and M. Luby, "Steepest Descent Can Take Exponential Time for Symmetric Connection Networks," *Complex Systems*, Vol. 2, pp. 191–196, 1988.

[12] F-C. Jeng and J.W. Woods, "Texture Discrimination Using Doubly Stochastic Gaussian Random Fields," *Proceedings: ICASSP*, pp. 1675–1678, 1989.

[13] R. Kindermann and J. L. Snell, *Markov Random Fields and Their Applications*, American Mathematical Society, 1980.

[14] S. Mitter, J. Marroquin, and T. Poggio, "Probabilistic Solution of Ill-Posed Problems in Computational Vision," *Proceedings: DARPA Image Understanding Workshop*, pp. 293–309, December 1985.

[15] J. Pearl, *Probabalistic Reasoning in Intelligent Systems.* Morgan Kaufman, Los Altos, California, 1988.

[16] T. Poggio et al "The MIT Vision Machine," P. H. Winston and S. A. Shellard, editors, *Artificial Intelligence at MIT: Expanding Frontiers.* MIT Press, Cambridge, Massachusetts, 1990.

[17] T. Poggio, E. B. Gamble, and J. J. Little, "Parallel Integration of Vision Modules," *Science*, Vol. 242, pp. 436–440, 1988.

[18] J. Rose and G. Steele, "C*: An Extended c Language for Data Parallel Programming," Technical Report PL87-5, Thinking Machines Corporation, Cambridge, Massachusetts, April 1987.

[19] D. B. Sher, "A Probabilistic Approach to Low-Level Vision," Technical Report 232, Department of Computer Science, University of Rochester, Rochesters, New York, October 1987.

[20] J. Subrahmonia, Y.P. Hung, and D. B. Cooper, "Model-based Segmentation and Estimation of 3D Surfaces from Two or More Intensity Images Using Markov Random Fields," In *Proceedings: Int. Conf. on Pattern Recognition*, pp. 390–397, 1990.

[21] M. J. Swain, "Parameter Learning for Markov Random Fields," Technical Report, Department of Computer Science, University of Rochester, Rochester, New York, August 1990.

[22] M. J. Swain, L. E. Wixson, and P. B. Chou, "Efficient Parallel Estimation for Markov Random Fields," M. Herion, R. Schacter, L. N. Kanal, and J. Lemmer, editors, *Uncertainty in Artificial Intelligence 5*, pp. 407-422, North-Holland, Amsterdam, 1990.

Stereo Matching

Stephen T. Barnard
Artificial Intelligence Center
SRI International
Menlo Park, California

1 Introduction

This chapter describes a system called CYCLOPS that uses simulated annealing to find a dense correspondence between two stereo images and hence to build a three-dimensional representation of a scene. The primary motivation for this work has been to devise a method for stereo matching that is practical for real applications, grounded in established theory of Bayesian inference and statistical physics, and consistent with biological and psychophysical knowledge of stereo vision in humans and other animals. CYCLOPS is presented here foremost as a practical application of Markov random fields (MRFs); the theory of MRFs will be treated only to the degree necessary to describe and interpret the algorithm. Nevertheless, some of the techniques used by CYCLOPS have some unusual, if not unique, features that theorists may find of interest. Among these are microcanonical annealing and multigrid annealing.

CYCLOPS poses stereo matching as a regularized optimization problem. It seeks the simplest correspondence between two images that leads to acceptable photometric error. Simulated annealing was chosen as the optimization method for several reasons, the most important being that it is robust, providing the means to escape from local minima. Simulated annealing employs a very simple algorithm to produce good average-case performance on large, nonlinear optimization problems. The disadvantage is that it requires a large amount of computation compared to deterministic methods, but this is mitigated by two factors: (1) the availability of parallel computers that can run simulated annealing algorithms very efficiently,

The work described in this chapter was supported under DARPA contracts MDA903-86-C-0084, DACA76-85-C-0004, and 89F737300. Use of the Connection Machine was provided by DARPA.

ISBN 0-12-170608-7

and (2) a very specialized representation, tailored to the properties of the stereo matching problem, that reduces the search space to a minimum.

The following section discusses some background facts about stereo vision. In particular, the essential geometrical relationships are defined, the essential problem (i.e., the *correspondence problem*) is posed, and three classes of computational approaches to solving it are discussed. Next we proceed with a detailed discussion of the CYCLOPS algorithm, beginning with a formulation of the correspondence problem as a combinatorial optimization. This section describes an MRF representation specifically tailored to the particular characteristics of stereo matching and implemented on a data-parallel SIMD computer (a Connection Machine). Some of the features of the system discussed in detail are multigrid annealing, microcanonical annealing, and three-pools representation. CYCLOPS is interpreted from two perspectives: Bayesian MAP estimation and physics. Analogies between CYCLOPS and stereo vision in higher mammals are suggested. Finally, implementation issues specific to SIMD parallelism are discussed, and the auxiliary functions of the system that support cartographic applications are briefly described.

2 Basics of Stereo Vision

Two eyes or cameras looking at the same objects from different perspectives provide the means to determine three-dimensional shape and position. Scientific investigation of this effect (called variously stereo vision, stereopsis, or single vision) has a rich history in psychology, biology, and, more recently, in the computational modeling of perception. The human visual ability to perceive depth is both commonplace and puzzling: one perceives three-dimensional relationships effortlessly, but the means by which one does so are largely unknown and hidden from introspection. Stereo vision, however, is one way to perceive depth that is relatively well understood from a computational standpoint. Stereo is an important method for machine perception because it leads to relatively direct measurements and, unlike monocular techniques, does not infer depth from weak and unverifiable photometric and statistical assumptions, nor does it require specific detailed models of objects. Once two stereo images are brought into point-to-point correspondence, recovering depth by triangulation is straightforward.

The geometric principle behind stereo vision, illustrated in Fig. 1, is quite simple. Assume that two cameras form images through left and right centers of perspective l and r, onto planes L and R. Furthermore, assume that the cameras are fixed upon point v, which is to say that the

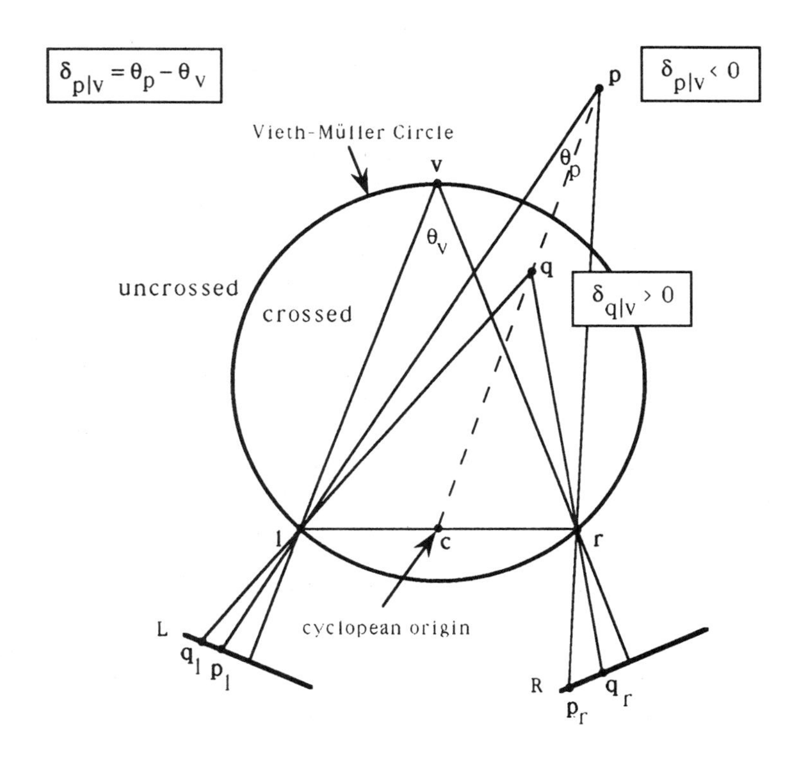

Figure 1: Basic stereo geometry.

two rays perpendicular to the image planes passing through the centers of perspective (the *principle rays*) intersect at v. Let θ_v be the angle between these principle rays. Now consider another point p projected onto image planes L and R as shown, and let the angle between these rays be θ_p. We say that the disparity of p with respect to v is $\delta_{p|v} = \theta_p - \theta_v$. The circle through l, r, and v (actually a sphere) is called the *Vieth-Müller circle.* Disparity is positive for points inside, negative for points outside, and zero for points on the circle. Positive and negative relative disparities are sometimes referred to as *crossed* and *uncrossed*, respectively.

Clearly, disparity is related to distance. Let $c = (l + r)/2$ be the midpoint between the focal points (the *cyclopean origin*), and p and q be two points on a line with c; if $|p-c| > |q-c|$ (i.e., p is farther from the observer than q) then $\delta_p < \delta_q$. The two projections of a single point in the scene are called *conjugate points*, and they determine the location of the point in the scene uniquely. This property of conjugate points establishes the importance of the *correspondence problem*: how can one determine which pairs of points in the two images correspond to actual points on surfaces in the scene? The scene is presumably made of light-reflecting surfaces, and individual markings on these surfaces will, in some sense, look about the same in the two images. In realistic imagery the correspondence problem is much harder than it may first appear, however, due to many complicating factors such as occlusion, periodic surface markings, surface areas with no markings at all, distortion of the surface markings in the images dues to perspective projection, sensor noise, optical distortion, etc.

Historically, the computational modeling of stereo vision has been driven by two motivations. The practical applications of automated stereo are so important, especially in cartography and robotics, that many engineering-oriented approaches have been tried. These often use "correlation" techniques: patches of intensities in one image are searched for in the other image by maximizing a measure of correlation or minimizing a measure of error. The other motivation is the desire to model biological stereo, and these approaches are typically feature-based: discrete local features (usually edges) are matched across images. A third approach, which could be called the optimization approach, represents a dense disparity map as the state variable of a system, usually defined on a two-dimensional grid. Stereo matching is then formulated as an optimization problem: find the best disparity map by maximizing an objective function that measures the "quality" of the state, or, equivalently, minimizing an "energy" function that measures the lack of quality. CYCLOPS is an example of an optimization method.

3 The CYCLOPS Algorithm

CYCLOPS uses a multigrid, stochastic-optimization algorithm to minimize a discrete version of the *stereo constraint equation*

$$\mathcal{E} = \int\int \left[(\mathcal{L}(x - \frac{\mathcal{D}}{2}, y) - \mathcal{R}(x + \frac{\mathcal{D}}{2}, y))^2 + \lambda |\nabla \mathcal{D}|^2 \right] dx dy \ , \tag{1}$$

where $\mathcal{L}$ and $\mathcal{R}$ are functions of image irradiance in the left and right views, $\mathcal{D} = \mathcal{D}(x, y)$ is a disparity map (a cyclopean map in this case), and λ is a regularization constant. The first term in the integrand is a measure of the photometric error associated with $\mathcal{D}$, while the second term (proportional to the squared magnitude of the gradient of disparity) is a measure of the first-order complexity of $\mathcal{D}$. By minimizing $\mathcal{E}$ one finds the flattest map with the least photometric error, with the trade-off between error and complexity specified by the regularization constant λ.

Note that in Eq. (1) disparity is a scalar function: corresponding points between $\mathcal{L}$ and $\mathcal{R}$ are assumed to have the same y coordinate, so disparity can be characterized as a shift in x alone; that is, there is no y disparity. This assumption involves no loss of generality, however, because the *epipolar constraint* provides a way to use camera-model information to remove y disparity (Fig. 2). Any point in three-dimensional space, together with the centers of projection of the two camera systems, defines a plane called an *epipolar plane*. The intersection of an epipolar plane with an image plane is called an *epipolar line*. Every point on an epipolar line in one image must correspond to a point on the corresponding epipolar line in the other image. The epipolar constraint therefore limits the searches for the matches to one-dimensional neighborhoods, as opposed to two-dimensional neighborhoods, with an enormous reduction in computational complexity. One can use the camera model to map the two images into a common rectification plane in which the epipolar lines are horizontal.

The discrete version of Eq. (1) is

$$E = \sum_{i,j} [\ (L(i - \left\lfloor \tfrac{D(i,j)}{2} \right\rfloor, j) - R(i + \left\lceil \tfrac{D(i,j)}{2} \right\rceil, j)^2 + \lambda \left| \nabla D(i,j) \right|^2], \tag{2}$$

where L and R are sampled versions of $\mathcal{L}$ and $\mathcal{R}$, D is a discrete, integer-valued disparity map, and the operator $|\nabla \cdot|^2$ is now proportional to a discrete approximation of the squared gradient magnitude

$$\left| \nabla D(i,j) \right|^2 \approx \sum_{(k,l) \in \mathcal{N}_{i,j}} [D(i,j) - D(k,l)]^2 \ , \tag{3}$$

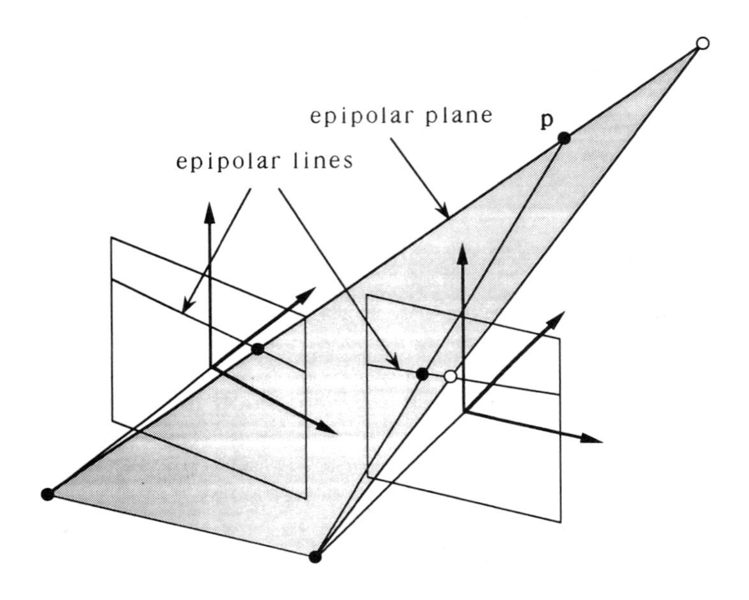

Figure 2: The epipolar constraint. Two camera systems with different orientations are shown. A point p in space and the two centers of projection define an epipolar plane (shaded) that interects the images in two epipolar lines. Any other point on the plane (for example, the point shown as an open circle) must project to image locations on the epipolar lines; therefore, the searches for corresponding points can be restricted to one dimension.

with $\mathcal{N}_{i,j}$ being the four nearest neighbors of grid location (i, j).

Simulated annealing provides a simple, yet subtle, method for dealing with combinatorial optimization problems such as minimizing Eq. (2) [8]. The essential concept is based on a thermodynamic computational metaphor: Treating the energy function (such as Eq. (2)) as the potential energy of a physical system, first heat the system to an equilibrium ensemble of high-energy states, and then gradually cool it, maintaining thermal equilibrium, until the temperature is zero and the ensemble of states "collapses" into a single minimum-energy state. Simulated annealing uses a Monte Carlo technique, originally developed in statistical physics, to generate a Markov chain of states (an MRF) sampled from the equilibrium ensemble of states at some temperature.

The most fundamental result of statistical physics states that at equilibrium the probability density of states is given by the *Boltzmann distribution*

$$P_{eq}(k) = \frac{\exp\left(\frac{-E_k}{\beta}\right)}{Z(\beta)}, \tag{4}$$

where $P_{eq}(k)$ is the probability of the system being in state k with energy E_k and β is the temperature of system. The function in the denominator is the *partition function*

$$Z(\beta) = \sum_k \exp\left(\frac{-E_k}{\beta}\right),$$

where the summation is over all states accessible to the system. If one wishes to measure macroscopic properties of some model system, such as the average energy at some temperature, Eq. (4) allows one to do so, in principle. One could simply generate many random states and weight them according to their probabilities. Unfortunately, since any random state is likely to have a very high energy, and hence a very low probability, this method will not work unless a prohibitive number of states are tried. This difficulty led to the development of *importance sampling* methods, in which states are not generated entirely at random but rather according to their probability. Metropolis *et al.* [10] described a simple algorithm for sampling an equilibrium ensemble that works as follows:

1. Begin in any state k. $t \leftarrow 0$.
2. $k' \leftarrow$ perturb(k) (Make a small change to k, typically changing only one degree of freedom.)
3. Let $\Delta E = E(k') - E(k)$.

4. If $\Delta E \leq 0$, then accept the change ($k \leftarrow k'$). Otherwise, accept the change with probability $\exp(-\Delta E/\beta)$.

5. $t \leftarrow t+1$. Go to step (2).

It can be shown that $P(k) \rightarrow P_{eq}(k)$ as $t \rightarrow \infty$. Simulated annealing uses the Metropolis algorithm (or perhaps a similar technique) by employing a *temperature schedule* to control the rate of change of β.

A simulated annealing algorithm using the standard Metropolis technique could be applied to the stereo matching problem directly:

1. Begin with a random disparity map D and a high temperature β.

2. Make a small change to D: $D' \leftarrow \text{perturb}(D)$.

3. Let $\Delta E = E(D') - E(D)$.

4. If $\Delta E < 0$, then accept the change ($D \leftarrow D'$). Otherwise, accept the change with probability $\exp(-\Delta E/\beta)$.

5. Decrease β according to some temperature schedule.

6. If $\beta = 0$, then done. Otherwise, go to step (2).

The perturbation function could be realized in a number of ways. For example, disparity could be randomly increased or decreased by one, or an entirely new disparity could be chosen from some fixed range. Various temperature schedules could be used; in general, the more slowly one cools the system the more likely one is to attain a final state close to a global optimum.

This direct approach will work if the range of disparity is not very large and if L and R are accurate measurements of image irradiance. For stereo data of realistic range and quality, however, it will either take too long to produce a useful answer, or it will produce incorrect disparities due to systematic errors in the data. Instead, CYCLOPS uses a modified simulated annealing algorithm that employs three more-or-less independent techniques to limit search and improve efficiency: multigrid annealing, the three-pools representation, and microcanonical annealing.

3.1 Multigrid Annealing

One of the most important features of stereo is that disparity scales linearly. If we reduce the size of the images by, say, a factor of 2, then the values in a disparity map are also reduced by a factor of 2. More precisely, given a disparity map $\mathcal{D}_0$ defined on the unit square, we can construct a series of reduced maps

$$\mathcal{D}_l(x, y) = \frac{1}{2}\mathcal{D}_{l-1}(2x, 2y) \quad 0 \leq x, y \leq \left(\frac{1}{2}\right)^l$$

that specify the correspondence between two sequences of reduced images,

$$\mathcal{L}_l(x, y) = \mathcal{L}_{l-1}(2x, 2y) \quad 0 \leq x, y \leq \left(\frac{1}{2}\right)^l$$

and

$$\mathcal{R}_l(x, y) = \mathcal{R}_{l-1}(2x, 2y) \quad 0 \leq x, y \leq \left(\frac{1}{2}\right)^l .$$

In the discrete domain, assuming for simplicity that L_0, R_0, and D_0 are defined on a $2^N \times 2^N$ grid, we can construct pyramid structures

$$\{L_l\}, \{R_l\}, \{D_l\} \quad 0 \leq l \leq N .$$

Each pyramid has $N + 1$ levels, with the grid size at level l being $2^{N-l} \times 2^{N-l}$.

The advantage of this representation is obvious. By starting at some very coarse level of resolution l_{max}, we can construct a coarse disparity map fairly quickly because the grid size and the range of disparity values are small. Using the linear scaling property of disparity, we can then interpolate the coarse map onto the next larger grid $l_{max} - 1$. This new map then becomes the starting point for another search. This process continues through the pyramid to the highest resolution:

1. Construct image pyramids $\{L_l\}$ and $\{R_l\}$, $0 \leq l \leq l_{max}$.
2. $l \leftarrow l_{max}$. $D_l \leftarrow 0$.
3. $D_l \leftarrow$ minimize(E_l).
4. If $l = 0$, then done. Otherwise, $D_{l-1} \leftarrow$ interpolate(D_l).
5. $l \leftarrow l - 1$. Go to step (3).

The multigrid algorithm is not only much more efficient than the direct approach, but it also allows one to correct for systematic errors in the data by using bandpass pyramids. This is important in real applications because raw image intensities are not usually invariant across stereo images for a variety of reasons, including different sensor biases and illumination effects. These sources of error, however, affect primarily the low-frequency components of the images. Using bandpassed images provides a kind of automatic gain control, similar to the effect of lateral inhibition in neural systems. The multigrid strategy of CYCLOPS starts by converting each epipolar-corrected image into a bandpass pyramid, in which the image is separated into band-limited components separated by one octave. A typical pair of cartographic stereo images and their associated bandpass pyramids are shown in Fig. 3.

There is a particularly efficient method of computing this structure (see Burt [2]) that works as follows. Suppose we have a digital image $I(i,j)$ with dimensions $2^N \times 2^N$. A low-pass, approximately Gaussian pyramid is defined recursively by

$$g_0(i,j) = I(i,j)$$

and

$$g_l(i,j) = \sum_{m=-2}^{2} \sum_{n=-2}^{2} w(m,n) g_{l-1}\left(i + m2^{l-1}, j + n2^{l-1}\right)$$

for l up to $\log N$. The array $w(m,n)$ is a symmetric, separable generating kernal chosen to be approximately Gaussian. Note that g_l is a weighted average over a 5×5 pattern of samples in g_{l-1} and that the sample spacing, 2^{l-1}, doubles across each level. Next, bandpass components are constructed as differences-of-Gaussians:

$$b_l(i,j) = g_l(i,j) - g_{l+1}(i,j) \quad l = 0, \cdots, (\log N) - 1 \; .$$

This structure is sometimes called a Laplacian pyramid because it provides a good approximation to the Laplacian of the Gaussian kernal, $\nabla^2 G$.

CYCLOPS constructs truncated bandpass pyramids from each epipolar-corrected image:

$$\{L_l\}, \{R_l\} \quad 0 \le l \le l_{max} \; .$$

The lowest level of resolution, l_{max}, is determined by the size of the original images and by a limit on the range of disparity between them. For typical aerial images with which we work, the lowest resolution in the pyramids would be 32×32. If the original images were 1024×1024, the pyramids would be six levels deep and the system would accommodate a disparity range of 127 pixels.

3.2 The Three-Pools Representation

To maximize the efficiency of the simulated-annealing search at each level, CYCLOPS uses a specialized representation of disparity that reduces the

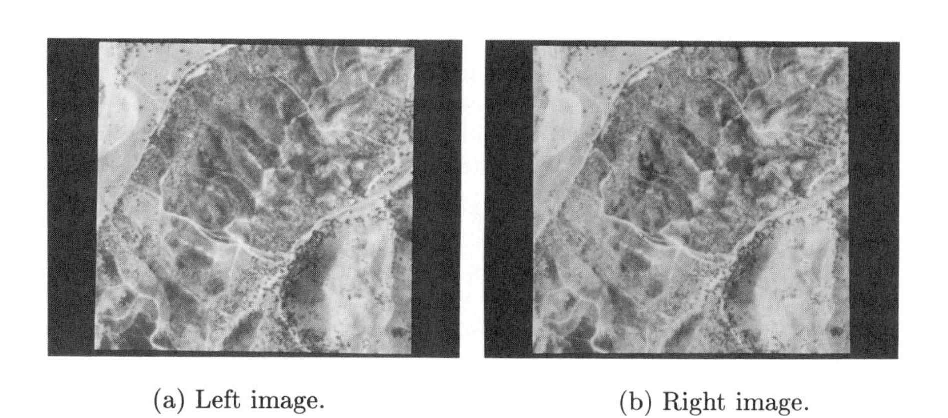

(a) Left image. (b) Right image.

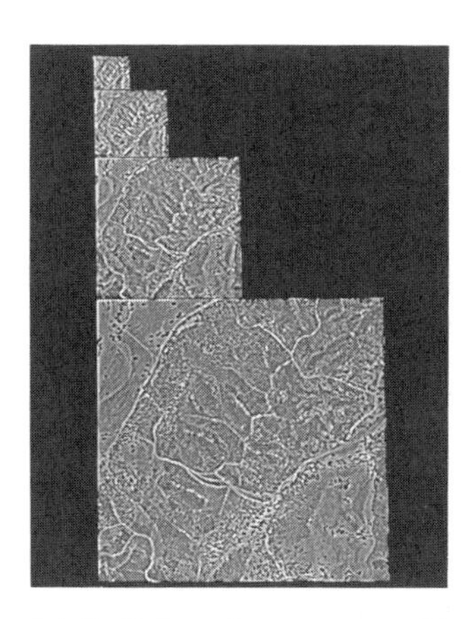

(c) Left bandpass pyramid.

(d) Right bandpass pyramid.

Figure 3: A cartographic example.

number of accessible states to a minimum. This is called the three-pools representation because it was originally suggested by a result from psychophysics: studies of human subjects with abnormal stereo vision indicate that there are three pools of disparity detectors, corresponding to positive (crossed), negative (uncrossed), and near-zero disparities [11].

The three-pools representation (illustrated in Fig. 4) works as follows. At each level of the resolution pyramid the disparity map D_l is the sum of a constant base disparity D_l^{base} (inherited by interpolation from the previous level) and a variable incremental disparity δ_l:

$$D_l(i,j) = D_l^{base}(i,j) + \delta_l(i,j),$$

with $\delta_l(i,j) \in \{-1,0,1\}$. Now δ_l is the state variable (the MRF at level l). Incremental disparity is interpreted as a correction to the coarse answer obtained from the previous level. In this way the system can handle a substantial range of disparity, but the search space at each level of the pyramid is kept as small as possible. The range of permissible disparities grows exponentially with the number of levels, with the range at level l given by

$$\left[-(2^{l_{max}-l+1}-1), 2^{l_{max}-l+1}-1\right] .$$

The interpolation is given by

$$D_l(i,j) = \frac{1}{2} \sum_{m\in\{0,1\}} \sum_{n\in\{0,1\}} D_l(\left\lfloor \frac{i+m}{2} \right\rfloor, \left\lfloor \frac{j+n}{2} \right\rfloor) .$$

3.3 Microcanonical Annealing

The method CYCLOPS uses for minimizing E_l, called microcanonical annealing, is somewhat different from the standard forms of simulated annealing. It has been adapted from a result in statistical physics by Creutz [4]. The advantage of microcanonical annealing is that it is particularly efficient on parallel computers with no floating-point hardware. It is also of some pedagogical interest because of its dual relationship to the standard techniques.

The Metropolis algorithm described previously is sometimes called the *heat bath* algorithm because it simulates a model immersed in a much larger system at temperature β (the heat bath) with which the model freely exchanges energy. The ensemble that the Metropolis algorithm samples is called the *canonical ensemble.* The Creutz algorithm, on the other hand, simulates a model thermally isolated from its surroundings, and therefore with constant energy — the microcanonical ensemble. The problem in designing such an algorithm is that a state transition must leave the system's

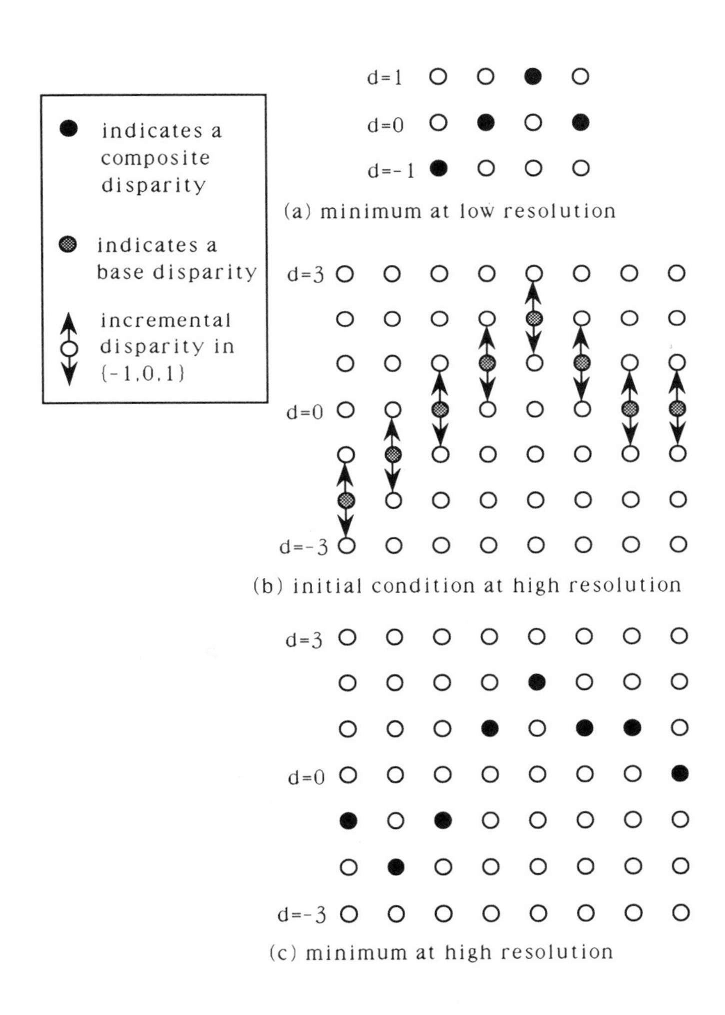

Figure 4: The three-pools representation.

total energy unchanged. Creutz accomplished this by augmenting the system with (at least) one degree of freedom, called a *demon*, that represents the kinetic energy of the system, E_D.

1. Begin in some state k with demon energy $E_D > 0$.
2. $k' \leftarrow \text{perturb}(k)$.
3. Let $\Delta E = E(k') - E(k)$.
4. If $\Delta E \leq E_D$, then accept the change $(k \leftarrow k')$ and subtract the energy from the demon $(E_D \leftarrow E_D - \Delta E)$. Otherwise, reject the change.
5. Go to step (2).

In this algorithm state transitions that lead to decreased potential energy are always accepted, while transitions that lead to increased potential energy are accepted only if the demon has enough energy to give up. E_D remains nonnegative.

Like the Metropolis algorithm, the Creutz algorithm generates a sequence of states with a probability density that converges in the limit to the Boltzmann distribution. Unlike the Metropolis algorithm, the Creutz algorithm does not require the evaluation of the exponential function, is naturally implemented with integer arithmetic, and requires the generation of far fewer random numbers. In the three-pools representation, for example, a state transition is generated simply by choosing a new incremental disparity, requiring only one random bit. The Metropolis step of accepting a state transition with probability $\exp(-\Delta E/\beta)$, which requires the generation of a high-precision random number, is avoided altogether.

Perhaps the most interesting property of microcanonical annealing is that, unlike the canonical algorithm, it doesn't use temperature as a control parameter. Instead, temperature is measured as a statistical feature of the system, the average demon energy

$$\beta = \langle E_D \rangle \ .$$

The parallel version of the Creutz algorithm uses a grid of demons (see Section 4.4) and the temperature is computed by averaging spatially over the grid. Heating and cooling are controlled by adding Δ_H or subtracting Δ_C from every demon after every update of the entire grid. At each level the system is heated until the temperature exceeds β_{max} and then cooled until it is less than β_{min}. The map is then interpolated to set the initial condition for the next level, and the heating and cooling processes are repeated, as described in Section 3.1.

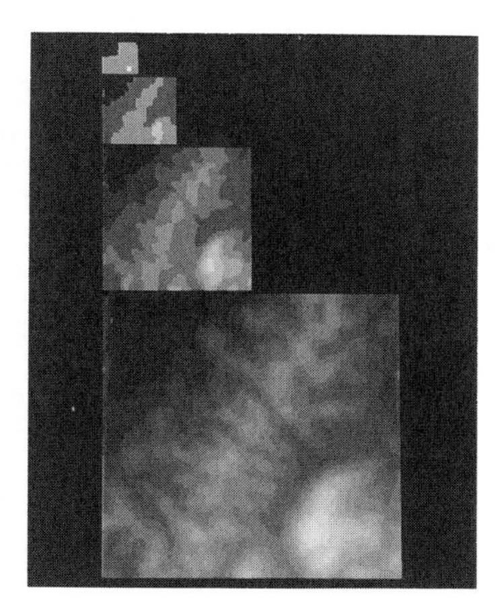

Figure 5: Disparity maps computed for the cartographic example.

3.4 Results

Figure 5 shows disparity maps at each level of resolution for the cartographic images shown earlier in Fig. 3. Gray values are assigned to disparities so that areas of higher elevation appear brighter. Also, the disparity maps are rescaled across levels to create comparably shaded images. The original images are of size 256 × 256. A four-level pyramid is used, with grids of size 32 × 32 up to 256 × 256. In the three-pools representation this is sufficient to accommodate disparities in the interval $[-15, 15]$. The annealing parameters were $\Delta_H = 10$, $\Delta_C = 3$, $\beta_{max} = 300$, and $\beta_{min} = 10$. Heating can proceed faster than cooling because the system reaches equilibrium much more quickly at higher temperatures. We find that a value of $\lambda = 64$ works well for images that are properly digitized into eight-bit pixels (using Eq. (3) for the gradient operator). This example required about four minutes to compute using a Connection Machine with 4K processors (the minimum configuration).

CYCLOPS is highly reliable for matching cartographic images, assuming that camera-model information is available to enforce the epipolar constraint. Figures 6 and 7 show the results for a quite different example — a ground-level, oblique view of a scene with abrupt depth discontinuities. All

(a) Left image.

(b) Right image.

(c) Left bandpass pyramid.

(d) Right bandpass pyramid.

Figure 6: A ground-level, oblique example with occlusion.

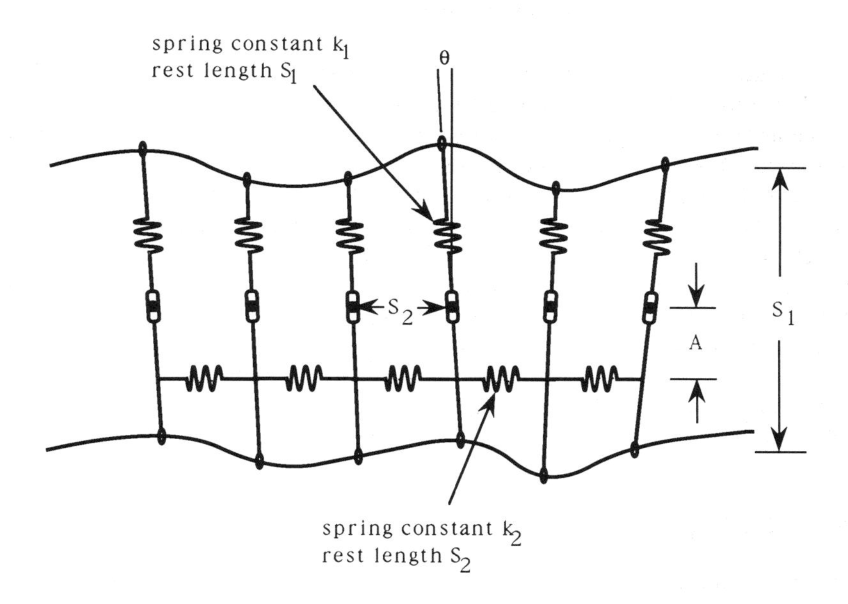

Figure 7: Disparity maps computed for the oblique example.

parameters are identical to those of the previous example, except that a five-level pyramid is used to accommodate the greater range of disparity in this pair of images. It is somewhat surprising that CYCLOPS has produced a reasonably good result, considering that it has no representation for occlusion, and that it computes a disparity for every grid point, even in areas that are not visible in one image or the other. We have experimented with coupled line processes for modeling occlusions, but so far the additional computational burden has not been justified by the marginally improved results.

It is known that stereo fusion in human vision is subject to a disparity-gradient limit [3], which requires that for fusion to occur the rate of change of disparity with respect to visual angle (a dimensionless number) must be bounded. Most observers have a limit of about 1. It has been shown that if an imaged surface has a disparity-gradient limit of less than two, then the left and right images are *topologically equivalent*; that is, all points in one image are visible in the other (i.e., the surface is not self-occluding), and furthermore the ordering of points is the same in both images [12]. This suggests that when we view stereoscopically a scene such as that in Fig. 6, we cannot simultaneously fuse both the foreground figure and the occluded background. Presumably, we use eye movements to bring the foreground and background into fusion separately, and then somehow integrate this information into a single, coherent perception of depth.

4 Interpretations of CYCLOPS

One of the interesting features of simulated annealing is that it can be interpreted and understood from different perspectives. Because the original concept of the Metropolis algorithm was developed for simulation of model systems in statistical physics, it is natural to look for physical analogies even when these methods are used for abstract optimization problems. More recent formalizations within the framework of Markov Random Fields lead to interpretations in terms of Bayesian inference. In addition, we can identify several points of similarity between CYCLOPS and the visual systems of humans and other "higher" mammals.

4.1 A Physical Analogy

The discussion of microcanonical annealing in Section 3.3 established one level of physical analogy by distinguishing the potential energy of Eq. (2) from the kinetic energy of the demons. Another kind of physical intuition can be obtained by considering the stereo constraint equation as the

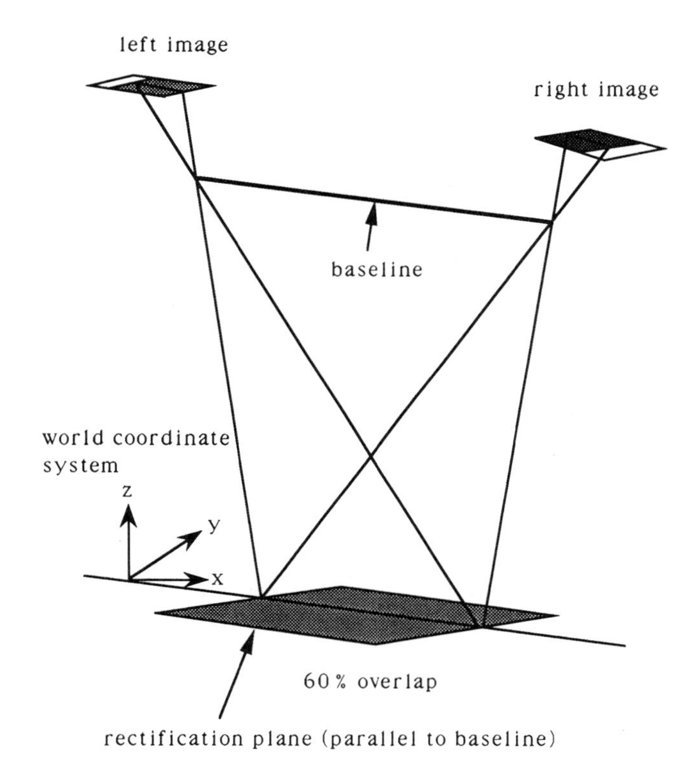

Figure 8: Spring model.

potential energy of the system of spring-loaded and spring-coupled levers illustrated in Fig. 8. This system consists of two surfaces $\mathcal{L}$ and $\mathcal{R}$, corresponding to the image irradiance functions, separated by a distance S_1. Between these surfaces is a grid of pivot points and spring-loaded and spring-coupled lever arms. The lever arms can rotate in the plane, but their endpoints are constrained to lie on the surfaces. The angle θ of a lever arm represents a disparity. It is easy to show that the potential energy in this system is proportional to the quantity $\mathcal{E}$ given by Eq. (1) [1]. The regularization constant is determined by the stiffness and the geometry of the system

$$\lambda = \left(\frac{k_2}{k_1}\right)\left(\frac{A}{S_1}\right) .$$

4.2 Bayesian MAP Estimation

A quite different interpretation of the computation is that of maximum *a posteriori* (MAP) estimation [7]. We observe $\{L, R\}$, which contains information about D. The Bayesian approach is to find D that maximizes $P(D|\{L, R\})$. Applying Bayes's Rule we have

$$P(D|\{L,R\}) = \frac{P(\{L,R\}|D)P(D)}{P(\{L,R\})}.$$

The probability in the the denominator, $P(\{L, R\})$, is constant, so the posterior probability $P(D|\{L, R\})$ is proportional to the numerator:

$$P(D|\{L,R\}) \propto P(\{L,R\}|D)P(D) . \tag{5}$$

The first factor in the right-hand side of Eq. (5), $P(\{L, R\}|D)$, is measured by the photometric error in the stereo constraint equation. (Given a disparity map, image pairs with less error are more likely.) The second factor, the prior probability $P(D)$, is measured by the variational term. (Disparity maps that are flatter are more likely.)

The disparity map is an MRF whose distribution of energies is guaranteed to be Boltzmann in the limit. Substituting the probabilities in Eq. (5) with Boltzmann distributions, we have

$$\begin{aligned} P(D|\{L,R\}) &\propto \exp\left(\frac{-E_1(L,R,D)}{\beta}\right)\exp\left(\frac{-E_2(D)}{\beta}\right) \\ &\propto \exp\left(\frac{-(E_1+E_2)}{\beta}\right) \\ &\propto \exp\left(\frac{-E}{\beta}\right), \end{aligned}$$

where E_1 and E_2 are the photometric error and the variational terms in E, respectively. Therefore, by minimizing E we maximize the posterior probability $P(D|\{L, R\})$.

4.3 Analogies to Biological Stereo Vision

Several nontrivial analogies can be made between CYCLOPS and current knowledge of the working of the human visual system.

4.3.1 Independent Spatial-Frequency Channels

There is considerable evidence that the human visual system has at least four, and possibly more, independent channels tuned to different spatial frequencies separated by about one octave [13]. The bandpass pyramids in CYCLOPS and similar structures in other models play the same role.

4.3.2 Sensitivity to Vertical Disparity

Human stereo vision can deal with a small amount of vertical disparity, but we resort to eye movements to align the visual fields as closely as possible [5, 6]. CYCLOPS uses information about the camera orientation to remove vertical disparity. Error in the camera model may cause some residual vertical disparity, but the simulated-annealing search is robust enough to handle this under reasonable conditions. Apparently, the search-space reduction of the epipolar constraint is so useful that vertical disparity is to be avoided as much as possible, but must be tolerated to degree, in both CYCLOPS and the human visual system.

4.3.3 Three Pools of Disparity

Studies of anomalous stereo vision suggest that there are three psychophysically distinct "pools" of disparity detectors, corresponding roughly to crossed (positive), uncrosssed (negative), and near-zero disparity relative to the vergence point [11]. CYCLOPS uses the same representation, coupled with several octaves of resolution hierarchy, to achieve efficiency without sacrificing dynamic range. At every level in the hierarchy, components of incremental disparity can assume only three local values: 1 (crossed), -1 (uncrossed), and zero. The base disparity, however, can grow by a factor of two across every level. The search space in any one level is therefore kept to a minimum, while the final composite disparity is allowed to have a substantial range.

4.3.4 Visual cortex

The first stereo computation in the human visual system occurs in the primary visual cortex. This part of the brain is remarkably uniform across its surface: a column of tissue about a millimeter square appears to characterize an elementary building block [9]. As a computational substrate, the visual cortex is then quite similar to a large grid of locally-connected processors. Since the cortical columns appear to be anatomically and functionally equivalent, they can be all be simulated by identical, locally interacting programs applied to different parts of the visual field, and therefore the visual cortex as a whole can be simulated efficiently by a fine-grained SIMD (single-instruction, multiple-data) architecture such as the Connection Machine.

4.4 Issues of Parallel Implementation

Implementing the CYCLOPS matching algorithm on the Connection Machine or on similar data-parallel computers is fairly straightforward because the simulated annealing algorithm is inherently parallel. Just as the molecules of a solid move simultaneously, the state variables of a simulated annealing algorithm can be updated in parallel. This section describes some of the techniques used in CYCLOPS.

One of the most important concepts in parallel computing is *virtualization.* The size of the problem may not, and usually will not, match perfectly the number of processors in the computer. For example, one may want to use a computer with 4K (64×64) processors to match stereo images of size 1024×1024. Virtualization is a way of having one physical processor do the work of a number of *virtual processors*, thereby presenting to the user a variable virtual geometry. The *LISP language of the Connection Machine provides very flexible and efficient tools for virtualization. This is especially important for multigrid algorithms, which must use several different virtual geometries at once, and which must transfer data between different geometries. The amount of time required to update an entire grid depends on the the *virtual-processor ratio*, which is the ratio of virtual processors to physical processors. (Ideally, the time should increase linearly with the virtual-processor ratio; in practice, however, the Connection Machine is relatively inefficient at low ratios.) The optimal granularity of the parallel system (that is, the ideal number of physical processors) is limited by the range of disparity the system must accommodate.

The microcanonical algorithm described in the previous section must be modified slightly for parallel implementation. The essential change is that CYCLOPS uses a grid of demons instead of just one. A state transition at a grid location is accepted or rejected according to the energy of the demon at that location. Temperature is measured by averaging the demon energies over the grid. It is important that the demons move about on the grid so that kinetic energy is transferred through the system effectively. CYCLOPS does this by applying a permutation to the demons after each iteration, moving each one to some random position, like shuffling a deck of cards. This operation is easy to do on the Connection Machine, which has a general mechanism for interprocessor communication. Some other parallel computers support only near-neighbor or row/column communication.

Other schemes for mixing the demons can be used on such systems. One method that works well is a kind of diffusion: simply replace the demon energies with an average of their four nearest neighbors.

It is important to realize that a state transition at one grid position must be made independently of the transitions of its neighbors. Given the neighborhood structure used in CYCLOPS (four nearest neighbors), this condition can be satisfied by dividing the grid into two disjoint "checkerboard" subsets:

$$S^{even} = \{(i,j) \mid \bmod_2(i+j) = 0\}$$

and

$$S^{odd} = \{(i,j) \mid \bmod_2(i+j) = 1\} .$$

The grid points S^{even} are the neighbors of S^{odd}, and *vice versa.* State transitions are made first at the locations in S^{even} and then at the locations in S^{odd}. Of course, this leads to some inefficiency because half the processors are idle at any time. It is possible to construct a virtualization in which a $2^N \times 2^N$ grid is represented by two smaller grids, each of size $2^{N-1} \times 2^N$, that correspond to S^{even} and S^{odd}. In principle, this should be more efficient because the virtual-processor ratio is half that of the more direct representation. On the Connection Machine, however, the complications this representation entails in the code add enough overhead to cancel almost all the theoretical improvement.

Constructing the bandpass pyramid in parallel requires a negligible amount of time compared to the annealing stages. At each level the local photometric errors (the terms in the first sum in Eq. (2)) at each grid point are tabulated into the three possible values they can attain. In this way the system avoids computing these squared differences more than once.

Since two stereo images will almost never overlap perfectly and there will be areas along the boundary that cannot be matched, handling the boundary condition of the photometric-error term can be tricky. It is possible for the algorithm to generate a state transition that matches a grid location in one image to a location off the grid in the other image. CYCLOPS uses a variable, stochastic boundary condition: if a disparity specifies an off-image match, the photometric error is taken to be equivalent to the current temperature. At high temperatures, therefore, the system will strongly resist accepting off-image matches, but as the temperature decreases such matches will be more likely to be accepted. In effect, there is a decreasing "pressure" forcing the matches into the interior of the images.

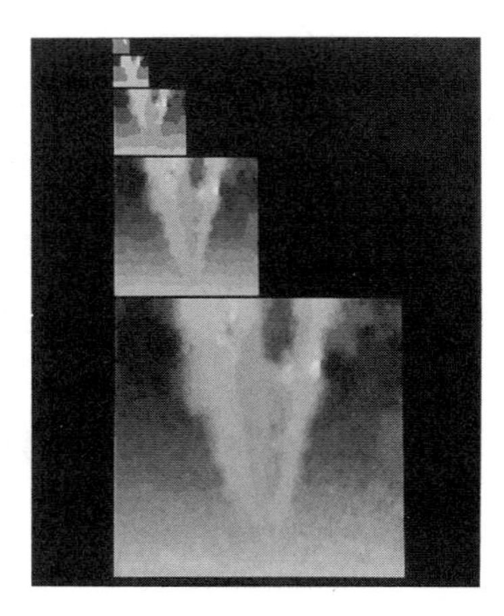

Figure 9: Epipolar correction.

4.5 Auxiliary Functions

Epipolar correction. CYCLOPS assumes that complete information about the camera models of the two stereo views is known. This model information includes, for each view, both the interior and exterior orientations. Interior orientation includes focal length, principal point, sampling rate, and possibly other parameters or even interpolation functions. Exterior orientation includes both the origin of the camera in an invariant world-coordinate system and an orthogonal transform matrix for translating between world and camera coordinates. This camera-model information is routinely compiled in standard cartographic practice. CYCLOPS uses the camera models to remove vertical disparity as shown in Fig. 9. A rectification plane is selected parallel to the baseline, passing through the origin of world coordinates, and approximately parallel to the two image planes. (The normal of the rectification plane bisects the angle between the principal rays of the two views.) The images are then backprojected onto this plane (a perspective transform) and then resampled using bilinear interpolation. In typical aerial imagery the views are spaced to give approximately 60% overlap in coverage. Images are chosen in the overlapping areas of the resampled images and given to the matching algorithm.

Subpixel estimates of disparity can be obtained by averaging over many states. When the finest level of resolution has been obtained, the system is not cooled to $\beta = 0$; instead, it is heated to a fairly high temperature, and a sequence of disparity maps is generated. The average disparity (over this sequence) quickly converges to a subpixel estimate.

4.5.1 Regular-grid Elevation Map

Once the disparities of the epipolar-corrected images are determined it is a relatively simple matter to compute the depth of each matched point. Using the camera model, rays from the centers of projection through the matched points are intersected to produce a grid of tuples in world coordinates:

$$\{x(i,j), y(i,j), z(i,j)\} \ .$$

Unfortunately, these tuples define an irregular mesh of points, while what is normally required in cartography is a grid of elevations at regular intervals of ground coordinates:

$$\{z(x,y) \mid x = x_0 + m\Delta x, y = y_0 + n\Delta y\} \ .$$

This is a standard resampling problem, equivalent to scan conversion in computer graphics. It is accomplished by triangulating the mesh of irregular points and then, in parallel, resampling each triangle on the regular grid. At the same time as the regular-grid elevation map is produced, the system creates a synthetic orthographic image of the terrain. CYCLOPS can also generate synthetic perspective views of the scene by texture mapping the ortho-image onto the grid of elevations and rendering the scene using a three-dimensional graphics package developed on the Connection Machine at SRI.

5 Summary

Markov Random Fields have proven to be an effective approach to the stereo matching problem. After posing the problem as an optimization of the stereo constraint equation, CYCLOPS uses a simulated annealing algorithm to approximate an optimal stereo correspondence. The algorithm uses a highly specialized representation tailored to stereo vision and is implemented efficiently on a parallel computer system. The robustness of the optimization algorithm yields consistent, high precision results, while the specialized representation and the parallel hardware lead to reasonable processing times for real applications. The solid theoretical basis for simulated annealing extends the approach beyond merely *ad hoc* heuristic search by clarifying the physical and logical meaning of the method.

The multigrid control structure of the CYCLOPS algorithm is crucial to its success. This structure allows the searches at higher resolutions to be guided by previous results at lower scales and, combined with the three-pools representation, drastically limits the search at any one scale without sacrificing the dynamic range of the system. Bandpass pyramids are effective for eliminating common systematic errors in stereo images. The significance of the microcanonical annealing algorithm lies primarily in that it eliminates the need for expensive floating-point hardware.

Currently the performance of CYCLOPS is adequate for cartographic applications, in which the need for high accuracy on large images is paramount. Applications that require real-time performance, such as vision for autonomous robots, await the development of cheaper, faster hardware. The technology of data-parallel computing is advancing rapidly, however, driven by the decreasing cost of memory and the increasing scale of circuit integration. I expect that in the near future we shall see many more practical applications of MRFs in a wide variety of problems.

Bibliography

[1] S. T. Barnard, *Stochastic stereo matching over scale*, International Journal of Computer Vision, 1989, pp. 17–22.

[2] P. J. Burt, *The Laplacian pyramid as a compact image code,* IEEE Trans. on Communications, 1983, pp. 532–540.

[3] P. J. Burt, and B. Julesz, *A disparity gradient limit for binocular fusion,* Science, 1980, pp. 615–617.

[4] M. Creutz, *Microcanonical Monte Carlo simulation*, Physical Review Letters, 1983, pp. 1411–1414.

[5] A. L. Duwaer and G. van den Brink, *What is the diplopia threshold?* Vision Res., 1981, pp. 295–309.

[6] A. L. Duwaer and G. van den Brink, *Diplopia thresholds and the initiation of vergence eye-movements,* Vision Res., 1981, pp. 1727–1737.

[7] D. Geman and S. Geman, *Stochastic relaxation, Gibbs distributions, and Bayesian restoration of images*, IEEE Proc. Pattern Analysis and Machine Intelligence, 1984, pp. 721–741.

[8] S. Kirkpatrick, C. D. Gelatt, and M. P. Vecchi, *Optimization by simulated annealing*, Science, 1983, pp. 671–680.

[9] D. H. Hubel and T. S. Weisel, *Brain mechanisms of vision*, in The Brain, W. H. Freeman and Co., New York, 1979.

[10] N. Metropolis, A. W. Rosenbluth, M. N. Rosenbluth, A. H. Teller, and E. Teller, *Equation of state calculations by fast computing machines*, J. of Chemical Physics, 1953, pp. 1087-1092.

[11] W. Richards, *Anomalous stereoscopic depth perception,* J. Opt. Soc. Amer., 1971, pp. 410–414.

[12] H. P. Trivedi and S. A. Lloyd, *The role of disparity gradient in stereo vision*, Perception, 1985, pp. 685–690.

[13] H. R. Wilson, and J. R. Bergen, *A four mechanism model for threshold spatial vision*, Vision Res., 1979, pp. 19–32.

3-D Analysis of a Shaded and Textural Surface Image

Rangasami L. Kashyap and Yoonsik Choe
School of Electrical Engineering
Purdue University, West Lafayette, Indiana

1 Introduction

An important task in computer vision is the recovery of 3-D scene information from single 2-D images. 3-D analysis of an image can be broken down into two main categories: *shape from shading* and *shape from texture*. *Shape from shading* technique uses the reflectance map, which shows scene radiance as a function of the surface gradient and the distribution of light sources, to extract 3-D surface information from image data [6, 15, 19]. On the other hand, *Shape from texture* analysis technique uses the texture pattern instead of shading to extract 3-D structure. Since texture gradients behave like intensity gradients, the shape of a surface can be inferred from the pattern of a texture on the surface by applying statistical texture analysis [23, 14, 22].

However, for describing a natural scene image, both these approaches have their own limitations. The *shape from shading* technique is applicable only under the assumption that the surface is smooth enough to have clear radiance information, while the *shape from texture* technique requires the surface to be relatively complex so that texture information can be extracted. Thus, neither technique is suitable to recover 3-D structure information from a natural scene, because both radiance and texture information coexist within the surface of a natural scene. Therefore, a robust technique is needed to handle this shortcoming. Recently, the fractal scaling parameter was introduced to measure the coarseness of the surface and was applied to represent the natural scene surface [20]. However, this fractal model is not enough to represent the real 3-D texture image, because even though two surfaces are estimated to have the same fractal scales, these surfaces can have different texture patterns.

In this paper, a 2-D orthographically projected *fractional differencing periodic* model, which is a composite model of *shape from shading* and *shape*

MARKOV RANDOM FIELDS
Theory and Applications

ISBN 0-12-170608-7

from texture, is developed to represent a 3-D surface image considering the scene image as the superposition of a smooth shaded image and a random texture image. The orthographical projection is adapted to take care of the nonisotropic distribution function due to the slant and tilt of a 3-D texture surface, and the fractional differencing periodic model is chosen because this model is able to simultaneously represent the coarseness and the pattern of the 3-D texture surface with the fractional differencing parameters c, d and the frequency parameters ω_1, ω_2, and it has the property of being flexible enough to synthesize both long-term and short-term correlation structures of random texture depending on the values of the fractional differencing parameter c and d. Since the object is described by a model involving several free parameters and the values of these parameters are determined directly from its projected image, it is possible to extract 3-D information and texture pattern directly from the given intensity values of the image without any preprocessing. Thus, the cumulative error obtained from each preprocessing can be minimized. For estimating the parameters, a hybrid method that uses both the least square and the maximum likelihood estimates is applied, and the estimation and the synthesis are done in frequency domain based on the local patch analysis. By using this model, the integrability problem that might occur in spatial domain analysis can be avoided, because only one inverse Fourier transform needs to be taken at the end of procedure to get the whole image.

The organization of this paper is as follows. In Section 2 we introduce the image model $z(l_1, l_2)$, which is obtained by superposing the deterministic function $x(l_1, l_2)$ and the random function $y(l_1, l_2)$, and the relationship between different directions of 3-D surface. In Sections 2.1, 2.2, and 2.3, an estimation scheme for the illumination direction, the modified reflectance map function for $x(l_1, l_2)$, and the orthographically projected fractional differencing periodic function for $y(l_1, l_2)$ are introduced. Section 3.1 outlines the estimation scheme for the parameters in the composite model. Section 3.2 then discusses some simulation results carried out to demonstrate the performance of the proposed algorithm Section 4 concludes the paper.

2 Model of 3-D Surface Image

The surface shape is usually defined in terms of the viewer's coordinate system. This system has axes l_1, l_2, l_3 with the l_3 axis in the viewing direction. The observed intensity function $z(l_1, l_2)$ of the local shape of a 3-D surface can be considered as the sum of a deterministic function $x(l_1, l_2)$ and a statistical random function $y(l_1, l_2)$, whose expected value is

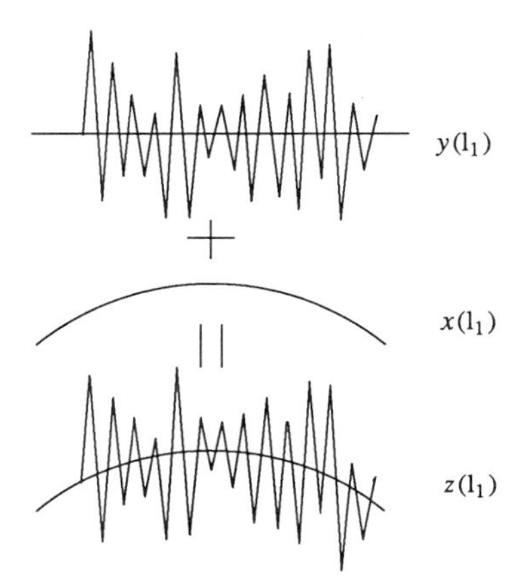

Figure 1: Superposition of a random and a deterministic function in 1-D case.

zero:

$$z(l_1, l_2) = x(l_1, l_2) + y(l_1, l_2) \tag{1}$$

and

$$E[z(l_1, l_2)] = x(l_1, l_2). \tag{2}$$

Thus, $x(l_1, l_2)$ can be simply estimated by smoothing, i.e., taking the average of intensity values in the proper size of window. Then $y(l_1, l_2)$ can be estimated by subtracting this function from the original image $z(l_1, l_2)$. Figure 1 shows this superposition in a simple 1-D case. Note that these intensity functions do not represent the actual shape of image, because the shading variations are caused by changes in surface orientation relative to the illumination direction **L**. Therefore, $x(l_1, l_2)$ and $y(l_1, l_2)$ need to be projected to the illumination direction to extract the shape information.

Let σ_L, τ_L be the slant and tilt of illumination direction **L** in the coordination system with l_1, l_2, l_3. The direction **L** induces a coordinate system with axes i_1, i_2, i_3, and this is derived from the following transformation:

$$\begin{pmatrix} i_1 \\ i_2 \\ i_3 \end{pmatrix} = \begin{pmatrix} \cos\sigma_L \cos\tau_L & \cos\sigma_L \sin\tau_L & -\sin\tau_L \\ -\sin\tau_L & \cos\tau_L & 0 \\ \sin\sigma_L \cos\tau_L & \sin\sigma_L \sin\tau_L & \cos\sigma_L \end{pmatrix} \begin{pmatrix} l_1 \\ l_2 \\ l_3 \end{pmatrix}. \tag{3}$$

Here, i_3 is the illumination direction **L**, and l_3 is the viewing direction.

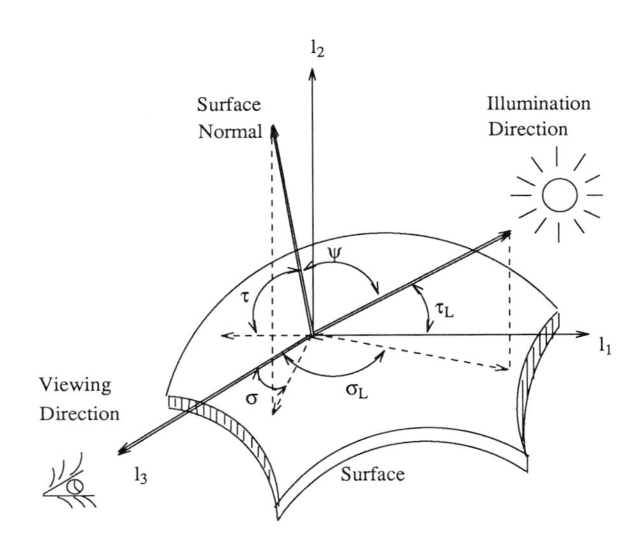

Figure 2: Three different directions on 3-D surface.

The surface normal, **N**, is another important direction of 3-D geometry. If we define a new coordinate system with axes, m_1, m_2, m_3, having the m_3-axis in the surface normal direction, these axes can be derived from the coordinate system of viewing direction, i.e., l_1, l_2, l_3, by using a similar coordination transform with different values of tilt τ and slant σ:

$$\begin{pmatrix} l_1 \\ l_2 \\ l_3 \end{pmatrix} = \begin{pmatrix} \cos\sigma\cos\tau & \cos\sigma\sin\tau & -\sin\tau \\ -\sin\tau & \cos\tau & 0 \\ \sin\sigma\cos\tau & \sin\sigma\sin\tau & \cos\sigma \end{pmatrix} \begin{pmatrix} m_1 \\ m_2 \\ m_3 \end{pmatrix}. \tag{4}$$

Figure 2 depicts the relations between these three different directions and the four different tilt and slant parameters, σ_L, τ_L, σ, and τ.

Let $x'(m_1, m_2)$ and $y'(m_1, m_2)$ be a deterministic and a random function respectively, defined on the surface normal image plane, i.e., $m_3 = 0$. Then, $z'(m_1, m_2)$, defined as the sum of these two functions, is merely the depth function z, observed from the surface normal. However, note that $y'(m_1, m_2)$ is not the rotated function of $y(l_1, l_2)$ but the projected function of $y(l_1, l_2)$ to the $m_1 - m_2$ plane so that they may satisfy the superposition property. Figure 3 depicts these relations between functions.

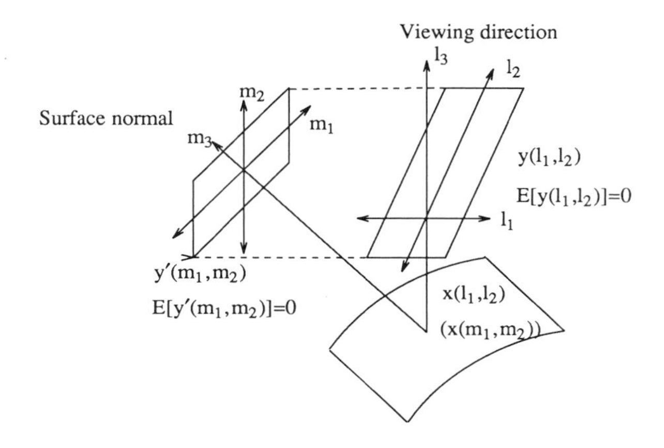

Figure 3: 3-D Geometry of the functions x, y, x', y'.

2.1 Estimation of Illumination Direction

The intensity function $z(l_1, l_2)$ does not represent the actual shape of the image, because the shading variations are caused by changes in surface orientation relative to the illumination direction **L**. Thus, the first step in estimating local surface orientation is the determination of the illumination direction for the surface. Then, $z(l_1, l_2)$ will be projected to the illumination direction to extract the shape information. Estimation methods of illumination direction **L** were suggested by Pentland [17] and Lee and Rosenfeld [15] in different ways.

Pentland's approach was based on the assumption that the surface normal of each local patch is isotropically distributed within a scene. Assuming that the distribution of the surface normal is known and the intensity value is measurable for different directions within the image, one could estimate the illumination direction **L** using a least square estimation procedure. On the other hand, Lee and Rosenfeld suggested a different statistical approach to estimate the illumination direction. For the slant of illumination direction σ_L, assuming that the slant σ is uniformly distributed over the Gaussian sphere surface, where a small circle corresponds to low slant and a large circle corresponds to high slant, the probability density function for the slant σ can be obtained by $(\sin \sigma)/2\pi$. Then, the expected values of image brightness $z(l_1, l_2)$ and $z^2(l_1, l_2)$ are determined by integrating over the illuminated region of surface.

Theorem 2.1 *The slant of illumination direction σ_L satisfies*

$$
\begin{aligned}
E[z] &= \frac{2\lambda\mu}{3\pi}(\sin\sigma_L + (\pi - \sigma_L)\cos\sigma_L) \\
E[z^2] &= \frac{(\lambda\mu)^2}{8}(1 + \cos\sigma_L)^2, \\
& \textit{when self-shadowed parts are included, and} \\
E[z] &= \frac{4\lambda\mu}{3\pi(1 + \cos\sigma_L)}(\sin\sigma_L + (\pi - \sigma_L)\cos\sigma_L) \\
E[z^2] &= \frac{(\lambda\mu)^2}{4}(1 + \cos\sigma_L), \\
& \textit{when self-shadowed parts are not included,}
\end{aligned}
$$

where $\lambda\mu$ is constant for the reflectance and the mathematical expectation is taken over the whole image.

The proof is in the appendices. □

Thus, the slant σ_L can be estimated from the previous equations by taking the average value of intensities over the whole image for $E[z]$ and $E[z^2]$. The tilt of illumination direction τ_L can be also estimated from Theorem 4.1 of [15].

$$\tau_L = \tan^{-1}\left(\frac{\hat{E}[\frac{\partial}{\partial l_2} z(l_1, l_2)]}{\hat{E}[\frac{\partial}{\partial l_1} z(l_1, l_2)]}\right). \tag{5}$$

Here, the estimated values $\hat{E}(\cdot)$ take over the whole image by taking the average value of the function.

Lee's method has several advantages over Pentland's method. First, calculation is much simpler, because calculations are required only once over the whole image. Second, we can calculate the slant of the illumination direction directly from two equations without knowing the value of $\lambda\mu$. Experimental results of Lee's method are known to be superior to Pentland's for the estimation of both tilt and slant. This has been confirmed by Lee and Rosenfeld [15] and Ferrie and Levine [4].

2.2 Model of the Deterministic Component x

As discussed before, a 3-D surface image can be considered to be the superposition of a texture image and a smooth shaded image. This smooth shaded image can be represented by the deterministic function $x(l_1, l_2)$. Thus, if the illumination direction values for the whole image are given, the surface orientation parameters slant, σ, and tilt, τ, can be estimated from $x(l_1, l_2)$, representing a smooth surface by *shape from shading* analysis.

Pioneering work on the inference of *shape from shading* was done by Horn [6] and his coworkers. To extract the 3-D shape function $H(\cdot, \cdot)$ from a single 2-D image, they used the reflectance map, which shows the intensity of the image as a function of the surface gradient and the illumination direction:

$$x(l_1, l_2) = \frac{p \cos \tau_L \sin \sigma_L + q \sin \tau_L \sin \sigma_L + \cos \sigma_L}{(p^2 + q^2 + 1)^{1/2}} \equiv R(p, q), \tag{6}$$

where

$$R(p, q) : \text{Reflectance map function}$$
$$p = \frac{\partial}{\partial l_1} H(l_1, l_2), q = \frac{\partial}{\partial l_2} H(l_1, l_2)$$
$$H(l_1, l_2) : \text{3-D shape function from the viewing direction,}$$
$$\text{and } \tau_L, \sigma_L : \text{Tilt, slant of the illumination direction.}$$

Here, from the relationship between the tilt τ, the slant σ of the surface and p, q,

$$\tau = \tan^{-1}\left(\frac{q}{p}\right), \quad \sigma = \cos^{-1} \frac{1}{\sqrt{p^2 + q^2 + 1}} \tag{7}$$

$$p = \tan \sigma \cos \tau, \quad q = \tan \sigma \sin \tau \ , \tag{8}$$

we can modify Eq. 6 to the function of σ, τ, σ_L, and τ_L as

$$\begin{aligned} x(l_1, l_2) \quad = \quad & \sin \sigma \cos \tau \cos \tau_L \sin \sigma_L \ + \ \sin \sigma \sin \tau \sin \tau_L \sin \sigma_L \\ & + \ \cos \sigma \cos \sigma_L. \end{aligned} \tag{9}$$

Construction of 3-D shape can be achieved by solving σ, τ in terms of $x(l_1, l_2)$ at each point and integrating those values. However, this approach needs the solutions of at least two difficult problems. First, this is an ill-posed problem because there are two unknown parameters and only one equation to be solved. Thus, we need an additional constraint to have the unique solution. Second, the final integrated shape can be different from the original shape, due to the cumulation of estimation errors.

To get the unique solution, the calculus of variation methods were used by minimizing the estimation error after adding one constraint for the smoothness [7] or integrability [5]. To handle the cumulative error, Pentland [19] suggested local shape analysis, which deals with only the local areas instead of a whole image. However, Pentland's technique has severe trouble in integrating all local area surface information. Recently, Pentland [17] developed another technique for solving the integrability problem. He suggested analysis in the frequency domain, instead of in the spatial domain. By using this method, the integrability problem can be avoided, because only one inverse Fourier transform needs to be taken at the end of a procedure. However, since the calculation of a convolution is required in the frequency domain to handle a simple multiplication operation in the spatial domain, calculation will be complicated. In this paper, these ill-posed problems and the integrability problems will be handled by an additional constraint from the texture pattern and frequency domain analysis, respectively. (It will be discussed in detail later.)

2.3 Model of the Random Component y

The random component $y(l_1, l_2)$ of the intensity function $z(l_1, l_2)$ can not be simply obtained by taking coordinate transformation to $y'(m_1, m_2)$, defined on the surface normal plane image. Because the random function $y(l_1, l_2)$ was assumed to be a 2-D random function, whose expectation value $E[y(l_1, l_2)]$ is zero, to satisfy the superposition properties, this function should rather be considered as the function obtained after projecting $y'(m_1, m_2)$ to the viewer's direction (as shown in Fig. 3). Thus, if we can model $y'(m_1, m_2)$ properly, $y(l_1, l_2)$ will be obtained by projecting this to the viewer's direction **L**.

2.3.1 Fractional Differencing Periodic Model for $\mathbf{y'(m_1, m_2)}$

The random function $y'(m_1, m_2)$, defined on the surface normal plane image, can be approximated by a 2-D random field model, which is distributed over the surface normal plane (as shown in Fig. 1). Note that since this model is based on a 2-D texture model, it will fit the plane surface more than any other shape of surface. Thus, we apply this model to the local shape of the image, because the local shape will be closer and closer to the plane surface when we take a smaller patch.

Among the various random field models, the fractional differencing periodic model is chosen to represent this random texture surface in this paper. The 1-D fractional differencing model was suggested by Hosking [8], generalizing the well-known ARIMA model of Box and Jenkins [1], which was originally designed to model a nonstationary random process. A typical

1-D first-order fractional differencing model is as follows:

$$y(l) = (1 - z)^{-\frac{c}{2}} \zeta(l), \tag{10}$$

where z is the delay operator in the direction of l, and $\zeta(l)$ is a white Gaussian noise.

This random process model has the property of being flexible to explain both the long-term and short-term correlation structure of a time series depending on the values of the fractional differencing parameter c, and it shares the basic properties with fractional Brownian motion defined by Mandelbrot and Van Ness [16].

Theorem 2.2 *When $c < 0$, fractional differencing process (Eq.10) has a short-term memory, and when $c > 0$, it has a long-term memory.*

The proof is in [8]. □

This long-term memory property was extended to the 2-D case and applied to detect the boundary of the mixed 2-D texture patterns of long-term persistence [13].

Theorem 2.3 *The fractional differencing process shares the basic properties with fractional Brownian motion.*

(*Proof*) Brownian motion is a continuous time stochastic process $B(t)$ with independent Gaussian increments. Its derivative is the continuous-time white noise process, which has constant spectral density. Fractional Brownian motion, $B_F(t)$, is a generalization of these processes. Then, fractional Brownian motion with parameter F, usually $0 < F < 1$, is equal to the $(\frac{1}{2} - F)$th fractional derivative of Brownian motion in Riemann-Liouville sense. The continuous-time fractional noise process is then the derivative of fractional Brownian motion; thus, it may also be thought of as the $(\frac{1}{2} - F)$th fractional derivative of continuous-time white noise. Therefore, the fractional differencing model (Eq. 10) is the discrete version of this continuous-time fractional white noise process, and it shares some properties with fractional Brownian motion [8]. □

In 3-D textural surface image, the fractional differencing parameter, which is the "fractal scaling parameter" in the terminology of Pentland [18, 20], indicates the roughness of the surface; that is, as the value of the fractal scaling parameter increases, the model represents a 3-D surface textured more roughly. However, Pentland's 2-D model with one fractal scaling parameter has limited modeling ability, because the fractal of the surface is assumed to be spatially isotropic. Thus, Pentland's model is not flexible enough to represent the wide range of different texture patterns encountered in practice, especially nonstationary random texture. In other words, this model can tell how rough the surface is but can not tell which pattern the surface is covered with. This limitation can be overcome by using the 2-D fractional differencing periodic model as follows [3]:

$$\begin{aligned} y'(m_1, m_2) &= (1 - 2\cos\omega_1 {z'_1}^{-1} + {z'_1}^{-2})^{-\frac{c}{2}} \\ &\quad \cdot (1 - 2\cos\omega_2 {z'_2}^{-1} + {z'_2}^{-2})^{-\frac{d}{2}} \zeta'(m_1, m_2) \end{aligned} \tag{11}$$

for $m_1, m_2 = 0,\ 1,\ \ldots,\ N - 1$.

The corresponding DFT of this function is

$$\begin{aligned} Y'(k_1, k_2) &= (1 - 2\cos\omega_1 e^{-j2\pi\frac{k_1}{N}} + e^{-j4\pi\frac{k_1}{N}})^{-\frac{c}{2}} \\ &\quad \cdot (1 - 2\cos\omega_2 e^{-j2\pi\frac{k_2}{N}} + e^{-j4\pi\frac{k_2}{N}})^{-\frac{d}{2}} W'(k_1, k_2), \end{aligned} \tag{12}$$

where z'_i is the delay operator associated with m_i, $\zeta'(m_1, m_2)$ is an i.i.d. Gaussian sequence, and $W'(k_1, k_2)$ is the corresponding DFT.

This model has four different parameters: c and d for the fractal scales and ω_1 and ω_2 for the frequencies of pattern in the direction of m_1 and m_2, respectively. Thus, this model represents the roughness of the surface and the pattern of the texture image at the same time even with the different values for the direction of m_1, m_2 separately. For example, consider the tree bark texture on the surface of a tree in Fig. 4-a. The 3-D structure of tree bark texture has different values of frequency and roughness in horizontal or vertical directions. Thus, it cannot be represented by either Pentland's model with only various fractal scales (Fig. 4-b,c), or second-order AR model with only directional frequencies (Fig. 4-d). Figure 5-a,b shows how well the fractional differencing periodic model fits to the tree bark texture.

2.3.2 Orthographically Projected Fractional Periodic Differencing Model for $y(l_1, l_2)$

The statistical model of the intensity function $y(l_1, l_2)$ cannot be simply obtained by rotating the coordinate axes because the expected values of both $y'(m_1, m_2)$ and $y(l_1, l_2)$ must be zero over the planes to satisfy the superposition properties. Therefore, such a function $y(l_1, l_2)$ that satisfies this requirement can be obtained by projecting the function $y'(m_1, m_2)$ orthographically to the viewer's image plane. Fig. 6 shows this projection.

A new coordinate system of the orthographically projected image from the viewing direction, l_1-l_2, can be obtained from the following two coordinate transformations:

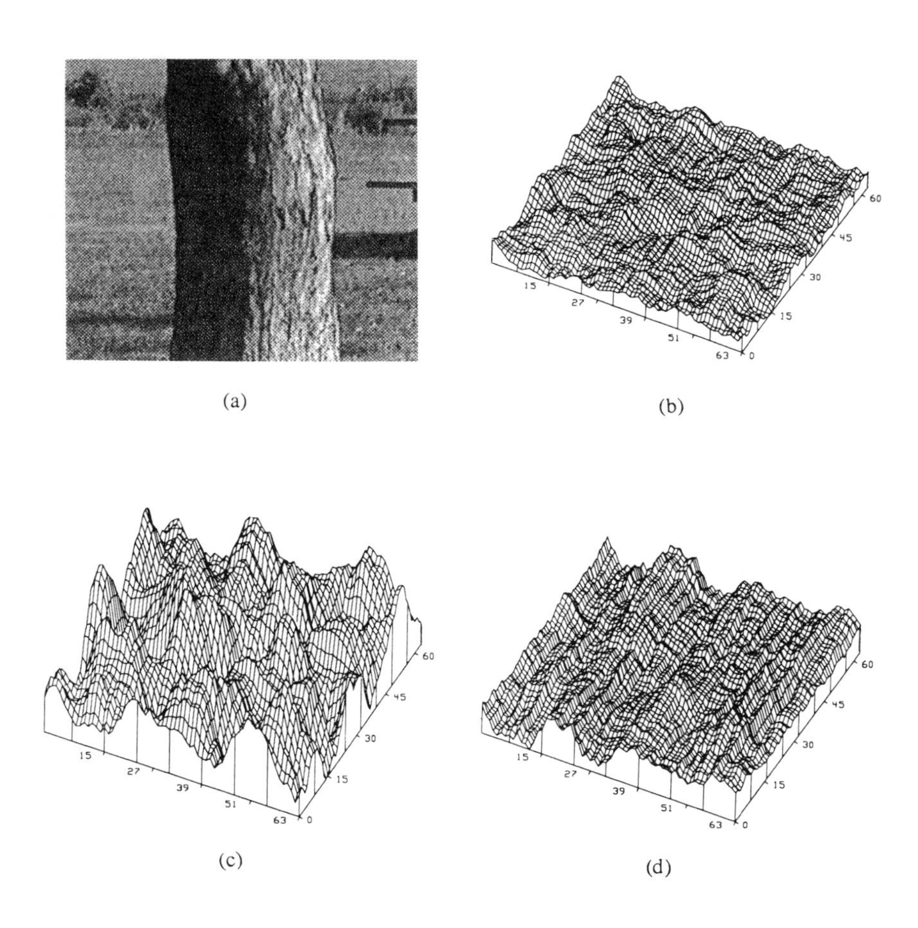

Figure 4: Synthesized surface shapes over 64 × 64 sized normal plane patch from (a) a tree image whose surface is covered by the tree bark texture; (b) Pentland's model with the fractal scale c = 0.9; (c) Pentland's model with the fractal scale $c = 1.5$; and (d) second-order AR model with the frequency $\omega_1 = 0.25$, $\omega_2 = 0.025$ in direction of m_1, m_2, respectively.

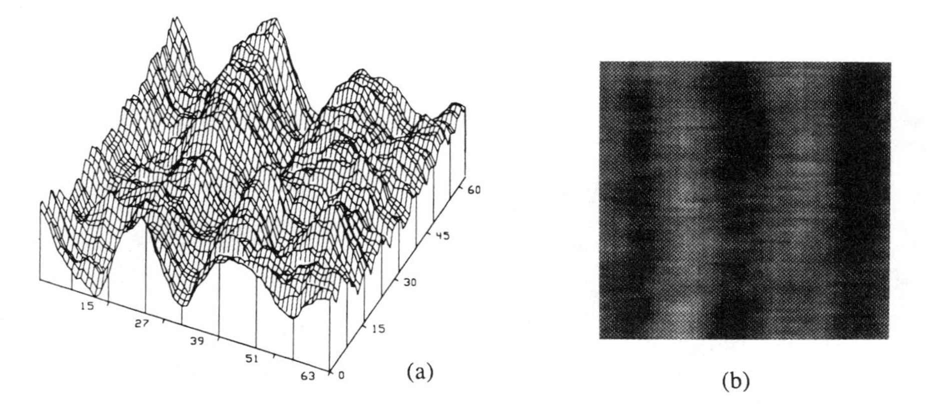

Figure 5: (a) Synthesized surface shape over 64 × 64 sized normal plane patch from fractional differencing periodic model with the frequency $\omega_1 = 0.25$, $\omega_2 = 0.025$ and the fractal scale $c = 1.5$, $d = 0.8$ in direction of m_1, m_2, respectively, and (b) corresponding surface image.

$$\begin{pmatrix} l_1' \\ l_2' \end{pmatrix} = \begin{pmatrix} \cos\tau & -\sin\tau \\ \sin\tau & \cos\tau \end{pmatrix} \begin{pmatrix} m_1 \\ m_2 \end{pmatrix} \tag{13}$$

and

$$\begin{pmatrix} l_1 \\ l_2 \end{pmatrix} = \begin{pmatrix} 1 & 0 \\ 0 & \cos\sigma \end{pmatrix} \begin{pmatrix} l_1' \\ l_2' \end{pmatrix}. \tag{14}$$

Here, τ is the angle between m_1 and l_1' axes and σ is the rotational angle based on l_1'-axis (Fig. 6).

Hence, the coordinate transformation of the orthographic projection between the m_1-m_2 system and l_1-l_2 system can be given as follows [9].

$$\begin{pmatrix} l_1 \\ l_2 \end{pmatrix} = \begin{pmatrix} \cos^2\tau + \cos\sigma\sin^2\tau & (1-\cos\sigma)\sin\tau\cos\tau \\ (1-\cos\sigma)\sin\tau\cos\tau & \sin^2\tau + \cos\sigma\cos^2\tau \end{pmatrix} \begin{pmatrix} m_1 \\ m_2 \end{pmatrix}. \tag{15}$$

Thus,

$$\begin{pmatrix} m_1 \\ m_2 \end{pmatrix} = \frac{1}{\cos\sigma} \begin{pmatrix} \sin^2\tau + \cos\sigma\cos^2\tau & (\cos\sigma - 1)\sin\tau\cos\tau \\ (\cos\sigma - 1)\sin\tau\cos\tau & \cos^2\tau + \cos\sigma\sin^2\tau \end{pmatrix} \begin{pmatrix} l_1 \\ l_2 \end{pmatrix}. \tag{16}$$

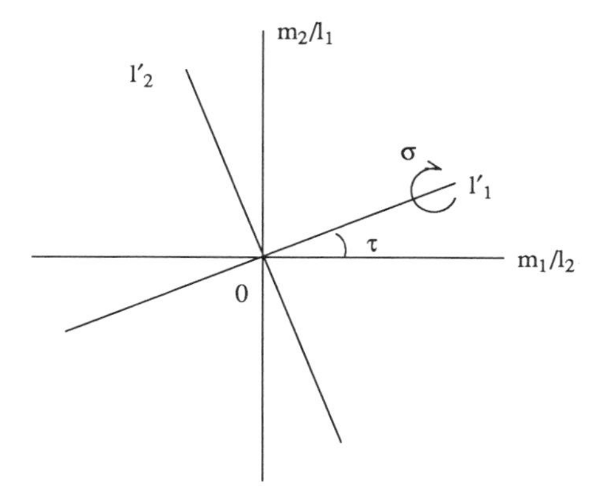

Figure 6: Coordinate transformation of the orthographic projection.

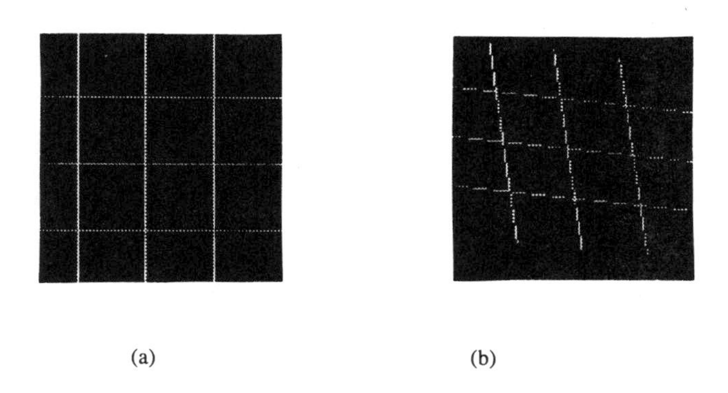

Figure 7: (a) A 2-D grid pattern image. (b) Image obtained after projecting the image (Fig. 7-a) orthographically, with $\sigma = \frac{\pi}{4}$ and $\tau = \frac{\pi}{8}$ in Eq. 15.

One grid pattern image (Fig. 7-a) was considered to demonstrate this orthogonal projection, and the coordinate transformation was taken to this image with $\sigma = \frac{\pi}{4}$ and $\tau = \frac{\pi}{8}$ (Fig. 7-b).

Therefore, by using the coordinate transform (Eq. 15), the model of intensity function $y(l_1, l_2)$ can be obtained from the fractional differencing periodic model of $y'(m_1, m_2)$ as follows. Let

$$y(l_1, l_2) = (1 - 2\cos\omega_1 z_1^{-1} + z_1^{-2})^{-\frac{c}{2}} (1 - 2\cos\omega_2 z_2^{-1} + z_2^{-2})^{-\frac{d}{2}} \zeta(l_1, l_2), \quad (17)$$

where z_1, z_2 are the delay operators, corresponding to l_1 and l_2, respectively. By the definition of DFT, $W(k_1, k_2)$ corresponding to white noise sequence $\zeta(l_1, l_2)$ is

$$W(k_1, k_2) = \sum_{m_1=0}^{N-1} \sum_{m_2=0}^{N-1} \zeta(l_1, l_2) \exp(-j\frac{2\pi}{N}(m_1 k_1 + m_2 k_2)) \quad (18)$$

and, from Eq. 15,

$$W(k_1, k_2) = \sum_{l_1=0}^{N-1} \sum_{l_2=0}^{N-1} \zeta(l_1, l_2) \exp(-j\frac{2\pi}{N\cos\sigma}[\{(\sin^2\tau + \cos\sigma \cdot \cos^2\tau)k_1 + (\cos\sigma - 1)\sin\tau\cos\tau k_2\}l_1 + \{(\cos\sigma - 1)\sin\tau \cdot \cos\tau k_1 + (\cos^2\tau + \cos\sigma\sin^2\tau)k_2\}l_2]). \quad (19)$$

Thus, as in [10], we can define

$$z_1 = z_1'^{\frac{\sin^2\tau + \cos\sigma\cos^2\tau}{\cos\sigma}} z_2'^{\frac{(\cos\sigma - 1)\sin\tau\cos\tau}{\cos\sigma}} \quad (20)$$

$$z_2 = z_1'^{\frac{(\cos\sigma - 1)\sin\tau\cos\tau}{\cos\sigma}} z_2'^{\frac{\cos^2\tau + \cos\sigma\sin^2\tau}{\cos\sigma}} \quad (21)$$

and

$$n_1 = \frac{1}{\cos\sigma}[(\sin^2\tau + \cos\sigma\cos^2\tau)k_1 + (\cos\sigma - 1)\sin\tau\cos\tau k_2], \quad (22)$$

$$n_2 = \frac{1}{\cos\sigma}[(\cos\sigma - 1)\sin\tau\cos\tau k_1 + (\cos^2\tau + \cos\sigma\sin^2\tau)k_2]. \quad (23)$$

Thus, from the orthographical projection, we can set

$$W(k_1, k_2) = W'(n_1, n_2). \quad (24)$$

Therefore, with these relations, we can have the projected version of $Y'(k_l, k_2)$, which is the DFT of $y(l_1, l_2)$ as follows:

$$Y(k_1, k_2) = (1 - 2\cos\omega_1 e^{-j2\pi\frac{n_1}{N}} + e^{-j4\pi\frac{n_1}{N}})^{-\frac{c}{2}} \cdot (1 - 2\cos\omega_2 e^{-j2\pi\frac{n_2}{N}} + e^{-j4\pi\frac{n_2}{N}})^{-\frac{d}{2}} W(k_1, k_2). \quad (25)$$

Note that the fractal scaling parameters c and d remain the same as the ones before projecting because of their scaling invariance property [8, 11].

3 Projected Texture on the 3-D Surface

As discussed in previous chapters, the intensity function of an image, $z(l_1, l_2)$, can be represented by the superposition of a deterministic function $x(l_1, l_2)$ and a random function $y(l_1, l_2)$. Therefore, if we can estimate all parameters, that is, σ, τ for the surface orientation and c, d, ω_1, ω_2 for the pattern of texture from the intensity function z directly, then we can get better estimates than the ones from the separate procedures for each x and y function. It is obvious that the estimation error from one procedure will cause another estimation error; thus, the errors will be cumulated. For that reason, in this chapter, a composite model of *shape from shading* and *shape from texture* will be discussed.

Consider the reflectance map function (Eq. 9) for x and the 2-D orthographically projected fractional differencing periodic model (Eq. 17) for y. Then the intensity function z can be represented by

$$z(l_1, l_2) = \epsilon[\sin\sigma\cos\tau\cos\tau_L\sin\sigma_L + \sin\sigma\sin\tau\sin\tau_L\sin\sigma_L + \cos\sigma\cos\sigma_L] + (1 - 2\cos\omega_1 z_1^{-1} + z_1^{-2})^{-\frac{c}{2}} \cdot (1 - 2\cos\omega_2 z_2^{-1} + z_2^{-2})^{-\frac{d}{2}} \zeta(l_1, l_2), \quad (26)$$

where ϵ is the normalization factor and z_1 and z_2 are the same as defined by Eq. 20 and Eq. 21, respectively.

Here, notice that the normalization factor ϵ should be multiplied to the reflectance map function $R(\sigma, \tau)$, because $R(\sigma, \tau)$ is the normalized function, which has a maximum value of 1. Because our model is based on the assumption that each patch is a slanted and tilted plane, the value of the deterministic function $x(l_1, l_2)$ is a constant over each patch; the best guess of the value of ϵ is the maximum value among the average intensities of the patches,

$$\epsilon = \max_{i=1,\cdots,M}[\frac{1}{N'^2}\sum_{l_1=1}^{N'}\sum_{l_2=1}^{N'} z_i(l_1, l_2)], \quad (27)$$

where M is the total number of patches in the whole image, N' is the size of patch, and $z_i(\cdot,\cdot)$ is the intensity function of the ith patch.

Therefore, the value of ϵ can be estimated from the whole image, before the actual procedure on each patch. Thus, we assume ϵ to be given, just as the illumination directions σ_L, τ_L.

The DFT of Eq. 26 will be

$$\begin{aligned} Z(k_1,k_2) &= X(k_1,k_2)+Y(k_1,k_2) \quad (28) \\ &= N^2\epsilon[\sin\sigma\cos\tau\cos\tau_L\sin\sigma_L+\sin\sigma\sin\tau\sin\tau_L\sin\sigma_L \\ &\quad +\cos\sigma\cos\sigma_L]\delta(k_1,k_2)+(1-2\cos\omega_1 e^{-j2\pi\frac{n_1}{N}} \\ &\quad +e^{-j4\pi\frac{n_1}{N}})^{-\frac{c}{2}}\cdot(1-2\cos\omega_2 e^{-j2\pi\frac{n_2}{N}} \\ &\quad +e^{-j4\pi\frac{n_2}{N}})^{-\frac{d}{2}}W(k_1,k_2), \quad (29) \end{aligned}$$

where n_1, n_2 are the same as defined by (Eqs. 22 and 23), and

$$\delta(k_1,k_2)=\begin{cases} 1, & \text{if } k_1,k_2=0 \\ 0, & \text{otherwise.} \end{cases} \quad (30)$$

3.1 Estimation of the Parameters c, d, $\omega_1,\omega_2,\sigma,\tau$

The estimation of parameters in this projected fractional differencing periodic model (Eq. 26) can be done in the frequency domain, modifying the techniques suggested in [12]. This frequency domain analysis has an advantage over the spatial domain analysis. Since we need only to take the inverse Fourier transform to get the whole image at the end of procedures, we can avoid the integrability problem that might occur in spatial domain analysis when the estimated surface orientation from each local shape has an error.

For estimation, all parameters can be estimated directly from the given data $Z(k_1,k_2)$, if we can obtain the likelihood function of $|Y(k_1,k_2)|$ from the relationship

$$Y(k_1,k_2)=Z(k_1,k_2)-X(k_1,k_2). \quad (31)$$

Then, since the noise sequence $\zeta'(m_1,m_2)$ is assumed to be a white Gaussian, $W(k_1,k_2)$ and $Y(k_1,k_2)$ follow the Rayleigh distribution as in the following theorem.

Theorem 3.1 *The modulus of the DFT of the noise sequence,* $\{|W(k_1,k_2)|,\ k_1=0,1,\cdots,N-1,\ k_2=0,1,\cdots,\lfloor N/2\rfloor\}$ *and*

$\{|Y(k_1,k_2)|,\ k_1 = 0,1,\cdots,N-1,\ k_2 = 0,1,\cdots,\lfloor N/2 \rfloor\}$ *are the white sequences with the following Rayleigh densities:*

$$f_{|W(k_1,k_2)|}(W) = \begin{cases} \frac{2W}{\rho N^2}\exp\{\frac{-W^2}{\rho N^2}\}, & W \geq 0 \\ 0, & otherwise, \end{cases}$$

$$f_{|Y(k_1,k_2)|}(Y) = \begin{cases} \frac{2s^2_{k_1,k_2}Y(k_1,k_2)}{\rho N^2}\exp\{\frac{-s^2_{k_1,k_2}Y^2(k_1,k_2)}{\rho N^2}\} & ,Y(k_1,k_2) \geq 0 \\ 0 & ,otherwise \end{cases}$$

where

$$s_{k_1,k_2} = |2\cos(\frac{2\pi n_1}{N}) - 2\cos\omega_1|^{\frac{c}{2}}|2\cos(\frac{2\pi n_2}{N}) - 2\cos\omega_2|^{\frac{d}{2}}.$$

The proof is in the Appendices. □

From these probabilistic properties, the estimation of parameters can be done by a hybrid method of *least square* and *maximum likelihood* estimations, which was suggested by Eom [3]. For LS estimation, if the values of the illumination direction σ_L, τ_L and the normalization factor ϵ are given, and parameters ω_1 ω_2, σ, τ are set, then $\theta = (c,d,\alpha)^T$ can be estimated by minimizing the following cost function.

$$\begin{aligned} J(\theta,\omega_1,\omega_2) &= \sum_{k_1=0}^{N-1}\sum_{k_2=0}^{\lfloor N/2\rfloor}(\log|Z(k_1,k_2) - N^2\epsilon[\sin\sigma\cos\tau\cos\tau_L\sin\sigma_L \\ &\quad + \sin\sigma\sin\tau\sin\tau_L\sin\sigma_L + \cos\sigma\cos\sigma_L]\delta(k_1,k_2)| + \alpha \\ &\quad + \frac{c}{2}\cdot\log|2\cos\frac{2\pi n_1}{N} - 2\cos\omega_1| + \frac{d}{2}\log|2\cos\frac{2\pi n_2}{N} \qquad (32) \\ &= -2\cos\omega_2|)^2\sum_{k_1=0}^{N-1}\sum_{k_2=0}^{\lfloor N/2\rfloor}(\log|Z(k_1,k_2) - N^2\epsilon[\sin\sigma\cos\tau \\ &\quad \cos\tau_L\sin\sigma_L + \sin\sigma\cdot\sin\tau\sin\tau_L\sin\sigma_L \\ &\quad + \cos\sigma\cos\sigma_L]\delta(k_1,k_2)| - \theta^T Q(k_1,k_2))^2, \qquad (33) \end{aligned}$$

where $\alpha = -E[\log|W(k_1,k_2)|]$, $\theta = (c,d,\alpha)^T$.

Here,

$$Q(k_1,k_2) = \begin{pmatrix} \frac{-1}{2}\log|2\cos\frac{2\pi n_1}{N} - 2\cos\omega_1| \\ \frac{-1}{2}\log|2\cos\frac{2\pi n_2}{N} - 2\cos\omega_2| \\ -1 \end{pmatrix}, \qquad (34)$$

Thus, the estimated values will be

$$[\hat{c}, \hat{d}, \hat{\alpha}]^T = (\sum_{k_1=0}^{N-1} \sum_{k_2=0}^{\lfloor N/2 \rfloor} Q(k_1, k_2) Q^T(k_1, k_2))^{-1} (\sum_{k_1=0}^{N-1} \sum_{k_2=0}^{\lfloor N/2 \rfloor}$$
$$\cdot Q(k_1, k_2) \log |Z(k_1, k_2) - N^2 \epsilon [\sin\sigma \cos\tau \cos\tau_L \sin\sigma_L$$
$$+ \sin\sigma \sin\tau \sin\tau_L \sin\sigma_L + \cos\sigma \cos\sigma_L] \delta(k_1, k_2)|). \tag{35}$$

Also, ML estimators of ω_1, ω_2, σ, τ can be calculated by maximizing the log-likelihood function $L(Y; \theta, \omega_1, \omega_2)$ with $\hat{\alpha}$ estimated previously.

$$L(Y; \theta, \omega_1, \omega_2) = \sum_{k_1=0}^{N-1} \sum_{k_2=0}^{\lfloor N/2 \rfloor} \log |\ Z(k_1, k_2) - N^2 \epsilon [\sin\sigma \cos\tau \cos\tau_L$$
$$\sin\sigma_L + \sin\sigma \sin\tau \sin\tau_L \sin\sigma_L + \cos\sigma \cos\sigma_L] \delta(k_1, k_2)|$$
$$- \sum_{k_1=0}^{N-1} \sum_{k_2=0}^{\lfloor N/2 \rfloor} \log(\frac{\rho N^2}{2}) + \frac{c}{2} N \sum_{k_1=0}^{N-1} \log |2(\cos\frac{2\pi n_1}{N} - \cos\omega_1)|$$
$$+ dN \sum_{k_2=0}^{\lfloor N/2 \rfloor} \log |2(\cos\frac{2\pi n_2}{N} - \cos\omega_2) - \frac{1}{\rho N^2} \sum_{k_1=0}^{N-1} \sum_{k_2=0}^{\lfloor N/2 \rfloor}$$
$$|2(\cos\frac{2\pi n_1}{N} - \cos\omega_1)|^c |2(\cos\frac{2\pi n_2}{N} - \cos\omega_2)|^d |Z(k_1, k_2)$$
$$- N^2 \epsilon [\sin\sigma \cos\tau \cdot \cos\tau_L \sin\sigma_L + \sin\sigma \sin\tau \sin\tau_L \sin\sigma_L$$
$$+ \cos\sigma \cos\sigma_L] \delta(k_1, k_2)|^2. \tag{36}$$

where ρ is a variance of $\zeta(l_1, l_2)$ and can be estimated by the following equation in mean square sense.

$$\hat{\rho} = \frac{1}{N^2} \exp(\gamma - 2\hat{\alpha} - \frac{\pi^2}{6N^2}), \tag{37}$$

where γ is Euler's constant ($= 0.5772157$) [3].

Therefore, the estimation scheme can be summarized as follows:

3.1.1 Estimation Algorithm

Step 1: Estimate the illumination direction σ_L, τ_L and the normalization factor ϵ from the whole image, by Theorem 2.1 and Eq. 27, respectively.

Step 2: For each patch, choose resonant frequencies ω_1, ω_2 in the range of $(0, \frac{\pi}{2})$ and the surface orientation parameters σ, τ in the range of $(-\frac{\pi}{2}, \frac{\pi}{2})$.

Step 3: With the given values of ω_1, ω_2, σ and τ, estimate c, d, and α by LS estimation algorithm (Eq. 35).

Step 4: Using $\hat{\alpha}$, compute the estimate of the variance of ρ, $\zeta(l_1, l_2)$, by equation (Eq. 37).

Step 5: Using the estimates $\hat{c}$, $\hat{d}$ and $\hat{\rho}$ found in Step 3 and 4, maximize the likelihood function given by (Eq. 36) with respect to ω_1, ω_2, σ, and τ.

Step 6: Using the estimates $\hat{\omega}_1$, $\hat{\omega}_2$, $\hat{\sigma}$ and $\hat{\tau}$, repeat Step 3 to Step 5 until the estimates have no significant change in successive iterations.

The results from computer simulation will be discussed in the next section.

3.2 Experimental Results

3.2.1 Experiment 1: *3-D Texture on a Sphere Surface*

In this experiment, a whole image that contains the shade and texture on a 3-D sphere surface was constructed using our proposed 3-D texture model, and this was compared with the images that were obtained by either applying the reflectance map function or by projecting the texture pattern only. Since our model is a composite of the *shape from shading* and the *shape from texture* models, the constructed image will look more natural than the ones from either the *shape from shading* or the *shape from texture* technique. For this experiment, the texture pattern was chosen as the fractional differencing periodic model (Eq. 17) with the parameter values, $\omega_1 = 0.2$, $\omega_2 = 0.2$, $c = 0.8$, and $d = 0.8$. From the given illumination direction σ_L, τ_L, and the surface orientations of sphere, σ, τ, the 3-D surface of a sphere covered by the chosen texture pattern was synthesized. Here, the illumination direction was chosen as $\sigma_L = -0.66$, $\tau_L = -0.66$, and the surface orientations $\sigma(i_1, i_2)$, $\tau(i_1, i_2)$ of the (i_1, i_2)th patch was given by

$$\sigma(i_1, i_2) = \begin{cases} \tan^{-1}\left(\frac{\partial H(l_1,l_2)/\partial l_2}{\partial H(l_1,l_2)/\partial l_1}\right)\Big|_{i_1,i_2}, & \text{if } H(l_1, l_2) \geq 63 \\ 0, & \text{otherwise} \end{cases} \tag{38}$$

and

$$\tau(i_1, i_2) = \begin{cases} \cos^{-1} \frac{1}{\sqrt{(\partial H/\partial l_1)^2|_{i_1,i_2} + (\partial H/\partial l_2)^2|_{i_1,i_2}, +1}} & \text{if } H(l_1, l_2) \geq 63 \\ 0, & \text{otherwise} \end{cases} \tag{39}$$

where the height function $H(l_1, l_2)$ is given by

$$H(l_1, l_2) = 15^2 - l_1^2 - l_2^2. \tag{40}$$

In this experiment, each 3-D texture pattern was synthesized on a 32 × 32 pixel sized planar patch, and a 16 × 16 pixel sized patch was taken from the center. The complete image of sphere (512 × 512) was obtained by adjoining these 16 × 16 pixel sized patches. Figures 8-a,b show the 3-D shape of a hemisphere and the corresponding sphere image obtained by the reflectance map function (Eq. 9) with the parameter values $\sigma_L = -0.66, \tau_L = -0.66, and \epsilon = 100$ Eq. 27, and Fig. 9 shows the orthographically projected texture image based on the given sphere surface with Eq. 17. Finally, with the composite model of the projected texture image and the reflectance map function Eq. 26, an image of 3-D texture on the surface of a sphere was synthesized Fig. 10.

Here, Fig. 8 and Fig. 9 illustrate that neither the *shape from shading* nor the *shape from texture* technique is suitable for representing the natural scene. Fig. 8 does not contain any detail information about the surface pattern, while Fig. 9 does not contain any shade. Thus, it is difficult to see the sphere in either Fig. 8 or Fig. 9. Figure 10, however, illustrates the sphere image with texture pattern on its surface clearly, thus showing our model's capability to represent the 3-D textural surface.

3.2.2 Experiment 2: *Parameter Estimation*

From this experiment, we want to show how accurate the estimated results are. Since each small patch is assumed as a tilted and slanted texture plane and the whole image is obtained by adjoining these patches, our proposed 3-D texture model will fit each small image patch and the surface orientation parameter will be estimated based on each patch. Thus, in this experiment, we will consider single patches of texture patterns that represent the tilted and slanted texture planes. From these patches the parameter values of the model (Eq. 26) will be estimated by the proposed estimation scheme and compared with the true values. For this experiment, three different 2-D texture patterns sized 64 × 64 were generated by Eq. 26 with the different values of parameters ω_1, ω_2, c, and d. Then, the projected images of the slanted and tilted texture planes were synthesized with the different values

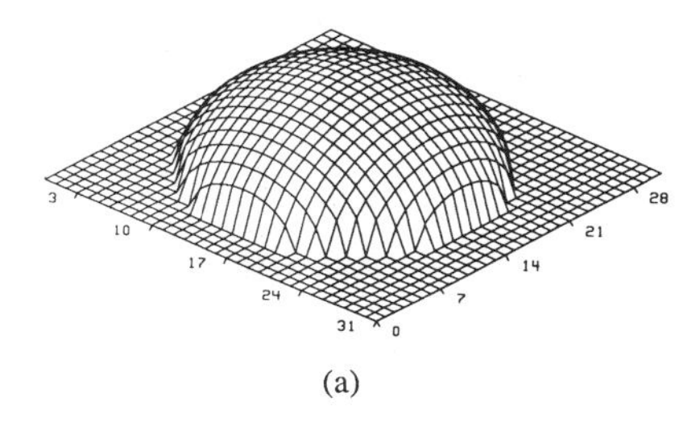

(a)

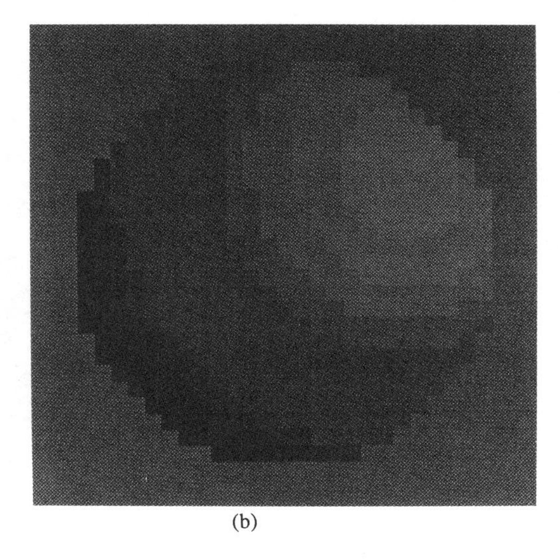

(b)

Figure 8: (a) Height function of a sphere obtained by Eq. 40. (b) Sphere image obtained by the reflectance map function (Eq. 9) with ϵ=100 (3.2), $\sigma_L = -0.66, \tau_L = -0.66$.

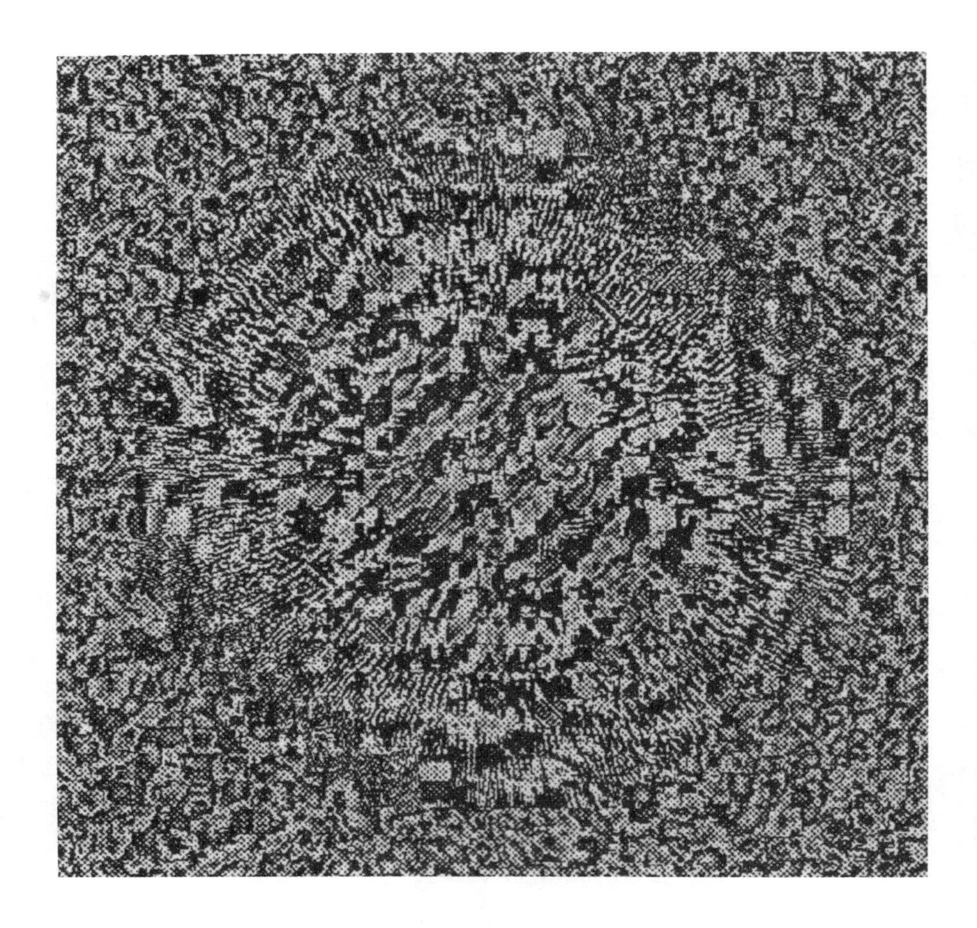

Figure 9: Image obtained after projecting the texture image orthographically to the sphere surface. (Background texture pattern is generated by the fractional differencing periodic model (Eq. 17) with $\omega_1 = 0.2$, $\omega_2 = 0.2$, c=0.8, d=0.8.)

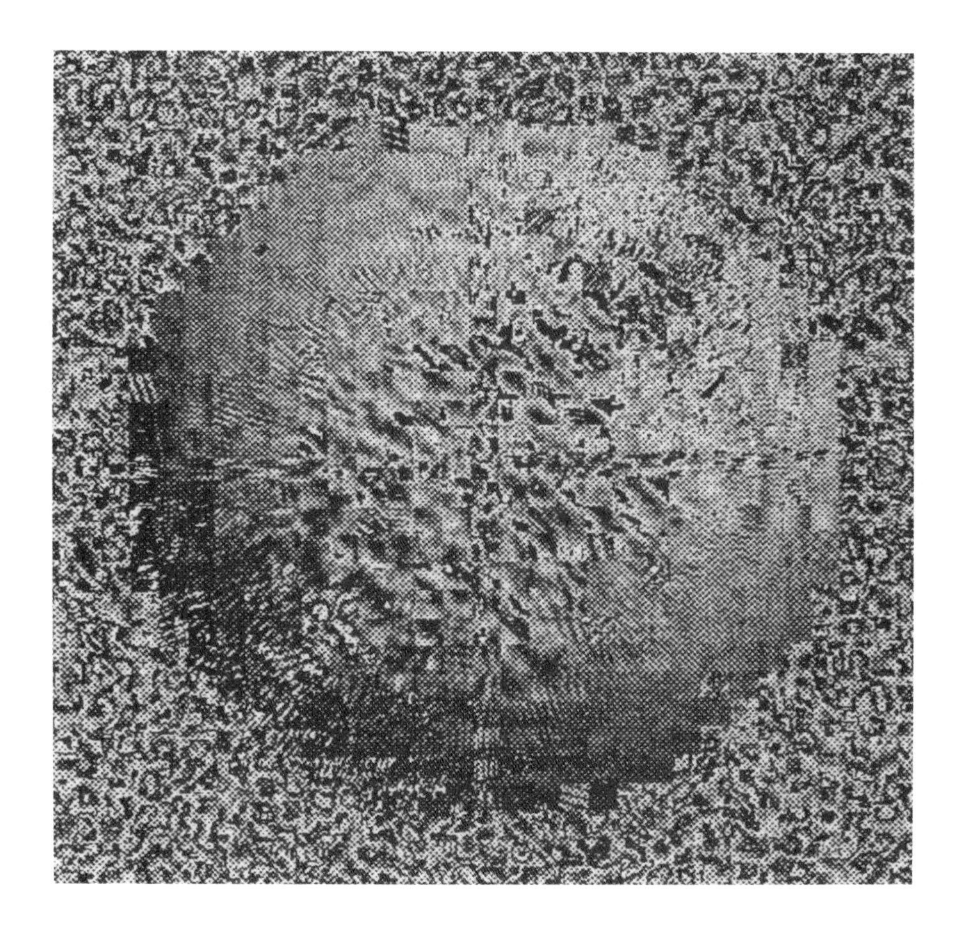

Figure 10: Image of 3-D texture on the surface of a sphere. (Image is generated by the composite model (Eq. 26) with $\sigma_L = -0.66$, $\tau_L = -0.66$.)

of the surface orientation σ, τ in Eq. 29, and zero mean white Gaussian noise with variance 10 was added to each image (Table 1). The values of the illumination direction σ_L, τ_L and the normalization factor ϵ was given by $\frac{\pi}{8}$, $\frac{\pi}{4}$, and 100, respectively. Thus, since the intensity value of a plane surface will be constant all over the patch, the deterministic part of Eq. 29 was considered as a constant and added to the random field part. For the estimation, the hybrid method of least square and maximum likelihood estimation, which was discussed in the previous section, was used to estimate parameters.

Figure 11-a shows one of the 2-D texture pattern that was generated by Eqs. 17-25 with the values of ω_1, ω_2, c, and d as 0.2, 0.2, 0.8, and 0.8, respectively, and Fig. 11-b shows the projected image of the tilted and slanted version of Fig. 11-a by $-\frac{\pi}{4}$ and $\frac{\pi}{8}$, respectively. As can be seen from Fig. 11-a and Fig. 11-b, due to the slanting, the distance between the dark spots along the normal direction of the tilted axis (the tilted axis is the northwest to southeast diagonal) reduces considerably (thus producing almost a continuous dark band along that direction in Fig. 11-b).

Here, 2-D texture model does not fit the pattern in Fig. 11-b, since this pattern has a nonisotropic random texture distribution due to the tilt and slant. Note that this synthesis technique does not require an interpolation after projecting. Because the white Gaussian random noise will still be white Gaussian noise after projecting, the random noise as input data can be generated for the projected model directly. From these three different 3-D texture patches, we have estimates close to the true values (Table 1). Also shown in the parentheses are absolute deviations from the true values. Therefore, we can say that the height function and the texture pattern of the whole image can be estimated properly from this local patch estimation process.

3.2.3 Experiment 3: *Tree Image*

A 512×512 real image of a part of a tree, which contains shade and texture on surface was considered (Fig. 4-a). From the whole image, the estimation values of the illumination direction σ_L, τ_L were determined to be $\hat{\sigma}_L = -1.3384$, $\hat{\tau}_L = -0.0426$, by applying Lee's method. Then, for this experiment, several local patches of size 32×32 were taken from the tree surface, and the parameter values were estimated by applying our projected fractional differencing periodic model Eq. 26 and estimation scheme (Eqs. 32-37). Average estimated values of the parameters were obtained as $\omega_1 = 0.486$, $\omega_2 = 0.053$, $c = 1.394$, and $d = 0.762$. A Synthesized tree surface image with the given illumination direction was constructed by projecting this texture pattern on a cylinder surface (Figure 12). This image looks

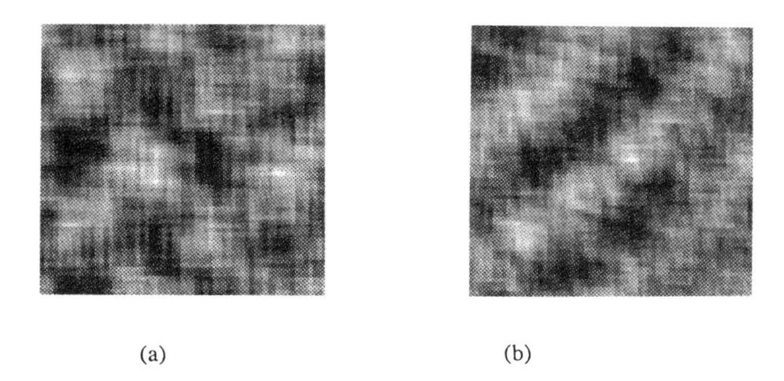

(a) (b)

Figure 11: (a) 2-D texture image obtained from Eq. 17 with $\omega_1 = 0.2$, $\omega_2 = 0.2, c = 0.8, d = 0.8$ (b) Projected image of the tilted and slanted version of Fig. 11(a) by $-\frac{\pi}{4}$ and $\frac{\pi}{8}$, respectively.

Table 1: True parameter values and the estimated parameter values of the projected fractional differencing periodic model (Eq.17): The random noise sequence as input data for each synthesized texture patch of size 64×64 was generated from white Gaussian noise with zero mean and variance 10. Estimated values were obtained from these synthesized images by equations Eq. 32-37).

Patches	True Values					
	ω_1	ω_2	c	d	σ	τ
Patch 1	0.2	0.2	0.8	0.8	0.4	0.78
Patch 2	0.2	0.2	0.4	0.8	0.78	0.78
Patch 3	0.2	0.4	0.6	0.8	0.2	0.6

Patches	Estimated Values					
	$\hat{\omega}_1$	$\hat{\omega}_2$	$\hat{c}$	$\hat{d}$	$\hat{\sigma}$	$\hat{\tau}$
Patch 1	0.204	0.208	0.803	0.807	0.393	0.744
	(0.004)	(0.008)	(0.003)	(0.007)	(0.007)	(0.036)
Patch 2	0.219	0.232	0.357	0.716	0.812	0.673
	(0.019)	(0.032)	(0.043)	(0.084)	(0.032)	(0.107)
Patch 3	0.191	0.378	0.620	0.589	0.299	0.598
	(0.009)	(0.022)	(0.020)	(0.211)	(0.099)	(0.002)

Figure 12: A synthesized tree image: Local patch size is 32×32, and each 3-D texture pattern was synthesized by the composite model (Eq. 26) with the illumination direction σ_L = -1.3384, τ_L = -0.0426 and ω_1 = 0.486, ω_2 = 0.053, c = 1.394, d = 0.762 for the random part. 16×16 pixel sized patch was taken from the center of it. The complete image of cylinder (512×512) was obtained by adjoining these 16×16 pixel sized patches.

very similar to the original tree image, and shows the ability of our model to represent a nonstationary texture pattern such as a tree bark on tree surface.

4 Conclusion

In this paper, a 2-D orthographically projected fractional differencing periodic model, which is a composite model of *shape from shading* and *shape from texture*, was developed to represent a 3-D surface image which contains information about both radiance and texture. This model has several advantages over the conventional approaches. First, as compared to the *shape from shading* techniques, this one always gives unique and more accurate solutions for the surface orientation parameter values, because of the additional constraint from the texture function part. Also, by using this analysis, the integrability problem that might occur in spatial domain analysis can be avoided, because only one inverse Fourier transform needs to be taken at the end of procedure to get the whole image. Second, as compared to the *shape from texture* techniques, the fractional differencing model has the property of being flexible to explain both the long-term and the short-term correlation structure of the texture pattern; thus, it has a superior ability of modeling different textures encountered in practice. The orthographical projection adds the additional flexibility to represent the 3-D rotated texture due to the slant and tilt of a surface normal plane. The estimation scheme for the parameters was based on the hybrid method of least square and maximum likelihood estimations. Thus, there may be more possible applications by combining with other techniques, such as region classification, segmentation, etc.

Appendices

Proof of Theorem 2.1

Let σ_L be the phase angle between the viewing direction $\mathbf{V}$ and the illumination direction $\mathbf{L}$, ψ be the incident angle between the illumination direction and the surface normal direction, and η be the emittance angle between the viewing direction and the surface normal direction. Then

$$\mathbf{V}\cdot\mathbf{L} = \cos\sigma_L, \mathbf{L}\cdot\mathbf{N} = \cos\psi, \text{and} \quad \mathbf{V}\cdot\mathbf{N} = \cos\eta. \tag{41}$$

Intensity function $x(l_1, l_2)$ can be represented by the following equation.

$$x(l_1, l_2) = \lambda\mu(\mathbf{L}\cdot\mathbf{N}) = \lambda\mu\cos\psi, \quad \text{when } \cos\psi \geq 0. \tag{42}$$

Then, for each surface patch, to determine the expected value of image intensity, we need to average $x(l_1, l_2)$, and, to determine the expected value of image intensity squared, we need to average $x^2(l_1, l_2)$ over the image of a hemisphere.

Define $\mathbf{O}$ as an unit vector orthogonal to the plane containing $\mathbf{L}$ and $\mathbf{V}$. Then, simple spherical trigonometry using a triangle with corners $\mathbf{V}$, $\mathbf{N}$, and $\mathbf{O}$, yields

$$\cos\eta = \cos\tau\cos\sigma, \tag{43}$$

and from a similar triangle with corners $\mathbf{L}$, $\mathbf{N}$, and $\mathbf{O}$, we can obtain

$$\cos\psi = \cos\tau\cos(\sigma - \sigma_L), \tag{44}$$

The illuminated half of the sphere runs from $\sigma = -\pi/2 + \sigma_L$ to $\sigma = +\pi/2 + \sigma_L$, while the visible half goes from $\sigma = -\pi/2$ to $\sigma = +\pi/2$, and the infinitesimal element of area is $\cos\tau d\sigma d\tau$.

The integrals that we are interested in are of the form

$$\int_{-\pi/2}^{\pi/2}\int_{-\pi/2+\sigma_L}^{\pi/2} f(\sigma,\tau)\cos\eta\cos\tau d\sigma d\tau, \tag{45}$$

where the $\cos\eta$ term compensates for the foreshortening due to the projection of the spherical surface onto the image. If we include self-shadowed areas in the computation of the average, we must divide the integral by the whole area, π, of the disc that is the projection of the hemisphere; on the other hand, if we do not include self-shadowed areas, we divide by the area $(\pi/2)(1+\cos\sigma_L)$ of the projection of the illuminated part of the hemisphere.

Thus, from Eq. 42 and Eq. 44,

$$\begin{aligned} E[x] &= \lambda\mu E[\cos\psi] && (46)\\ &= \lambda\mu E[\cos\tau\cos(\sigma-\sigma_L)]. && (47)\end{aligned}$$

Here, to get the value of $E[\cos\psi]$, we need to evaluate the integral

$$\begin{aligned} B_1 &= \int_{-\pi/2}^{\pi/2}\int_{-\pi/2+\sigma_L}^{\pi/2} (\cos\eta\cos\psi)\cos\tau d\sigma d\tau && (48)\\ &= \int_{-\pi/2}^{\pi/2}\int_{-\pi/2+\sigma_L}^{\pi/2} (\cos\tau\cos\sigma)(\cos\tau\cos(\sigma-\sigma_L))\cos\tau d\sigma d\tau \\ &= \int_{-\pi/2}^{\pi/2}\cos^3\tau d\tau\int_{-\pi/2+\sigma_L}^{\pi/2}\cos(\sigma-\sigma_L)\cos\sigma d\sigma \\ &= \frac{2}{3}(\sin\sigma_L + (\pi-\sigma_L)\cos\sigma_L). && (49)\end{aligned}$$

Therefore,

$$E[x] = \frac{2\lambda\mu}{3\pi}(\sin\sigma_L + (\pi-\sigma_L)\cos\sigma_L), \tag{50}$$

or

$$E[x] = \frac{4\lambda\mu}{3\pi(1+\cos\sigma_L)}(\sin\sigma_L + (\pi-\sigma_L)\cos\sigma_L), \tag{51}$$

depending on whether we average over all image regions, including self-shadowed parts, or not.

Similarly, to get the value of $E[x^2]$, we need to evaluate the integral

$$B_2 = \int_{-\pi/2}^{\pi/2} \int_{-\pi/2+\sigma_L}^{\pi/2} \cos\eta \cos^2\psi \cos\tau d\sigma d\tau \tag{52}$$

$$= \int_{-\pi/2}^{\pi/2} \int_{-\pi/2+\sigma_L}^{\pi/2} (\cos\tau\cos\sigma)(\cos\tau\cos(\sigma-\sigma_L))^2 \cos\tau d\sigma d\tau$$

$$= \int_{-\pi/2}^{\pi/2} \cos^4\tau d\tau \int_{-\pi/2+\sigma_L}^{\pi/2} \cos^2(\sigma-\sigma_L)\cos\sigma d\sigma$$

$$= \frac{\pi}{8}(1+\cos\sigma_L)^2. \tag{53}$$

Therefore,

$$E[x^2] = \frac{(\lambda\mu)^2}{8}(1+\cos\sigma_L)^2, \tag{54}$$

or

$$E[x^2] = \frac{(\lambda\mu)^2}{4}(1+\cos\sigma_L), \tag{55}$$

depending on whether we average over all image regions, including self-shadowed parts, or not. □

Proof of Theorem 3.1

Let $W'(k_1, k_2)$ be the DFT of a white noise sequence $\zeta'(m_1, m_2)$ with a normal distribution, that is,

$$\zeta'(m_1, m_2) \sim N(0, \rho), \quad \text{for } m_1, m_2 = 1, 2, \cdots, N. \tag{56}$$

Then, $W'(k_1, k_2)$ will follow another normal distribution as follows [2].

$$W'(k_1,k_2) \sim N(N \cdot E[\zeta'(m_1,m_2)],\ \rho N^2) \tag{57}$$

$$\sim N(0, \rho N^2) \tag{58}$$

and

$$\text{Re}[W'(k_1,k_2)], \text{Im}[W'(k_1,k_2)] \sim N(0, \frac{\rho N^2}{2}), \tag{59}$$

where $\text{Re}[W']$ and $\text{Im}[W']$ are the real and the imaginary parts of $W'(k_1, k_2)$, respectively.

Define new coordinate systems, (l_1, l_2) and (n_1, n_2), which can be obtained from the coordinate transformations (Eqs. 15, 22, 23). Thus, based on the new coordinate system, a new noise sequence $\zeta(l_1, l_2)$ which has the corresponding DFT, $W(k_1, k_2)$, can be defined. As defined before, since the density function of Re[W'] or Im[W'] follows a normal distribution and the coordinate transformation are the linear transformation, the density function of $\zeta(l_1, l_2)$,Re[W] will follow the same normal distributions to the ones of $\zeta'(m_1, m_2)$, Re[W']. That is,

$$\zeta(l_1, l_2) \sim N(0, \rho), \quad \text{for } l_1, l_2 = 1, 2, \cdots, N \tag{60}$$

and

$$\text{Re}[W(n_1, n_2)], \text{Im}[W(n_1, n_2)] \sim N(0, \frac{\rho N^2}{2}). \tag{61}$$

Now, consider the density function of $|W(n_1, n_2)|$. Since $W(n_1, n_2)$ is complex, defining W be $W(n_1, n_2)$,

$$|W(n_1, n_2)| = \sqrt{\text{Re}^2[W] + \text{Im}^2[W]}. \tag{62}$$

Define $\phi = \tan^{-1}(\frac{\text{Im}[W]}{\text{Re}[W]})$. Then, we can have the following relations:

$$\text{Re}[W] = |W(n_1, n_2)| \cos \phi \tag{63}$$

$$\text{Im}[W] = |W(n_1, n_2)| \sin \phi, \quad \text{for } |W| > 0. \tag{64}$$

Thus, the joint probability density of $|W'|$ and ϕ can be obtained from the Jacobian, and the joint probability density of Re[W'] and Im[W'] is as follows:

$$\begin{aligned} f_{|W|,\phi}(|W|, \phi) &= \frac{1}{\text{Jacobian}} f_{\text{Re}[W],\text{Im}[W]}(\text{Re}[W], \text{Im}[W]) \\ &= |W| f_{\text{Re}[W]}(|W| \cos \phi) \cdot f_{\text{Im}[W]}(|W| \sin \phi) \\ &= \frac{|W|}{\pi \rho N^2} e^{\frac{-(\text{Re}^2[W] + \text{Im}^2[W])}{\rho N^2}}, \quad \text{for } |W| > 0, \end{aligned} \tag{65}$$

where

$$\text{Jacobian} = \begin{vmatrix} \frac{\partial |W|}{\partial \text{Re}[W]} & \frac{\partial |W|}{\partial \text{Im}[W]} \\ \frac{\partial \phi}{\partial \text{Re}[W]} & \frac{\partial \phi}{\partial \text{Im}[(W]} \end{vmatrix} \tag{66}$$

$$= \frac{1}{|W|}. \tag{67}$$

Therefore, since $|W|$ and ϕ are independent, and the density function of ϕ, $f_\phi(\phi)$ is $\frac{1}{2\pi}$,

$$f_{|W|\phi}(|W|, \phi) = f_{|W|}(|W|) \cdot f_\phi(\phi), \tag{68}$$

and

$$f_{|W|}(W) = \begin{cases} \frac{2W}{\rho N^2} e^{\frac{-W^2}{\rho N^2}}, & W \geq 0 \\ 0, & \text{otherwise.} \end{cases} \tag{69}$$

Now, set

$$s_{k_1,k_2} = |2\cos(\frac{2\pi n_1}{N}) - 2\cos\omega_1|^{\frac{c}{2}} |2\cos(\frac{2\pi n_2}{N}) - 2\cos\omega_2|^{\frac{d}{2}}. \tag{70}$$

Then, from the definition of n_i, (Eqs. 22, 23), we can have

$$|Y(k_1, k_2)| = s_{k_1,k2}^{-1} |W(k_1, k_2)| \tag{71}$$

$$= s_{k_1,k_2}^{-1} |W'(n_1, n_2)|. \tag{72}$$

Thus, the density function of $|Y(k_1, k_2)|$ can be represented by the following equations:

$$f_{|Y(k_1,k_2)|}(Y) = s_{k_1,k_2} f_{|W(k_1,k_2)|}(s_{k_1,k_2} Y(k_1, k_2)) \tag{73}$$

$$= \begin{cases} \frac{2s_{k_1,k_2}^2 Y(k_1,k_2)}{\rho N^2} e^{\frac{-s_{k_1,k_2}^2 Y^2(k_1,k_2)}{\rho N^2}} & Y(k_1, k_2) \geq 0 \quad \square \\ 0, & \text{otherwise.} \end{cases} \tag{74}$$

Bibliography

[1] G. E. P. Box and G. M. Jenkins, *Time Series Analysis: Forecasting and Control,* Holden-Day, 1969.

[2] D. R. Brillinger, *Time Series, Data Analysis and Theory, Expanded Edition*, Holden-Day Inc., 1981.

[3] K. - B. Eom, *Robust Image Models With Application* Ph.D Dissertation, Purdue University, West Lafayette, Indiana, 1986.

[4] F. P. Ferrie and M. D. Levine, "Where and Why Local Shading Analysis Works," *IEEE Trans. Pattern Analysis and Machine Intelligence,* vol.PAMI-11, no. 2, pp. 198–206, Feb. 1989.

[5] R. T. Frankot and R. Chellappa, "A Method for Enforcing Integrability in Shape from Shading Algorithms," *IEEE Trans. Pattern Analysis and Machine Intelligence*, vol. 10, no. 4, pp. 439–451, July 1988.

[6] B. K. P. Horn, *Robot Vision*, McGraw-Hill, MIT Press, 1986.

[7] B. K. P. Horn and M. J. Brooks, "The Variational Approach to Shape From Shading," *Computer Vision, Graphics, and Image Processing*, vol. 33, pp. 174–208, 1986.

[8] J. R. M. Hosking, "Fractional Differencing," *Biometrika*, vol. 68, pp. 165–176, 1981.

[9] K. Kanatani, "Detection of Surface Orientation and Motion from Texture by a Stereological Technique," *Artificial Intelligence*, vol. 23, pp. 213–237, 1984.

[10] H. Kang and J. K. Aggarwal, "Design of Two-Dimensional Recursive Filters by Interpolation," *IEEE Trans. Circuit and Systems,* vol. CAS-24, pp. 281–291, 1977.

[11] R. L. Kashyap, "Image Models," in *Handbook of Pattern Recognition and Image Processing,* Academic Press, Inc., pp. 281–310, 1986.

[12] R. L. Kashyap and K. - B. Eom, "Estimation In Long-Memory Time-Series Model," *Journal of Time Series Analysis*, vol. 9, pp. 35–41, 1988.

[13] R. L. Kashyap and K–B, Eom, "Texture Boundary Detection Based On The Long Correlation Model," *IEEE Trans. Pattern Analysis and Machine Intelligence*, vol. PAMI-11, no. 1, pp. 58–67, Jan. 1989.

[14] J. R. Kender, "Shape From Texture: An Aggregation Transform That Maps A Class Of Textures Into Surface Orientation," *Proc. 6th IJCAI*, pp. 475–480, 1979.

[15] C–H. Lee and A. Rosenfeld, "Improved Methods of Estimating Shape From Shading Using the Light Source Coordinate System," *Artificial Intelligence*, vol. 26, pp. 125–143, 1985.

[16] B. B. Mandelbrot and J. W. Van Ness, "Fractional Brownian Motions, Fractional Noises and Application," *SIAM Rev.*, vol. 10, pp. 422–437, 1968.

[17] A. P. Pentland, "Finding The Illumination Direction," *J. Opt. Soc. America,* vol. 72, pp. 448–455, 1982.

[18] A. P. Pentland, "Fractal-Based Description of Natural Scenes," *IEEE Trans. Pattern Analysis and Machine Intelligence,* vol. PAMI-6, pp. 661–674, 1984.

[19] A. P. Pentland, "Local Shading Analysis," *IEEE Trans. Pattern Analysis and Machine Intelligence,* vol. PAMI-6, pp. 170–187, March 1984.

[20] A. P. Pentland, "Shading Into Texture," *Artificial Intelligence* vol. 29, pp. 147–170, 1986.

[21] A. P. Pentland, "The Transform Method for Shape From Shading," *Vision Science Technical Report 106,* M.I.T. Media Lab., July 1988.

[22] K. A. Stevens, "The Visual Interpretation Of Surface Contours," *Artificial Intelligence*, vol. 17, pp. 47-73, 1981.

[23] A. P. Witkin, "Recovering Surface Shape and Orientation From Texture," *Artificial Intelligence*, vol. 17, pp. 17-45, 1981.

Shape from Texture Using Gaussian Markov Random Fields

Fernand S. Cohen and Maqbool A. S. Patel
Department of Electrical and Computer Engineering
Drexel University
Philadelphia, Pennsylvania

1 Introduction

The inference of the shape of a 3-D object from its image (or images) has been the concern of many researchers for the past two decades. Towards that end, many monocular cues have been exploited. Horn and his co-workers [1, 2] have done pioneering work on the inference of shape from shading. They have used the reflectance map to extract 3D surface information from image data. Pentland [3] and Brooks and Horn [4] used a parametrized reflectance map to recover the light source from the image when certain assumptions were made. Gidas and Torreao [5] presented a Bayesian/geometric framework for reconstructing 3-D shapes from shading information using Gibbs distribution for modeling shapes. Bolle and Cooper [6] used 2-D quadric polynomial approximation to surface image, where the approximation was constrained by the 3-D object surface shape, to classify an image patch into one of a set of candidate surfaces. Others bypassed the approach based on the irradiance equation (shape from shading), and considered inferring shape by considering special contours on the surface [7, 8], or by considering the distribution of the contours of constant image intensity as a function of 3-D surface shape [9]. Shape information can also be extracted from multiple images. A powerful Bayesian/geometric framework for reconstructing 3-D shapes and obtaining depth information from multiple images is presented in [10]. They circumvented the correspondence problem necessary for estimating the depth by modeling the 3-D world by 3-D parametrized surfaces, and estimated these parameters

This work was supported by the National Science Foundation under grant number IRI-8913958.

ISBN 0-12-170608-7

directly from two or more images. In [11], this idea is extended to include the segmentation of the image into surface patches.

Shape parameters can also be calculated from regular patterns on the surface or from texture gradients. Gibson [12] first proposed the texture density gradient as the primary basis for surface perception by humans. Bajcsy and Lieberman [13] used the two-dimensional Fourier spectrum to detect texture gradient as a cue to depth information.Witkin [14] studied the problem of recovering the local orientation of known 3-D surface markings by studying how the marking features are transformed by the projection operation. Davis *et al.* [15] developed an efficient algorithm for recovering local orientations of known 3-D surface markings. Ikeuchi [16] also exploited regular patterns on the surface to extract shape information. He used a spherical perspective projection imaging model and studied the local distortions of a repeated texture pattern due to the the image projection. Kender [17] obtained the orientation of a 3-D planar surface from the images of parallel edges drawn on it. It relies on an abstract representation (the normalized textural property map) of the effects of surface orientation on a particular texture property, and prestored information about a particular texture pattern. Ohta and his co-workers [18] obtained the 3-D planar surface orientation by computing a 2-D affine transformation that approximates the distortion of the texel patterns under perspective projection. This semi-perspective transformation was also used by Brown *et al.* [19] for extracting shape from texture, contour, and motion. Brzakovic [20] obtained the orientation for planar surfaces with regular or globally regular directional textures under a perspective camera model. Finally, Choe and Kashyap [21] also dealt with the problem of modeling and synthesis of texture on 3-D surfaces using a long-correlation periodic model for the texture combined with shape from shading technique.

Texture is usually viewed from one of the two approaches, either structural or statistical. We look at textured images as realizations or samples from parametric probability distributions on the image space. We use the Gaussian Markov random fields (GMRF) to model the texture. The reason for that choice is given in Section 3. The properties of the Markov random fields (MRF) and their use in filtering, image modeling, and segmentation has been treated in various papers [22–34]. They were shown to be a compact representation for a variety of textures.

2 Problem Statement and Assumptions

The goal of this chapter is to model images that result from the projective distortions of *homogeneous* textures laid on 3-D surfaces, as they are seen

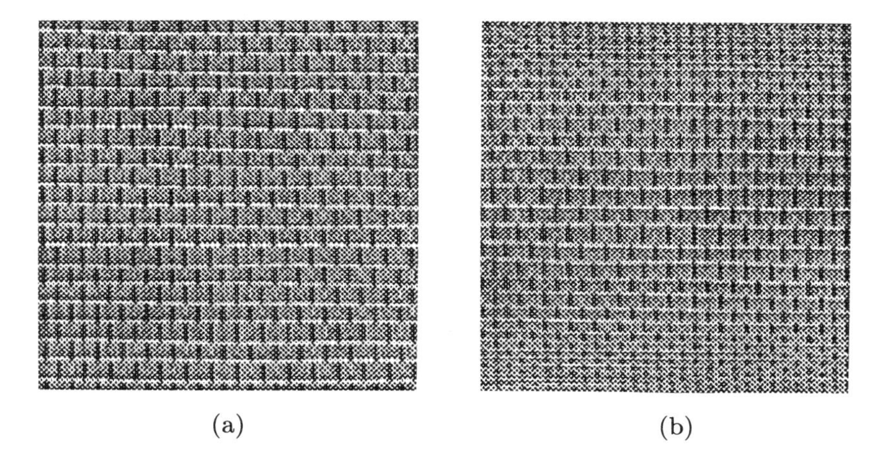

(a) (b)

Figure 1: Texture Rendering.

by a camera. The surface texture is obtained by laying down a *homogeneous* texture (thought of as a rubber planar sheet) on the surface. (See Fig. 1 for an example.) The homogeneous texture on the rubber planar sheet is referred to as the *parent texture* and is modeled by a stationary GMRF. Figure 1 shows the process of rendering the parent texture on the 3-D surface. The parent homogeneous texture is shown in Fig. 1a. Figure 1b shows an image of the parent texture after it has been laid onto a cylinder without any local distortions. This image of the cylinder was generated by a ray tracing algorithm [38] under an orthographic camera model.

The projected texture is described by a Gaussian Random Field, whose probability density function is an explicit function of the parent texture parameters, the surface shape, and the camera model.

Throughout this chapter, a camera with an orthographic projection geometry model is assumed. We refer the reader to [37, 38] for the treatment of the problem under a semi-perspective camera model. The camera optical axis is along the z-axis, and the image plane parallel to the x–y plane. A viewer-centered coordinate system is adopted.

The basic modeling concepts are used in extracting shape information from texture. We address the problem of how to recognize and recover the shape of a 3-D *homogeneous-textured* surface from a single image of the surface patch.

This chapter is a summary of the works that appear in [35 – 38].

3 2-D Homogeneous Texture Model

For modeling the homogeneous planar texture, we use GMRFs. The reasons for that choice are : (i) they are synthesis models and are capable of producing textures that match and capture well many man-made as well as natural micro- and-macro-textures [26, 28, 32, 33]; (ii) the parameters (or their associated sufficient statistics) of the model are estimable from data samples, and the appropriateness of the model can be assessed objectively by hypothesis testing; (iii) minimum error classifiers are readily built.

3.1 Gaussian Markov Random Field Model Structure

Let $g_o(i,\ j)$ be the intensity at pixel $\boldsymbol{r} = (i,\ j)$, and let $g(i,\ j) = (g_o(i,\ j) - \mu)$, where μ is the mean of $g_o(i,\ j)$. The GMRF is a special case of the general class of MRF [22–34]. It is a 2-D non-causal autoregressive (AR) process described by the difference equation,

$$g(\boldsymbol{r}) = \sum_{\boldsymbol{v} \varepsilon D_p} \beta_{\boldsymbol{r} - \boldsymbol{v}} g(\boldsymbol{v}) + n(\boldsymbol{r}), \tag{1}$$

where $\beta_{\boldsymbol{r} - \boldsymbol{v}} = \beta_{-(\boldsymbol{r} - \boldsymbol{v})}$, and D_p is a neighbor set that is shown in Fig. 2 for $p = 5$. $\{n(\boldsymbol{r})\}$ is a Gaussian noise sequence with zero mean and autocorrelation function given by:

$$R_n(\boldsymbol{r}, \boldsymbol{v}) = \begin{cases} \varphi^2 , & \text{if } \boldsymbol{v} = \boldsymbol{r} \\ -\varphi^2\ \beta_{\boldsymbol{r} - \boldsymbol{v}} , & \text{if } \boldsymbol{v} \varepsilon\ D_p \\ 0 , & \text{otherwise} \end{cases} . \tag{2}$$

The power spectral density associated with $g(m,\ n)$ is given by (see [22]):

$$S_g(\Omega_1,\ \Omega_2)\ =\ \varphi^2/(1 - \sum_{(k,l)\varepsilon D_p} \beta_{k,l} \exp\{-j(k\Omega_1 + l\Omega_2)\}). \tag{3}$$

$S_g(\Omega_1, \Omega_2)$ is continuous, and can be computed anywhere for $-\pi < \Omega_1, \Omega_2 \leq \pi$.

The GMRF is parametrized by a parameter set $\boldsymbol{\gamma} = (\mu, \varphi^2, \boldsymbol{\beta})$, with $\boldsymbol{\beta} = (\beta_{10}, \beta_{01}, \beta_{11}, \beta_{1,-1}, ..)$. Let $\boldsymbol{g}_o = \{g_o(m,\ n),\ -N/2 \leq m,\ n \leq N/2 - 1\}$ be an $N \times N$ image obtained from the infinite extent GMRF $g_o(m,n)$. The joint density function of $\boldsymbol{g}_o$ is $p(\boldsymbol{g}_o|\gamma) \sim \eta(\boldsymbol{U}, [\Sigma])$, where $\boldsymbol{U} = \mu\mathbf{1}$ is the mean vector, and $[\Sigma]$ is the covariance matrix of $\boldsymbol{g}_o$. Under a toroidal lattice approximation [24-26], the covariance matrix $[\Sigma]$ is block-circulant. Let $\boldsymbol{G} = \{G(m,n),\ -N/2 \leq m,\ n \leq N/2 - 1\} =$ 2D N-DFT$\{\boldsymbol{g}_o - \boldsymbol{U}\}$, where $G(m,n)$ is defined as:

5	4	3	4	5
4	2	1	2	4
3	1	X	1	3
4	2	1	2	4
5	4	3	4	5

Figure 2: D_p Neighborhood.

$$G(m,n) = \sum_{k=-N/2}^{N/2-1} \sum_{l=-N/2}^{N/2-1} (g_o(k,l) - \mu)\exp\{-j(2\pi/N)[mk+nl]\}. \quad (4)$$

Let $R_o = \{(m,n) : -(N/2-1) \leq m, \ n \leq N/2\}$, $R_{1R} = \{(m,n) : (0,0), (0,N/2), (N/2,0), (N/2,N/2)\}$, $R_1 = \{(m,n) : 0 \leq m \leq N/2, \ 0 \leq n \leq N/2\} \cup \{(m,n) : -(N/2-1) \leq m \leq -1, \ 1 \leq n \leq N/2-1\}$, $R_2 = R_1 - R_{1R}$, $R_{-2} = \{(-m,-n) : (m,n)\varepsilon R_2\}$. Since $g(m,n)$ is real, it follows that $G(m,n)$ exhibits a conjugate symmetry, and, therefore, only the set R_1 defined before determines the complete DFT [39] . Moreover, it is easy to verify that $G(m,n)$ is real for $(m,n)\varepsilon R_{1R}$, and is complex for $(m,n)\varepsilon R_2$. $\boldsymbol{G}$ is a white [24–26] zero mean complex Gaussian field, with $p(\boldsymbol{G}|\gamma)$ given by [35]:

$$\begin{aligned} p(\boldsymbol{G}|\gamma) \;=\; & \prod_{(m,n)\varepsilon R_2} \frac{1}{\pi N^2} S_g(m,n) \cdot \exp\{-\sum_{(m,n)\varepsilon R_2} |G(m,n)|^2/N^2 S_g(m,n)\} \\ & \cdot \prod_{(m,n)\varepsilon R_{1R}} (1/2\pi N^2 S_g(m,n))^{1/2} \cdot \\ & \exp\{-\sum_{(m,n)\varepsilon R_{1R}} |G(m,n)|^2/2N^2 S_g(m,n)\}, \end{aligned} \quad (5)$$

where $S_g(m,n)$ is obtained from the continuous power spectral density $S_g(\Omega_1,\Omega_2)$ by evaluating it at $\Omega_1 = 2\pi m/N$, and $\Omega_2 = 2\pi n/N$, and is given by

$$S_g(m,n) = \varphi^2/(1 - 2\sum_{(k,l)\epsilon D'_p} \beta_{k,l}\cos\{(2\pi/N)[mk+nl]\}). \tag{6}$$

D'_p is the non-symmetric half-plane neighborhood of the set D_p in Fig. 2. Under a toroidal structure, the likelihood function of $\boldsymbol{g}$ is given by:

$$\begin{aligned} p(\boldsymbol{g}|\gamma) &= \prod_{-N/2\leq m,n\leq N/2-1} (1/2\pi N^2 S_g(m,n))^{1/2} \\ &\cdot \exp\{-\sum_{-N/2\leq m,n\leq N/2-1} |G(m,n)|^2/2N^2 S_g(m,n)\}. \end{aligned} \tag{7}$$

4 Gaussian Random Field under Linear Transformation

In this section, we show what happens to a homogeneous GMRF texture model parametrized by $\boldsymbol{\gamma}$ under a *linear* transformation of the image plane. In particular, we compute a bona-fide likelihood function associated with the image data that results from subjecting the parent homogeneous textured plane to a linear transformation (e.g. rotation and scale). The resulting texture model is a Gaussian random field that is parametrized by $\boldsymbol{\gamma}$ and the parameters of linear transformation.

4.1 Likelihood Function for the Texture Data under Linear Transformation

Let $g_c(x,y)$ be a continuous infinite extent image that is looked upon as a realization from a wide-sense stationary Gaussian random field, whose Power Spectral Density (PSD) $S_c(\omega_1,\omega_2)$ is band-limited. Let $g_o(m,n) = g_c(x = \Delta_x m,\ y = \Delta_y n) = g_c(m,n)$ be the sampled infinite extent image obtained from $g_c(x,y)$ by sampling at a period $\boldsymbol{\Delta}_s = (\Delta_x, \Delta_y) = (1,1)$, which is higher than the Nyquist rate, i.e., it is assumed here that $\boldsymbol{\Delta}_s$ is such that no aliasing occurs in the PSD. Let $g_o(m,n)$ be modeled by a GMRF parametrized by a parameter set $\boldsymbol{\gamma}$, with its PSD $S_g(\Omega_1,\Omega_2)$ given in Eq. (3). Since there is no aliasing, it can be easily shown that for $-\pi < \Omega_1, \Omega_2 \leq \pi$,

$$S_g(\Omega_1,\Omega_2) = S_c(\omega_1 = \Omega_1, \omega_2 = \Omega_2), \tag{8}$$

where $S_c(\omega_1,\omega_2)$ is the continuous PSD of the continuous image $g_c(x,y)$. Let $[\boldsymbol{T}]$ be any non-singular linear transformation matrix. Let $g_s(x,y) =$

$g_c(u, v)$, where $(u\ v) = [\boldsymbol{T}]^{-1}(x\ y)$, be the continuous transformed image, and let $S_s(\boldsymbol{\omega})$ be its PSD. $S_s(\boldsymbol{\omega})$ (where $\boldsymbol{\omega} = (\omega_1,\ \omega_2)$) is given by:

$$S_s(\omega) = \det\{[\boldsymbol{T}]^t\} S_c([\boldsymbol{T}]^t \omega). \tag{9}$$

Let $g_I'(m, n) = g_s(x = \Delta_x m,\ y = \Delta_y n) = g_s(x = m, y = n) = g_s(\boldsymbol{r}) = g_c([\boldsymbol{T}]^{-1}\boldsymbol{r})$, with $\boldsymbol{r} = (m, n)$, be the sampled transformed image. Let $S_I'(\boldsymbol{\Omega})$ where $\boldsymbol{\Omega} = (\Omega_1, \Omega_2)$, be the PSD associated with $g_I'(m, n)$. For $-\pi < \Omega_1, \Omega_2 \leq \pi$, $S_I'(\boldsymbol{\Omega})$ is periodic with period 2π in each direction and is given as:

$$S_I'(\boldsymbol{\Omega}) = \det\{[\boldsymbol{T}]\} \sum_{k=-\infty}^{\infty} \sum_{l=-\infty}^{\infty} S_g([\boldsymbol{T}]^t(\boldsymbol{\Omega} - 2\pi \boldsymbol{r}_{kl})). \tag{10}$$

In general, not all the terms in the sum in Eq. (10) contribute nonzero values to $S_I'(\boldsymbol{\Omega})$ in the frequency band $-\pi < \Omega_1, \Omega_2 \leq \pi$. (See Fig. 3 for an example.) Hence, Eq. (10) reduces to:

$$S_I'(\boldsymbol{\Omega}) = \det\{[\boldsymbol{T}]\} \sum_{k=-K}^{K} \sum_{l=-L}^{L} S_g([\boldsymbol{T}]^t(\boldsymbol{\Omega} - 2\pi \boldsymbol{r}_{kl})), \tag{11}$$

where K and L are the first integers above which $S_g([\boldsymbol{T}]^t(\boldsymbol{\Omega} - 2\pi r_{kl}))$ for $|k| > K$, $|l| > L$ has no nonzero values inside the region $-\pi < \Omega_1, \Omega_2 \leq \pi$. In Fig. 3, for example, we can see that K and L are confined to 1, and that $S_g([T]^t(\boldsymbol{\Omega} - 2\pi \boldsymbol{r}_{11}))$ and $S_g([\boldsymbol{T}]^t(\boldsymbol{\Omega} - 2\pi \boldsymbol{r}_{-1,-1}))$ terms are not contributing in the sum (11) in the $-\pi < \Omega_1, \Omega_2 \leq \pi$ range. Let $\{g_{\mathrm{I}}(m, n) : -N/2 \leq m,\ n \leq N/2 - 1\} = \{g_{\mathrm{I}}'(m, n) : -N/2 \leq m,\ n \leq N/2 - 1\}$ be the $N \times N$ windowed image, and let $G_I(m, n)$ be the 2-D N-DFT $\{g_{\mathrm{I}}(m, n)\}$, then under a toroidal structure as in Section 3,

$$\mathrm{VAR}[G_I(m, n)] = S_I(m, n) = N^2 S_I'(\Omega_1 = 2\pi m/N, \Omega_2 = 2\pi n/N). \tag{12}$$

Hence, using Eq. (11), Eq. (12) becomes:

$$S_I(m, n) \approx N^2 \det\{[\boldsymbol{T}]\} \sum_{k=-K}^{K} \sum_{l=-L}^{L} S_g([\boldsymbol{T}]^t(\boldsymbol{\Omega} - 2\pi \boldsymbol{r}_{kl})), \tag{13}$$

where $S_g(\boldsymbol{\Omega})$ is given in Eq. (3). The $G_I(m, n)$'s, for $-N/2 \leq m,\ n \leq N/2 - 1$, form a white Gaussian, zero mean, random field, with $\mathrm{VAR}[G_I(m, n)] = S_I(m, n)$ given in Eq. (13).

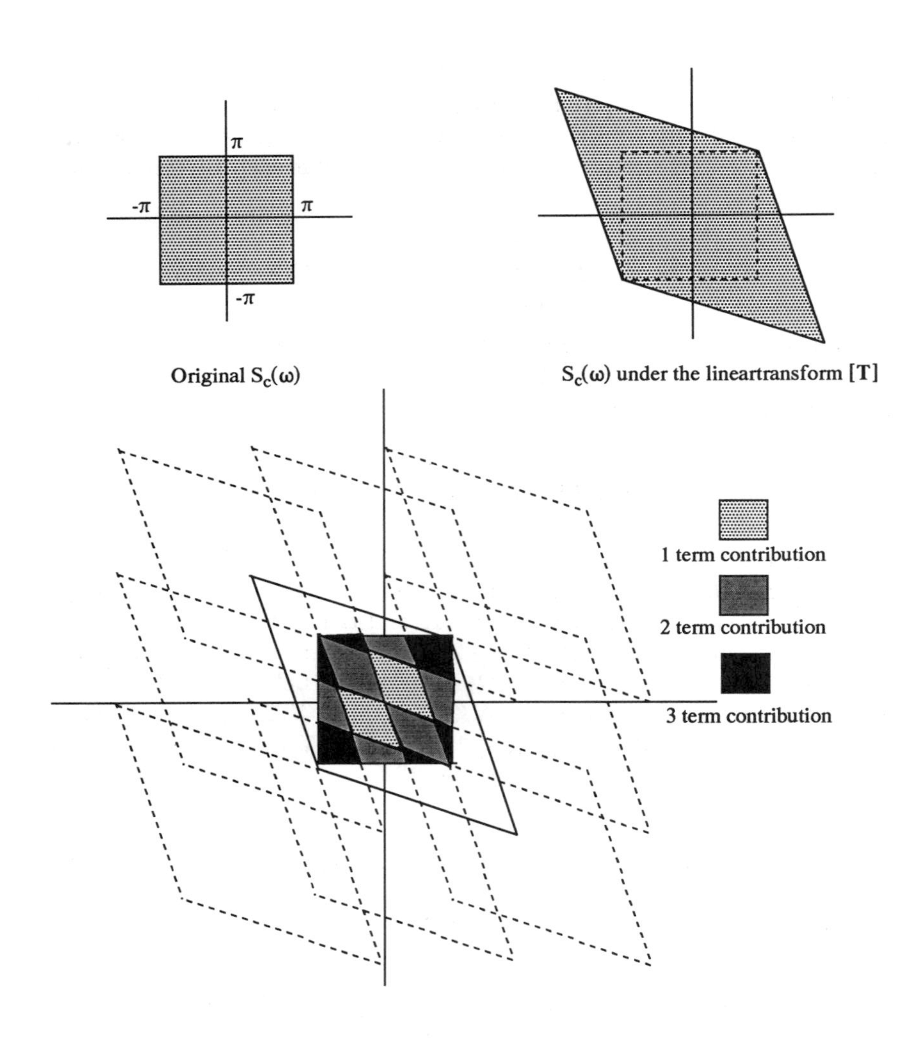

Figure 3: Terms in Eq. (12) that contribute nonzero values to $S_I'(\mathbf{\Omega})$ in the frequency band $-\pi < \Omega_1, \Omega_2 \leq \pi$.

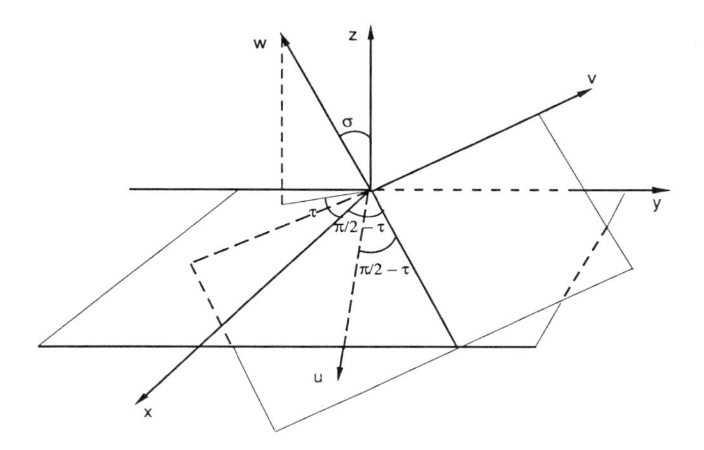

Figure 4: Imaging geometry.

The likelihood function of the $\boldsymbol{G}_I = \{G_I(m,n)\text{'s, for } -N/2 \le m,n \le N/2-1\}$ is given by

$$p(\boldsymbol{G}_I|\gamma,\sigma,\tau) = \prod_{(m,n)\varepsilon R_2} (1/\pi S_I(m,n))$$
$$\exp\{-\sum_{(m,n)\varepsilon R_2} |G_I(m,n)|^2/S_I(m,n)\}\cdot \prod_{(m,n)\varepsilon R_{1R}} (1/2\pi N^2 S_I(m,n))^{1/2}$$
$$\exp\{-\sum_{(m,n)\varepsilon R_{1R}} |G_I(m,n)|^2/2N^2 S_I(m,n)\} \tag{14}$$

Remark: *In this derivation it is not necessary to know the bandwidth of the band-limited PSD* $S_c(\omega_1,\omega_2)$ *associated with the continuous random field* $g_c(x,y)$ *as long as the PSD* $S_g(\Omega_1,\Omega_2)$ *of the sampled field* $g_o(m,n)$ *(modeled here by a GMRF with parameter set* γ*) is a non-aliased version of* $S_c(\omega_1=\Omega_1,\omega_2=\Omega_2)$*, for* $-\pi<\Omega_1,\Omega_2\le\pi$*, (i.e., as long as* $S_g(\Omega_1,\Omega_2) = S_c(\omega_1=\Omega_1,\omega_2=\Omega_2)$*, for* $-\pi<\Omega_1,\Omega_2\le\pi$*).*

4.2 Orthographic Projection Camera Model

One example of a linear transformation $[\boldsymbol{T}]$ is the orthographic projection $[\boldsymbol{O}]$. Let σ and τ, shown in Fig. 4, be the slant angle and the tilt angle of the planar surface relative to the viewer coordinate system.

If the surface is parallel to the image frame and has no slant or tilt, then the texture in the image coincides with the texture on the surface. The effects of the slant and tilt of the plane on the image texture under orthographic projection can be easily derived as follows. Consider a textured plane that coincides with the image plane (i.e., the uvw object plane coordinates and the xyz image coordinates coincide, and the normals to the planes are along w and z). To realize a slant of σ and tilt of τ of the object plane relative to the image plane, we first rotate the object plane by an angle $(\pi/2 - \tau)$ about the w axis. This results in a rotation of the uv axes, with the w axis remaining the same. We then rotate by an angle of $-\sigma$ the object plane about that new u axis. Here, the vw axes are changing, and the u axis remains unchanged. These two rotations realize the desired slant and tilt; however, the texture on the plane has been rotated by an angle of $(\pi/2 - \tau)$. To undo this effect, we rotate the object plane about the new w axis by an angle $(\tau - \pi/2)$. The rotations are shown in Fig. 4. A point with coordinates $(u\ v\ 0)^t$ relative to the uvw coordinates system will have its xyz coordinates given by:

$$\begin{bmatrix} x \\ y \\ z \end{bmatrix} = \begin{bmatrix} \sin\{\tau\} & \cos\{\tau\} & 0 \\ -\cos\{\tau\} & \sin\{\tau\} & 0 \\ 0 & 0 & 1 \end{bmatrix} \begin{bmatrix} 1 & 0 & 0 \\ 0 & \cos\{\sigma\} & \sin\{\sigma\} \\ 0 & -\sin\{\sigma\} & \cos\{\sigma\} \end{bmatrix}, \cdot \begin{bmatrix} \sin\{\tau\} & -\cos\{\tau\} & 0 \\ \cos\{\tau\} & \sin\{\tau\} & 0 \\ 0 & 0 & 1 \end{bmatrix} \begin{bmatrix} u \\ v \\ 0 \end{bmatrix}. \tag{15}$$

As the point on the object plane is orthographically projected on the image plane, its image coordinates $(i\ j)^t$ are simply $(x\ y)^t$, which are given in Eq. (15). Hence,

$$\begin{bmatrix} i \\ j \end{bmatrix} = [\boldsymbol{T}] \begin{bmatrix} u \\ v \end{bmatrix} = \begin{bmatrix} 1-\rho\cos^2\{\tau\} & \rho\sin\{\tau\}\cos\{\tau\} \\ \rho\sin\{\tau\}\cos\{\tau\} & 1-\rho\sin^2\{\tau\} \end{bmatrix} \begin{bmatrix} u \\ v \end{bmatrix}, \tag{16}$$

where $\rho = (1-\cos\{\sigma\})$. Eq. (16) is also reported in [40]. Three consecutive operations are performed to arrive at the slanted and tilted images, which are summed up in the factorization of $[\boldsymbol{T}]$ as:

$$[\boldsymbol{T}] = [\boldsymbol{R}]^{-1}[\boldsymbol{S}][\boldsymbol{R}] = [\boldsymbol{O}] = [\boldsymbol{O}^t], \tag{17}$$

where $[\boldsymbol{R}]$ is the 2×2 rotation matrix by an angle of $(\pi/2 - \tau)$, $[\boldsymbol{S}]$ is a 2×2 scaling diagonal matrix having its diagonal elements, 1 and $\cos\{\sigma\}$, respectively, and $[\boldsymbol{R}]^{-1}$ is a 2×2 rotation matrix by an angle of $(\tau - \pi/2)$.

The effect of the orthographic projection is a scaling (compression) by $\cos\{\sigma\}$ of the texture in the direction normal to the rotation axis determined by τ (Fig. 4), i.e., the texture appears uniformly compressed by $\cos\{\sigma\}$ in the direction normal to the rotation axis determined by τ. Because of the linearity of the transform in Eq. (16), the resulting texture is homogeneous.

4.3 On the Markovianity of the Transformed GMRF

In this section, we comment on the Markovianity [35] of a texture modeled by a GMRF when it undergoes a linear transformation. As we have seen in Section 3, a GMRF $g(m,n)$ has a PSD given in Eq. (3). Let $\gamma_g(\Omega_1,\Omega_2) = \varphi^2/S_g(\Omega_1,\Omega_2)$, where $\gamma_g(\Omega_1,\Omega_2)$ is given as:

$$\begin{aligned}\gamma_g(\Omega_1,\Omega_2) &= \left(1 - \sum_{(k,l)\varepsilon D_p} \beta_{k,l}\exp\{-j(k\Omega_1 + l\Omega_2)\}\right) \\ &= \sum_{(k,l)\varepsilon D_p\cup(0,0)} a_{k,l}\exp\{-j(k\Omega_1 + l\Omega_2)\}, \end{aligned} \tag{18}$$

where $a_{0,0} = 1$, and $a_{k,l} = -\beta_{k,l}$, for $(k,\ l) \neq (0,\ 0)$ and $(k,\ l)\varepsilon D_p$, and $a_{k,l} = 0$, otherwise. Then a necessary condition for GMRF to be of finite order is to have $\{a_{k,l}\}$ sequence to be of finite extent. $\gamma_g(\Omega_1,\Omega_2)$ in Eq. (18) can be thought of as the discrete-time Fourier transform of the finite-extent sequence $\{a_{k,l}\}$, with the $a_{k,l}$'s given as the inverse discrete-time Fourier transform of $\gamma_g(\Omega_1,\Omega_2)$, i.e.,

$$a_{k,l} = \frac{1}{4\pi^2}\int_{-\pi}^{\pi}\int_{-\pi}^{\pi} \gamma_g(\Omega_1,\Omega_2)\exp\{j(k\Omega_1 + l\Omega_2)\}d\Omega_1 d\Omega_2. \tag{19}$$

$a_{k,l}$ has a nonzero value only for $(k,l)\varepsilon D_p \cup (0,0)$. As shown in Section 4.1, the PSD of the transformed infinite-extent image $g'_I(m,n)$ is given in Eq. (11). In terms of $\gamma_g(\Omega_1,\Omega_2)$ in Eq. (18), $S'_I(\Omega_1,\Omega_2)$ is given as:

$$S'_I(\Omega_1,\Omega_2) = (\varphi^2\det\{[\boldsymbol{T}]\})\sum_{k=-K}^{K}\sum_{l=-L}^{L}\frac{1}{\gamma_g([\boldsymbol{T}](\boldsymbol{\Omega} - 2\pi\boldsymbol{r}_{kl}))}. \tag{20}$$

Let $\gamma'_I(\Omega_1,\Omega_2) = (\varphi^2\det\{[\boldsymbol{T}]\})/S'_I(\Omega_1,\Omega_2)$, then from Eq. (20), it follows that

$$\gamma_I'(\Omega_1, \Omega_2) = \left[\sum_{k=-K}^{K} \sum_{l=-L}^{L} \frac{1}{\gamma_g([\boldsymbol{T}]([\boldsymbol{\Omega}] - 2\pi \boldsymbol{r}_{kl}))} \right]^{-1}. \tag{21}$$

The inverse discrete-time Fourier transform of $\gamma_I'(\Omega_1, \Omega_2)$ is in general not of finite extent, and hence would not correspond to a GMRF of finite order.

5 MLE of Surface Shape Parameters

5.1 Estimating Plane Orientation

When γ is known, the MLE for σ and τ are obtained by maximizing Eq. (14) with respect to σ and τ. For an image whose size is relatively large, then the true values for σ and τ can be obtained and Cramer–Rao bounds can be achieved [35]. When γ is not known *a priori*, then the problem becomes ill-posed and unidentifiable, and neither the original texture nor σ and τ can be recovered just from the projective relation. Any value of σ and τ defines a possible surface orientation and a possible reconstruction of the unprojected texture. That means that there will be a family of possible texture/orientation pairs that can generate the image texture. (See Fig. 5.) The projection can be thought of as many-to-one mapping.

5.2 Generalization to More Complex Regular Surfaces

To generalize the model to non-planar surfaces, we view the 3-D surfaces as made up of local planar patches tangential to the surface at the point of consideration. Each of these planar patches is projected onto the image plane according to the assumed camera model projection. As one can imagine, the smaller the planar patch is, the better the approximation to the surface. On the other hand, the bigger the window is, the more reliable are the estimates for the orientation. Hence, one has to select the window size carefully based on the surface curvature and how reliable the estimates are. Under this planar facet model, a given image is partitioned into relatively small non-overlapping windows and the orientation of the plane in each of these windows is obtained. Surface shape estimation can be obtained from the set of computed unit normals $\boldsymbol{N}_k$'s. There are many ways of estimating surface shape from the unit normals [41]. In this chapter, we adopt a window-based approach wherein the image data is partitioned into non-overlapping or sliding square windows (say, M windows) that are

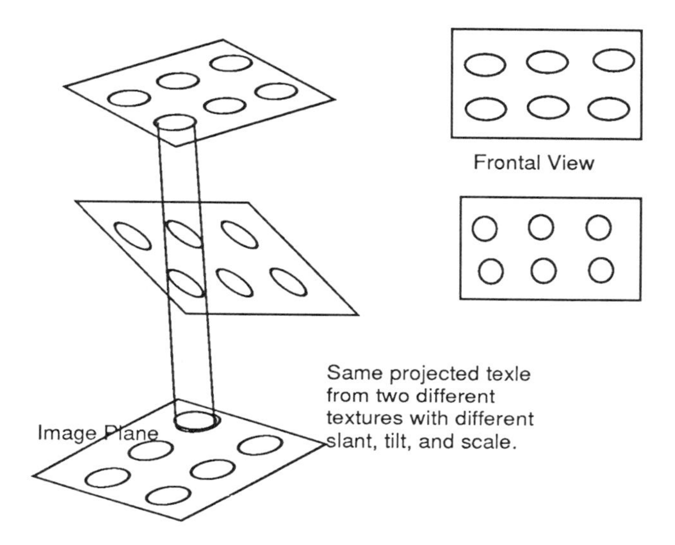

Figure 5: Same projected texel from two different textures with different slant, tilt, and scale.

assumed to emanate from locally planar patches. For each window, the plane orientation parameters are computed as explained in Section 5.1.

6 Surface Patch Classification from Surface Normals

The problem here is to classify a given surface patch into one of c possible surfaces denoted as $\omega_1, \omega_2, \ldots, \omega_c$, from the set of computed unit normals $\boldsymbol{N}_k$'s. In spherical coordinate systems, the unit normal $\boldsymbol{N}_k = (1 \; \sigma_k \; \tau_k) = (1 \; \boldsymbol{n}_k)$. Hence, we can cast the classification problem in terms of the $\boldsymbol{n}_k$'s. Let $\boldsymbol{n} = (\boldsymbol{n}_1, \boldsymbol{n}_2, \ldots, \boldsymbol{n}_M)$ be the set of estimated window plane normals. To classify the image patch to one of c classes, we use the Bayesian decision rule,

$$\max_{\omega_c}\{p(\omega_c|\boldsymbol{n})\} = \max_{\omega_c}\{p(\boldsymbol{n}|\omega_c)p(\omega_c)/p(\boldsymbol{n})\}, \tag{22}$$

which reduces to (23) for equal class prior $p(\omega_c)$,

$$\max_{\omega_c}\{p(\boldsymbol{n}|\omega_c)\}, \tag{23}$$

where $p(\boldsymbol{n}|\omega_c)$ is given by:

$$p(\boldsymbol{n}|\omega_c) = \int_{\boldsymbol{\Gamma}_c} p(\boldsymbol{n}, \boldsymbol{\Gamma}_c|\omega_c)d\boldsymbol{\Gamma}_c = \int_{\boldsymbol{\Gamma}_c} p(\boldsymbol{n}|\boldsymbol{\Gamma}_c, \omega_c)p(\boldsymbol{\Gamma}_c|\omega_c)d\boldsymbol{\Gamma}_c, \tag{24}$$

where $\boldsymbol{\Gamma}_c$ are the parameters associated with class ω_c. Eq. (24) is, in general, not available except for the case of reproducible densities [42]. To obtain a closed-form expression for $p(\boldsymbol{n}|\omega_c)$,we make use of the following approximation [42, 43]:

$$p(\boldsymbol{n}|(\boldsymbol{\Gamma}_c, \omega_c) \approx p(\boldsymbol{n}|\boldsymbol{\Gamma}_c^*, \omega_c)\exp\{-(1/2)(\boldsymbol{\Gamma}_c^* - \boldsymbol{\Gamma}_c)^t[\Psi(\boldsymbol{\Gamma}_c^*)](\boldsymbol{\Gamma}_c^* - \boldsymbol{\Gamma}_c)\}, \tag{25}$$

where $\boldsymbol{\Gamma}_c^*$ is the maximum likelihood of $\boldsymbol{\Gamma}_c$, and $[\Psi(\boldsymbol{\Gamma}_c^*)]$ is the Hessian matrix with elements a_{mn},

$$a_{mn} = -\partial^2 \log\{p(\boldsymbol{n}|\boldsymbol{\Gamma}_c^*, \omega_c)\}/\partial\Gamma_m\partial\Gamma_n, \tag{26}$$

and Γ_m is the mth component of $\boldsymbol{\Gamma}_c^*.\boldsymbol{\Gamma}_c^*$ converges with probability 1 to $\boldsymbol{\Gamma}_c$ as $M \to \infty$. Assuming a uniform improper $p(\boldsymbol{\Gamma}_c|\omega_c)$ yields :

$$\begin{aligned} \log\{p(\boldsymbol{n}|(\boldsymbol{\Gamma}_c, \omega_c)\} \approx\ & \log\{p(\boldsymbol{n}|(\boldsymbol{\Gamma}_c^*, \omega_c)\} - m(c)\log\{1/2\pi\} \\ & - \log\{\det\{[\Psi(\boldsymbol{\Gamma}_c^*)]\} + \log\{p(\boldsymbol{\Gamma}_c^*|\omega_c)\}, \end{aligned} \tag{27}$$

where $m(c)$ is the cardinality of $\boldsymbol{\Gamma}_c$.

Using Eq. (27) and assuming a uniform improper $p(\boldsymbol{\Gamma}_c|\omega_c)$ for all $\boldsymbol{\Gamma}_c$ and ω_c, the decision rule in Eq. (23) simplifies to:

$$\max_{\omega_c}\{\log\{p(\boldsymbol{n}|\boldsymbol{\Gamma}_c^*, \omega_c)\} - m(c)\log\{1/2\pi\} - \log\{\det\{[\Psi(\boldsymbol{\Gamma}_c^*)]\}\} \tag{28}$$

The decision rule in Eq. (28) has been introduced by Kashyap [44] in a different context. Their concern was to decide on the *best* order of an autoregressive process. A decision rule similar to Eq. (28) was also used by Bolle and Cooper [6] for classifying an image patch into one of c possible quadric intensity polynomials.

In our analysis, we have assumed that the σ and τ components associated with $\boldsymbol{n}$ have white Gaussian perturbation that are independent of each other, i.e.,

$$\begin{aligned}\sigma_k(x_k, y_k) &= \hat{\sigma}_k(x_k, y_k) + w_1(x_k, y_k), \\ \tau_k(x_k, y_k) &= \hat{\tau}_k(x_k, y_k) + w_2(x_k, y_k),\end{aligned} \tag{29}$$

where $w_m(x_k, y_k) \sim \eta(0, \nu^2), m = 1, 2$. $\hat{\sigma}_k(x_k, y_k)$ and $\hat{\tau}_k(x_k, y_k)$ in Eq. (29) are the true surface slant and tilt, respectively, evaluated at the centroid (x_k, y_k) of the kth window. $\hat{\boldsymbol{n}}_k = \{\hat{\sigma}_k(x_k, y_k), \hat{\tau}_k(x_k, y_k)\}$ is a function of the parameters of the underlying surface. Its equation can be derived in a straightforward manner from the equations of the polynomial surface (e.g., a plane, cylinder, or sphere). If the polynomial surface (of order p) is described in the implicit form as

$$f(x, y, z) = \boldsymbol{\Gamma}^t \boldsymbol{r}_p = 0, \tag{30}$$

where $\boldsymbol{\Gamma}$ are the polynomial parameters $\boldsymbol{r}_p = (x^p, y^p, z^p, \ldots, x, y, z, 1)$ is the set of monomials. The normal vector to the surface at point (x, y, z) is obtained by taking the gradient of $f(x, y, z)$ with respect to x, y, and z. The unit normal is given as $\boldsymbol{N} = (N_x, N_y, N_z) = \boldsymbol{\nabla} f(x, y, z)/|\boldsymbol{\nabla} f(x, y, z)|$, where $\boldsymbol{\nabla} = (\partial/\partial x, \partial/\partial y, \partial/\partial z)$. In spherical coordinates $\boldsymbol{N} = (1, \sigma, \tau) = (1, \boldsymbol{n})$, with

$$\hat{\boldsymbol{n}} = (\hat{\sigma} = \cos^{-1}\{\sqrt{1 - N_x^2 - N_y^2}\},\ \hat{\tau} = \tan^{-1}\{N_y/N_x\}). \tag{31}$$

In light of the model in Eq. (28), the conditional class likelihood $p(\boldsymbol{n}|\boldsymbol{\Gamma}_c, \omega_c)$ is given as :

$$p(\boldsymbol{n}|\boldsymbol{\Gamma}_c, \omega_c) = \prod_{k=1}^{M} p(\boldsymbol{n}_k|\boldsymbol{\Gamma}_c, \omega_c), \tag{32}$$

where

$$p(\boldsymbol{n}_k|\boldsymbol{\Gamma}_c, \omega_c) \sim \eta(E\{\boldsymbol{n}_k|\boldsymbol{\Gamma}_c, \omega_c\}, \nu^2[\boldsymbol{I}]), \tag{33}$$

where $E\{\boldsymbol{n}_j|\boldsymbol{\Gamma}_c, \omega_c\}$ is the expected value of the normal at the centroid of window j under the assumption that the surface is ω_c, and that the surface is parametrized with the parameter vector $\boldsymbol{\Gamma}_c$. This is computed as in Eq. (31).

7 Surface Patch Classification—Unknown Texture Case

As in the planar case, if the parameter set γ of the parent texture is known, the problem becomes well-defined and the true orientations of each patch can be estimated. In the real world, however, the parameter set γ of the parent texture is seldom known *a priori*; and hence this problem of orientation estimation becomes ill-posed as described earlier. To alleviate this problem, one can select a window (say, window 1) and maximize the likelihood given in Eq. (14) with respect to γ and $[\boldsymbol{O}]$ to obtain a possible parameter set and orientation pair, i.e.,

$$\max_{\gamma,[\boldsymbol{O}]} \{p(\boldsymbol{G}_1|\gamma,[\boldsymbol{O}])\}, \tag{34}$$

or, equivalently,

$$\max_{\gamma,\sigma,\tau} p(\boldsymbol{G}_1|\gamma,\sigma,\tau), \tag{35}$$

where $\boldsymbol{G}_1$ is the discrete Fourier transform (DFT) of the data in window 1. Let $\gamma^*(1)$ and $[\boldsymbol{O}(1)]^*$ (corresponding to $(\sigma^*(1),\ \tau^*(1))$) be the MLE for γ and $[\boldsymbol{O}]$, respectively. Since all the windows originated from the same parent texture, we impose $\gamma^*(1)$ on the rest of the windows, and obtain the orientations of all the windows in the image patch, i.e.,

$$\max_{[\boldsymbol{O}(J)],J\neq 1} \{p(\boldsymbol{G}_J|\gamma^*(1),[\boldsymbol{O}(J)])\} \tag{36}$$

for any window J in the image patch. Apart from the computational complexity associated with this method, it would fail in many cases. To illustrate the problem associated with this method, consider the texture shown in Fig. 6a. The texture given in Fig. 6a is the parent, while the textures in Figs. 6b and 6c were obtained from the parent texture by subjecting the parent texture to an orientation of $(\sigma = 60^o,\ \tau = 45^o)$, and an orientation of $(\sigma = 60^o, \tau = -45^o)$, respectively.

Consider the texture in Fig. 6b. We estimate the parent parameter set γ^* and the orientation (say, $[\boldsymbol{O}]^*$) for this texture by maximizing the likelihood given in Eq. (14). One possible pair of $\{[\boldsymbol{O}^*],\gamma]^*\}$ could be as shown in Fig. 6d, with $[\boldsymbol{O}]^*$ corresponding to $\sigma^* = 0$ and $\tau^* = 0$. Note that the texture in Fig. 6d can be obtained from the texture in Fig. 6b through an identity projection, i.e., $\sigma = 0$ and $\tau = 0$. As pointed out earlier, γ^* and $[\boldsymbol{O}]^*$ are not necessarily the true ones. Consider the texture in Fig. 6c. Recalling the fact that an orthographic projection of a texture

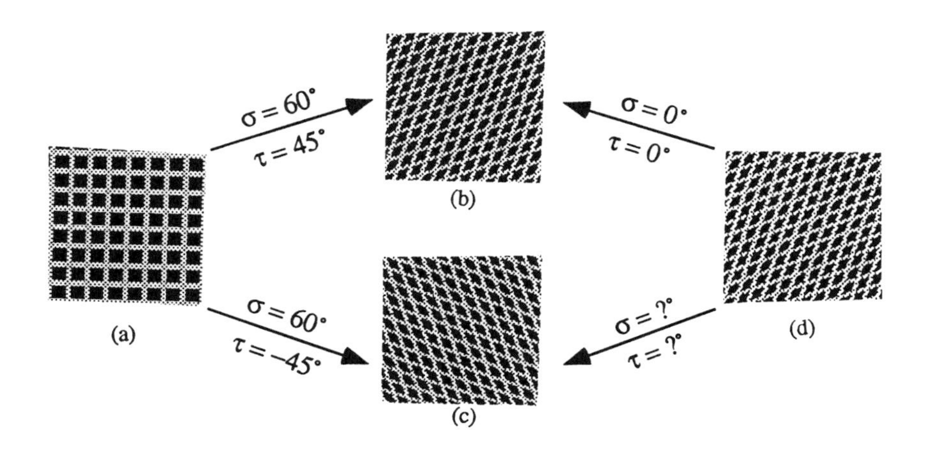

Figure 6: Consistency and uniqueness of the parent texture.

always results in a compression of the texture by an amount $\cos\{\sigma\}$ in the τ direction, one can see, therefore, that we cannot obtain the texture in Fig. 6c by compressing the texture in Fig. 6d in any direction . This implies that there exists no orthographic projection that will transform the texture in Fig. 6d to the texture shown in Fig. 6b. One can obtain the texture in Fig. 6c by rotating the texture in Fig. 6d by an amount $\pi/2$; but the rotation is not an orthographic projection! We overcome this problem by making use of what we call *stereo windows*. This method is illustrated in the following section.

7.1 Stereo Windows

Pick any two windows in the image and call them 1 and 2 for illustration. The idea here is to find a parent texture, which under $[\boldsymbol{O}(1)]$ will yield the texture in window 1 and under $[\boldsymbol{O}(2)]$ yields the texture in window 2. This is illustrated in Fig. 7. To achieve this, we maximize the likelihood function,

$$\max_{[\boldsymbol{O}(1)],[\boldsymbol{O}(2)]} \{p(\boldsymbol{G}_1|\gamma_1,[\boldsymbol{O}(1)])p(\boldsymbol{G}_2|\gamma_2,[\boldsymbol{O}(2)])\}, \tag{37}$$

subject to the constraint that $\gamma_1 = \gamma_2$, where $\boldsymbol{G}_1$ and $\boldsymbol{G}_2$ are the DFT of the data in the two windows under consideration. This maximization will yield the parent texture parameter set γ^* and the two orientations, $[\boldsymbol{O}(1)]^*$ and $[\boldsymbol{O}(2)]^*$.

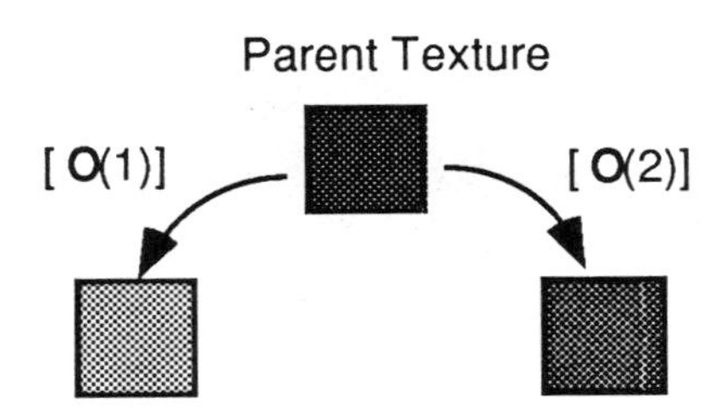

Figure 7: Stereo windows.

There are two cases of interest with regard to the uniqueness of the parent texture. Assume that there exist two parent textures that under orthographic projection give rise to the textures in windows 1 and 2, respectively. The first of these two parent textures with the appropriate orientations, $[\boldsymbol{O}(1)]$ and $[\boldsymbol{O}(2)]$, is shown in Fig. 7. Let the orientations associated with the other parent texture be $[\boldsymbol{O}(3)]$ and $[\boldsymbol{O}(4)]$, respectively. Then the linear transform $[\boldsymbol{T}]$ that relates the texture in window 1 to that in window 2 is given by:

$$[\boldsymbol{T}] = [\boldsymbol{O}(1)][\boldsymbol{O}(2)]^{-1} = [\boldsymbol{O}(3)][\boldsymbol{O}(4)]^{-1}, \tag{38}$$

where (See Eq. 17)

$$[\boldsymbol{O}(1)][\boldsymbol{O}(2)]^{-1} \quad = \quad [\boldsymbol{R}(1)]^t[\boldsymbol{S}(1)][\boldsymbol{R}(1)][\boldsymbol{R}(2)]^t[\boldsymbol{S}(2)]^{-1}[\boldsymbol{R}(2)]$$

and

$$[\boldsymbol{O}(3)][\boldsymbol{O}(4)]^{-1} = [\boldsymbol{R}(3)]^t[\boldsymbol{S}(3)][\boldsymbol{R}(3)][\boldsymbol{R}(4)]^t[\boldsymbol{S}(4)]^{-1}[\boldsymbol{R}(4)]. \tag{39}$$

After a few algebraic manipulations, we arrive at the following result:

$$[\boldsymbol{R}(1-3)]^t[\boldsymbol{S}(1)][\boldsymbol{R}(1-3)][\boldsymbol{R}(3)][\boldsymbol{R}(4)]^t[\boldsymbol{R}(2-4)]^t[S(2)]^{-1} \\ [\boldsymbol{R}(2-4)] = [\boldsymbol{S}(3)][\boldsymbol{R}(3)][\boldsymbol{R}(4)]^t[S(4)]^{-1}. \tag{40}$$

From Eq. (40), these two results follow :

(i) When the two windows have the same tilt, i.e., $\tau(1) = \tau(2)$ and $\tau(3) = \tau(4)$, Eq. (40) reduces to:

$$[\boldsymbol{S}(1)][\boldsymbol{S}(2)]^{-1} = [\boldsymbol{S}(3)][\boldsymbol{S}(4)]^{-1}, \tag{41}$$

which translates to:

$$\cos\{\sigma(1)\}/\cos\{\sigma(2)\} = \cos\{\sigma(3)\}/\cos\{\sigma(4)\}. \tag{42}$$

Eq. (42) means that there exists an infinite number of parent textures that satisfy Eq. (42) that yield, under orthographic projection, the textures in windows 1 and 2. This is true as long as the two textures have the same tilts;

(ii) When the windows have different tilts, then it follows that

$$\begin{aligned} [\boldsymbol{R}(1-3)]^t[\boldsymbol{S}(1)][\boldsymbol{R}(1-3)] &= [\boldsymbol{S}(3)], \\ [\boldsymbol{R}(2-4)]^t[\boldsymbol{S}(2)]^{-1}[\boldsymbol{R}(2-4)] &= [\boldsymbol{S}(4)]^{-1}, \end{aligned} \tag{43}$$

which implies that

$$[\boldsymbol{O}(1)] = [\boldsymbol{O}(3)] \text{ and } [\boldsymbol{O}(2)] = [\boldsymbol{O}(4)]. \tag{44}$$

The result in Eq. (44) implies the uniqueness of the parent texture as long as the tilts in the two windows are different.

The procedure described above can be thought of as some type of a regularization process [45]. Ways of speeding up the computations for estimating γ^* from Eq. (36) are discussed in [38]. Once we have the parent texture, we impose this parent texture on the rest of the windows to obtain their orientations. The problem is now reduced to surface classification and shape extraction from a known parent texture.

7.2 Classification for the Case of Plane, Cylinder, and Sphere

In this chapter, we have restricted our analysis to planar, cylindrical, and spherical surfaces. In terms of the two cases we have discussed earlier, both the planar and the cylindrical surfaces fall under case 1, whereas the spherical surface falls under case 2. In the case of a plane, we cannot do much, except to recognize that it is a plane. Note that all the windows for a plane have the same σ and τ. We select a window $\boldsymbol{I}$ in the image, and maximize the likelihood in Eq. (14) with respect to the orientation $[\boldsymbol{O}]$ and the parameter set γ . Since the MLE for the orientation $[\boldsymbol{O}]$ and the parameter set γ are not guaranteed to be the true ones, one can reduce the computational burden by imposing an arbitrary orientation, say, $\sigma = 0$

and $\tau = 0$, and obtain a parameter set $\boldsymbol{\gamma}$ for the parent texture. This is done by maximizing the likelihood (14) with respect to $\boldsymbol{\gamma}$ while imposing the orientation that corresponds to $\sigma = 0$, $\tau = 0$, i.e.,

$$\max_{\boldsymbol{\gamma}}\{p(\boldsymbol{G}_I|\boldsymbol{\gamma}, \sigma = 0, \tau = 0)\}. \tag{45}$$

Let $\boldsymbol{\gamma}^*$ be the MLE for $\boldsymbol{\gamma}$. Imposing $\boldsymbol{\gamma}^*$ on the rest of the windows should result in the same orientation for all the windows.

The cylindrical case is more complicated. All the windows for a cylindrical patch have the same tilt, but different slants. Again, as we did in the planar case, we impose an arbitrary orientation on a selected window "k", and obtain a parameter set $\boldsymbol{\gamma}^*(k)$ for this window. This $\boldsymbol{\gamma}^*(k)$ is imposed on the rest of the windows. We select the first window in Fig. 8 ($k = 1$) and obtain the $\boldsymbol{\gamma}^*(1)$ for this window by imposing $\sigma = 0$ and $\tau = 0$. The texture in window 5 cannot be obtained by a compression of the texture in window 1. It can only be obtained by an expansion of that texture. This implies that we should allow for $[\boldsymbol{O}]$ and $[\boldsymbol{O}]^{-1}$ in the orientation estimation process. If $[\boldsymbol{O}]$ corresponds to a scaling of the texture by an amount $1/\cos\{\sigma\}$ along τ, then $[\boldsymbol{O}]^{-1}$ corresponds to the scaling of the texture by an amount $\cos\{\sigma\}$ along the direction τ. Let $s = (1/\cos\{\sigma\}$ or $\cos\{\sigma\})$. We maximize the likelihood (14) with respect to s and τ, i.e.,

$$\max_{s,\tau}\{p(\boldsymbol{G}_j|\gamma = \boldsymbol{\gamma}^*(1), s, \tau\}. \tag{46}$$

A scale ($s < 1.0$) corresponds to the case of compression, i.e., orientation $[\boldsymbol{O}]$, and the case of the scale ($s > 1.0$) corresponds to the case of expansion, i.e., orientation $[\boldsymbol{O}]^{-1}$. Note here that, although we were not able to recover the true orientations and the parent texture, we will still be able to recognize the surface as a cylinder and estimate its radius and axis orientation. This is due to the fact that the relative orientation between the window normals is preserved. We need to go through a normalization procedure to obtain the (σ, τ) distribution from (s, τ) distribution. We select the window i, which has the largest value of $s = s(i)$ (this corresponds to the window which has the largest expansion, i.e., that is the nearest to the camera.) We normalize the rest of the scales $s(j)$ with respect to $s(i)$. This is done by replacing the scale $s(j)$ by $s(j)/s(i)$ for each window j. Then the orientation of the window j is given in terms of $(\cos^{-1}\{(s(j)/s(i)\}, \tau)$.

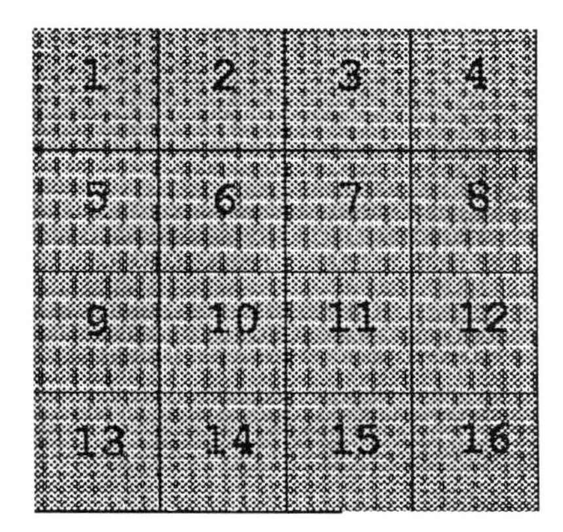

Figure 8: Cylinder image partitioned into 16 windows.

8 Experiments

We ran experiments on the 256×256 cylindrical and spherical patches shown in Figs. 9 and 10. The true radii (in pixels) were 130 and 190 for the cylindrical and spherical surfaces, respectively. These experimental images were obtained as follows. First, the parent textures from the Brodatz album [46] were scanned, and later these textures were mapped onto the cylindrical and the spherical surfaces by a ray-tracing like algorithm [38]. The 256×256 image patches shown in Figs. 9 and 10 were extracted out of these images. The images were partitioned into 16 windows, each of size 64×64, for which the planar facet orientations were estimated. Since the parent textures were available to us, we estimated the parameters of these parent textures by fitting eighth-order GMRF models.

In the first part of the experiment, we assumed the knowledge of the parent texture parameters. In all four cases, the image patch was correctly classified and the estimated radii were close to the true ones. Tables 1a and 2a show the expected values for $\cos\{\sigma\}$ and τ evaluated at the window centroids, for the cylinder and sphere, respectively. Tables 1b and 2b show the estimated values of $\cos\{\sigma\}$ and τ for the cylinder and the sphere image patches shown in Figs. 9a and 10a, respectively. Based on these estimated values, the radius of the cylinder was estimated to be 110, and the orientation of the axis was found to be along the y-axis. The sphere radius was estimated to be 168.

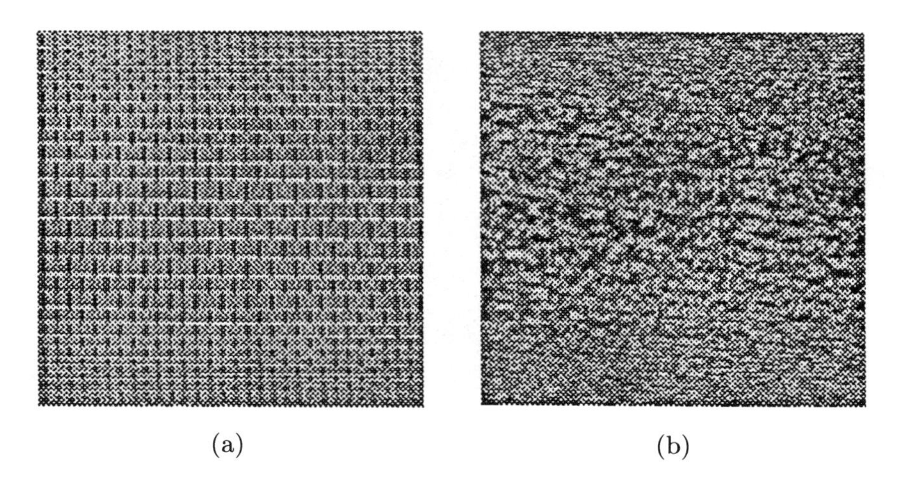

(a) (b)

Figure 9: (a) Cylindrical patch (texture D14). (b) Cylindrical patch (texture D24).

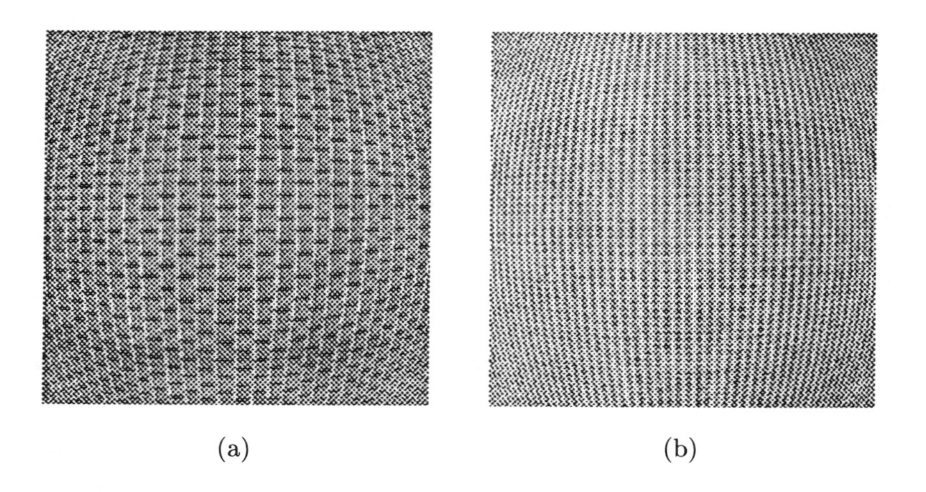

(a) (b)

Figure 10: (a) Spherical patch (texture D14). (b) Spherical patch (texture D21).

Window j^{th} coordinate

$i \backslash j$	1	65	129	193
1	(0.67,0.0)	(0.67,0.0)	(0.67,0.0)	(0.67,0.0)
65	(0.97,0.0)	(0.97,0.0)	(0.97,0.0)	(0.97,0.0)
129	(0.97,0.0)	(0.97,0.0)	(0.97,0.0)	(0.97,0.0)
193	(0.65,0.0)	(0.65,0.0)	(0.65,0.0)	(0.65,0.0)

Window i^{th} Coordinate

cos{σ} and τ (in degrees) distribution
Expected Values for Cylinder
(a)

Window j^{th} coordinate

$i \backslash j$	1	65	129	193
1	(0.67,0.0)	(0.67,0.0)	(0.67,0.0)	(0.67,0.0)
65	(0.97,0.0)	(0.97,0.0)	(0.97,0.0)	(0.97,0.0)
129	(0.97,0.0)	(0.97,0.0)	(0.97,0.0)	(0.97,0.0)
193	(0.65,0.0)	(0.65,0.0)	(0.65,0.0)	(0.65,0.0)

Window i^{th} Coordinate

cos{σ} and τ (in degrees) distribution
Estimated Values for Cylinder
Known Texture Case
(b)

Window j^{th} coordinate

$i \backslash j$	1	65	129	193
1	(0.41,-0.46)	(0.39,0.233)	(0.41,0.5)	(0.41,0.5)
65	(0.93,-0.01)	(0.88,-0.09)	(0.90,-0.02)	(0.92,1.2)
129	(0.9,-0.29)	(0.99,-0.01)	(1.0,-0.02)	(0.82,-0.05)
193	(0.39,-1.35)	(0.42,0.53)	(0.39,-9.3)	(0.41,0.5)

Window i^{th} Coordinate

cos{σ} and τ (in degrees) distribution
Estimated Values for Cylinder
Unknown Texture Case
(c)

Table 1.

Window j^{th} coordinate

$i \backslash j$	1	65	129	193
1	(0.65,45.0)	(0.83,18.1)	(0.83,-19.2)	(0.65,-46.6)
65	(0.83, 71.9)	(0.97,45.0)	(0.97,-46.8)	(0.82,-72.3)
129	(0.83,-70.8)	(0.97,-43.21)	(0.97,45.0)	(0.82,71.21)
193	(0.65,-44.4)	(0.82,-17.72)	(0.82,18.79)	(0.65,45.0)

Window i^{th} Coordinate

cos{σ} and τ (in degrees) distribution
Expected Values for Sphere
(a)

Window j^{th} coordinate

$i \backslash j$	1	65	129	193
1	(0.50,51.3)	(0.78,22.7)	(0.73,-28.4)	(0.47,-48.6)
65	(0.77, 84.6)	(1.0,0.88)	(1.0,-0.92)	(0.71,-69.05)
129	(0.77,-53.6)	(1.0,1.0)	(1.0,-0.8)	(0.79,84.65)
193	(0.51,-38.4)	(0.76,-31.1)	(0.81,18.86)	(0.50,52.0)

Window i^{th} Coordinate

cos{σ} and τ (in degrees) distribution
Estimated Values for Sphere
Known Texture case
(b)

Window j^{th} coordinate

$i \backslash j$	1	65	129	193
1	(0.52,49.9)	(0.90,18.1)	(0.85,-35.4)	(0.50,-52.1)
65	(0.82, 78.2)	(1.0,-0.84)	(0.99,-1.7)	(0.7,-65.1)
129	(0.67,-70.25)	(1.0,1.0)	(1.00,0.98)	(0.62,74.21)
193	(0.51,-50.4)	(0.83,-28.72)	(0.83,20.79)	(0.50,53.0)

Window i^{th} Coordinate

cos{σ} and τ (in degrees) distribution
Estimated Values for Sphere
Unknown Texture case
(c)

Table 2.

In the second part of the experiment, we assumed that the parent texture parameters were unknown. Again, the image patches were correctly classified. The estimated values for the normalized $\cos\{\sigma\}$ and τ for each of the 64×64 windows is shown in Table 1(c) for the cylindrical surface of Fig. 9a. For the spherical surface shown in Fig. 10a, the estimated values of the $\cos\{\sigma\}$ and τ for each of the 64×64 windows is shown in Table 2c. Based on these estimates, the estimated radius for the cylinder was 105, and for the spherical surface, it was 160. Note that when the slant angle σ is zero, the tilt angle τ is undefined. This means that for slant angles approaching zero, the estimates for τ become unreliable. This explains the discrepancy between the estimated values of τ as compared to their expected values for the four windows at the center of the image in the sphere case.

9 Discussion and Conclusion

We presented a novel and unifying approach for the problem of shape from texture. Unlike prior approaches to shape from texture-which basically used special-purpose algorithms designed to exploit particular features of the texture, like directionality; pattern regularities; texel shape, size, and spacing; and special markings to extract shape information — our method proposes a general framework that is suited to all these textural features. The GMRFs are successful in modeling textures with directionality like in the pressed calf-leather (Texture D24); regular coarse texture like the woven aluminum wire (Texture D14); micro-texture like the French canvas (Texture D21); etc. The same decision–theoretic and estimation approach was applied to all the textures. Shape parameter estimation was posed as a maximum likelihood estimation problem whose performance can be assessed, and its asymptotic properties of efficiency and unbiasedness are most desirable. Minimum error bounds for the orientation estimates can be derived. A minimum error Bayes classifier was used to classify an image patch into one of C possible surfaces.

Bibliography

[1] B. K. P. Horn, "Obtaining Shape from Shading Information," *Psychology of Computer Vision*, P. H. Winston, Ed., McGraw-Hill, New York, 1975.

[2] B. K. P. Horn and R. W. Sjoberg, "Calculating the Reflectance Map," *Applied Optics*, Vol. 18, June 1979.

[3] A. P. Pentland, "Finding the Illuminant Direction," *Journal of the Optical Society of America*, Vol. 72, April 1982.

[4] M. J. Brooks and B. K .P. Horn, "Shape and Source from Shading," *Proc. IJCAI Conf.*, Los Angeles, CA, August 1985.

[5] B. Gidas and J. Torreao, "A Bayesian/Geometric Framework for Reconstructing 3D Shapes in Robot Vision," *Proc. SPIE Conf., High-Speed Computing II*, Vol. 1058, Los Angeles, CA, January 1989.

[6] R. M. Bolle and D. B. Cooper, "Bayesian Recognition of Local 3D Shape by Approximating Image Intensity Functions with Quadric Polynomials," *IEEE Trans. PAMI*, Vol. 6, July 1984.

[7] F. S. Cohen and J.-Y. Wang,"3-D Recognition and Shape Estimation from Image Contours", *Proc. 1992 IEEE Conf. Computer Vision and Pattern Recognition*, Urbana Champaign, Illinois, June 1992.

[8] K. A. Stevens, " The Visual Interpretation of Visual Contours," *Artificial Intelligence*, Vol. 17, August 1981.

[9] R. D. Rimey and F. S. Cohen, "A Maximum Likelihood Approach for Segmenting Range Data," *IEEE Trans. RA*, April 1988.

[10] B. Cernushi-Frias, D. B. Cooper, Y. Hung, and P. Belhumeur, "Toward a Model-Based Bayesian Theory for Estimating and Recognizing Parametrized 3-D Objects Using Two or More Images taken from Different Positions," *IEEE Trans. PAMI*, Vol. 11, October 1989.

[11] D. B. Cooper, Y. Hung, and J. Subrahmonia, "General Model-Based 3D Surface Estimation, Recognition, and Segmentation from Multiple Images," *Proc. DARPA Image Understanding Workshop*, Pittsburgh, PA, September 1990.

[12] J. J. Gibson, *The Perception of the Visual World*, Houghton Mifflin, Boston 1950.

[13] R. Bajcsy and L. Lieberman, "Texture Gradients as a Depth Cue," *Computer Vision, Graphics and Image Processing*, Vol. 5, 1976.

[14] A. P. Witkin, "Recovering Surface Shape and Orientation from Texture," *Artificial Intelligence*, Vol. 17, 1981.

[15] L. S. Davis, L. Lanos, and S. M. Dunn, "Efficient Recovery of Shape from Texture," *IEEE Trans. PAMI*, Vol. 5, 1983.

[16] K. Ikeuchi, "Shape from Regular Patterns," *Artificial Intelligence*, Vol. 22, 1984.

[17] J. R. Kender, "Shape from Texture : A Computational Paradigm," *Proc. Image Understanding Workshop*, May 1979.

[18] Y. Ohta, K. Maenobu, and T. Sakai, "Obtaining Surface Orientation from Texels under Perspective Projection," *Proc. IJCAI Conf.*, 1980.

[19] C. Brown, J. Aloimonos, M. Swain, P. Chou, and A. Basu, "Texture, Contour, Shape, and Motion ," *Pattern Recognition Letters*, Vol. 5, North-Holland, Amsterdam, 1987.

[20] D. Brzakovic, *Computer Based 3D Description from Texture*, Ph. D. dissertation, Department of Electrical Engineering, University of Florida, Gainesville, FL, 1984.

[21] Y. Choe and L. R. Kashyap, "A Model and Synthesis of Texture on 3-D Surface," *Proc. 26th Annual Allerton Conf. on Communication, Control, and Computing*, Urbana, IL, September 1988.

[22] J. W. Woods, "Two-Dimensional Discrete Markov Random Fields," *IEEE Trans. IT*, Vol. 18, March 1972.

[23] J. E. Besag, "Spatial Interaction and the Statistical Analysis of Lattice Systems," *Journal of the Royal Statistical Society*, Series B, Vol. 36, 1974.

[24] J. E. Besag and P. Moran, "On the Estimation and Testing of Spatial Interaction in Gaussian Lattices," *Biometrika*, Vol. 62, 1975.

[25] R. L. Kashyap and R. Chellappa, "Estimation and Choice of Neighbors in Spatial Interaction Models of Images," *IEEE Trans. IT*, Vol. 29, January 1983.

[26] R. Chellappa and R. L. Kashyap, "Texture Synthesis Using 2D Noncausal Autoregressive Models," *IEEE Trans. ASSP*, Vol. 33, February 1985.

[27] G. R. Cross and A. K. Jain, "Markov Random Field Texture Models," *IEEE Trans. PAMI*, Vol. 5, January 1983.

[28] F. S. Cohen, "Markov Random Fields for Image Modelling and Analysis," Chap. 10 in *Modelling and Applications of Stochastic Processes*, U. Desai, Ed., Kluwer Academic Press, Boston, MA, 1986.

[29] S. Geman and D. Geman, "Stochastic Relaxation, Gibbs Distributions and the Bayesian Restoration of Images," *IEEE Trans. PAMI*, Vol. 6, November 1984.

[30] H. Derin and H. Elliot, "Modeling and Segmentation of Noisy and Textured Images Using Gibbs Random Fields," *IEEE Trans. PAMI*, Vol. 9, January 1987.

[31] R. Cristi, "Markov and Recursive Least Squares Methods for the Estimation of Data with Discontinuities," *IEEE Trans. ASSP*, Vol. 38, November 1990.

[32] F. S. Cohen and D. B. Cooper, "Simple Parallel Hierarchical and Relaxation Algorithms for Segmenting Noncausal Markov Random Fields," *IEEE Trans. PAMI*, Vol. 9, March 1987.

[33] Z. Fan and F. S. Cohen, "Textured Image Segmentation as a Multiple-Hypothesis Test," *IEEE Trans. Systems and Circuits*, June 1988.

[34] F. S. Cohen and Z. Fan, "Maximum Likelihood Unsupervised Textured Image Segmentation," *Computer Vision Graphics and Image Processing: GraphicalModels and Image Processing*, Vol. 54, May 1992.

[35] F. S. Cohen, Z. Fan, and M. Patel, "Classification of Rotated and Scaled Textured Images Using Gaussian Markov Random Field," *IEEE Trans. PAMI*, Vol. 13, February 1991.

[36] F. S. Cohen and M. Patel "Modeling and Synthesis of Images of 3-D Textured Surfaces," *Computer Vision, Graphics and Image Processing: GraphicalModels and Image Processing*, Vol. 53, November 1991.

[37] M. Patel and F. S. Cohen "Local Surface Shape Estimation of 3-D Textured Surfaces using Gaussian Markov Random Fields and Stereo-Windows" *Proc. 1992 IEEE Conf. Computer Vision and Pattern Recognition*, Urbana Champaign, Illinois, June 1992.

[38] M. Patel, *Shape Extraction of 3D Textured Scenes from Image Data*, Ph.D. Dissertation, Drexel University, Electrical and Computer Engineering Department, PL, 1991.

[39] A. K. Jain, *Fundamentals of Digital Image Processing*, Prentice Hall, Englewood Cliffs, NJ, 1988.

[40] K. Kanatani, "Detection of Surface Orientation and Motion from Texture by Stereological Technique," *Artificial Intelligence* , Vol. 23, 1984.

[41] P. J. Besl and R. C. Jain, "Intrinsic and Extrinsic Surface Characteristics," *Proc. CVPR, 1985.*

[42] S. Zacks, *Parametric Statistical Inference*, Pergamon, 1981.

[43] D. B. Cooper, "When Should a Learning Machine Ask for Help?" *IEEE Trans. on IT*, Vol. 12, July 1974.

[44] R. L. Kashyap, "Optimal Choice of AR and MA Parts in Autoregressive Moving Average Processes," *IEEE Trans. PAMI*, Vol. 4, March 1982.

[45] B. K. P. Horn, *Robot Vision*, MIT Press, Cambridge, MA, and McGraw-Hill Co., NY, 1986.

[46] P. Brodatz, *Textures*, Dover Publications, New York , 1966.

The Use of Markov Random Fields in Estimating and Recognizing Objects in 3D Space

David B. Cooper[†], **Jayashree Subrahmonia**[†], **Yi-Ping Hung**[§] **and Bruno Cernuschi-Frias**[‡]

[†]Laboratory for Engineering Man/Machine Systems
Division of Engineering, Brown University

[‡]Facultad de Ingenieria,
Universidad de Buenos Aires, and CONICET, Argentina

[§]Institute of Information Science
Academia Sinica, Nankang, Taipei, Taiwan

1 Introduction

In the search for a general approach to computer vision, the path that we have followed for a number of years has been to adopt a decision-theoretic framework. The problems of interest to us have been those of object recognition in a cluttered occluding environment, and scene sensing and understanding. The approach involves formulating these problems in terms of Bayesian inferencing. This provides meaningful performance functionals. Then inferencing can be realized as maximum likelihood or maximum *a posteriori* probability estimation, or as minimum probability of misclassification recognition, or as a combination of these. Many tools for parallel or sequential processing and for making meaningful approximations and developing bounds then become available. The approach requires the use of models for the objects and scenes of interest and for the data generation. We use a combination of algebraic, geometric and probabilistic models. This provides for a very powerful and interesting domain of structures. The sensed data that we use are images and range data, the latter consisting of points in 3D space. One of the things we will show in this

ISBN 0-12-170608-7

paper is our approach to developing expressions for the joint probabilities of these data sets, i.e., p (sensed data | 3D model). The other thing that is necessary for the Bayesian approach is the specification of the *a priori* probability for an object or a scene, i.e., p (object) or p (scene). The focus of this paper is on some examples of and some ideas on the use of Markov Random Fields [27, 1, 18, 10, 13, 16, 14, 20] for this purpose.

2 Preliminary Comments on Models Used in this Paper

We present ideas on four levels of modeling.

Level 1: The problem of interest here is the estimation of 3D surfaces that are highly variable but with depth variation that is smooth except for occasional discontinuities. The Markov Random Field (MRF) used as an *a priori* model for these surfaces is one with a small neighborhood structure that permits smooth variation and incorporates a line process that permits occasional depth discontinuities. In the surface reconstruction process, the purpose of this field is largely to act as a smoother (i.e., regularization).

Level 2: The problem of interest here is the simultaneous segmentation of three-space into regions in each of which one 3D primitive surface model applies, and the estimation of the parameters of these primitive surfaces. Hence, the problem is combined primitive model recognition, primitive model parameter estimation, and model-based segmentation of images and three-space. This modeling can be handled in various ways. Our level 2 is the simplest model. Our images are assumed to be views of groups of primitive 3D surfaces, e.g., spheres, cylinders, planes and unrestricted polynomials in x, y, z, or other polynomials restricted in some way. Hence, the models used for primitives are true 3D models. However, since we assume that images are taken from a restricted range of positions, e.g., the angles between the optical axes of the cameras in their most extreme positions are not more than 20 or 30 degrees, there will be much occlusion of surfaces. Hence, we assume no *a priori* knowledge of the positions of primitives with respect to one another, and only crude knowledge of the number of each type of primitive seen, the sizes of the primitive patches seen, and of the primitive parameters. The way the relations among the primitives enter in this simplest model is through a minimal specification of the regions in an image, each of which views one primitive surface. We use a simple MRF for this purpose, which labels each pixel, or a small block of pixels, with the 3D surface being viewed at that location.

Level 3: The problem here is an extension of that in level 2. In addition to 3D polynomial surfaces, we permit 3D textures, i.e., high frequency stochastic structures in x, y, z. For example, the depth map, z as a function of x, y, for a workstation keyboard or for books, journals, and reports in a bookcase, looks like a stochastic process having some colored spectrum. Stochastic depth maps are also appropriate for the foliage for small indoor trees such as a ficus or a dracaena. Alternatively, for these trees a volumetric MRF might be more appropriate, and this would certainly seem to be the case for larger indoor trees. As we will see, the inferencing for the volumetric model will be much more computation-intensive.

Level 4: The problem here is to use the same primitive surfaces as in level 2, but now to control shape directly and much more carefully by putting a MRF on the parameters for all of the primitives. For example, a simplest case might be to model a car having a trunk with nine planes—one each for the engine compartment front, the hood, the windshield, the roof, the rear window, the rear deck, the trunk back, and each of the car sides. Each plane can be viewed as a node in a nine-node graph, and there are three parameters associated with each node. The three parameters associated with a node are the three parameters specifying the plane for the node. Hence, a car in this class is a point in a 27-dimensional sample space, and the class of all such cars is described by a probability distribution over this 27-dimensional space. We propose a MRF for this purpose. More generally, we use higher order polynomials in x, y, z, as primitives. For our modeling purposes, we describe them by their location, orientation and shape or by their geometric invariants. The plane is a degenerate primitive in the sense that it has only location and partial orientation—it does not have shape parameters.

Levels 1 through 4 have direct extensions to complex scenes. In the paper, we describe the models in some detail, and present experimental results in using them for 3D surface inferencing from images.

Other recent unusual MRF applications are to 3D surface shape estimation from texture [9], relations among objects in outdoor scenes [22], 2D and 3D pointwise perturbations of templates [8] and 3D blob estimation in Single Photon Emission Computed Tomography [16].

3 The Fundamental Equations: Joint Probability of Two or More Images And Of Prior Information

3.1 Bayesian Problems of Interest

Among the problems of interest are the following:

1. Maximum *a posteriori* probability (MAP) estimation of a 3D parameterized surface from a sequence of images.
2. Joint MAP recognition of 3D surface type and surface parameter estimation.
3. Minimum probability of error recognition of 3D surface type.

It turns out that for these objectives, it is also necessary to estimate the parameters specifying the pattern on the object surface, i.e., the intensity of the reflected light at each point seen on the object surface. Consider a sequence of images $I_1(.), I_2(.), \ldots, I_N(.)$ *. Let $\mathbf{a}$ denote the vector of parameters for the object surface model, and let $\boldsymbol{\alpha}$ be the vector of parameters that includes the surface parameters and the parameters that specify the pattern seen on the surface (for example, if the object is a plane, then $\mathbf{a}$ has three components for specifying plane orientation and location, whereas if the object is a sphere, then $\mathbf{a}$ has four components).

The solution to the first problem is

$$\max_{\boldsymbol{\alpha}} p(\mathbf{I}_1, \ldots, \mathbf{I}_N, \boldsymbol{\alpha}) = \max_{\boldsymbol{\alpha}} p(\mathbf{I}_1, \ldots, \mathbf{I}_N \mid \boldsymbol{\alpha}) p(\boldsymbol{\alpha}). \tag{1}$$

The vector $\mathbf{I}_n$ has as its components the image intensities at the pixels in image n. If l denotes the label for the object surface model, the solution to the second problem is

$$\max_{l, \boldsymbol{\alpha}} p(\mathbf{I}_1, \ldots, \mathbf{I}_N \mid l, \boldsymbol{\alpha}) p(\boldsymbol{\alpha} \mid l) p(l). \tag{2}$$

The solution to the third problem is

$$\max_{l} \int_{-\infty}^{\infty} p(\mathbf{I}_1, \ldots, \mathbf{I}_N \mid l, \boldsymbol{\alpha}) p(\boldsymbol{\alpha} \mid l) p(l) d\boldsymbol{\alpha}. \tag{3}$$

Equations (2) and (3) are used when the data can arise from more than one model, and joint model determination and parameter estimation or just

*Regular font denotes scalars, and bold font denotes vectors

model determination is desired. Note that in (2), $\max_{\boldsymbol{\alpha}} p(\mathbf{I}_1, \ldots, \mathbf{I}_N \mid l, \boldsymbol{\alpha})$ can be appreciably larger when l specifies *cylinder* than when l specifies *plane*, since a cylinder can always approximate the image data better than can a plane. For purposes of model recognition, this is usually countered by the presence of $p(\boldsymbol{\alpha} \mid l)$ in equations (2) and (3) because, e.g., if a *plane* is present in the data, $p(\boldsymbol{\alpha} \mid l)$ should be larger for $l = plane$ than for $l = cylinder$. The solution to (3) has smaller recognition of surface type error than the solution to (2).

More generally, there will be many objects in a scene. Assume the number of objects in a scene can vary. Then the generalization of the preceeding is that what must be estimated is the number of surfaces present, their types, and their parameter values. These generalizations are the following:

4. MAP estimation of the number of surfaces present, the surface types, and their parameter values.

5. Minimum probability of error recognition of the number of surfaces present and their types.

Suppose there are K surfaces in the scene. Let $\mathbf{L}^t = (l_1, l_2, \ldots, l_K)$ denote the labels for these K surfaces, and $\boldsymbol{\alpha}^t = (\boldsymbol{\alpha}_1^t, \boldsymbol{\alpha}_2^t, \ldots, \boldsymbol{\alpha}_K^t)$ denote the associated parameter vectors. Then the solution to the fourth problem is

$$\max_{K,\mathbf{L},\boldsymbol{\alpha}} p(\mathbf{I}_1, \ldots, \mathbf{I}_N \mid \boldsymbol{\alpha}, \mathbf{L}) p(\boldsymbol{\alpha}, \mathbf{L}, K). \tag{4}$$

Two comments are appropriate here. First, there are a number of difficulties in implementing (4) to produce physically meaningful results. One problem is that $\boldsymbol{\alpha}$ is a continuous random vector whereas $\mathbf{L}$ and K are discrete and finite. One solution is to discretize $\boldsymbol{\alpha}$. Then, we are dealing with a probability function. This discretization and the discretization of the $\mathbf{I}_n$ are governed by physical considerations; the most appropriate quantization sizes are neither obvious nor automatic. There are also other possible approaches that must be explored in dealing with MAP estimation involving both continuous and discrete random parameters.

The second comment is that for the most part, if only a portion of a geometric surface (e.g., a plane or a cylinder) is present in the scene, the portion will often be determined by intersections with other surfaces. For example, a closed can is the intersection of a cylinder with two parallel planes, the top and the bottom. Hence, the portions of the cylindrical and two planar surfaces involved in the can model are determined by intersections of surfaces of infinite extent. Otherwise, $\boldsymbol{\alpha}, \mathbf{L}$ do not determine the extent of the surfaces in 3D, though, for purposes of (1)-(4), this is often not necessary.

Note that a minimum description length (MDL) solution [19] can be thought of as a special case of (4). An alternative to the discretization of $\boldsymbol{\alpha}$ is to assign a cost to each value of $\boldsymbol{\alpha}$ and then to find a true Bayesian solution, i.e., minimum expected cost to the joint estimation of $\boldsymbol{\alpha}$ and recognition of $\mathbf{L}$ and K. The solution to (4) is such a solution when the cost associated with each value of $\boldsymbol{\alpha}$ is 1 and $p(\mathbf{L}, K)$ is a uniform distribution. Finally, the solution to the fifth problem is

$$\max_{K,\mathbf{L}} \int_{-\infty}^{\infty} p(\mathbf{I}_1, \ldots, \mathbf{I}_N \mid \boldsymbol{\alpha}, \mathbf{L}) p(\boldsymbol{\alpha}, \mathbf{L}, K) d\boldsymbol{\alpha}. \tag{5}$$

Equations (1)-(5) are simple to compute. We now briefly discuss the functions involved.

3.2 Probability Distribution Models For The Data

There are a number of models possible for $\mathbf{I}_n$. In [12], a polynomial contour model is used. That is, the useful data in an image is assumed to be curves across which the image intensity changes rapidly. We have used 2D polynomials to represent these curves; the 3D surface pattern is parameterized by the coefficients of these polynomials. The data used there is not the original images but rather edge-maps, i.e., an image is replaced by the result of running a simple edge detector over the image. In the present paper, we assume $I_n(\mathbf{u})$ is some true noiseless pixel intensity at location $\mathbf{u}$ in the nth image, $\mu_n(\mathbf{u})$ plus white Gaussian noise having mean zero and variance σ^2. We assume a pinhole camera model and a Lambertian surface for the object. Then a point P on the object surface is seen with equal brightness in all the images. Hence $p(\mathbf{I}_1, \mathbf{I}_2, \ldots, \mathbf{I}_N \mid \boldsymbol{\alpha})$ in (1) can be written approximately as

$$\prod_{n=1}^{N} (2\pi\sigma^2)^{-d/2} exp\{-\frac{1}{2\sigma^2} \sum_{\mathbf{u}\in\mathbf{D}_n} [I_n(\mathbf{u}) - \mu_n(\mathbf{u})]^2\}, \tag{6}$$

where $\mathbf{D}_n$ is the set of pixels in the nth image, and d is the number of pixels in $\mathbf{D}_n$. See [5, 6] for details. The $\mu_n(\mathbf{u})$ are functionally related. This can be seen as follows:

Consider Fig. 1 illustrating the geometry for the two images, $\mathbf{I}_1$ and $\mathbf{I}_2$. Take any point in image 1. Since the two cameras are calibrated, the surface point P seen at $\mathbf{s}$ lies along the backward projected line from $\mathbf{s}$ through the camera lens center and the equation of this line in three-space

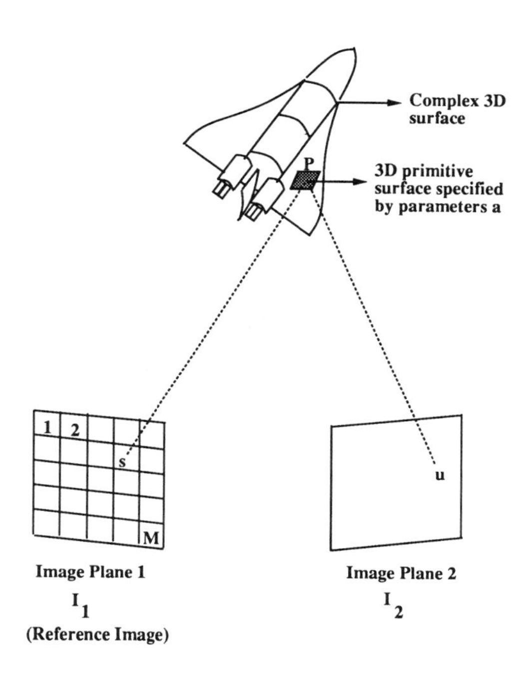

Figure 1: Functional relation between a pair of images taken by calibrated cameras.

can be determined from the image. Assume the surface point seen lies on the surface specified by the parameter vector $\mathbf{a}$. Then the point P seen is the intersection of the backward projected ray determined by $\mathbf{s}$ with the surface specified by $\mathbf{a}$. This point is seen in image 2 at the intersection of the image plane at camera 2 with the forward projected ray from P, through the lens center of camera 2. Hence, we denote this point in $\mathbf{I}_2$ by $\mathbf{u}(\mathbf{s}, \mathbf{a})$. Because of the Lambertian assumption, if $\mathbf{a}$ is the true parameter value, we have

$$\mu_n(\mathbf{u}(\mathbf{s}, \mathbf{a})) = \mu_1(\mathbf{s}). \tag{7}$$

Consequently, an approximation to (6) is

$$\prod_{n=1}^{N} (2\pi\sigma^2)^{-d/2} exp\{-\frac{1}{2\sigma^2} \sum_{\mathbf{s}\in\mathbf{D}_1} [I_n(\mathbf{u}(\mathbf{s}, \mathbf{a})) - \mu_1(\mathbf{s})]^2\}. \tag{8}$$

The values $\mu_1(\mathbf{s})$ are *a priori* unknown. Hence, they must be treated as *a priori* unknown parameters and must be estimated along with $\mathbf{a}$ or must be treated as nuisance parameters and integrated out. They specify the pattern on the 3D surface. The *a priori* unknown parameters $\boldsymbol{\alpha}$ then consist of the 3D surface parameters $\mathbf{a}$, the pattern parameter vector $\boldsymbol{\mu}_1$ having components $\mu_1(\mathbf{s})$, $\mathbf{s} \in \mathbf{D}_1$, and σ^2. When N=2, equation (8) has some very nice properties. In particular, if the $\mu_1(\mathbf{s})$ are independent, uniformly-distributed random variables, then

$$\begin{aligned}&\max_{\boldsymbol{\mu}_1} p(\mathbf{I}_1, \mathbf{I}_2 \mid \mathbf{a}, \boldsymbol{\mu}_1)p(\boldsymbol{\mu}_1)p(\mathbf{a}) = \\ &(2\pi\sigma^2)^{-d/2} exp\{-\tfrac{1}{4\sigma^2} \sum_{\mathbf{s}\in\mathbf{D}_1} [I_1(\mathbf{s}) - I_2(\mathbf{u}(\mathbf{s},\mathbf{a}))]^2\}\times \\ &p(\boldsymbol{\mu}_{1MAP})p(\mathbf{a}),\end{aligned} \tag{9}$$

where $\mu_{1MAP}(\mathbf{s}) = \frac{1}{2}[I_1(\mathbf{s})+I_2(\mathbf{u}(\mathbf{s},\mathbf{a}))]$ is the MAP estimate of $\mu_1(\mathbf{s})$ given $\mathbf{a}$, and $p(\boldsymbol{\mu}_{1MAP})$ is constant over a rectangular solid. Note that $\mu_{1MAP}(\mathbf{s})$ is also the maximum likelihood estimate for $\mu_1(\mathbf{s})$. For $3 \leq N$, the situation is more complex, but can be treated effectively [17]. Equation (9) is to be used in the maximization (2) when $p(\boldsymbol{\mu}_1)$ is the uniform probability density function and N=2.

One other equation [4] that facilitates the practical computation of equations (1)-(5) is

$$\begin{aligned}&p(\mathbf{I}_1, \ldots, \mathbf{I}_N \mid \boldsymbol{\alpha}) \approx \\ &p(\mathbf{I}_1, \ldots, \mathbf{I}_N \mid \hat{\boldsymbol{\alpha}}_N) exp\{-\tfrac{1}{2}(\boldsymbol{\alpha} - \hat{\boldsymbol{\alpha}}_N)^t \Phi_N (\boldsymbol{\alpha} - \hat{\boldsymbol{\alpha}}_N)\},\end{aligned} \tag{10}$$

where $\hat{\boldsymbol{\alpha}}_N$ is the maximum likelihood estimate (MLE) of $\boldsymbol{\alpha}$ based on the image data $\mathbf{I}_1, \ldots, \mathbf{I}_N$, and Φ_N is the second derivative matrix having i, jth component

$$-\frac{\partial^2}{\partial\alpha(i)\partial\alpha(j)} ln\, p(\mathbf{I}_1, \ldots, \mathbf{I}_N \mid \hat{\boldsymbol{\alpha}}_N),$$

where these derivatives are computed on equation (8). Hence all the useful information about $\boldsymbol{\alpha}$ is summarized in the quadratic form in the exponent of equation (10). When (10) is used for $p(\mathbf{I}_1, \ldots, \mathbf{I}_N \mid l, \boldsymbol{\alpha})$ in (3), the result is

$$\begin{aligned}&\max_l p(\mathbf{I}_1, \ldots, \mathbf{I}_N \mid l, \hat{\boldsymbol{\alpha}}_N)\times \\ &(2\pi)^{q/2} \mid \Phi_N \mid^{-1/2} p(\hat{\boldsymbol{\alpha}}_N \mid l)p(l),\end{aligned}$$

where q is the number of components in $\boldsymbol{\alpha}$, and $\mid \Phi_N \mid$ is the determinant of Φ_N. Note that this should give a more accurate recognition than (2).

We now discuss how problems 1-5 can be solved and show examples of experiments for the solutions to two problems.

4 A Computationally Practical Approach To Surface Estimation

In order to use parallel processing to achieve real time operation, we partition an image, e.g., the first, into square windows, assume each window is a view of a single 3D surface and first estimate these surfaces independently in parallel. Within each window, parallel processing can also be used. Hence, for illustrative purposes, if we consider $\mathbf{D}_1$ in (8) as a single window, initial estimation of the surface seen in the window is realized by maximizing (8) with respect to $\mathbf{a}$. This is maximum likelihood estimation (MLE). In addition to these windows being well suited to parallel processing, a window also constitutes a good-sized chunk of data so that an estimate having useful accuracy can be obtained as a starting point for further processing. Then, these initial estimates can be used as a starting point for global estimation of the entire complex surface seen in the images. Since gradient descent is the optimization technique used in most of our solutions, some care must be exercised to deal with the multimodal functions involved. We handle this at the individual window estimation stage, as discussed in [6, 17].

A rough indication of the computation cost involved is that, for a 32×32 window, when 10 iterations are required to convergence, the running time on a SUN 3/60 is a few seconds. Usually, the algorithm can be successfully run on a subset of the pixels in a window, substantially reducing the running time. With inexpensive DSP technology, program optimization and parallel implementation, the running time can be reduced by orders of magnitude.

5 Map 3D Surface Reconstruction

Our approach to the estimation of highly variable surfaces is to model the surface as a stochastic process and then do MAP estimation. The surface estimation is then given by (1) which for two images becomes

$$\max_{\boldsymbol{\alpha}} p(\mathbf{I}_1, \mathbf{I}_2 \mid \boldsymbol{\alpha}) p(\boldsymbol{\alpha}). \tag{11}$$

Vector $\mathbf{a}$ here models the surface. The simplest surface model is the depth map. One way to model the depth map is to partition the first image into M windows and model the surface depth map as being constant over each window. Thus $\mathbf{a}$ here is the vector $(a^1, a^2, \ldots, a^M)$ where a^k denotes the surface depth in window k. The window size is chosen by the designer and can consist of one pixel if the finest possible resolution is needed.

We model $p(\mathbf{a})$ as a MRF on the 2D lattice where a lattice point is the center of a window in the first image. Only the first order neighborhood system is considered here, as this produced good results in practice. However, this method can be extended to higher order neighborhood systems. The free boundary condition is used. Using the MRF-Gibbs equivalence [1], $p(\mathbf{a})$ can be expressed as the Gibbs distribution

$$p(\mathbf{a}) = \frac{1}{Z} exp \left\{ - \sum_{\mathbf{c} \in \mathbf{C}} V_{\mathbf{C}}(\mathbf{a}) \right\}, \tag{12}$$

where Z is the normalizing constant, $V_{\mathbf{C}}(\mathbf{a})$ is the contribution of clique $\mathbf{c}$ to the energy of $\mathbf{a}$ and $\mathbf{C}$ is the set of all cliques with respect to the chosen neighborhood system. The cliques here are a pair of adjacent lattice points in the horizontal and vertical directions.

An appropriate MRF model for encoding the prior knowledge about the smoothness of the 3D surface is one with clique potentials:

$$V_{\mathbf{C}}(\mathbf{a}) = \lambda_0 \parallel a^k - a^j \parallel^2 \tag{13}$$

where $\mathbf{c} = \{k, j\} \in \mathbf{C}$.

A potential can be treated as a penalty for having a surface with depth a^k at site k and a^j at site j. The constant λ_0 controls the amount of smoothing. If the surface seen within one region is known to be smoother than one seen in another region, its corresponding λ_0 will be larger, i.e., the penalty incurred by the difference between neighboring $\mathbf{a}$'s in the region known to be smoother will be larger. The distance measure, $\parallel . \parallel$, was chosen to be the Euclidean distance, but one can use other distance measures as well.

When the scene contains discontinuities in depth variation, the estimation using the above smoothness model tends to smooth out the discontinuities that are important for scene analysis. The smoothing effect is especially severe when the 3D measurement data is sparse. To remedy this problem, we encode the discontinuities into the MRF models. A possible method is to model the discontinuities with a line process [11, 16, 7], and then couple the line process with the original smooth field $\mathbf{a}$ by designing an appropriate neighborhood system and its corresponding potentials. This results in a coupled MRF, which provides a piecewise-smooth description of 3D surfaces with occasional discontinuities at the surface boundaries.

Let l_{kj} denote the line element between windows k and j. The variable l_{kj} takes values 1 or 0. The value $l_{kj} = 1$ indicates a discontinuity between windows k and j and $l_{kj} = 0$ indicates that there is no discontinuity between windows k and j. Let $\mathbf{E}$ denote the vector containing all the line elements. The unknowns in the problem now are $\mathbf{a}$ and $\mathbf{E}$. We estimate them by maximizing the posterior probability, $p(\mathbf{a}, \mathbf{E}|\mu_{1MAP}, \mathbf{I}_2, \mathbf{I}_2)$, which is proportional to $p(\mathbf{I}_1, \mathbf{I}_2|\mathbf{a}, \mathbf{E}, \mu_{1MAP})p(\mathbf{a}|\mathbf{E})p(\mathbf{E})$.

We model $p(\mathbf{E})$ as a MRF on 2D lattice where a lattice point lies at the center of the boundary between two windows. For the experiments illustrated in Figure 8, we found that using only a zeroth order neighborhood system was adequate. Thus, each clique consists of just one site. Since $\mathbf{E}$ has a Gibbs distribution, $p(\mathbf{E})$ is specified by the potential $V_{\mathbf{c}}(\mathbf{E})$ for all cliques, $\mathbf{c}$. We use the following potential function for the clique $\mathbf{c}$ consisting of the site between windows k and j:

$$V_{\mathbf{c}}(\mathbf{E}) = \lambda_1 l_{kj}$$

The parameter λ_1 controls the frequency of occurrence of edges in this apriori model for $\mathbf{E}$.

Use of this model for $p(\mathbf{E})$ produces reasonably good results as can be seen in Fig. 5. If, however, it is desired to have discontinuities that lie along law curvature curves, a larger neighborhood system needs to be used, e.g., a MRF where the neighborhood of a lattice point is two layers of surrounding lattice points, and the interaction parameters are chosen to encourage roughly straight lines.

Consider an example where the discontinuities lie along vertical and horizontal curves. We model $p(\mathbf{E})$ by a MRF with a neighborhood system for which the cliques consist of either single sites or pairs of adjacent sites. The potential function for a clique consisting of a single site has a form similar to the one given above. For the clique $\mathbf{c}$ consisting of a pair of adjacent sites (corresponding to the edge elements $l_{k_1 l_1}$ and $l_{k_2 l_2}$), we assign the potential function as follows. If $l_{k_1 j_1}$ and $l_{k_2 j_2}$ correspond to a pair of horizontal or a pair of vertical edge sites, assign a small clique potential if $l_{k_1 j_1}$ and $l_{k_2 j_2}$ have the same value and a large clique potential if they have different values. If $l_{k_1 j_1}$ corresponds to a vertical edge and $l_{k_2 j_2}$ corresponds to a horizontal edge, or vice versa, assign a large clique potential in order not to favor corners. Thus, by choosing a first order neighborhood system and appropriate clique potentials, we can build an a priori model for discontinuities that favors long vertical and horizontal curves. By choosing an even higher order neighborhood system, one can model discontinuities that lie along low curvature curves in arbitrary directions.

The potential function for clique $\mathbf{c} = \{k, j\}$ in the field of $\mathbf{a}$ given $\mathbf{E}$ can now be written as

$$V_{\mathbf{c}}(\mathbf{a}) = \lambda_0(1 - l_{kj})\|a^k - a^j\|^2 \tag{14}$$

For convenience, we have considered the $\mathbf{a}$ and $\mathbf{E}$ fields separately and the combined field is a coupled MRF. However, in a general framework, we can consider a single MRF field directly for the **a** and **E** vectors. The graph for this field is the combined graph of the **a** and **E** fields. The potential function then for a clique consisting of sites k and j of the **a** field and site l_{kj} of the **E** field is $V_{k,j}(\mathbf{a}, \mathbf{E}) = \lambda_0(1 - l_{kj})\|a^k - a^j\|^2 + \lambda_1 l_{kj}$. The total energy function is the same for both ways of viewing the resulting MRF.

We now focus on the data term, $p(\mathbf{I}_1,\mathbf{I}_2 \mid \boldsymbol{\mu}_{1MAP}\mathbf{a}, \mathbf{E})$. Let $\mathbf{I}_1^m$ denote the image intensity seen in the mth window in image 1 and $\mathbf{I}_2^m$ be its corresponding image data seen in image 2. Since the noise in an image is white and independent of the noise in the other image,

$$p(\mathbf{I}_1, \mathbf{I}_2 \mid \boldsymbol{\mu}_{1MAP}, \mathbf{a}, \mathbf{E}) = \prod_{m=1}^{M} p(\mathbf{I}_1^m, \mathbf{I}_2^m \mid \boldsymbol{\mu}_{1MAP}^m, \mathbf{a}, \mathbf{E}) \tag{15}$$

which, from Section 3, is

$$p(\mathbf{I}_1, \mathbf{I}_2|\boldsymbol{\mu}_{1MAP}^m, \mathbf{a}, \mathbf{E}) = (2\pi\sigma^2)^{-\frac{1}{2}} \prod_{m=1}^{M} \exp\{-\frac{1}{4\sigma^2} \sum_{\mathbf{s}\in\mathbf{D}^m} [I_1(\mathbf{s}) - I_2(\mathbf{u}(\mathbf{s}, a^m))]\} \tag{16}$$

Combining the prior term given above and the data term, the posterior probability of **a** and **E** given $\boldsymbol{\mu}_{1MAP}$ given the image date $\mathbf{I}_1$ and $\mathbf{I}_2$ can be written as

$$p(\mathbf{a}, \mathbf{E}|\boldsymbol{\mu}_{1MAP}\mathbf{I}_1, \mathbf{I}_2) \propto p(\mathbf{I}_1, \mathbf{I}_2|\mathbf{a}, \mathbf{E}, \boldsymbol{\mu}_{1MAP})p(\mathbf{a}|\mathbf{E})$$

$$\propto \exp\left\{(-\tfrac{1}{4\sigma^2} \textstyle\sum_{m=1}^{M} \sum_{s\in\mathbf{D}^m} [I_1(\mathbf{s}) - I_2(\mathbf{u}(\mathbf{s}, a^m))]^2) - \sum_{k,j} V_{k,j}(\mathbf{a}, \mathbf{E})\right\} \tag{17}$$

where the potential $V_{k,j}(\mathbf{a}, \mathbf{E}) = \lambda_0(1 - l_{kj})\|a^k - a^j\|^2 + \lambda_1 l_{kj}$.

Our goal is to find the parameters **a** that maximize this posterior probability. The global maximization can be implemented by simulated annealing based on the Gibbs sampler [11]. However, the computational cost incurred by the above stochastic relaxation is very large. So, we use deterministic relaxation techniques introduced by Cohen and Cooper [10], and later called ICM (Iterative Conditional Means technique) [2]. Since the technique is deterministic, it guarantees only a local maximum. Therefore, it is important to have good initial estimates. In our case, we use the maximum likelihood estimates (MLE) for the surface depth as the initial estimate.

If an image is a view of a few regions in each of which the surface has a certain smoothness that is different than that for the other regions, the surface estimation problem can be considered to be that of joint surface

Figure 2: First image.

estimation, the segmentation of the image into regions of different smoothness and the estimation of the appropriate smoothing constant λ for each region.

5.1 Experimental Results

In the set of experiments shown, a sequence of images taken by a SONY XC-39 CCD camera mounted on a PUMA robot were used. The object to be estimated is a cylindrical cereal box lying on a planar table top. Figures 2 and 3 show two 242×256 pixel images of the cereal box. The first image is divided into windows of size 16×16 pixels. In the experiment, each surface patch seen in a small window is modeled by a planar patch with orientation fixed to the direction of the optical ray passing through the window's center. Hence, the only surface parameter to be estimated for each window is the depth of the surface patch, i.e., the distance from the lens center to the center of the surface patch. This set of depth parameters constitutes a 2D array called the depth map. In this case, the general surface reconstruction algorithm reduces to an efficient algorithm for depth reconstruction.

The depth map after the maximum likelihood estimation is shown in Fig. 4. This is the initial estimate used for doing the MAP estimation. The MLEs provide good initial estimates for doing the deterministic relaxation so that the final results are close to the global maximum. Fig. 5 shows the result after the MAP estimation using the line process. The algorithm does a good job of smoothing and preserving depth discontinuities.

The computational cost on a SUN 3/60 is: for the initial MLE, a fraction of a second for each window since only depth and not orientation is estimated for each window; this is typically followed by 3-4 iterations for

Figure 3: Second image.

Figure 4: Surface reconstruction after maximum likelihood estimation.

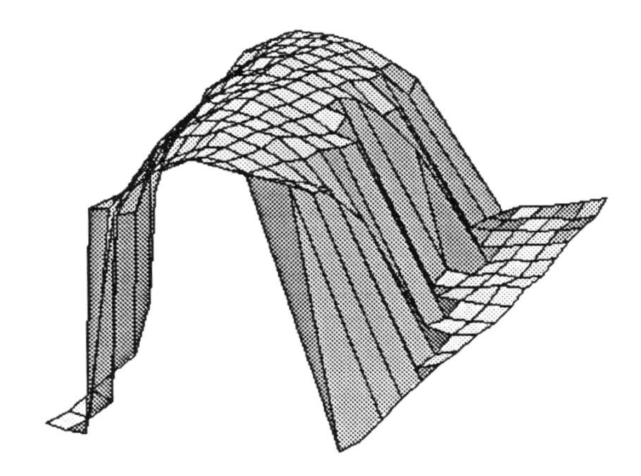

Figure 5: Surface reconstruction after MAP estimation.

the MAP estimation.

6 Map Estimation of the Number of Surfaces, their Types and Parameter Values

The solution to problem 4 is given in equation (4). In this section, we assume that $p(\mathbf{L}, K)$ has a uniform distribution, and assume very little prior information about $\boldsymbol{\alpha} \mid \mathbf{L}, K$. The problem of designing probability distributions for interesting classes of $\boldsymbol{\alpha}$ is taken up in Section 8. To avoid having to make the choices concerning (4) discussed in Section 3.1, we have implemented (5). The implementation of (5) also returns a MAP estimate for $\boldsymbol{\alpha}$.

Though we have programs to estimate a single cylinder or sphere from a pair of images, we have implemented the segmentation of images into regions, each looking at a single primitive surface, only for planar primitives.

The experiments were run on pairs of images taken by cameras having about 20° between their optical axes. Then, the cameras are looking at the scene from roughly the same directions, much occlusion in the scene is possible and considerable computational cost can be incurred in determining all the planes in the scene. In order to arrive at a solution that is "computationally practical for real time processing," we have implemented the following, which is not guaranteed to provide the optimal accuracy solution, but which appears to provide a usefully accurate solution at reasonable computational cost.

The search strategy here is:

1. Partition image 1 into M $Q \times Q$ pixel windows. Using these windows and image 2, estimate the 3D planar patch seen in each window. Do these independently in parallel.

2. Cluster the M windows into K clusters, where all the windows in a cluster are assumed to be viewing the same plane. The clustering is based on the three-parameter planar patch vectors estimated in step 1. This initial clustering is not based on the image data because of the computational cost involved. Thus, the clustering is poor, especially if Q, the window size, is small.

3. Using the result of step 2 as a starting point, come up with a best joint clustering of the windows and estimation of the K planes. Because of the occlusion and interpretation problems that we referred to, we handle this stage by putting a MRF on the labels for the windows. The primary purpose of this MRF is to discourage the formation of small primitive patches in the surface reconstruction, but the field can be designed to favor simple geometries. We use this field to guide the joint MAP estimation of $\boldsymbol{\alpha}$ and the segmentation of the images, rather than using 3D consistency. This provides the K approximately best 3D planes in the surface reconstruction.

4. Given the results in step 3, the asymptotically Bayesian behavior discussed in Section 3.2 can be used in equation (5) to determine the roughly optimal K, i.e., the number of 3D planes present.

We now proceed to give the details. For two images, consider

$$\max_{\boldsymbol{\alpha},\mathbf{L}} p(\mathbf{I}_1, \mathbf{I}_2 \mid \boldsymbol{\alpha}, \mathbf{L}) p(\boldsymbol{\alpha}, \mathbf{L} \mid K). \tag{18}$$

Suppose the reference image $\mathbf{I}_1$ is partitioned into M windows, and t^m denotes the label associated with the surface patch seen in the mth window, i.e., $t^m = l_i$, where l_i is the label of the ith surface that is seen in the mth window. Then, the first problem can be stated as getting the MAP estimate of $(\mathbf{T} = (t^1, \ldots, t^M)$ and $\boldsymbol{\alpha}) \mid K$, i.e., finding the $\mathbf{T}$ and $\boldsymbol{\alpha}$ that maximize

$$where p(\mathbf{I}_1, \mathbf{I}_2 \mid \boldsymbol{\alpha}, \mathbf{T}) p(\boldsymbol{\alpha}, \mathbf{T} \mid K), \tag{19}$$

$p(\boldsymbol{\alpha}, \mathbf{T} \mid K) = p(\boldsymbol{\alpha} \mid \mathbf{T})p(\mathbf{T} \mid K)$. We model $\mathbf{T} \mid K$ with a MRF by choosing an appropriate neighborhood system and its associated potential function on all the cliques. Unless prior information is available, we take

$$p(\boldsymbol{\alpha} \mid \mathbf{T}) = \prod_{k=1}^{K} p(\boldsymbol{\alpha}_k \mid \mathbf{T}),$$

and $\boldsymbol{\alpha}_k$ is uniformly distributed over some region. Note, we assume the $p(\boldsymbol{\alpha}_k \mid \mathbf{T})$ are independent of one another, and that all $p(\boldsymbol{\alpha}_k \mid \mathbf{T})$ are the same. The prior distribution for $\mathbf{T} \mid K$ can be written as follows:

$$p(\mathbf{T} \mid K) = \frac{1}{Z} exp\{-U(\mathbf{T})\}, \tag{20}$$

where Z is a constant and the energy function is of the form

$$U(\mathbf{T}) = \sum_{\mathbf{c} \in \mathbf{C}} V_{\mathbf{c}}(\mathbf{T}), \tag{21}$$

with $V_{\mathbf{c}}(\mathbf{T})$ being the contribution of the clique $\mathbf{c}$ to the energy of $\mathbf{T}$. In this paper, the first order neighborhood system is used, where the cliques are individual sites and pairs of adjacent horizontal and vertical sites, and the potential for a clique $\mathbf{c} = \{m, j\}$ is chosen to be:

$$V_{\{m,j\}}(\mathbf{T}) = V_{\{m,j\}}(t^m, t^j) = \begin{cases} 0 & if \quad t^m = t^j \\ \beta/4 & if \quad t^m \neq t^j. \end{cases} \tag{22}$$

However, the potential functions for the cliques need not be restricted in any way, and can be designed to formulate a wide variety of behaviors. The preceding MRF is homogeneous. It tends to generate blob-like regions. Blob size increases with increasing β. It is the simplest field, and was designed solely to discourage segmentation into small regions or regions with wiggly boundaries. In this particular type of application, we assumed that the segmentation is determined largely by the data, i.e., that there are enough pixels per surface patch to determine the parameters of the surface patch, and the MRF for the labeling plays a secondary role.

We now focus on the data term in Equation (19), $p(\mathbf{I}_1, \mathbf{I}_2 | \boldsymbol{\mu}_{1MAP}, \mathbf{a}, \mathbf{T})$. This is similar to (15) except that $\mathbf{a}^m$ denotes the parameter vector that describes the 3D surface patch seen in the mth window. Then, $\mathbf{a}^m$ is the surface parameter vector associated with the surface label t^m.

$$p(\mathbf{I}_1, \mathbf{I}_2 | \boldsymbol{\mu}_{1MAP}, \mathbf{a}, \mathbf{T}) = \prod_{m=1}^{M} p(\mathbf{I}_1^m, \mathbf{I}_2^m | \boldsymbol{\mu}_{1MAP}^m, \mathbf{a}, \mathbf{t}^m) \tag{23}$$

which is

$$p(\mathbf{I}_1, \mathbf{I}_2 | \boldsymbol{\mu}_{1MAP}, \mathbf{a}, \mathbf{T}) \propto \tag{24}$$

$$\prod_{m=1}^{M} \exp\{-\frac{1}{4\sigma^2} \sum_{\mathbf{s} \in \mathbf{D}^m} [I_1(\mathbf{s}) - I_2(\mathbf{u}(\mathbf{s}, \mathbf{a}^m))]^2\} \tag{25}$$

The MAP estimation of **a** and the image segmentation becomes the maximization of

$$p(\mathbf{I}_1, \mathbf{I}_2 | \boldsymbol{\mu}_{1MAP}, \mathbf{a}, \mathbf{T}) p(\mathbf{a}|\mathbf{T}) p(\mathbf{T}|K) \tag{26}$$

Stochastic relaxation [15] for maximizing equation (25) is computationally too costly. Deterministic relaxation, such as the iterative conditional means (ICM) [10, 2] is computationally simple but is only guaranteed to find a local maximum. It is necessary to have fairly decent initial estimates for **a** and **T** to get good final estimates. Therefore, the first two stages of the algorithm are computational reasonable operatoins for getting the initial estimates for **a** and **T**.

For step 4 of the algorithm, the best K is obtained by computing (5) for each K. The MRF for $\mathbf{T} \mid K$ is not used here, because it models only local behavior useful in the clustering. Hence, in step 4 we use $\hat{\boldsymbol{\alpha}}_k$ and $\hat{t}^m$ from step 3 and the asymptotic integration described at the end of Section 3.2. Let Φ_{Kk} denote the Φ matrix for the region in image 1 that has been determined as viewing the kth plane under the assumption that K planes are in the scene. Let δ_{Kk} be the number of pixels in this region. Then, for large δ_{Kk}, $|\Phi_{Kk}|^{-1/2}$, where $|.|$ denotes the determinant, can be approximated by constant $\times (\delta_{Kk})^{-5/2}$ [3], where q, the number of parameters needed to specify a plane, is equal to 3. Hence, we choose K as the maximizer of

$$\prod_{k=1}^{K} (\delta_{Kk})^{-5/2} exp\{-\frac{1}{4\sigma^2} \sum_{\mathbf{S} \in region K_k} [I_1(\mathbf{s}) - I_2(\mathbf{u}(\mathbf{s}, \mathbf{a}^k))]^2\}$$

where K_k denotes the kth class when the total number of classes is K.

More generally, when the l_k can be different types of primitives, we would proceed analogously.

6.1 Experimental Results

Experimental results using real images are shown in Figs. 6 through 9. All the experiments are run using only planes as primitives. Here, both planar orientation and depth are estimated.

Figure 6: First image.

Figure 7: Second image.

For the experiment shown, the scene consists of three cylinders (c1, c2 and c3) and two planes (p1 and p2). In this example, the two planes are well fit by the primitive model well whereas the three cylinders cannot be well approximated by a single primitive planar patch each. The purposes of this experiment were first to see whether planar surfaces are found when nonplanar surfaces are also present, and second to explore the quality of the planar approximations to nonplanar surfaces. Figs. 6 and 7 show the two images used as input. Two of the cylinders and one of the planes have shiny surfaces. Patches of size 32 $\times$ 32 pixels are used to do the local surface estimation. For the final MAP segmentation and reestimation using a MRF, windows of size 16 $\times$ 16 pixels were used. The algorithm comes up with eight classes. Two of the classes correspond to the two planes, p1 and p2, and are well fit by the model. The cylinder c1 is approximated by one plane, c3 by two and c2 by three. The reason for this is that c1 and c3 have small curvatures and only a small portion of each of these cylinders is seen in both images. Cylinder c2, however, has large curvature and most of it is seen in both images. Figure 8 shows the segmentation of the portion of the reference image that is seen in both of the input images. Here each region of constant intensity corresponds to one plane found. The surface reconstruction after the fine segmentation is shown in Fig. 9.

Additional details of the algorithm and the highest confidence first test used are given in [23, 7], respectively.

In the experiments run, the MLE for the individual windows is followed by K-means clustering of the 3-component parameter vectors for the planes, which takes a few seconds, and then segmentation and parameter estimation using the MRF and based on the highest confidence first test, which takes around 20-30 minutes on a SUN 3/60. The final segmentation using the MRF is relatively computationally costly because the raw image data is used. This was used to achieve maximum accuracy. Approximations that would reduce required computation should be explored.

We are presently exploring the limits to the smallest surfaces that can be found with this approach.

7 3D Texture Models

There are regions in 3D in which the surfaces present cannot be modeled by a small number of low degree polynomial surface patches and space curves. Rather, a large number of small surface and curve patches must be used. Examples are an electronic chassis, an indoor leafy plant, books and journals in a bookcase, the keyboard for a workstation, etc. Trying to understand such a region by estimating and then interpreting these primitives

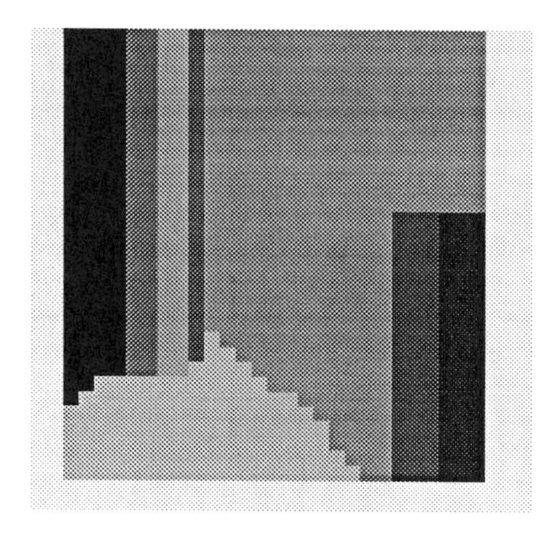

Figure 8: Segmentation of the first image.

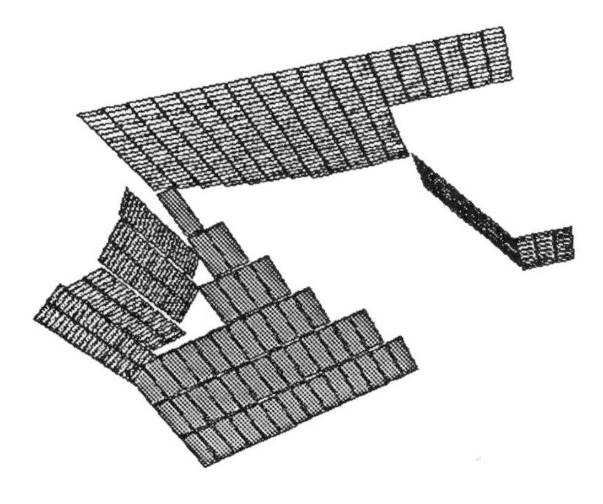

Figure 9: Surface reconstruction after MAP estimation.

may well be computationally infeasible. There is useful statistical regularity in some of these regions, so our approach is to model such structure as a 3D texture and estimate, segment, and recognize it using a Bayesian stereo or range data image understanding system.

An example of 3D textures on which we have run experiments is a set of books on a bookshelf, varying in depth and inclined in some direction. This is the view that a mobile robot would have. Our approach to modeling such a texture is as a textured depth map about a mean plane, and the texture variations are perpendicular to the plane. The vector $\mathbf{a} = \{a^1, a^2, \ldots, a^M\}$ denotes the depth map, where a^m is the depth for window m. Again, we emphasize that the texture variation is modeled perpendicular to its mean plane rather than perpendicular to the viewing direction. We cast this problem into a MRF formulation by modeling the parameter a^k as a Gaussian Markov Random Field (GMRF) on a 2D lattice where a lattice point is the center of a window in the first image. However, different MRF models can be used to model a wide range of 3D textures.

In our initial experiments, we used a first order neighborhood system. The cliques for this neighborhood system are the singleton sites and pairs of adjacent sites. The singleton cliques are assigned zero potential. $\beta_{1,0}$ and $\beta_{0,1}$ denote the interaction between adjacent sites in the horizontal and vertical directions.

The parameters, γ, say, of the GMRF are estimated by maximizing the joint likelihood of both data and $\mathbf{a}$, given γ, as a function of $\mathbf{a}$ and γ. This joint likelihood of the two images in the stereo pair, $\mathbf{I}_1$ and $\mathbf{I}_2$, can be written using Bayes rule as

$$p(\mathbf{I}_1, \mathbf{I}_2, \mathbf{a} \mid \gamma) \propto p(\mathbf{I}_1, \mathbf{I}_2 \mid \mathbf{a}) p(\mathbf{a} \mid \gamma). \tag{27}$$

The first term is the data term. The model that we use for image intensity observation is that of some true intensity corrupted by white Gaussian noise having variance σ^2. Then, from the analysis in Section 3.2, the data term can be written as

$$p(\mathbf{I}_1, \mathbf{I}_2 \mid \mathbf{a}, \mu_{1MAP}) \propto \prod_{m=1}^{M} exp\{-\tfrac{1}{4\sigma^2} \sum_{\mathbf{s} \in \mathbf{D}^m} [I_1(\mathbf{s}) - I_2(\mathbf{u}(\mathbf{s}, a^m))]^2\}. \tag{28}$$

The second term in (23) is the distribution of $\mathbf{a}$.

7.1 Experimental Results

Experimental results are shown in Figs. 10-13. Figures 10 and 11 are the data sets. These are images of a 3D scene consisting of a cylinder, which

Figure 10: First image.

Figure 11: Second image.

can be modeled as a polynomial function, and two 3D textures, i.e., the set of books inclined in the horizontal direction and the set of books inclined in the vertical direction. The different steps of the experiment are the following:

First, texture MRFs are estimated from learning data, a different set of images. This learning was supervised and was done as follows.

1. A pair of images to be used as learning data were taken. They viewed a region of horizontally oriented books and a region of vertically oriented books. These orientations were slightly different than those in Fig. 10 and 11 in order to get a feel for the robustness of our approach. Then a depth map was estimated. This was done by partitioning the first image into 16×16 blocks and using the 3D surface estimation algorithm [6] to estimate a 3D plane for each block. Only the depth of each plane is used in stage 2.

2. The estimated depth map from stage 1 was then used in a supervised learning mode to estimate the MRF parameters for the two depth texture models. A mean plane is estimated for each texture region and then the texture parameters are estimated. For the two textures shown, the values of the β parameters were found to be (0.18, 0.018) and (0.02, 0.2) where the first entry corresponds to $\beta_{1,0}$ and the second corresponds to $\beta_{0,1}$. The sum of the β parameters has to be strictly less than 0.5 to guarantee that the field model is Gaussian. The value of σ_d, the variance of the GMRF, was found to be 0.4 inches for both of the textures.

3. Using the learned texture models from stage 2, a hierarchical segmentation algorithm is run to segment the 3D data in Figs. 10 and 11 into regions, in each of which the data can be approximated either as a single polynomial function or a 3D texture. At the moment, the only polynomial functions that we use are planes, so a few of them are required for curved surfaces.

The hierarchical segmentation algorithm is a Bayesian stereo model-based algorithm that segments the image pair into regions of decreasing size, checking each time to see whether a 3D depth texture model or a 3D polynomial surface model is appropriate for a region. If there is at least one appropriate model, it chooses the best. Otherwise, it continues and examines smaller regions. (See [10] for a similar approach to the segmentation of a single textured image.)

The surface reconstruction after segmentation is shown in Figs. 12 and 13. The two figures are the reconstruction of the same surface, but viewed

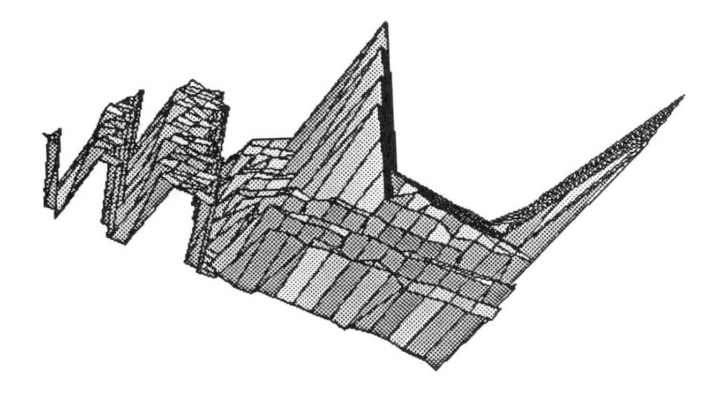

Figure 12: One view of the final surface reconstruction.

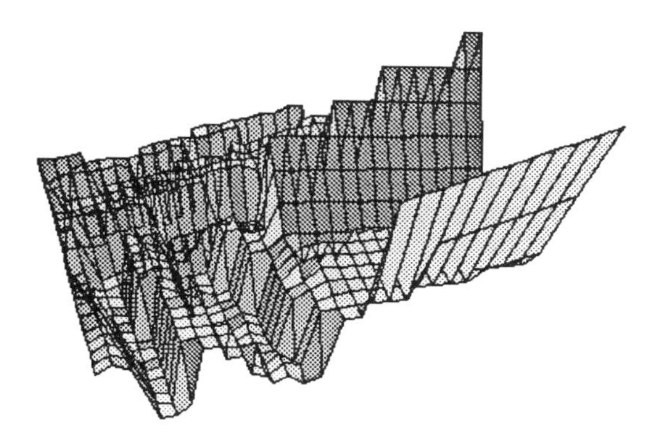

Figure 13: Second view of the final surface reconstruction.

from two different angles. The cylindrical portion is approximated by three planes, since we use only planes as primitives here. Details of the algorithm to fit polynomial functions to the data are given in Section 6 and in [23]. The algorithm finds one class for each of the 3D textures. Figure 12 shows the reconstruction of one of the textures clearly and Figure 13 shows the reconstruction of the other clearly. The algorithm stops after one step of the hierarchy since all the regions get correctly classified at this point.

The preliminary experiments conducted have been solely for the purpose of investigating the feasibility of 3D depth texture segmentation in a stereo system. The next step is to develop a joint $\mathbf{a}$ and γ estimation approach that functions at reasonable computational cost, and to explore the extent to which a system could function in more of an unsupervised mode with less prior information about texture-model parameter values.

8 Probability Measures for the Objects in a Class

Whereas probability measures have been used in the past to measure shape pointwise as stochastic processes [11, 8], or for relative object location [22], we are not aware of any attempt of measuring shape in terms of complex primitives.

We turn to the level 4 problem posed in Section 2, i.e., the design of a mathematically consistent and physically meaningful MRF for the parameters for all the primitives modeling an object in a class. To begin, consider the car modeling problem posed there. At a coarse level, a car having a trunk is approximated by nine planes (See Fig. 14). This can be represented by a nine node graph (See Fig. 15), with a 3D vector, the parameters for a plane, associated with each node.

To model all of the surface points, it is not only necessary to specify the planes but also to specify the portions of each plane that are used. The problem is automatically taken care of here, because the region of each plane that is used is determined by the intersections with other planes. A small problem arises at the bottom of the car. A tenth plane should be introduced to bound from below the portions of the front, rear, and side planes used in the model. We do not bother with that in the present discussion, but it may have to be considered in a real inferencing system.

From among the different parameterizations of a plane, we use the following. A plane can be specified by giving its orientation in three-space, i.e., a normal vector, and a point that the plane must pass through as shown in Fig. 16. In this figure a 3D plane is represented by its intersection with

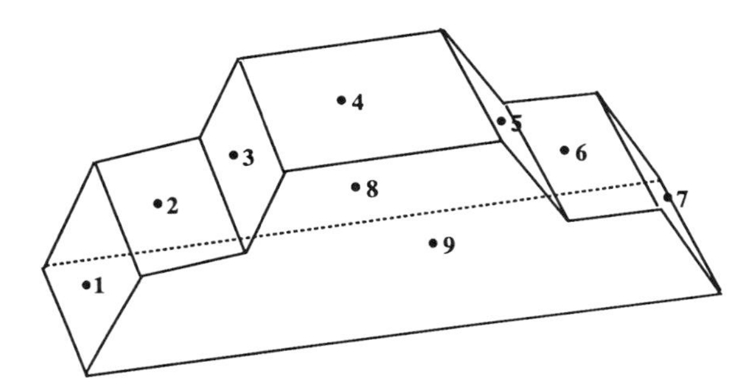

Figure 14: Planar approximation of a car

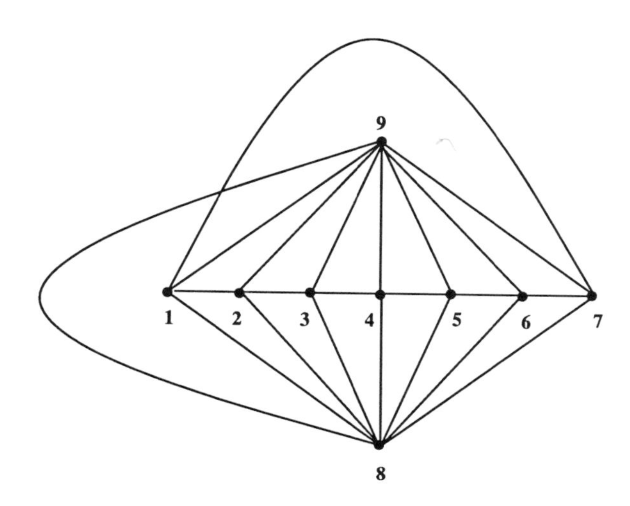

Figure 15: Graph representation for a car

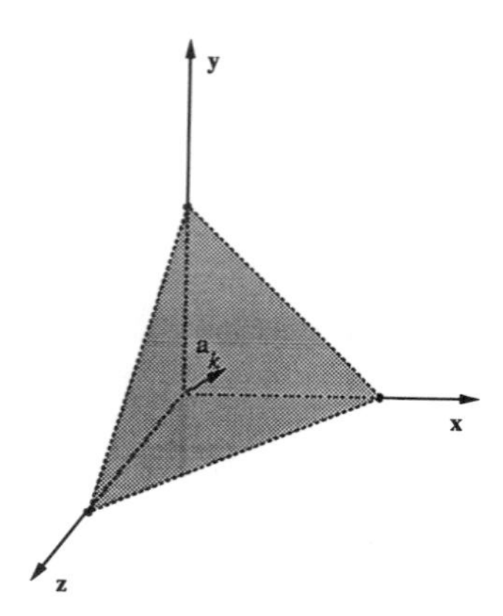

Figure 16: Parameterization of a plane.

the three coordinate planes, i.e., the xy, xz and yz planes, and $\mathbf{a}_k$ denotes the normal vector to the plane. This can be realized by specifying a vector $\boldsymbol{\gamma}$ in three-space with its tail at the origin and such that the plane is normal to $\boldsymbol{\gamma}$ and passes through the head of $\boldsymbol{\gamma}$. Let the parameter vectors $\mathbf{a}_k$ for the nine planes in the car model be these representations. We position a car within its local coordinate system by giving it a center and an orientation. Perhaps simplest is to put the center of mass, i.e., first joint moment of the car, at the origin and orient the car such that the eigenvectors of the second central moment matrix (scatter matrix) of the surface points are parallel to the local coordinate axes. For the following discussion, let the a_{k_1} axis (where a_k is the first component of $\mathbf{a}_k$) be orthogonal to the vertical plane of symmetry, i.e., the vertical plane running from front to back that bisects the car.

Now we want to put a joint distribution on these nine 3D vectors. Let the $\mathbf{a}_k$ be 3D vectors specified in this local coordinate system for the cars. Hence the distribution for the vector $\mathbf{a}^t \equiv (\mathbf{a}_1^t, \mathbf{a}_2^t, \ldots, \mathbf{a}_9^t)$ is given with respect to this local coordinate system and the local coordinate system can be described by a distribution within a world coordinate system. The local coordinate system is specified by six parameters, namely, the location of the origin of the coordinate system and the three unit vectors for the coordinate system orientation. Equivalent to the unit vectors is the transformation that rotates the local coordinate system from an initial alignment with the world coordinate axes.

Within the local coordinate system, a Gaussian distribution or a mixture of a few Gaussians would seem to make sense for $\mathbf{a}$. In the former case, it means that the class of **a**'s can be viewed as perturbations of some basic template. Determining useful parameters for a 27 $\times$ 27 covariance matrix appears to be an imposing task. The other extreme is to treat the $\mathbf{a}_k$ as independent of one another. Here, the one standard deviation contour would be some small ellipsoid about a mean vector. This model is appealing, but perhaps not accurate enough for object class recognition and object recognition within the class. We propose a MRF with the neighborhood system shown by the arcs in the graph in Fig. 15. In this simplest case, there are pairwise interactions between a node and its neighbors. The four neighbors of a surface are the surfaces that intersect the surface or constrain the surface significantly in some sense.

The joint distribution is singular because for $k = 1, 2, \ldots, 7$, a_{k_1}, i.e., the first component of $\mathbf{a}_k$, is 0. Also, because of the symmetry about the plane of axes 2 and 3, there is a 1-1 map relating $\mathbf{a}_8$ and $\mathbf{a}_9$. Hence, within the local coordinate system for the class, there are only 17 parameters to specify the nine planes. We have chosen cliques to be individual nodes and pairs of adjacent nodes in the graph. The simplest clique energy for a pair of nodes is $\mathbf{a}_j^t B_{jk} \mathbf{a}_k$ where B_{jk} is a 3×3 matrix of interaction parameters between nodes j and k. For $k = 1, 2, \ldots, 7$, $a_{k_1} = 0$, so the only nonzero entries in B_{jk} are in the lower right 2×2 block on the diagonal. Similarly, the structures of the blocks B_{kk} and of the blocks B_{jk}, where one of the j and k belongs to $\{1, 2, \ldots, 7\}$ and the other to $\{8, 9\}$, are clear.

8.1 Quadric Surfaces

Modeling with 3D quadric surface patches [5] is more interesting than with planes, because in addition to having position, they also have shapes. The general nondegenerate quadric in x, y, z is either an ellipsoid, a hyperboloid of one sheet, or a hyperboloid of two sheets.

The quadric

$$\begin{aligned} &\phi(x, y, z) = \\ &a_{200}x^2 + a_{020}y^2 + a_{002}z^2 + a_{110}xy + a_{101}xz + \\ &a_{011}yz + a_{100}x + a_{010}y + a_{001}z + a_{000} = 0 \end{aligned} \tag{29}$$

has a center, which is the solution of $\nabla\phi(x, y, z) = 0$, an orientation, which is the set of the three orthonormal eigenvectors of the matrix of coefficients of the second degree terms in $\phi(x, y, z)$, and a shape which is determined by the three eigenvalues of the matrix of coefficients of second degree terms.

Note that the location and orientation parameters can be thought of as an intrinsic coordinate system for the quadric, with the origin of the intrinsic coordinate system being displaced from the object coordinate system origin by the vector of location parameters, and the orientation of the intrinsic coordinate system specified by three rotations, one about each coordinate axis of the object coordinate system. Hence, a general quadric is specified by six parameters that determine its intrinsic coordinate system, three shape parameters that determine an ellipsoid or one of two types of hyperboloids within the intrinsic coordinate system, and a_{000}, a size parameter (a_{000} must be provided since otherwise the shape eigenvalues are known only up to a multiplicative factor.) For this nondegenerate case, there is a 1-1 map from the geometric parameters onto the space of coefficients in (25). A probability distribution could be put on the polynomial coefficients in (25) within the object coordinate system. However, there is then no simple interpretation of the 3D geometry controlled by the distribution. Alternatively, a probability distribution for a class of objects composed of quadric patches or quadric and planar patches can be put on the geometric parameters for the primitives, which seems to be more meaningful.

8.2 Higher Degree Polynomials and Invariants

Putting probability measures on higher degree polynomials is more complicated. We are exploring a number of approaches to this, built around geometric invariants. By geometric invariants, we mean functions of the polynomial coefficients, for the surface, that are independent of the coordinate system used [5, 24, 25, 26].

9 Conclusions

A number of 3D surface models of increasing complexity were discussed in this paper, and algorithms using three of them for 3D surface estimation, model recognition, and model-based segmentation, based on two or more images taken by calibrated cameras in different positions, were implemented. We feel that MAP 3D variable surface estimation is computationally attractive and should produce close to optimal accuracy. Bayesian segmentation of images and 3D space into regions, in each of which one primitive surface model applies (only 3D planes in our experiments), appears to be computationally practical and to have the potential for highly accurate results. The use of 3D depth texture models in segmentation for scene understanding is possible and potentially highly useful, but at this time it is not clear that this can be done to a desired accuracy in real time. Finally, an approach to putting MRFs on collections of polynomial primitives to model classes of complex 3D objects was outlined. We believe that this is a rich and potentially very important subject area to be explored.

10 Acknowledgments

This work was partially supported by NSF Grant #IRI-8715774 and NSF-DARPA Grant #IRI-8905436.

Bibliography

[1] J. Besag. Spatial Interaction and the Statstical Analysis of Lattice Systems. *Journal of Royal Statistical Society*, B **36**:192–236, 1974.

[2] J. Besag. On the Statistical analysis of dirty pictures. *Journal of Royal Statistical Society*, B **48**:259–302, 1986.

[3] R.M. Bolle and D.B. Cooper. Bayesian Recognition of Local 3D Shape by Approximating Image Intensity Functions with Quadratic Polynomials. *IEEE Transactions on Pattern Analysis and Machine Intelligence*, pages 418–429, July 1984.

[4] R.M. Bolle and D.B. Cooper. On Optimally Combining Pieces of Information, with Applications to Estimating 3D Complex-Object Position from Range Data. *IEEE Transactions on Pattern Analysis and Machine Intelligence*, pages 619–638, September 1986.

[5] B. Cernuschi-Frias. *Orientation and Location Parameter Estimation of Quadric Surfaces in 3D from a Sequence of Images*, Ph.D. thesis, Brown University, May 1984, UMI Press.

[6] B. Cernuschi-Frias, D.B. Cooper, Y.P. Hung, and P.N. Belhumeur. Toward a Model-based Bayesian Theory for Estimating and Recognizing Parameterized 3D Objects Using Two or More Images Taken from Different Positions. *IEEE Transactions on Pattern Analysis and Machine Intelligence*, pages 1028–1052, October 1989.

[7] P.B. Chou. *The Theory and Practice of Bayesian Image Labelling.* Technical Report 258, Computer Science Dept., University of Rochester, August 1988.

[8] Y. Chow, U. Grenander, and D. M. Keenan. *HANDS: A Pattern Theoretic Study of Biological Shapes.* Division of Applied Mathematics, Brown University, 1988.

[9] F. Cohen, M. Patel. Shape from Texture Using Gaussian Markov Random Fields, in *Markov Random Fields: Theory and Applications*, R. Chellappa and A.K. Jain, (Eds.), Academic Press, New York, NY 1992.

[10] F. Cohen, D.B. Cooper, J.F. Silverman, and E.B. Hinkle. Simple Parallel Hierarchical and Relaxation Algorithms for Segmenting Textured Images Based on Noncausal Markovian Random Field Models. In *Proceedings, IEEE Seventh International Conference on Pattern Recognition*, pages 1104–1107, July 1984.

[11] D.B. Cooper and F.P. Sung. Multiple Window Parallel Adaptive Boundary Finding in Computer Vision. *IEEE Transactions on Pattern Analysis and Machine Intelligence*, pages 299–316, May 1983.

[12] D.B. Cooper, Y.P. Hung, and G. Taubin. A New Model-Based Stereo Approach for 3D Surface Reconstruction Using Contours on the Surface Pattern. In *Proceedings, IEEE International Conference on Computer Vision*, pages 74–83, December 1988.

[13] H. Derin and H. Elliot. Modelling and Segmentation of Noisy and Textured Images using Gibbs Random Fields. *IEEE Transactions on Pattern Analysis and Machine Intelligence*, pages 39–55, January 1987.

[14] R.C. Dubes and A.K. Jain. Random Field Models in Image Analysis. *Journal of Applied Statistics*, **16**, 1989.

[15] S. Geman and D. Geman. Stochastic Relaxation, Gibbs Distributions and the Bayesian Restoration of Images. *IEEE Transactions on Pattern Analysis and Machine Intelligence*, pages 721–741, November 1984.

[16] S. Geman and K. Manbeck. Isotropic Priors for Single Photon Emission Computed Tomography, in *Markov Random Fields: Theory and Applications*, R. Chellappa and A.K. Jain, (Eds.), Academic Press, New York, NY 1992.

[17] Y.P. Hung, D.B. Cooper, and B. Cernuschi-Frias. *Asymptotic Bayesian Surface Estimation Using an Image Sequence.* Technical Report LEMS-73, Brown University, June 1990.

[18] R.L. Kashyap and R. Chellappa. Estimation and Choice of Neighbors in Spatial-Interaction Models of Images. *IEEE Transactions on Information Theory*, pages 60–72, January 1983.

[19] Y.G. Leclerc. Image and Boundary Segmentation via Minimum-Length Encoding on the Connection Machine. In *Proceedings, DARPA Image Understanding Workshop*, pages 1056–1069, May 1989.

[20] B.S. Manjunath, T. Simchony and R. Chellappa. Stochastic and Deterministic Networks for Texture Segmentation. *IEEE Transactions on Acoustics, Speech, and Signal Processing*, pages 1039–1049, June 1990.

[21] J. Marroquin, S. Mitter, and T. Poggio. Probabilistic Solution to Ill-Posed Problems in Computer Vision. *Journal of the Americal Statistical Association*, **82**:76–89, March 1987.

[22] J.W. Modestino and J. Zhang. A Markov Random Field Model-based Approach to Image Interpretation. In *Proceedings, IEEE Computer Society Conference on Computer Vision and Pattern Recognition*, pages 458–465, June 1989.

[23] J. Subrahmonia, Y.P. Hung, and D.B. Cooper. Model-based Segmentation and Estimation of 3D Surfaces from Two or more Intensity Images Using Markov Random Fields. In *Proceedings, Tenth International Conference on Pattern Recognition*, pages 390–397, June 1990.

[24] G. Taubin and D.B. Cooper. *Recognition and Positioning of 3D Piecewise Algebraic Objects Using Euclidean Invariants.* IBM Technical Report #RC 16211, Yorktown Heights, NY, October 1990.

[25] G. Taubin. *Estimation of Planar Curves and Nonplanar Space Curves Defined by Implicit Equations, with Applications to Edge and Range Image Segmentation. IEEE Transactions on Pattern Analysis and Machine Intelligence*, pages 1115–1138, November 1991.

[26] G. Taubin. *Recognition and Positioning of Rigid Objects Using Algebraic and Moment Invariants.* Ph.D. thesis, Brown University, May 1991.

[27] J.W. Woods. Two-dimensional Discrete Markovian Random Fields. *IEEE Transactions on Information Theory*, pages 232–240, March 1972.

A Markov Random Field Model-Based Approach To Image Interpretation

J. W. Modestino[‡] and J. Zhang[†]

[‡]Electrical, Computer and Systems Engineering Department
Rensselaer Polytechnic Institute
Troy, New York

[†]Department of Electrical Engineering and Computer Science
University of Wisconsin
Milwaukee, Wisconsin

1 Introduction

Image interpretation is the process of understanding the meaning of an image through identifying significant objects in the image and analyzing their spatial relationships. The need for image interpretation can be found in many diverse fields of science and engineering. For example, a major application of image interpretation is in remote-sensing, or aerial/satellite photointerpretation, which is widely used in geological surveys and military air reconnaissance [1]–[3]. Image interpretation also plays an important part in biomedical science and particle physics, where many of the experimental results are recorded in the form of photographs [4],[5].

Traditionally, the task of image interpretation is performed by well-trained and experienced *human experts*. However, analyzing a complex image is quite labor-intensive. Hence, much of the research in image processing has been directed towards constructing *automated* (computerized) image interpretation systems. Recent research in intelligent robots has created yet another need for automated image interpretation. In this case, the robots need to understand what they *see* with imaging sensors to be able to perform intelligent tasks in complex environments [6],[7]. Here, the robots have to rely entirely on automated image interpretation.

Most of the existing image interpretation techniques involve two major operations *low-level* and *high-level* processing. In low-level processing, the

ISBN 0-12-170608-7

representation of an image is transformed, through image processing operations, such as edge detection and region segmentation, from a *numerical* representation, as an array of pixel intensities, to a *symbolic* representation, as a set of spatially related *image primitives*, such as edges and regions. Various features are then extracted from the primitives. These features may include: the lengths of significant edges, average intensities of regions, shape and/or texture descriptors, etc. Also extracted would be the spatial relationship between the image primitives. In high-level processing, image domain knowledge is used to assign object labels, or interpretations, to the primitives and construct a description as to "what is present in the image." In the rest of this chapter, we often refer to the object labels as interpretations and the overall interpretations for *all* the primitives in the image as the interpretation of the image.

The main approach in early research on image interpretation was that of classification [4],[5], [8], in which isolated image primitives are classified into a finite number of object classes according to their feature measurements. However, since low-level processing often produces erroneous or incomplete primitives, and noise in the image may often cause measurement errors in the features, the performance of image interpretation systems using the classification approach is quite limited. The main problem here is that the rich knowledge of the spatial constraints between objects, used by human experts, has not been used in the high-level processing.

To solve this problem, most of the recent techniques have adopted the *knowledge-based*, or *expert system* [9], [10], approach. In this approach, domain knowledge, and especially spatial constraints, are used in high-level (some also in low-level) processing. Hence, an ambiguous object may be recognized as the result of successful recognition of its neighboring objects. Even more fundamentally, an object can be recognized from combining the feature information from several spatially related image primitives. Finally, low-level processing errors may be corrected, or at least mitigated, through feedback from high-level processing to low-level image processing.

The early work in knowledge-based image interpretation has been summarized in Nagao and Matsuyama [11], Binford [12], and Ohta [13], among others. Recently, a number of more sophisticated experimental systems have been constructed for different application domains, such as high-altitude aerial photographs [11], [14], [15], [16]–[18]; airport scenes [19], [20], [21] and outdoor scenes [13], [22]–[24]. Many of these systems are still undergoing continuous improvements through architecture modification and domain extension. New ideas and systems are constantly emerging, as can be seen in recent PRCV and SPIE conferences and workshops, and several pertinent technical reports [24], [25], [26]. While success has been demonstrated to various degrees in these systems, developing a *general*, *domain-independent* and *systematic* method for constructing knowledge-based image interpretation systems is still an open problem [22].

In this chapter, we describe a general, domain-independent, stochastic model-based approach to the image interpretation problem. In this approach, the interpretation labels to be assigned to the primitives of an image are modeled as a Markov random field (MRF) defined on the spatial adjacency graph formed by the extracted primitives, where the randomness is used to model the uncertainty in the assignment of the labels. As a result, the domain knowledge, whatever it may be, can be systematically represented in terms of the clique functions associated with the underlying Gibbs probability distribution function (pdf) describing the MRF. Under the MRF modeling assumption, image interpretation is then formulated as the optimization problem of maximizing the *a posteriori* probability of interpretation given domain knowledge and feature measurements. Then, simulated annealing is used to find the optimal set of interpretation labels. In this chapter, we present a special region-based version of this approach; that is, the primitives are segmented regions and for the sake of simplicity, we do not include feedback from high-level to low-level processing. However, research is currently ongoing to include use of linear edge segments as primitives as well as high-to-low-level feedback [27], [28].

2 The MRF Model-Based Approach to Image Interpretation

The MRF model, as an extension of the one-dimensional Markov process, has recently attracted much attention in the image processing and computer vision community. The main advantage of the MRF model is that it provides a general and natural model for the interaction between spatially related random variables, and there is a relatively flexible optimization algorithm, simulated annealing, that can be used to find the *globally* optimal realization that, in this case, corresponds to the maximum *a posteriori* (MAP) interpretation. Up to now, the success of MRF models has been demonstrated mostly in low-level image processing applications, such as region segmentation [29]–[36] and edge detection [37], where they are defined on two-dimensional (2-D) lattices on which the images are represented as 2-D arrays. For example, in stochastic model-based image segmentation, the pixels are classified into a finite number of statistical *classes* and the MRF is used to model the spatial distribution of pixel classes, or region distributions [29]–[36]. However, as demonstrated by Kinderman and Snell

[38], the MRF can be defined, in general, on graphs for which the 2-D lattice is a special case. In what follows, we will briefly review the concepts associated with the MRF defined on graphs and show how this can be applied to the image interpretation problem. More comprehensive treatments of MRFs can be found in [38], [39].

2.1 The MRF Model on Graphs

Let $\mathbf{G} = \{\mathbf{R}, \mathbf{E}\}$ be a graph, where

$$\mathbf{R} = \{R_1, R_2, \ldots, R_N\} \tag{1}$$

is the set of nodes represented by R_i, $i = 1, 2, \ldots, N$; $\mathbf{E}$ is the set of edges connecting them. Suppose that there exists a *neighborhood system* on $\mathbf{G}$, denoted by

$$\mathbf{n} = \{n(R_1), n(R_2), \ldots, n(R_N)\}, \tag{2}$$

where $n(R_i)$, $i = 1, 2, \ldots, N$, is the set of all the nodes in $\mathbf{R}$ that are neighbors of R_i, such that:

1. $R_i \notin n(R_i)$, and
2. if $R_j \in n(R_i)$, then $R_i \in n(R_j)$.

Let

$$\mathbf{I} = \{I_1, I_2, \ldots, I_N\} \tag{3}$$

be a family of random variables defined on $\mathbf{R}$. Then, $\mathbf{I}$ is called a *random field*, where I_i is the random variable associated with R_i. Notice that the random variables I_i's here can be numerical as well as symbolic, e.g., interpretation labels. We say $\mathbf{I}$ is an MRF on $\mathbf{G}$ with respect to the neighborhood system $\mathbf{n}$ if and only if:

1. $P[\mathbf{I}] > \mathbf{0}$, for all realizations of $\mathbf{I}$;
2. $Pr[I_i | I_j, \text{all } R_j \neq R_i] = P[I_i | I_j, R_j \in n(R_i)]$,

where $P[\cdot]$ and $P[\cdot|\cdot]$ are the joint and conditional pdf's, respectively. Intuitively, the MRF is a random field with the property that the statistic at a particular node depends only on that of its neighbors.

An important feature of the MRF model defined before is that its joint pdf has a general functional form, known as the *Gibbs distribution*, which is defined based on the concept of *cliques* [38], [39]. Here, a clique associated with the graph $\mathbf{G}$, denoted by c, is a subset of $\mathbf{R}$ such that it contains either a single node or several nodes that are *all* neighbors of each other. If we denote the collection of all the cliques of $\mathbf{G}$ with respect to the neighborhood system $\mathbf{n}$ as $\mathbf{C}(\mathbf{G}, \mathbf{n})$, the general functional form of the pdf * of the MRF can be expressed as the following Gibbs distribution:

$$P[\mathbf{I}] = Z^{-1}\exp[-U(\mathbf{I})], \tag{4a}$$

where

$$U(\mathbf{I}) = \sum_{c \in \mathbf{C}(\mathbf{G},\mathbf{n})} V_c(\mathbf{I}) \tag{4b}$$

is called the *Gibbs energy function* and the $V_c(\mathbf{I})$'s are called clique functions defined on the corresponding cliques $c \in \mathbf{C}(\mathbf{G}, \mathbf{n})$. Finally,

$$Z = \sum_{all\ \mathbf{I}'} \exp[-U(\mathbf{I}')], \tag{4c}$$

is the normalization factor to make (4a) a valid pdf. Notice that the preceding MRF pdf is quite rich in that the clique functions can be arbitrary as long as they depend only on the nodes in the corresponding cliques. Due to this unique structure, in which the global and local properties are related through cliques, the MRF model-based approach to image interpretation provides potential advantages in knowledge representation, learning, and optimization, as will be discussed in more detail later. More importantly, this method provides a useful mathematical framework for the study of image interpretation procedures.

2.2 The MRF Model-Based Formulation

As described in Section 1, for the time being we restrict the image interpretation problem to that of labeling *segmented regions*. Suppose, for a given image, there are N disjoint regions resulting after segmentation,† denoted by $\mathbf{R} = \{R_1, R_2, \ldots, R_N\}$. Then $\mathbf{R}$ can be represented by a set of nodes in a connected graph, called the *adjacency graph*, denoted by $\mathbf{G} = \{\mathbf{R}, \mathbf{E}\}$ where the edge set $\mathbf{E}$ is such that a node R_i is connected to another node R_j if and only if the corresponding regions are spatially adjacent. A neighborhood system, denoted by $\mathbf{n}$, can also be defined on the adjacency graph.

*Actually, this is a probability mass function (pmf) due to the discrete nature of $\mathbf{I}$, although we will not make this distinction in what follows and continue to use the term pdf.

†Clearly, the number N of segmented regions is a random variable depending upon the image as well as the segmentation procedure.

For simplicity, in what follows we define the neighbors of a node to be the nodes that are connected to it directly by an edge of $\mathbf{G}$, i.e., only spatially adjacent regions are neighbors. Now, given the neighborhood system, we can also find the cliques for the adjacency graph. As an illustration, we have shown in Fig. 1 the adjacency graph and all its cliques for a particular synthetic conceptual image. This image is intended to represent a car on a road between two fields with the sky as a background. In forming the adjacency graph, we assume perfect segmentation of the image objects.

As described in Section 1, image interpretation is the process of assigning object labels to the segmented regions according to domain knowledge and feature measurement information (or *measurements*, in short) made on these regions. From the preceding graphical formulation, the interpretation of the image can be represented as a random vector $\mathbf{I}(\mathbf{R}) = \{I_1, I_2, \ldots, I_N\}$, defined on the adjacency graph $\mathbf{G}$, where we use $\mathbf{I}(\mathbf{R})$ to emphasize the relationship between interpretation and the symbolic representation in terms of segmented regions. Here, I_i, $i = 1, 2, ..., N$, is the interpretation label for node R_i while $I_i \in L$ and $L = \{L_1, L_2, \ldots, L_M\}$ is the set of all the interpretation labels. Hence, $\mathbf{I}(\mathbf{R})$ is a random field. Let us denote the domain knowledge as $\mathbf{K}$ and all the measurements made on the segmented regions as $\mathbf{X}(\mathbf{R})$. Now, we can define image interpretation as the following *optimization problem*: for a given $\mathbf{R}$, find $\mathbf{I}_0(\mathbf{R})$, such that

$$\mathbf{I}_0(\mathbf{R}) = \arg \max_{\mathbf{I} \in \{L\}^N} P[\mathbf{I}(\mathbf{R}) \mid \mathbf{K}, \mathbf{X}(\mathbf{R})], \tag{5}$$

where $P[\cdot|\cdot,\cdot]$ is the *a posteriori* pdf of the interpretation given the domain knowledge and measurements, while $\{L\}^N$ is the set of all possible interpretation vectors of length N. The formulation of Eq. (5) is also known as the maximum *a posteriori* (MAP) formulation.

Two problems must be solved in applying the above MAP approach to image interpretation. Specifically, we need an explicit expression for the conditional pdf in Eq. (5) and an optimization method to avoid the computationally explosive nature of exhaustive combinatorial search. Feldman and Yakimovsky [40], and Faugeras and Price [14], [15] have considered similar formulations to that of Eq. (5) and proposed heuristic expressions for the *a posteriori* pdf using the marginal pdf's of single and joint pdf's of pairs of interpretation labels. They have also used different relaxation schemes to find local optimal solutions, some of which have also been studied in [54]–[57]. On the other hand, the MRF model discussed above appears to provide a natural solution to the preceding two problems. More specifically, assume that $\mathbf{I}(\mathbf{R})$ forms an MRF. Then, the pdf appearing in Eq. (5) is the Gibbs distribution,

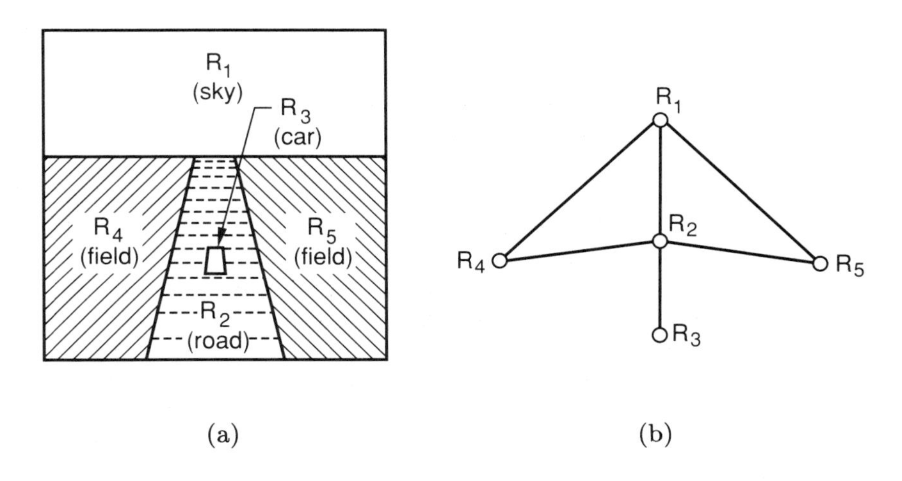

Node or Region	Associated Cliques
R_1	$\{R_1\}, \{R_1,R_2\}, \{R_1,R_4\}, \{R_1,R_5\}, \{R_1,R_2,R_4\}, \{R_1,R_2,R_5\}$
R_2	$\{R_2\}, \{R_2,R_3\}, \{R_2,R_4\}, \{R_2,R_5\}$
R_3	$\{R_3\}$
R_4	$\{R_4\}$

(c)

Figure 1: A synthetic conceptual image for image interpretation. (a) The synthetic conceptual image. (b) The adjacency graph of the conceptual image. (c) Cliques of the adjacency graph.

$$P[\mathbf{I}(\mathbf{R}) \mid \mathbf{K}, \mathbf{X}(\mathbf{R})] = Z^{-1}\exp[-U\big(\mathbf{I}(\mathbf{R})\,;\,\mathbf{K}, \mathbf{X}(\mathbf{R})\big)], \tag{6a}$$

with energy function,

$$U\big(\mathbf{I}(\mathbf{R})\,;\,\mathbf{K}, \mathbf{X}(\mathbf{R})\big) = \sum_{c \in \mathbf{C}(\mathbf{G}, \mathbf{n})} V_c\big(\mathbf{I}(\mathbf{R})\,;\,\mathbf{K}, \mathbf{X}(\mathbf{R})\big), \tag{6b}$$

where the $V_c(\cdot;\cdot,\cdot)$'s are the clique functions. Indeed, as will be seen in the subsequent sections, through imposing a neighborhood system and the aforementioned Markov property 2, the MRF model-based formulation provides a general and systematic approach for knowledge representation and knowledge acquisition through appropriate construction of the clique functions. For the optimization strategy, the simulated annealing procedure can be used to find the globally optimal interpretation for the image. In addition, the approach of [14], [15], [40], can be shown to be special cases with certain neighborhood structures and clique functions. Finally, when used in the context of image interpretation, the MRF model suggests that the interpretation for a particular region, given those of all other regions, depends only on the interpretations of its neighboring regions. This is often a reasonable assumption in practical applications. For example, the identification of a region as a car might depend on whether its neighboring regions are a road but has little to do with the identity of the regions spatially far removed from it. In the rest of the chapter, we will model the interpretation vector $\mathbf{I}(\mathbf{R})$ as an MRF.

3 The Design of Clique Functions

In the MRF model-based formulation of the preceding section, it is clear that the optimal interpretation, $\mathbf{I}_0(\mathbf{R})$, should be the one that minimizes the energy function, or has the minimum energy. For a given image, the optimal interpretation depends on how the energy function is defined. In general, we would like the optimal interpretation obtained under the MRF assumption to be the one that is most *consistent* with the measurements, and domain knowledge. For example, in aerial photointerpretation, suppose we know that a car has small area and would usually be on a road. An interpretation with a car having large area or in the sky should obviously be considered *not* optimal. This type of consistency requirement can be achieved by properly selecting the energy functional or, rather, the corresponding clique functions. It will be seen in the following that, by using the MRF model, the domain knowledge can be organized easily and systematically as clique functions to provide a proper energy functional such that the consistency between the interpretation, the measurements and domain knowledge is maintained.

Without loss of generality, we assume that all the clique functions are non-negative. Then, a general principle for the selection of a clique function is the following:

Design Rule:

> If the interpretation of the regions (or region for a *singleton* clique) in a clique tends to be consistent with the measurements and domain knowledge, the clique function decreases, resulting in a decrease in the energy function; otherwise, the clique function increases, resulting in a corresponding increase in the energy function.

In this way, an interpretation for the image that is most consistent with the measurements and domain knowledge will have the minimum energy, or achieve the optimum. Based on this principle, we now propose a general approach to defining clique functions from domain knowledge. We first consider the clique functions for single-node cliques, and then extend the result to the case of multiple cliques.

3.1 Clique Functions for Single-Node Cliques

Let c be an arbitrary single-node clique with one node, R. Let the corresponding clique function be denoted by $V_c\big(I(R)\ \mathbf{K}, \mathbf{X}(R)\big)$; it depends only upon the single node R, its interpretation $I = I(R)$, and the measurements $\mathbf{X}(R)$ on the corresponding segmented region R, as well as the domain knowledge represented by $\mathbf{K}$. Suppose that $\mathbf{X}(R)$ has m components, $X_1(R), X_2(R), \ldots, X_m(R)$, representing measurement values of m well-defined *features* of R, e.g., *average gray level, area, standard deviation of gray levels*, etc. Assuming the components of $\mathbf{X}(R)$ are independent, we can define a clique function for clique c as:

$$V_c\big(I(R)\,;\, \mathbf{K}, \mathbf{X}(R)\big) = \sum_{i=1}^{m} p_c^{(i)}(I(R), \mathbf{K}) B_c^{(i)}\big(I(R); \mathbf{K}, X_i(R)\big), \qquad (7)$$

where $B_c^{(i)}(\cdot;\cdot,\cdot)$, $i = 1, 2, \ldots, m$, are called *basis functions* for the corresponding clique function. These quantities are functions of the ith feature measurement, $X_i(R)$, parameterized by the interpretation $I(R)$ and, of course, depend upon the domain knowledge, $\mathbf{K}$. The $p_c^{(i)}(I, \mathbf{K})$'s are a set of nonnegative numbers that can be conveniently normalized so that

$$\sum_{i=1}^{m} p_c^{(i)}(I, \mathbf{K}) = 1, \tag{8}$$

and are *weights* associated with the basis functions. Here, $p_c^{(i)}(I, \mathbf{K})$ not only depends on i but also on the interpretation $I(R)$ as well as $\mathbf{K}$.

Now the problem of designing clique functions becomes that of designing the basis functions of the features and determining their weights. We first consider the design of the basis functions. Without loss of generality, we assume that all the basis functions are nonnegative. Then, the consistency principle for designing clique functions (in the previous Design Rule) applies to the design of the basis functions. Here, it is sufficient to consider the design of a particular basis function for a single-node clique c, denoted by $B_c\big(I(R); \mathbf{K}, X(R)\big)$, where, for notational simplicity, the index i has been dropped. According to the consistency requirement between interpretation, measurements, and domain knowledge, we want the basis function to be small when $I(R)$, $X(R)$ are *consistent* according to $\mathbf{K}$; otherwise, it should be large. One way to achieve this is to take a probabilistic approach. In particular, we consider the *a posteriori* pdf $P_c[I(R) \mid \mathbf{K}, X(R)]$. This is the probability that, based on the domain knowledge, $\mathbf{K}$, and the measurement, $X(R)$, the interpretation of the node R should be $I(R)$. By definition, the probability $P_c[I(R) \mid \mathbf{K}, X(R)]$ is such that for $I(R)$ consistent with the measurements and domain knowledge, it is large; otherwise it is small. Hence, a non-increasing function of this pdf can be used as a basis function. For example, the logarithm of the pdf has been suggested [38] as a reasonable basis function for general MRFs. In this case, we can define:

$$B_c\big(I(R); \mathbf{K}, X(R)\big) = -\alpha_c \log P_c[I(R) \mid \mathbf{K}, X(R)], \tag{9}$$

where α_c is a positive weighting constant and $-log(\cdot)$ is a monotonically decreasing function. Another way of selecting the basis function is to use

$$B_c\big(I(R); \mathbf{K}, X(R)\big) = \alpha_c\big(1 - \beta_c P_c[I(R) \mid \mathbf{K}, X(R)]\big), \tag{10}$$

where α_c and β_c are positive constants and $\beta_c P_c[I(R) \mid \mathbf{K}, X(R)] < 1$. Usually, we want the normalization constants α_c and β_c, to be such that $0 \leq B_c\big(I(R); \mathbf{K}, X(R)\big) \leq 1$.

To find the pdf $P_c[I(R) \mid \mathbf{K}, X(R)]$, Bayes's conditional pdf formula can be used; that is,

$$\begin{aligned} P_c[I(R) \mid \mathbf{K}, X(R)] &= P_c[X(R) \mid \mathbf{K}, I(R)] \\ &\quad \cdot P_c[\mathbf{K}, I(R)] P^{-1}[\mathbf{K}, X(R)], \end{aligned} \tag{11}$$

where the first term is the *likelihood functional* of the measurement conditioned on the interpretation, which can be found easily under proper modeling assumptions, and the second term is the prior pdf of the interpretations, which can be determined from *a priori* information, or heuristically. Finally, the last term is the inverse of the pdf of $X(R)$, which does not depend on I and hence can be dropped in the basis function. For the sake of simplicity, we assume the prior probability to be a constant; that is, the interpretations are equally likely *a priori*, then the second term can also be dropped.

To further illustrate what we mean by $P_c[I(R) \mid \mathbf{K}, X(R)]$ and how a basis function can be defined from it, consider an example of a single-node clique, c. Suppose we have the following *knowledge*:

1. The node could be sky, field, car, road, denoted by interpretation labels L_s, L_f, L_c and L_r.
2. The average gray level, a feature of the preceding objects, should be close to G_s, G_f, G_c, and G_r, respectively, and $G_s > G_r > G_c > G_f$.
3. The distribution of the measured average gray level of the node, conditioned on each label (sky, field, car, road), is Gaussian; that is, if the measured average gray level $X(R) = G$, then

$$P[X(R) \mid \mathbf{K}, I(R) = L_\delta] = \frac{1}{\sqrt{2\pi}\sigma_\delta} exp\left[-\frac{(G - G_\delta)^2}{2\sigma_\delta^2} \right], \tag{12}$$

where $\delta = s, f, c, r$. Then, from Eq. (9), a possible basis function could be of the form,

$$B_c\big(I(R)\,;\, \mathbf{K}, X(R)\big) = \alpha_c \left(\frac{1}{2} \log 2\pi\sigma_\delta^2 + \frac{(G - G_\delta)^2}{2\sigma_\delta^2} \right). \tag{13}$$

Plots of several basis functions, including the one proposed by Modestino [41], are shown in Fig. 2. Notice that these functions are all *window-like* functions. When domain knowledge about a feature can be expressed in terms of a nominal value, as is in the example before, the specification of the corresponding basis function can be greatly simplified to that of merely constructing one of these window functions. The piecewise linear basis function is particularly interesting in that it is very easy to compute and, as will be seen in later sections, it is relatively robust against measurement errors or image noise. Hence, we give it a special notation, $g(x; a_1, a_2, b_1, b_2)$ where x is the independent variable and a_1, a_2, b_1, b_2 are the four *corner points*, with $a_1 \leq a_2 \leq b_1 \leq b_2$. Similar functions have

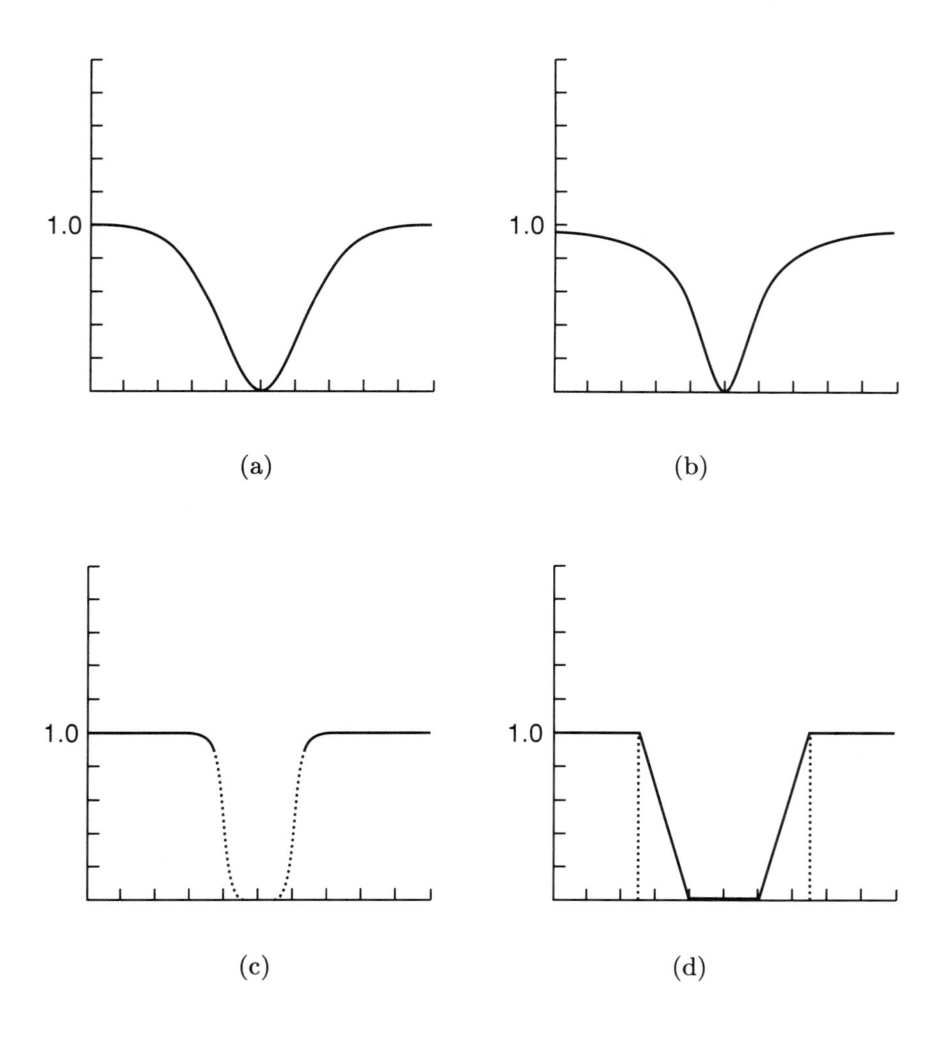

Figure 2: Examples of different basis functions. (a) Gaussian function. (b) Modestino's function, basic form: $y^2/(1+y^2)$. (c) Extension of Modestino's function, $y^n/(1+y^n), n = 8$. (d) The piecewise linear function.

been used in [22] for a rule-based image interpretation system and in the applications of fuzzy set theory [42].

3.2 Clique Functions for Multiple-Node Cliques

The extension of the clique function design procedure from the case of single-node cliques of Section 3.1 to the case of multiple-node cliques is quite straightforward. Here, we still design clique functions through designing a set of basis functions, as indicated in Eq. (7). However, the designing of the basis function is slightly more complicated here in that we may have two types of basis functions. The first type is the basis function for feature measurements, as in the case for single-node cliques. The features in this case could be quantities such as *mutual boundary length, contrast*, etc. Basis functions for these feature measurements can be designed in the same way as that in Section 3.1, using the window functions of Fig. 2. The second type of basis functions is that for spatial constraints. The constraints in this case could be statements such as "a car should be on (neighboring to) the road," "a car should never be in the sky," etc. In this case, we can still use the probabilistic approach in the spirit of Eqs. (9)–(10). For example, consider an arbitrary clique c with multiple nodes denoted by $\mathbf{R}_c$ and interpretations $\mathbf{I}_c(\mathbf{R}_c)$. Let $P_c[\mathbf{I}_c(\mathbf{R}_c) \mid \mathbf{K}]$ be the probability that the combination of interpretations $\mathbf{I}_c(\mathbf{R}_c)$ is valid according to domain knowledge. For example, we might have:

$$
\begin{aligned}
P_c[\mathbf{I}_c(\mathbf{R}_c)|\mathbf{K}] &= 1 \text{ if } \mathbf{I}_c \text{ is a valid combination,} \\
&\quad\ \text{according to domain knowledge;} \\
&= 0 \text{ if } \mathbf{I}_c \text{ is not a valid combination,} \\
&\quad\ \text{according to domain knowledge.}
\end{aligned} \tag{14}
$$

Similar to Eqs. (9)–(10), we can define the basis function as:

$$
B_c(\mathbf{I}_c(\mathbf{R}); \mathbf{K}) = \alpha_c(1 - P_c[\mathbf{I}_c(\mathbf{R}) \mid \mathbf{K}]). \tag{15}
$$

3.3 The Selection of the Weights for the Basis Functions

The weights of the basis functions in Eqs. (7)–(8) control the contributions of the individual basis functions to the value of a clique function. For simplicity, we may make them all equal. In our current experiments, we start with this simple scheme and then, if a feature is too unreliable for a particular object type, we will reduce the corresponding weight. In addition, adjustments are also made by trial and error through examining interpretation results on representative training images. A more sophisticated approach, currently under investigation, is to select a weight based on how powerful the corresponding feature is for object recognition and discrimination. For example, a useful indication of whether a given feature is good for object discrimination can be obtained from the *inter-cluster distances* [43], where the clusters are formed by the measurements of the feature from different objects. Similarly, a useful indication of whether a feature is good for recognizing a particular type of object, say, type I, can be obtained from the *intra-cluster standard deviation* [43], where the cluster is formed by the measurements of the ith feature on many objects of type I.

3.4 Remarks

To conclude this section, we note several interesting points. First, through the design of clique functions, we have a systematic approach for representing spatial knowledge; that is, for organizing the domain knowledge into a set of well-defined clique functions. This approach also provides guidelines as to what kind of knowledge one would want for the purpose of image interpretation; basically, knowledge concerning objects spatially related as members of different types of cliques.

Secondly, under the current neighborhood system assumptions, there are at most four different clique types, as shown in Fig. 3, which contain at most four nodes. This is due to the fact that the adjacency graph associated with the segmented regions is a *planar* graph, a graph without overpassing edges while a clique containing five or more nodes causes overpassing of edges in the graph. Hence, the design of clique functions is relatively simple due to the small number of different clique types.

Finally, as has been pointed out in Section 2, the Gibbs distribution is a very rich distribution in that, as long as the clique functions depend only on the corresponding cliques, their form can be somewhat arbitrary. The general guidelines provided in this section on designing clique functions are based on the considerations of the consistency requirement in the image interpretation problem. While they provide useful insights into the image interpretation problem and offer practical solutions, they are not necessarily the only or the best choice.

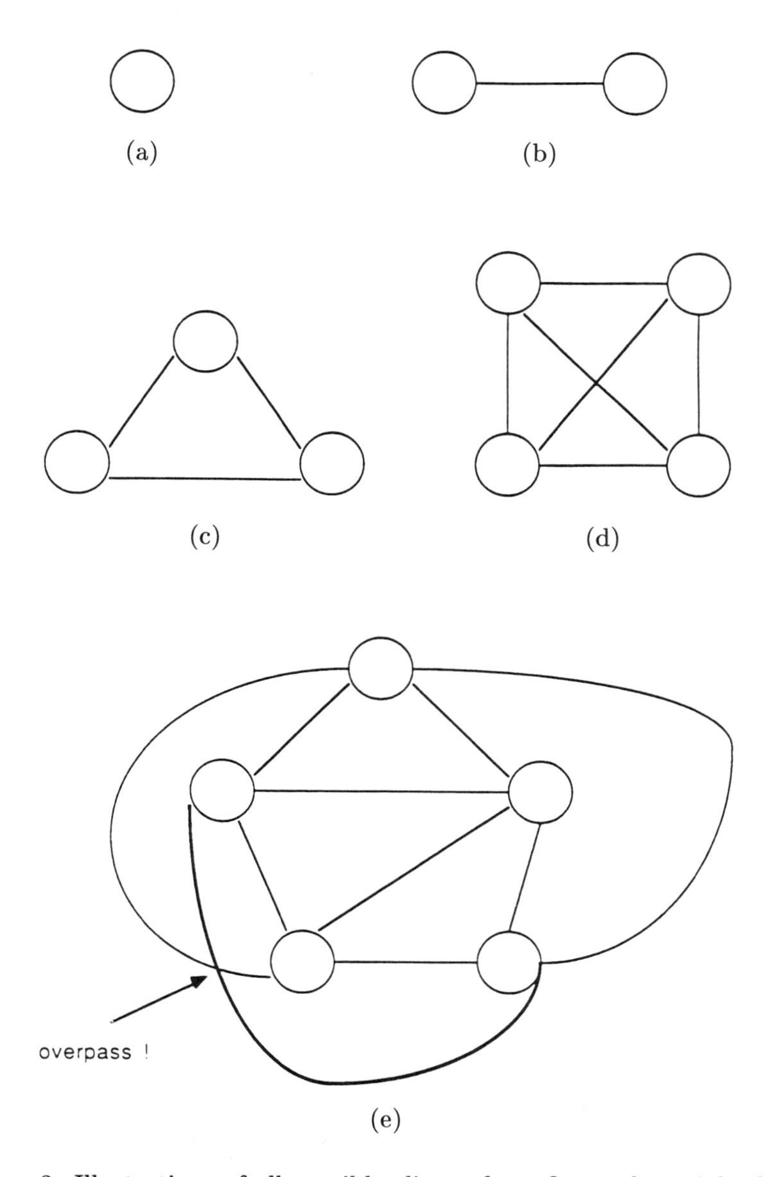

Figure 3: Illustrations of all possible cliques for a first-order neigborhood system. (a) Single-node clique. (b) Two-node clique. (c) Three-node clique. (d) Four-node clique. (e) Five nodes can not form a clique.

4 Implementation through Simulated Annealing

In the last section, image interpretation is formulated as a MAP estimation problem. Under the MRF modeling assumption, this becomes the problem of minimizing a properly defined energy function such that the interpretation obtained is most consistent with measurements and domain knowledge. The simplest optimization method is an exhaustive search procedure. This, however, results in an exponential complexity of $O(M^N)$, where M is the number of labels and N is the number of nodes in the adjacency graph. An alternative is the simulated annealing algorithm, a stochastic iterative optimization procedure, that will find the global maximum of the pdf of the MRF, or the minimum of the energy function, without excessive computation [29], [48]. The simulated annealing algorithm has been widely used in various applications involving combinatorial optimization, such as VLSI layout [44], channel coding [45], and image segmentation [46], [47].

For convenience, we describe this algorithm here in the context of a minimization problem. Let the function to be minimized be $E(\mathbf{x})$, where $\mathbf{x}$ is the indepdendent variable. This algorithm can be loosely described as follows:

Simulated Annealing

1. Select an initial *temperature* parameter T_0 and randomly choose an initial variable $\mathbf{x}_0$. Iteration begins.

2. At step k, perturb $\mathbf{x}_k$ by $\hat{\mathbf{x}}_{k+1} = \mathbf{x}_k + \Delta\mathbf{x}$ and compute $\Delta E = E(\hat{\mathbf{x}}_{k+1}) - E(\mathbf{x}_k)$.

3. If $\Delta E < 0$, accept the change; that is, $\mathbf{x}_{k+1} = \mathbf{x}_k + \Delta\mathbf{x}$. If $\Delta E > 0$, accept the change only with probability $p = e^{-\Delta E/T}$.

4. If there is a considerable drop in energy, or enough iterations, lower the temperature.

5. If the energy becomes stable and the temperature is very low, stop; otherwise, go back to Step 2.

For image interpretation, the implementation is straightforward when the definition of the method of perturbation and the annealing schedule (i.e., how the temperature is lowered) are decided. We first order all the nodes arbitrarily as node $1, 2, ..., N$. Then, an iteration is defined as *one* visit to *all* the nodes according to this order. When a node is visited, a perturbation of the interpretation vector is performed through generating a new label for this node from a uniform distribution of all the possible interpretations, $L_1, L_2, \ldots, L_M$, where M is the number of different labels, or from the conditional probability distribution of the MRF (i.e., the Gibbs sampler of [29]). In our experiments, we found that the two perturbation methods provide roughly the same results in terms of convergence to the optimal interpretations, while the Gibbs sampler is more complicated in computation structure, so we have mainly made use of the first approach for perturbation. Finally, for the annealing schedule, the temperature is lowered after each iteration according to $T_{l+1} = \alpha T_l$, where $0.5 < \alpha < 1$, which we have used successfully in an MRF model-based MAP image segmentation procedure [46]. In particular, we have selected $T_0 = 1$ and $\alpha = 0.92$ for all our experiments.

5 Interpretation of Synthetic Images

To gain some understanding of the MRF approach to real-world image interpretation, it makes sense to investigate this approach under the somewhat ideal situation in which we have acceptable segmentations and relatively strong features and spatial constraints in the domain knowledge. If the MRF model-based approach works well here, then it is reasonable to expect that it will be effective for real-world image interpretation, provided we can produce good segmentations (or are able to deal with poor ones) and find strong features and spatial constraints. We can create such an ideal situation through generating synthetic images and studying the performance of the MRF model-based approach in interpreting them. In this section, we describe some of the experimental results on synthetic images taken from a more complete study [41], [46].

The synthetic images used in this experiment are variations of the conceptual image of Fig. 1, which contains such objects as sky, road, field, and car, all of which appear as regions of *constant* gray levels. The assumed domain knowledge associated with this image is shown in Table 1 where object features (with precise definitions in Table 2 and spatial constraints are stated. In the experiments [41], [46] we have found that, compared to other basis functions of Fig. 2, the piecewise-linear basis functions were more effective for object recognition and less sensitive to segmentation error. Hence, all the results for image interpretation described here have been obtained using clique functions composed of this type of basis function. In Table 3, these clique functions are shown in terms of their basis functions and weights for the synthetic image (corner points a_1, a_2, b_1, b_2, weight p, and α_c for the basis function for spatial constraints). Finally, the segmentation algorithm used here is a Gaussian model-based segmentation algorithm [49] that has been quite effective for aerial photograph segmentation.

The experiments on the synthetic image described in this section contain three parts. First, the *ideal image* of Fig. 1 is interpreted with results shown in Fig. 4. In this case, the extracted segmentation is perfect, as shown in Fig. 4b, since all the regions have constant gray levels. The interpretation result in Fig. 4c, in which different gray levels indicate different object labels (as shown in Fig. 4d), shows that all the regions are correctly identified. This suggests that when the image is well-segmented and the domain knowledge is sufficient, the MRF model-based approach is effective. As reported in [46], starting from a random initial interpretation vector, the simulated annealing converged within 25 iterations. In this result, and the rest of the results of this section, the clique functions of Table 3 have been used, and the number of iterations for the simulated annealing is set to 25.

In the second part of the experiments, the ideal image of Fig. 1 is corrupted by additive white Gaussian noise to generate degraded images of different signal-to-noise ratios (SNRs). These images then are presented for interpretation. Here, the added noise should result in errors in segmentation and feature measurements. This is used to study the performance of the MRF approach under moderately imperfect segmentation. In [46], experiments have been performed for images with SNR of 20dB, 10dB, and 3dB with similar results. Hence, the results are only shown in Fig. 5 for the 3dB case. Again, all the objects are correctly identified. Here, the car has a very small area and its identification might be most seriously affected by segmentation error. However, since the road can be well identified, the car can still be identified partly due to the spatial constraints between them. To be able to deal with more serious segmentation errors, such as the case in which the car is split into several regions, the clique functions have to be expanded to allow the merging of regions. This is currently under investigation [27].

In the last part of the experiments, we have considered the case where there are *unknown* objects in the image for which no information is available in the knowledge base. There are two main causes for unknown regions, e.g., the region belongs to an unknown object or the region belongs to a known object but the feature measurements are far from the nominal values of all known objects due to segmentation errors. In both cases, it is more desirable to assign an *unknown* label to such regions than to risk making a mistake. Hence, we have added an *unknown* label to the set of object labels. In this part, an *unknown* object that is similar to a car is placed in the right *field* region of the images used in the preceding experiments and

	Region Knowledge	
Object Type	Area (No. of pixels)	Average Gray Level
Car	≤ 800	$\doteq 150$
Sky	≥ 25000	$\doteq 200$
Road	$\doteq 11700$	$\doteq 100$
Field	≥ 13500	$\doteq 50$

a.) a single region

			Mutual Knowledge
Object Type	Boundary Length	Contrast	Spatial Constraints
Sky, Car	0	–	Impossible combination
Field, Car	0	–	impossible combination
Sky, Road	$\doteq 56$	$\doteq 100$	valid combination
Car, Road	$\doteq 120$	$\doteq 50$	valid combination
Sky, Field	$\doteq 100$	$\doteq 150$	valid combination
Road, Field	$\doteq 180$	$\doteq 150$	valid combination

b.) two regions

	High - Order Knowledge
Object Types	Spatial Constraints
Sky, Car, Field	Impossible combination
Sky, Car, Road	impossible combination
Sky, Road, Field	valid combination
Road, Car, Field	impossible combination

c.) three regions

Table 1: Summary of Assumed Knowledge for the Conceptual Image.

1. Area: A = the number of pixels in the region R.
2. Average Gray Level: $$G = \frac{1}{A} \sum_{(i,j) \in R} x(i,j),$$ where $x(i,j)$'s are gray levels of the pixels in R.
3. Standard Deviation of Gray Levels: $$S = \left(\frac{1}{A} \sum_{(i,j) \in R} (x(i,j) - G)^2 \right)^{1/2}.$$
4. Compactness: $$C = \frac{P^2}{4\pi A} - 1,$$ where P is the perimeter of R defined as P = the number of boundary points of R.
5. Partial Compactness: C = sample standard deviation of r, where r is the random measurement of the radius of region R as shown in text.

a.) Features for a Single Region R.

1. Boundary Length: $$B_{i,j} = \frac{P_i + P_j}{2}.$$
2. Contrast: $$C_{i,j} = \lvert G_i - G_j \rvert.$$

b.) Features for Two Adjacent Regions R_i and R_j.

Table 2: Definitions of Several Region-Based Features.

	Features	
	Area	Average Gray Level
Label	a_1, a_2, b_1, b_2, p	a_1, a_2, b_1, b_2, p
Car	750, 790, 810, 850, 0.5	165, 175, 185, 187, 0.5
Sky	11500, 11600, 11900, 12000, 0.5	85,95,105,115,0.5
Road	19900, 20000, 65536, 65536, 0.5	193, 195, 205, 215, 0.5
Field	13300, 13400, 13900, 14000, 0.5	0, 0, 55, 65, 0.5

a.) cliques of a single node

	Features		
	Boundary Length	Contrast	Spatial Constraints
Labels	a_1, a_2, b_1, b_2, p	a_1, a_2, b_1, b_2, p	$\alpha_{c,p}$
Sky, Car	–	–	1.0, 1.0
Field, Car	–	–	1.0, 1.0
Sky, Road	0, 0, 55, 65, 0.5	85,95,105,115,0.5	0.0, 1.0
Car, Road	105, 115, 125, 135, 0.5	0, 0, 85, 95, 0.5	0.0, 1.0
Road, Field	140, 150, 1000, 1000, 0.5	35, 45, 55, 65, 0.5	0.0, 1.0
Sky, Field	85, 95, 115, 0.5	140, 150, 257, 257, 0.5	0.0, 1.0
all others	–	–	0.5, 1.0

b.) cliques of two nodes

Labels	$\alpha_{c,p}$
Sky, Car, Field	1.0, 1.0
Field, Car, Road	1.0, 1.0
Sky, Road, Field	0.0, 1.0
Road, Car, Field	1.0, 1.0
all others	0.5, 1.0

c.) three regions

Table 3: Linear Basis Functions for the Synthetic Image.

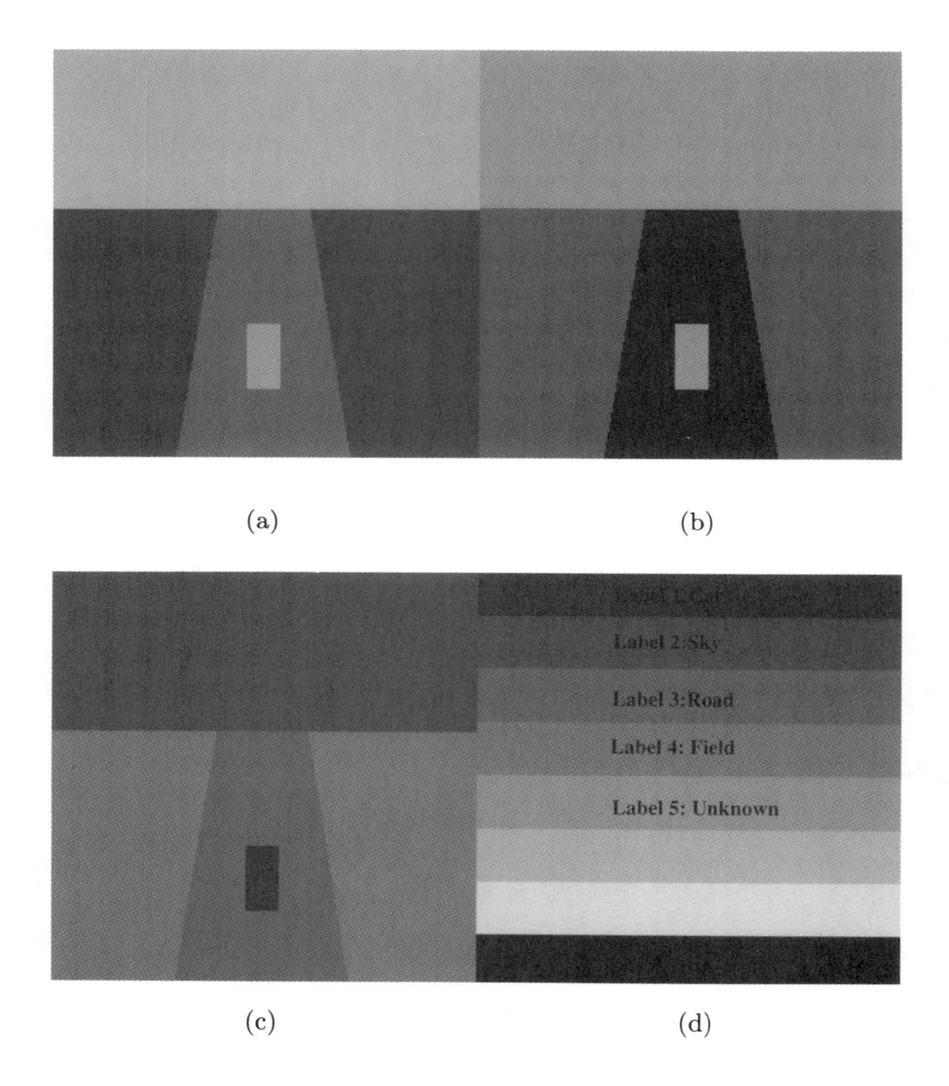

Figure 4: Interpretation of the ideal image. (a) The original image. (b) The segmented image. (c) The interpreted image. (d) Gray levels: Labels.

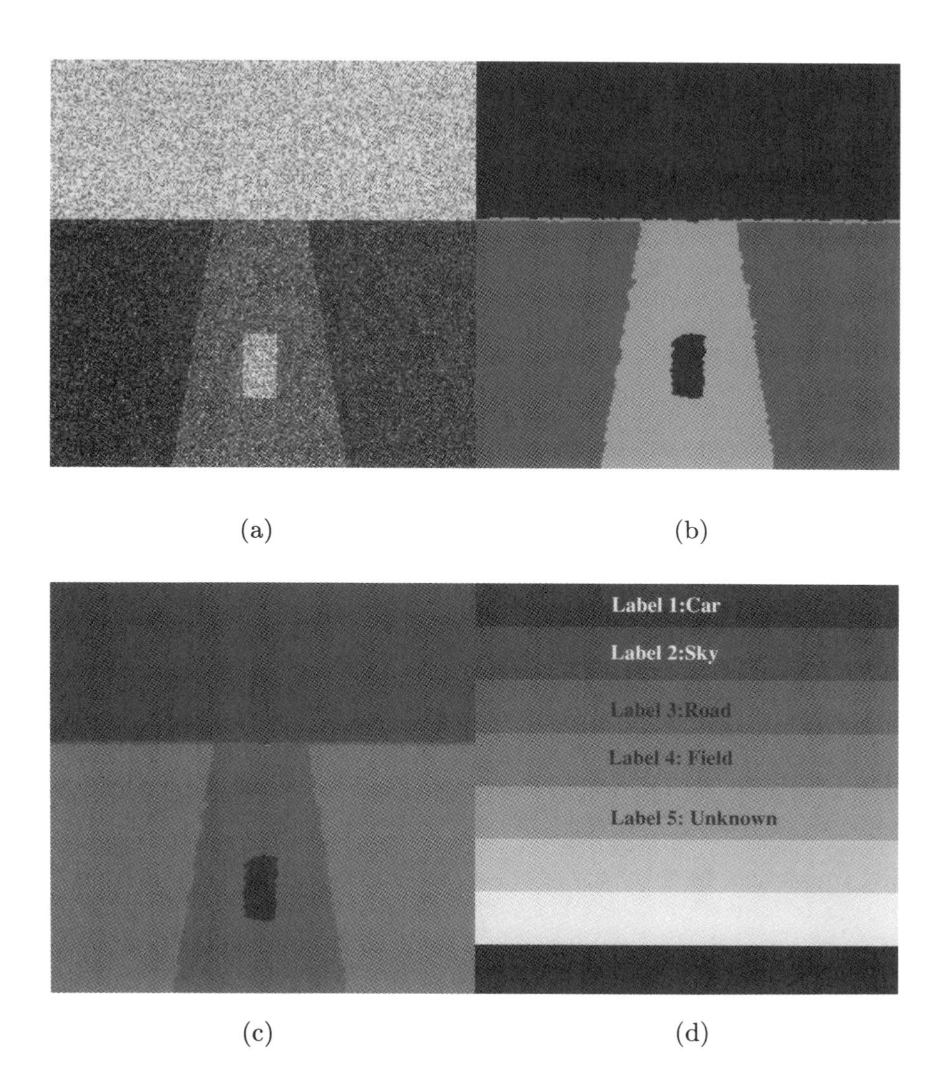

Figure 5: Interpretation of the 3dB SNR image. (a) The original image. (b) The segmented image. (c) The interpreted image. (d) Gray levels: Labels.

the resulting image has been reinterpreted [46]. Here, we have only shown the results for the ideal image and the image with 3dB SNR in Figs. 6 and 7, respectively. All the regions have been identified correctly and the unknown region is not identified as a car, since that will violate the spatial constraint in the knowledge base. This example, to some extent, shows the flexibility of the MRF approach.

6 Interpretation of Real-World Images

To test the practical applicability of the proposed MRF approach for image interpretation, experiments have been performed on real-world images; in particular, aerial photographs, which have been digitized to 256 gray level images of 256×256 pixels. Experimental results obtained here also provide further insights and useful guidelines on how to effectively apply this general approach in practice. Since we have no control over the generation process of the images involved, the task of interpreting real-world images is much more complicated and difficult than that of interpreting synthetic images. We have proceeded in two steps; namely, knowledge acquisition through constructing clique functions of the MRF model and interpretation using simulated annealing.

6.1 Knowledge Acquisition

This is the process of gathering information about the objects of interest in real-world images. This information is usually represented in terms of features and spatial constraints. For example, we might like to obtain information about cars, such as "a car has an average area of 800 pixels" and "a car is always on (neighboring to) a road." Here, the area is a feature, while the neighborhood relationship is a spatial constraint. Knowledge acquisition is also a *selection* process in which we select certain features and constraints from *all* the features and constraints we know about the objects to form a knowledge base. The selection process is necessary, since some of the features and constraints are not essential to interpretation, while they only add system complexity and computational burden. In the selection of the features and constraints, we want to select the ones that are most powerful for object recognition and discrimination. At this point, this selection is performed heuristically through trial-and-error. As a future goal, a general approach to solve this problem, such as the one described in Section 3.3, needs to be found.

There are many sources for knowledge acquisition. For aerial photointerpretation, one source of information/knowledge is from map information

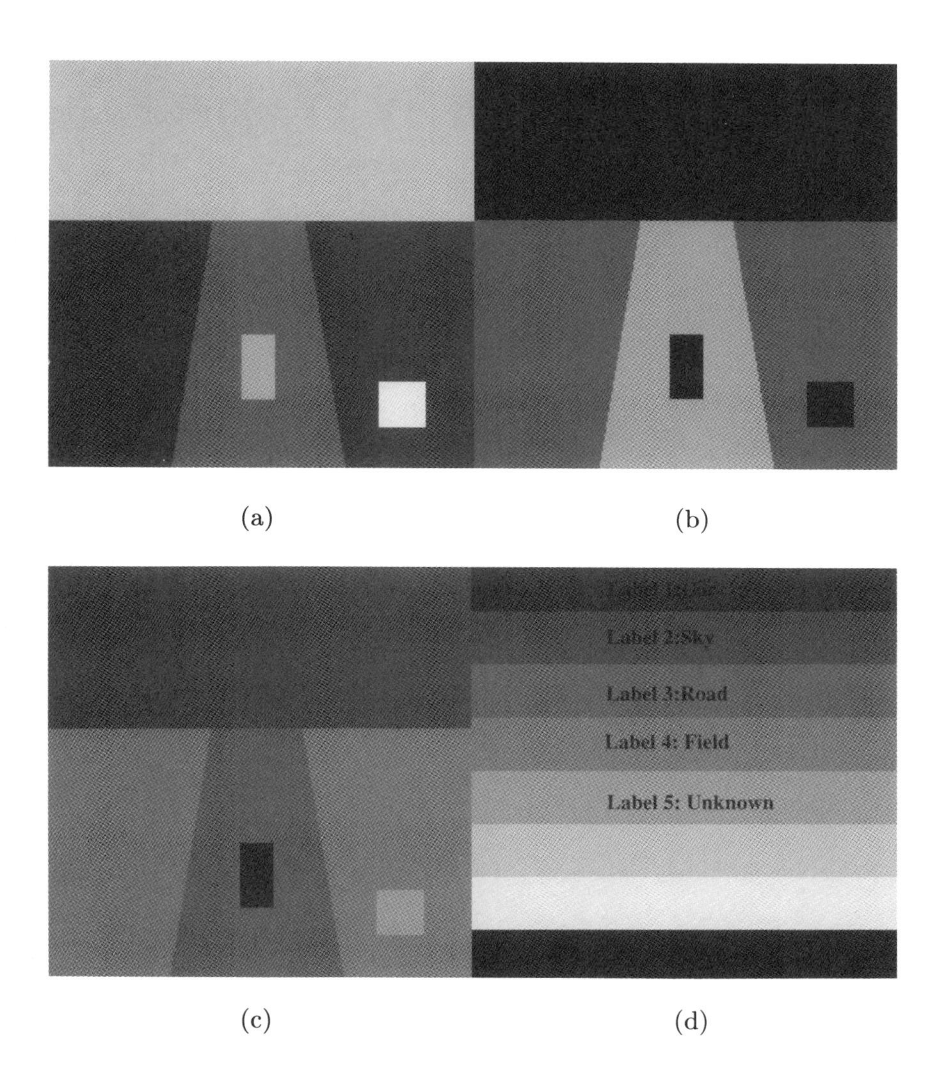

Figure 6: Interpretation of the ideal image with an unknown object. (a) The original image. (b) The segmented image. (c) The interpreted image. (d) Gray levels: Labels.

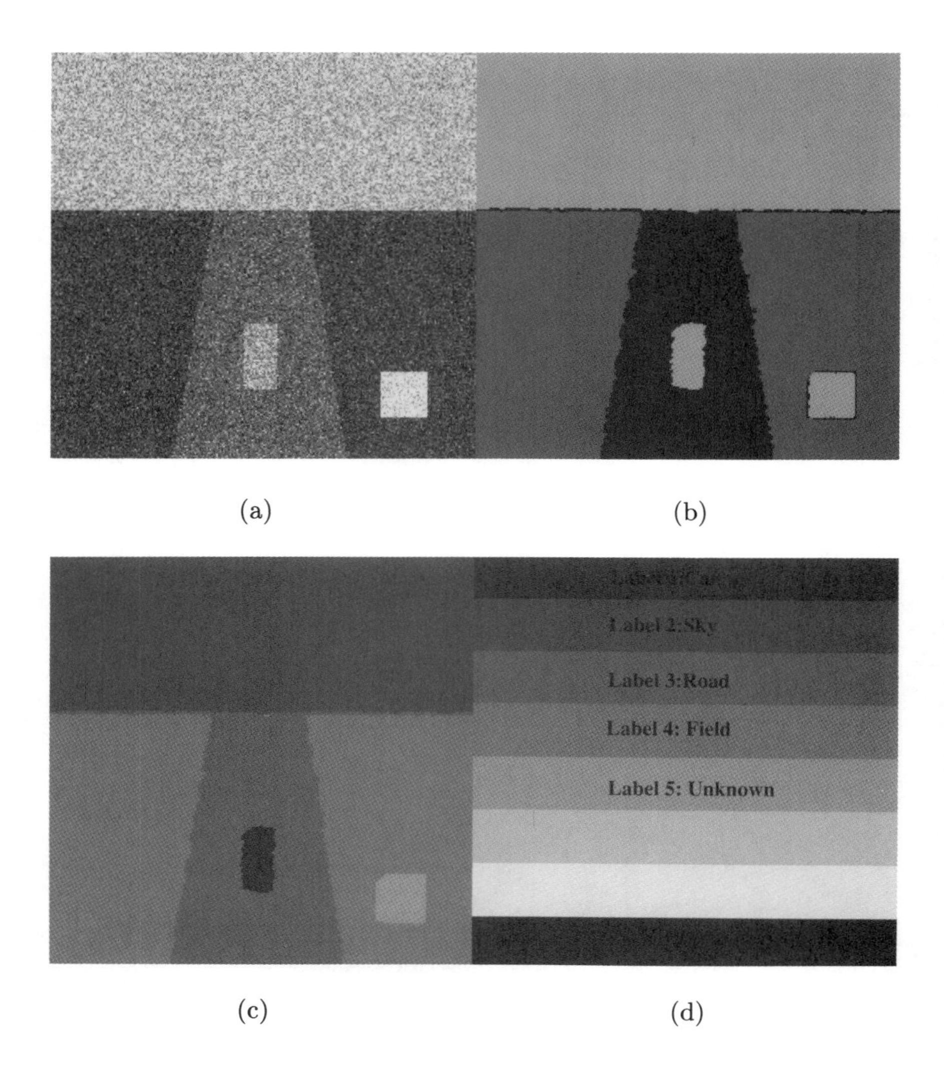

Figure 7: Interpretation of the 3dB SNR image with an unknown object. (a) The original image. (b) The segmented image. (c) The interpreted image. (d) Gray levels: Labels.

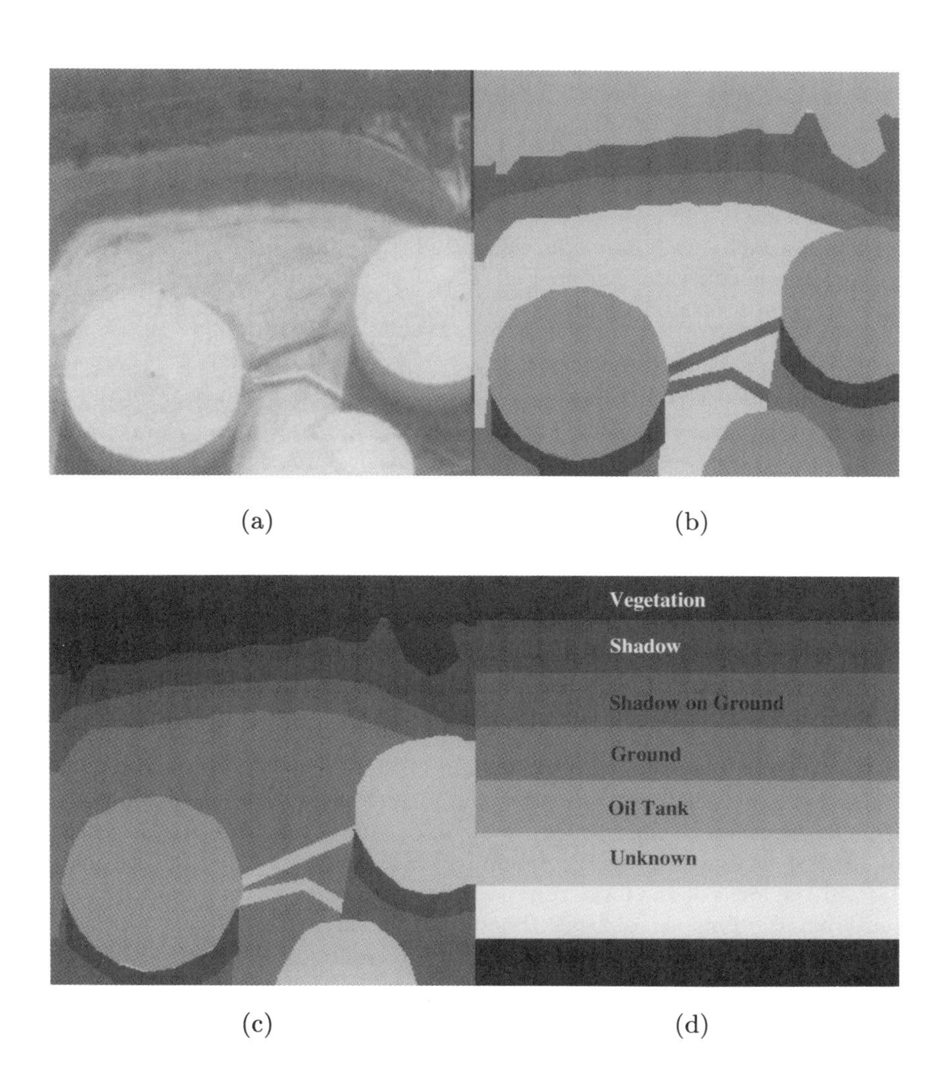

Figure 8: Interpretation of the training image. (a) The original image. (b) The segmented image. (c) The interpreted image. (d) Gray levels: Labels.

and characteristics of the man-made objects in the area being photographed [20]. This approach is used often in constructing practical photointerpretation systems and it usually involves a large amount of information. A less complicated alternative is the training approach. In this approach, a number of representative or training images are first segmented and interpreted by human experts; knowledge is then extracted from these segmented and interpreted images. The training approach is relatively simple, while not sacrificing generality in principle; hence, it is very useful in experimental studies such as this. In this work, we make exclusive use of the training approach.

For simplicity, only one training image is used in this experiment— an aerial photograph as shown in Fig. 8. After human expert segmentation and interpretation performed using an interactive display-segmentation facility at RPI [51], it has been found to contain mainly the following types of objects: (1)vegetation region (VEGE); (2)shadow (SHD1); (3)shadow on the ground (SHD2); (4)ground (GRND); and (5)oil tank (OLTK). The feature measurements selected for the objects include: area, average gray level, compactness of an object, and contrast between two regions. (See the definitions in Table 2.) The constraints selected for the objects include a number of neighboring relationships. For the clique functions, we have again used the piecewise-linear basis functions for the features and basis functions of Eq. (15) for spatial constraints. In particular, the corner points of the basis function for a given feature of a given object is determined from observing the *maximum* and *minimum* of the measurements of that feature on this type of object identified by the human interpreter. *Guard intervals* are introduced around the the maximum and minimum to determine the exact values of the four corner points. This, as well as the selection of the weights, has been performed heuristically, since there are relatively few objects of each type. When the number of objects of different types is large, the process of determining the corner points can be automated [50]. In Table 4, we have shown the domain knowledge learned from the training data in the form of the basis functions from which the clique functions are constructed. This set of basis functions and subsequent clique functions has been used to obtain all the experimental results to be described in this section.

6.2 Interpretation Using Simulated Annealing

In this experiment, interpretation has been performed on two test images. The first one is the training image itself. This is used to verify the correctness and effectiveness of the knowledge obtained in the form of clique functions from the training stage. Here, we have performed interpretation on the training image, using both manual segmentation and computer segmentation, as shown in Figs. 8 and 9, respectively. Since the computer segmentation provides comparable quality to that of the manual segmentation, the subsequent interpretation results are both quite good. Here, most of the regions are correctly identified, except that some of the oil tanks, which appear only partly in the image, are labeled as unknown objects. The reason for this is that in the knowledge base, oil tanks are characterized as circular objects, as reflected from the definition of the compactness in Table 2 and the corner points for the corresponding basis functions in Table 4. In other words, oil tanks that are only partly in the image were treated as unknown objects in the training stage and hence, it is not surprising that they have been interpreted as such in the interpretation stage. To recognize these *partial* oil tanks, more powerful shape features should be used, as described next.

The second test image, as shown in Fig. 10a, is *cut* from a larger image (2048×2048), from which the training image is obtained, and contains similar objects to those in the training image. This image is used to test the usefulness of the knowledge and clique functions obtained from the training image. Notice that in the original image, the gray tone and texture of several oil tanks are so close to those of their surroundings that they are very hard to extract even by human eyes; it is then not surprising that the computer segmentation of several regions corresponding to oil tanks is rather poor, as shown in Fig. 9c, especially in their shapes. As a result, in the interpretation results shown in Fig. 10c, most of the regions are correctly identified except for these oil tank regions. In fact, the compactness feature failed to be effective, since the regions corresponding to the oil tanks have very noisy boundaries and some of them are only partially in view. To solve this problem, Yu has proposed a more robust shape feature, called partial compactness, for the oil tanks [52]. This feature is based on the idea of obtaining a large number of estimates of the radius of a segmented region from random points on the boundary of the region, as illustrated in Fig. 11. Here, we have shown a set of three random points, A, B, and C, on the boundary of a segmented region. The vertical equal division lines of AB and BC intersect at O, resulting in a random estimate of the radius r. In this way, every set of three random boundary points provides a random estimate of radius. If a region is reasonably circular, or partially circular, the variance of the random estimates tends to be small. It has been shown by Yu [52] that this feature is quite powerful for recognizing both circular and partially circular objects, and is relatively robust to noisy boundaries. An interpretation of the second image is performed,

	Features			
	Area	Average Gray Level	Compactness	Partial Compactness
Label	a_1, a_2, b_1, b_2, p	a_1, a_2, b_1, b_2, p	a_1, a_2, b_1, b_2, p	a_1, a_2, b_1, b_2, p
Vegetation	11500, 12000, 13000, 13500, 0.5	120, 123, 127, 130, 0.5	–	–
Shadow 1	–	135, 140, 150, 155, 1.0	–	–
Shadow 2	0, 0, 6500, 7000, 0.5	105, 110, 140, 145, 0.5	–	–
Ground	–	185, 190, 200, 205, 1.0	–	–
Oil tank	4500, 4500, 64000, 64000	200, 205, 256, 256, 0, 33	0.0, 0.0, 0.95, 0.95, 1.0	0.0, 0.0, 10.0, 10.0, 1.0

a.) basis function for cliques of a single node

	Features	
	Contrast	Spatial Constraints
Labels	a_1, a_2, b_1, b_2, p	$\alpha_{c,p}$
Vegetation and Shadow 1	10, 15, 20, 30, 1.0	0.0, 1.0
Shadow 1 and Shadow 2	0.5, 30, 35, 1.0	0.0, 1.0
Shadow 1 and Ground	35, 40, 60, 65, 1.0	0.0, 1.0
Shadow 1 and Oil Tank	45, 50, 75, 80, 1.0	0.0, 1.0
Shadow 2 and Ground	55, 60, 90, 95, 1.0	0.0, 1.0
Shadow 2 and Oil Tank	75, 80, 105, 110, 1.0	0.0, 1.0
Ground and Oil Tank	0.5, 25, 30, 1.0	0.0, 1.0
Vegetation and Oil Tank	impossible combination	1.0, 1.0
Oil Tank and Oil Tank	impossible combination	1.0, 1.0
all other cases	–	0.5, 1.0

b.) basis function for features and spatial constraints for cliques of two nodes

Table 4: Knowledge and Clique Function for the Aerial Photos.

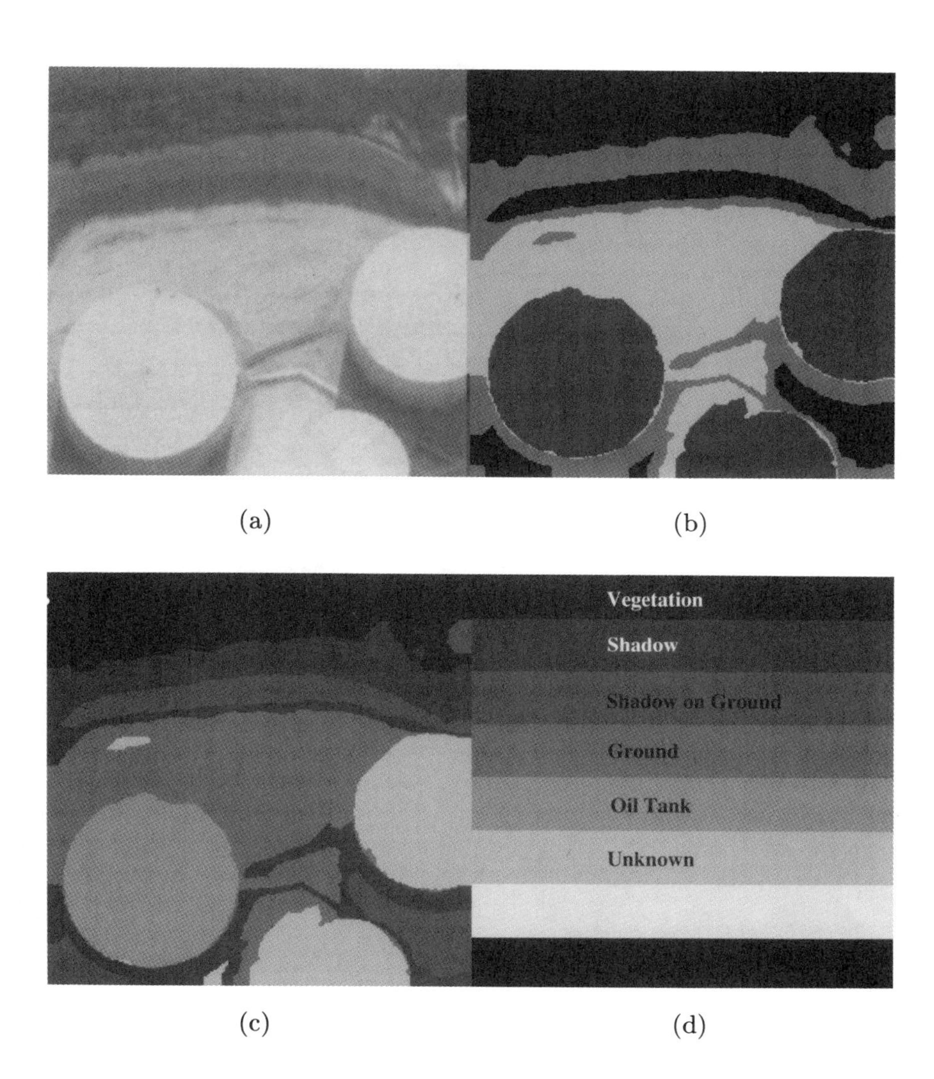

Figure 9: Interpretation of the training image with computer segmentation. (a) The original image. (b) The segmented image. (c) The interpreted image. (d) Gray levels: Labels.

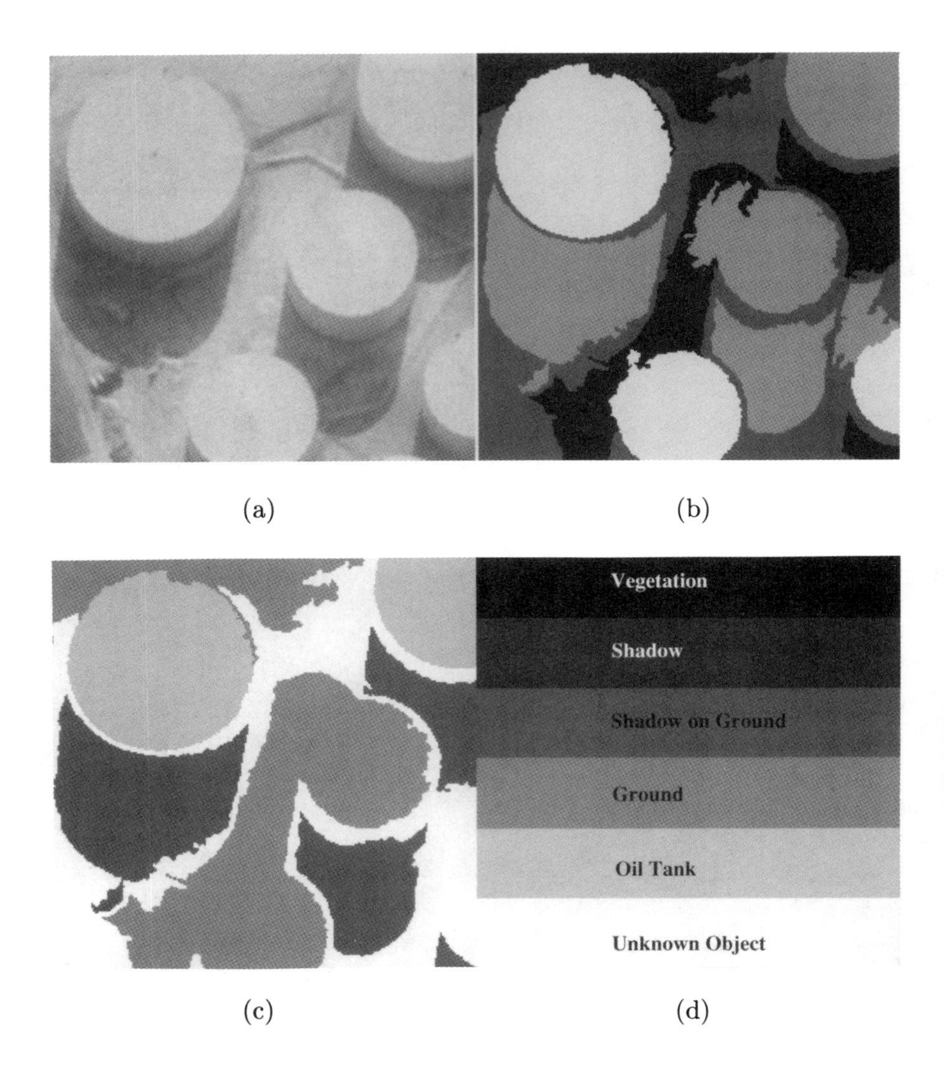

Figure 10: Interpretation of the test image with the compactness feature. (a) The original image. (b) The segmented image. (c) The interpreted image. (d) Gray levels: Labels.

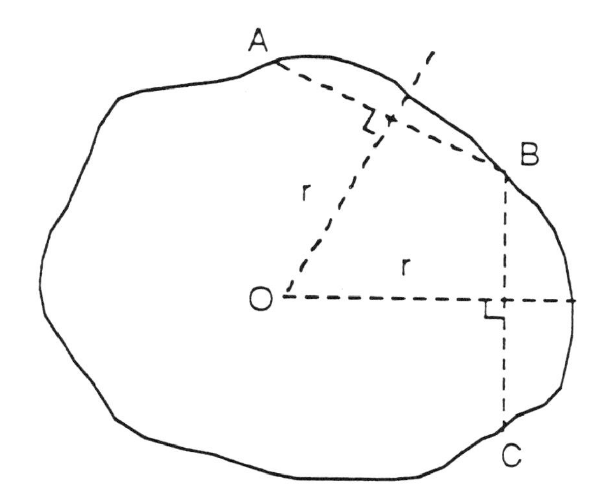

Figure 11: Illustration of the partial compactness feature.

again, using the knowledge- base and clique functions of Table 4, except with the compactness replaced by Yu's partial compactness (also shown in Tables 2 and 4). As can be seen from the interpretation results shown in Fig. 12c, improvements are obtained in that all the oil tanks are correctly identified. This justifies the points made in the experiments on interpreting synthetic images, that the MRF model-based formulation is quite powerful and feature selection is the crucial problem in applying it to real-world image interpretation.

6.3 Remarks

In concluding this section, we want to point out that the purpose of the set of simple experiments performed here is to understand some of the fundamental problems in knowledge-based image interpretation, such as knowledge representation, the selection of features and constraints, and the interaction between interpretations of different regions. While this should provide useful insights into how to build practical systems, the knowledge base and clique functions used here are not sufficient yet as a practical system. For example, the features used in these experiments are meant to be neither sufficient nor the best to use, as would have been done in build-

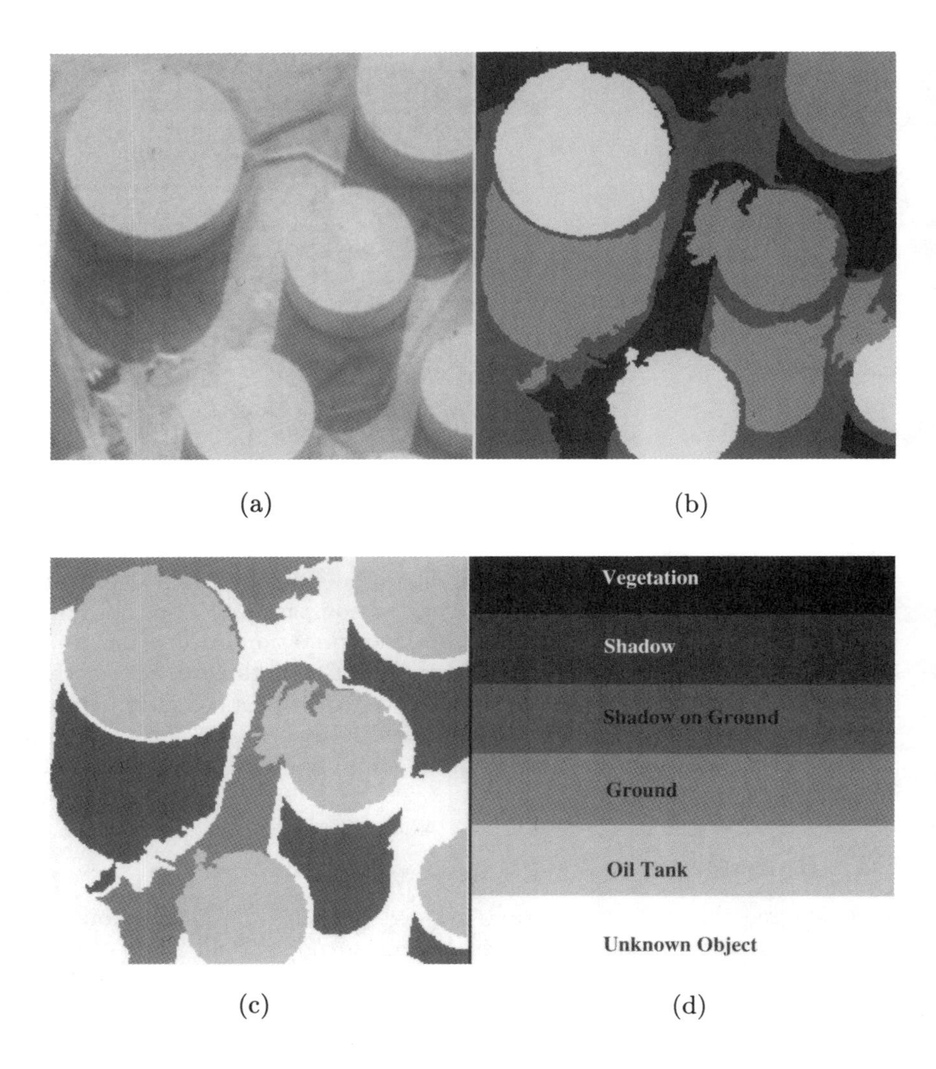

Figure 12: Interpretation of the test image with the partial compactness feature. (a) The original image. (b) The segmented image. (c) The interpreted image. (d) Gray levels: Labels.

ing practical systems; but rather, they are used here to demonstrate useful concepts and important issues related to the performance of knowledge-based image interpretation systems. Finally, a practical system should also incorporate the capability of recognizing complex objects composed of simpler ones; for example, recognizing a runway area from road-like regions containing a number of airplanes. It appears that the MRF model-based formulation can be used here to construct a hierarchical representation for objects of different complexity. In this representation, the regions corresponding to simple objects form a low-level MRF, while regions corresponding to complex objects that are collections of simple object regions form a high-level MRF. The optimal interpretation is the one that maximizes the pdf of this hierarchical MRF. In fact, a similar hierarchical MRF model has been used successfully for image segmentation using the pyramid image structures [53].

7 Summary and Conclusions

In this chapter, we have described an MRF model-based approach for automated image interpretation demonstrated as a region-based approach. In this approach, an image is first segmented into a collection of disjoint regions of certain homogeneous image properties. These regions, together with their spatial relationships, form an adjacency graph, and image interpretation is achieved through assigning object labels, or interpretations, to the regions using domain knowledge and measurements of features and spatial relationships extracted from them. In this approach, the interpretations are modeled as an MRF on the adjacency graph, and the image interpretation problem is formulated as that of finding the *best* realization of the MRF given the domain knowledge and measurements. Through the MRF model, this approach provides a systematic methodology for organizing and representing domain knowledge through properly designed clique functions associated with the pdf of the underlying MRF, which are to be designed in such a way that the optimal interpretation found is most consistent with domain knowledge and measurements. In particular, we have proposed a structure for the clique functions as a weighted sum of basis functions of features and spatial constraints. Finally, the simulated annealing algorithm is used to find the globally optimal interpretation.

To study the efficacy of the MRF model-based approach, image interpretation experiments have been performed on both synthetic images and real-world aerial photographs. In the experiments, we have found the piecewise-linear basis function provides robust performance for the interpretation in the presence of measurement errors caused by imperfect segmentation. We have also found that selecting powerful features and constraints is very crucial to real-world image interpretation; when such features and constraints are used, most of the objects in the image are correctly recognized.

Although the results here are still preliminary, they do suggest several promising directions for future research work. Specifically, future research should include the following three immediate research tasks. First of all, a more general approach is needed to determine the weights and corner points for the piecewise-linear basis functions used to construct the clique functions. Secondly, work should be done to incorporate other primitives, such as linear edge segments, and high-level to low-level feedback, such as the split-and-merge of original segmented regions during interpretation, into the MRF model-based approach, as described in Section 2. Some preliminary results for these two tasks have already been obtained [27], [28]. Finally, the MRF model-based approach needs to be tested on more diverse real-world images, such as additional aerial photographs.

Bibliography

[1] T. E. Avery, *Interpretation of Aerial Photographs*, Burgess Publishing Company, Minneapolis, Minnesota, 1977.

[2] S.A. Drury, *Image Interpretation in Geology*, Allen and Unwin, Ltd., London, UK, 1987.

[3] D.P. Paine, *Aerial Photography and Image Interpretation for Resource Management*, John Wiley and Sons, Inc., New York, 1981.

[4] A. Rosenfeld and A.C. Kak, *Digital Picture Processing*, Academic Press, New York, 1976.

[5] A. Rosenfeld and A.C. Kak, *Digital Picture Processing*, 2nd Ed., Academic Press, New York, 1982.

[6] B. Horn, *Robot Vision*, MIT Press, Cambridge, Massachusetts, 1986.

[7] *Proc. of the 6th Int. Conf. on Robot Vision and Sensor Control*, Paris, France, 1986.

[8] R.A. Schowengerdt, *Techniques for Image Processing and Classification in Remote Sensing*, Academic Press, New York, 1983.

[9] N.J. Nilsson, *Principles of Artificial Intelligence*, Tioga Publishing Company, Palo Alto, California, 1980.

[10] R.D. Keller, *Expert System Technology: Development and Application*, Lourdon Press, Englewood Cliffs, New Jersey, 1987.

[11] M. Nagao and T. Matsuyama, *A Structural Analysis of Complex Aerial Photographs*, Plenum Press, New York, 1980.

[12] T. Binford, "Survey of Model-Based Image Analysis Systems,", *The Int. J. of Robotics Research*, Vol. 1, No. 1, pp. 587–633, 1982.

[13] Y. Ohta, *Knowledge-Based Interpretation of Outdoor Natural Color Scenes*, Pitman Advanced Publishing Program, Boston, 1985.

[14] O.D. Faugeras and K.E. Price, "Semantic Description of Aerial Images Using Stochastic Relaxation," *IEEE Trans. Pattern Anal. Machine Intel.*, Vol. PAMI-3, pp. 633–642, Nov., 1981.

[15] K.E. Price, "Relaxation Matching Techniques: a Comparision," *IEEE Trans. Pattern Anal. Machine Intel.*, Vol. PAMI-7, pp. 617–623, Sept., 1985.

[16] V.S.S. Hwang, "Evidence Accumulation for Spatial Reasoning in Aerial Image Understanding," Ph.D. Thesis, Univ. of Maryland, College Park, 1984.

[17] V.S.S. Hwang, T. Matsuyama, L. Davis, and A. Rosenfeld, "Evidence Accumulation for Spatial Reasoning in Aerial Image Understanding," Report CS-TR-1300, Univ. of Maryland, College Park, 1983.

[18] R. Prasannappa, L. Davis, and V.S.S. Hwang, "A Knowledge-Based Vision System for Aerial Image Understanding," Report CS-TR-1758, Univ. of Maryland, College Park, Jan., 1987.

[19] R. Brooks, "Symbolic Reasoning Among 3-Dimensional Models and 2-Dimensional Images," *Artificial Intelligence*, Vol. 17, pp. 285–394, 1981.

[20] D.M. McKeown, W.A. Harvey, and J. McDermott, "Rule-Based Interpretation of Aerial Imagery," *IEEE Trans. Pattern Anal. Machine Intel.*, Vol. PAMI-7, pp. 570–585, Sept., 1985.

[21] D.M. McKeown and W.A. Harvey, "Automating Knowledge Acquisition for Aerial Image Interpretation," Technical Report CMU-CS-87-102, Carnegie-Mellon Univ., Pittsburgh, Philadelphia, Jan., 1987.

[22] E.M. Riseman and A.R. Hanson, "A Methodology for the Development of General Kowledge-Based Vision Systems," Technical Report COINS 86-27, Univ. of Massachusetts, Amherst, July, 1986.

[23] A.R. Hanson and E.M. Riseman, "From Image Measurements to Object Hypothesis," Technical Report COINS 87-129, Univ. of Massachusetts, Amherst, Dec., 1987.

[24] B. Draper, R. Collins, J. Brolio, A.R. Hanson, and E.M. Riseman, "The Schema System", Technical Report COINS 88-76, Univ. of Massachusetts, Amherst, Sept., 1988.

[25] E.M. Riseman and A.R. Hanson, "Summary of Image Understanding Research at the University of Massachusetts," Technical Report COINS 88-32, Univ. of Massachusetts, Amherst, May, 1988.

[26] "Technical Reports of the Computer Vision Laboratory, 1986–1988", Univ. of Maryland, College Park, 1988.

[27] J. Zhang, "Merging of Segmented Regions Through Simulated Annealing," RPI Technical Report, Troy, New York, in preparation.

[28] J. Zhang, "Consistent Combination of Local Interpretation for Image Analysis," RPI Technical Memo, Troy, New York, May, 1988.

[29] S. Geman and D. Geman, "Stochastic Relaxation, Gibbs Distribution and the Bayesian Restoration of Images," *IEEE Trans. Pattern Anal. Machine Intel.*, Vol. PAMI-6, pp. 721–741, Nov., 1984.

[30] C.W. Therrien, T.F. Quatieri, and D.E. Dudgeon, "Statistical Model-Based Algorithms for Image Analysis," *Proc. IEEE*, Vol. 74, April, 1986.

[31] J. Zhang and J.W. Modestino, "Markov Random Fields with Applications to Texture Classification and Discrimination," *Proc. The 20th Annual Conf. on Information Science and Systems*, Princeton University, Princeton, New Jersey, March, 1986.

[32] H. Derin and H. Elliot, "Modelling and Segmentation of Noisy and Textured Images Using Gibbs Random Fields," *IEEE Trans. Pattern Anal. Machine Intel.*, Vol. PAMI-9, pp. 39–55, Jan., 1987.

[33] F.S. Cohen and D.B. Cooper, "Simple, Parallel, Hierachical and Relaxation Algorithms for Segmenting Non-Causal Markovian Random Field Models," *Proc. IEEE Pattern Anal. Machine Intel.*, Vol. PAMI-9, pp. 195–219, March, 1987.

[34] J. Besag, "On the Statistical Analysis of Dirty Pictures," *J. Royal Stat. Soc. B.*, Vol. 48, pp. 259–302, 1986.

[35] J. Zhang and J.W. Modestino, "Unsupervised Image Segmentation Using a Gaussian Model," to be submitted to *IEEE Trans. PAMI.*

[36] J. Marroquin, S. Mitter, and T. Poggio, "Computer Vision," *J. Amer. Stat. Association*, Vol. 82, pp. 76–89, March, 1987.

[37] P.B. Chou and C.M. Brown, "Multi-Modal Segmentation Using Markov Random Fields," *Proc. IJCAI*, pp. 663–670, 1987.

[38] R. Kinderman and J.L. Snell, *Markov Random Fields and Their Applications*, RI: Amer. Math. Soc., Providence, Rhode Island, 1980.

[39] J. Besag, "Spatial Interaction and the Statistical Analysis of Lattice Systems," *J. Roy. Statist. Soc.*, Series B., Vol. 36, pp. 192–226., 1974.

[40] J.A. Feldman and Y. Yakimovsky, "Decision Theory and Artificial Intelligence: I. A Semantic-Based Region Analyzer," *Artificial Intelligence*, Vol. 5, pp. 325–348, 1974.

[41] J.W. Modestino, "A Hierarchical Region-Based Approach to Automated Photointerpretation," RPI report, Troy, New York, March 1987.

[42] L. Zadeh, "Approximate Reasoning Based on Fuzzy Logic," *Proc. 6th IJCAI*, pp.1004–1010, 1979.

[43] J.T. Tou and R.C. Gonzalez, *Pattern Recognition Principles*, Addison-Wesley Publishing Company, Reading, Massachusetts, 1974.

[44] S. Kirkpatrick, C.S. Gelatt, and M. P. Vecchi, "Optimization by Simulated Annealing," *Science*, Vol. 220, pp. 671–680, May, 1983.

[45] A.A. El Gamal, L.A. Hemachandra, I. Shperling, and V.K. Wei, "Using Simulated Annealing to Design Good Codes," *IEEE Trans. Infor. Theory*, Vol. IT-33, pp. 116–123, Jan., 1987.

[46] J. Zhang, "Two-Dimensional Stochastic Model-Based Image Analysis," Ph.D. Thesis, Rensselaer Polytechnic Institute, Troy, New York, Aug., 1988.

[47] C.S. Won and H. Derin, "Segmentation of Noisy Textured Images Using Simulated Annealing," *Proc. ICASSP*, pp. 563–566, 1987.

[48] B. Gidas, "Nonstationary Markov Chains and Convergence of the Annealing Algorithm," *J. Statist. Phys.*, Vol. 39, pp. 73–131, 1985.

[49] J. Zhang and J.W. Modestino, "Image Segmentation Using a Gaussian Model," *Proc. Conf. on Info. Sci. and System*, Princeton University, Princeton, New Jersey, March, 1988.

[50] J. Zhang, "Learning in the MRF Model-Based Approach for Image Interpretation," in preparation.

[51] J. Kanai, "Interpretation of Real Images Using the MRF Model-Based Method," RPI Report, Troy, New York, September, 1987.

[52] L.Y. Yu, "New Results on Image Segmentation and Interpretation Using the MRF Model," RPI Report, Troy, New York, June, 1988.

[53] C. Bouman and B. Liu, "Segmentation of Textured Images Using A Multiple Resolution Approach," *Proc. ICASSP*, New York, pp. 1124–1127, March, 1988.

[54] A. Rosenfeld, R.A. Hummel, and S.W. Zucker, "Scene Labeling by Relaxation Operations," *IEEE Trans. Syst., Man, Cybern.*, Vol. SMC-6, pp. 420–433, 1976.

[55] R.M. Haralick and L.G. Shapiro, "The Consistent Labeling Problem: Part I," *IEEE Trans. Pattern Anal. Machine Intel.*, Vol. PAMI-1, pp. 173–184, April, 1979.

[56] R.M. Haralick and L.G. Shapiro, "The Consistent Labeling Problem: Part II," *IEEE Trans. Pattern Anal. Machine Intel.*, Vol. PAMI-2, pp. 193–203, May 1980.

[57] R.A. Hummel and S.W. Zucker, "On the Foundation of Relaxation Labeling Process," *IEEE Trans. Pattern Anal. Machine Intel.*, Vol. PAMI-5, pp. 267–287, May, 1983.

A Markov Random Field Restoration of Image Sequences

T.J. Hainsworth and K.V. Mardia
Department of Statistics,
The University of Leeds,
Leeds LS2 9JT, U.K.

1 Introduction

There has been much interest in the problem of analyzing and restoring degraded images of some true scene. Many methods have been proposed for the segmentation of a single 2-dimensional image: those utilizing local spatial/contextual dependencies between the memberships of neighboring pixels [e.g. Besag, 1986; Geman and Geman, 1984; Kent and Mardia, 1987, 1988; Mardia and Hainsworth, 1988] generally produce better results than non-spatial methods [e.g. Ridler and Calvard, 1978].

There is now growing interest in the analysis of multiple images of a scene. Here we must take into account not only the spatial dependencies between neighboring pixels in the same frame, but also the temporal relationship between a pixel in one frame and its "parent" from a previous frame. Thus, the segmentation should reflect the spatial characteristics of the current frame moderated by the requirement for consistency of the current segmentation with those in previous frames. Our aim here is to obtain the most consistent segmentation for the sequence.

We shall consider the extension of some single-frame segmentation methods to the multiple-frame case, where the image sequences are of the following two types:

(i) "slowly-varying" sequences;
(ii) "quickly-varying" sequences.

By slowly varying sequences we mean that there is relatively little object movement and only a relatively small number of pixels change class, whereas by quickly varying sequences we mean that the object is subjected

$$
\begin{array}{ccccccc}
\circ & \times & \circ & \times & \circ & \times & \circ \\
\times & & \times & & \times & & \times \\
\circ & \times & \circ & \times & \circ & \times & \circ \\
\times & & \times & & \times & & \times \\
\circ & \times & \circ & \times & \circ & \times & \circ
\end{array}
$$

Figure 1: Configuration of pixel sites ($\circ$) and associated edge sites ($\times$).

to significant translation, rotation and scale change so that there could be a large number of pixels changing class.

In the second case, a stage of image *registration* will be required in addition to the segmentation stage, since there will be significant phase differences between the frames in sequences of quickly varying images.

We shall first consider the case of slowly varying image sequences. We follow a Bayesian approach for the recursive reconstruction of image sequences along the lines of Green and Titterington (1987), where the priors take the form of Gibb's distributions. In spirit, this work is an extension of Geman and Geman (1984) and Besag (1986) for image sequences; *cf* Wright (1989) who uses a similar approach for the "fusion" of multispectral images. Section 2 introduces our notation. In Sections 3 and 4, we consider the use of particular energy/potential functions in the Gibb's distribution. Section 3 describes a stochastic relaxation approach to image reconstruction, as in Geman and Geman (1984). Section 4 provides details of an Iterated Conditional Modes (ICM) approach in the context of the segmentation of a sequence of k images (k-ICM).

In Section 5, we describe a maximum likelihood approach for *registration* of fast-varying sequences of images under an assumption of rigid body motion.

In Section 6, we briefly consider extensions of some other related single-frame segmentation methods to the multiple frame case. In Sections 7 and 8, we consider the application of the methods to real and synthetic examples.

2 Extension of Geman & Geman Approach

2.1 Notation

Let $\mathbf{X}_t^L \equiv (X_{ti}^L)$ denote the true image at time t, where X_{ti}^L is the membership (label/color) of pixel i, $i = 1, \ldots, N$, lying on a rectangular lattice, at time t. Suppose that between each pair of adjacent pixels there exists a notional *edge site* ; see Figure 1. Let $\mathbf{X}_t^E \equiv (X_{tij}^E)$ be the associated edge map, where X_{tij}^E is the state of the edge site between pixels i and j at time t. We shall consider a binary edge process, for which $X_{tij}^E = 1$ if there is an edge between pixels i and j in frame t and $X_{tij}^E = 0$ otherwise. The purpose of the edge sites is to allow for discontinuities in the *pixel image* which must surely arise at inter-population boundaries. We shall write $\mathbf{X}_t \equiv (\mathbf{X}_t^L, \mathbf{X}_t^E)$ to denote the composite pixel/edge true scene.

Let $\mathbf{Y}_t \equiv (Y_{ti})$, be the observed image at time t where Y_{ti} is the observed gray-level (signal) at pixel site i, $i = 1, \ldots, N$ at time t. The state of the edge sites is unobserved, but we aim to make inferences about the edge map from the neighboring observations on the pixel lattice.

Let $\mathbf{X}_{\leq t} \equiv \{\mathbf{X}_s : s \leq t\}$ be the set of true scenes up to and including time t. Similarly, we define $\mathbf{X}_{<t} \equiv \{\mathbf{X}_s : s < t\}$. Let $\mathbf{X}_{\leq t}^{(k)} \equiv \{\mathbf{X}_s : t-k < s \leq t\}$ and $\mathbf{X}_{<t}^{(k)} \equiv \{\mathbf{X}_s : t-k < s < t\}$. We shall use the corresponding notation for $\{\mathbf{X}_t^L\}$, $\{\mathbf{X}_t^E\}$ and $\{\mathbf{Y}_t\}$.

We shall also require notation $\mathbf{X}_{t\backslash i}^L \equiv \{\mathbf{X}_{tj}^L : j = 1, \ldots, N ; j \neq i\}$ to denote the pixel image at the time t excluding pixel i. Similarly, let $\mathbf{X}_{t\backslash ij}^E \equiv \{X_{trs}^E : (r, s) \neq (i, j)\}$ denote the edge map at time t excluding the edge site between pixels i and j.

We aim to estimate $\mathbf{X}_t = (\mathbf{X}_t^L, \mathbf{X}_t^E)$ given provisional estimates of $\mathbf{X}_{<t} \equiv (\mathbf{X}_{<t}^L, \mathbf{X}_{<t}^E)$ and observations $\mathbf{Y}_{<t}$. Our recursive formulation follows Green and Titterington (1987). The MAP (maximum a posteriori) solution is that $\mathbf{X}_t$ which maximizes the posterior distribution,

$$P(\mathbf{X}_t | \mathbf{X}_{<t}, \mathbf{Y}_{\leq t}) \propto P(\mathbf{Y}_{\leq t}, \mathbf{X}_{\leq t}) \, P(X_{\leq t}) \quad .$$

It is computationally infeasible to obtain the MAP solution directly. We shall consider some locally based solutions based on stochastic optimization techniques under some simplifying assumptions.

Assumptions

We first provide a set of assumptions which are designed to reflect our prior beliefs on the nature of the evolution of the image sequence. We make the following *temporal* assumptions:

$$P(\mathbf{X}_t \mid \mathbf{X}_{<t}) \equiv P(\mathbf{X}_t \mid \mathbf{X}_{<t}^{(k)}) \tag{2.1}$$

$$\text{and} \quad P(\mathbf{Y}_t \mid \mathbf{X}_{\leq t}, \mathbf{Y}_{<t}) \equiv P(\mathbf{Y}_t \mid \mathbf{X}_t^L) \quad . \tag{2.2}$$

Assumption (2.1) is a Markov assumption about the sequence of true scenes, whereas (2.2) assumes that the observed image depends only on the labels of the pixels for that frame.

We shall also restrict (2.2) further by assuming conditional (spatial) independence between the observations in a given frame,

$$P(\mathbf{Y}_t \mid \mathbf{X}_t^L) = \prod_{i=1}^{N} P(Y_{ti} \mid X_{ti}^L) \quad . \tag{2.3}$$

We shall make a further Markov assumption regarding the nature of the true scene *within* a frame. Let $\delta i(t)$ denote the neighborhood of pixel i at time t, with

$$\delta i(\leq t; k) \equiv \{\delta i(s) : t - k < s \leq t\} \quad .$$

For example, for the $k = 2$ frame situation we could choose to use a second order neighborhood in "3D" space (viewing the "stacked" sequence of 2D frames as a 3D image) as shown in Figure 2. We shall make use of this neighborhood scheme in Section 4, where we consider a k-frame ICM approach for the reconstruction of the pixel sites alone. A simpler neighborhood structure is used in Section 3 for the simultaneous reconstruction of pixel and edge classes using stochastic relaxation.

We make the *spatio-temporal* assumption that

$$P(X_{ti}^L \mid \mathbf{X}_{\leq t} \setminus X_{ti}^L) \equiv P(X_{ti}^L \mid \mathbf{X}_{\delta i(\leq t;k)}) \quad , \tag{2.4}$$

where $\mathbf{X}_{\delta i(\leq t;k)}$ are the neighboring pixel/edge sites of pixel i in the k frames up to and including frame t. We shall see that the spatio-temporal Markov assumption (2.4) follows when we use a Gibbsian prior for $\mathbf{X}_{\leq t}$. Under assumptions (2.1)-(2.4), it can be shown that

$$P(X_{ti}^L \mid \mathbf{Y}_{\leq t}, \mathbf{X}_{\leq t} \setminus X_{ti}^L) \propto P(X_{ti}^L \mid \mathbf{X}_{\delta i(\leq t;k)})\, P(Y_{ti} \mid X_{ti}^L) \quad . \tag{2.5}$$

This suggests the possibility of pixel-image reconstruction using local updates based on (2.5). An edge-update rule is obtained by making a corresponding assumption to (2.4) for the conditional prior of X_{tij}^E given $X_{\leq t} \setminus X_{tij}^E$.

```
            o                        o   ×   o   ×   o
            ×                        ×       ×       ×
o     ×     o_i    ×     o           o   ×   o_i ×   o
            ×                        ×       ×       ×
            o                        o   ×   o   ×   o

        δi(t-1)                            δi(t)
```

Figure 2: A possible neighborhood configuration $\delta i(\leq t; 2)$ for a sequence of two frames. ($\circ$ $\equiv$ pixel sites, $\times$ $\equiv$ edge sites).

2.2 Parameter Estimation

We shall seek to estimate $\mathbf{X}_t$ given a provisional estimate of $\mathbf{X}_{<t}$ and suitable estimates of population noise parameters. In the following derivations we assume known population parameters. However, in practice these parameters are estimated by maximum likelihood conditional on the current provisional estimate of the membership X_t^L.

3 Map Estimation by Simulated Annealing

We shall now consider a particular Gibbsian model formulation extending Geman and Geman (1984) along the lines of Wright (1989). Note that in our model we assume that the true scene consists of p (known) distinct classes, but our main interest is in the special case of $p = 2$ classes, as appropriate for the problem of object segmentation.

Under the simplifying assumption (2.2), the conditional posterior takes the form

$$P(\mathbf{X}_t \mid \mathbf{X}_{<t},\ \mathbf{Y}_{\leq t}) \ \propto \ P(\mathbf{Y}_t|\mathbf{X}_t^L)\ P(\mathbf{X}_{\leq t}) \quad . \tag{3.1}$$

The MAP solution is that value of $\mathbf{X}_t$ which maximizes the posterior probability (3.1). Following Geman and Geman (1984) we shall seek an estimate for the MAP solution based on stochastic relaxation.

As usual we shall assume that the signals $Y_{si},\ s \leq t,\ i = 1, \ldots, N$ are conditionally independent Gaussian random variables given the membership: $Y_{ti} \sim IN[\mu(X_{ti}),\ \sigma^2]$; *i.e.*,

$$P(\mathbf{Y}_t \mid \mathbf{X}_t) = (2\pi\sigma^2)^{-N/2}\, exp\{-\frac{1}{2\sigma^2} \sum_{i=1}^{\mathbf{N}} [Y_{ti} - \mu(X_{ti})]^2\} \quad .$$

We shall consider only the case of fixed population parameters, which is reasonable in the context of our real application. However, the model is easily extended to deal with parameters which vary with time. Further, since our reconstructions seek to integrate over the labels from different frames, the method proposed here is also suitable for application to multi-sensed data or images in which the noise parameters / characteristics vary with time.

It is known that the underlying scene is modeled as a Markov Random Field (MRF) if and only if the prior probability takes the form of Gibb's distribution. We shall use a Gibbsian prior of the form,

$$P(\mathbf{X}_{\leq t}) = exp\{-U(X_{\leq t})\} \;/\; Z \quad , \tag{3.2}$$

where Z is an appropriate normalizing constant and $U(X_{\leq t})$ is an appropriate energy (cost) function for the true image sequence computed as the sum of certain potential functions. The penalty function $U(X_{\leq t})$ should reflect our prior beliefs on the local structure of the true scene. In particular, we want $U(\mathbf{X}_{\leq t}) \equiv U(\mathbf{X}^{(k)}_{\leq t})$.

We shall use an energy function of the form,

$$\begin{aligned} U(\mathbf{X}_{\leq t}) = {} & \beta_1\, Cost_1\,(\mathbf{X}^L_t|\mathbf{X}^E_t) + \beta_2\, Cost_2\,(\mathbf{X}^E_t) + \\ & \beta_3\, Cost_3\,(\mathbf{X}^L_t|\mathbf{X}^L_{<t}) + \beta_4\, Cost_4\,(\mathbf{X}^E_t|\mathbf{X}^E_{<t}), \end{aligned} \tag{3.3}$$

where $Cost_i(\ )$, $i = 1, \ldots, 4$ are appropriate cost functions realized as the sum of certain image potentials and $\beta_i \geq 0$, $i = 1, \ldots, 4$ are appropriate weighting parameters. Note that $Cost_1(.)$ and $Cost_2(.)$ are applied to the current frame t, with $Cost_3(.)$ and $Cost_4(.)$ used to regulate consistency of the current frame with previous frames. We have assumed that the current pixel classes depend on the past only through the past pixel classes and that the current edges depend on the past only through the past edge states. However, a more general form of (3.3) could include additional terms to model the dependence of $\mathbf{X}^L_t$ on $\mathbf{X}^E_{<t}$ and of $\mathbf{X}^E_t$ on $\mathbf{X}^L_{<t}$.

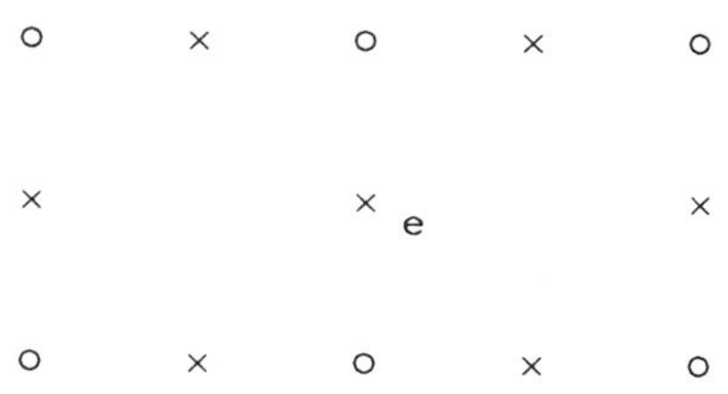

∘	×	∘	×	∘
×		× e		×
∘	×	∘	×	∘

Figure 3: Neighborhood system for edge sites: the central edge site (e) has 6 neighboring edges (∘ pixels, × edges)

For within-frame contributions, we shall use the usual forms of cost function:

$$Cost_1(\mathbf{X}_t^L \mid \mathbf{X}_t^E) = \sum_{i \sim j} (1 - X_{tij}^E)[1 - 2\delta(X_{ti}^L - X_{tj}^L)] \ , \quad (3.4)$$

$$Cost_2(\mathbf{X}_t^E) = \sum_{c \epsilon \mathcal{C}_t} V(c) \ , \quad (3.5)$$

where $\delta(x) = 1$ if $x = 0$; $= 0$ otherwise. The summation in (3.4) is over all neighboring pairs of pixels $i \sim j$ in frame t, the summation in (3.5) is over all cliques of edge sites c which lie in the set $\mathcal{C}_t$ of all such cliques for frame t, and $V(c)$ is the potential associated with clique c. A clique is a set of mutual neighbors; we use the neighborhood system in which each edge site has 6 nearest neighbors as shown in Figure 3. The potentials $V(c)$ are low for "likely" configurations and "high" for "unlikely" configurations. We shall use the potentials (Wright, 1989) for cliques of only size 4 given in Figure 4; these were found to work better than the potentials ($V_0 = 0,\ V_1 = 2.7,\ V_{2c} = 1.8,\ V_{2s} = 0.9,\ V_3 = 1.8,\ V_4 = 2.7$) of Geman and Geman (1984) for images containing edges at a variety of orientations. [See Silverman *et al*, 1990, for a consideration of the problem of appropriate specification of edge potentials.]

The effect of $Cost_1(.)$ is to penalize against having neighboring pixel sites with different colors (labels) if there is no intermediate edge: if there

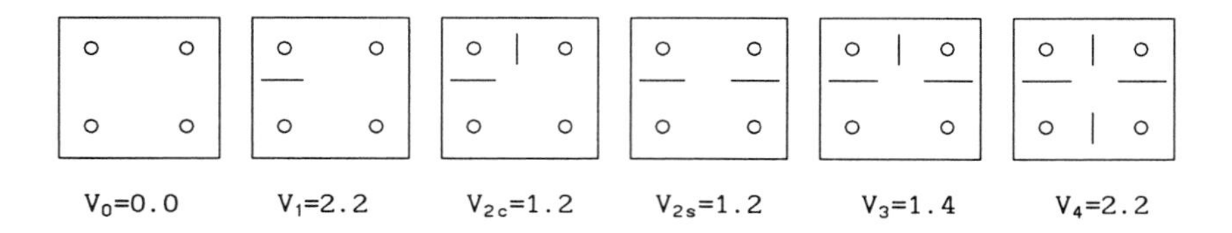

Figure 4: Potential energies for all possible edge clique configurations up to reflections and rotations.

is an edge between the pixels, then we allow a difference in color (discontinuity) to exist between the pixels at no added cost. $Cost_2(.)$ supports the use of the most likely edge site configurations, i.e. those with lowest potential. Note that if we were to choose the edge classes to minimize $Cost_2(.)$ alone, then the image would contain no edges. In practice, however, the balance of conflicting costs leads to more sensible segmentations.

We shall consider the use of the following cost functions to promote consistency between the segmentations of the image sequence:

$$Cost_3(\mathbf{X}_t^L \mid \mathbf{X}_{<t}^L) = \sum_{i=1}^{N} \gamma(X_{ti}^L;\ X_{<t,i}^L) \tag{3.6}$$

and

$$Cost_4(\mathbf{X}_t^E \mid \mathbf{X}_{<t}^E) = \sum_{i \sim j} \gamma(X_{tij}^E;\ X_{<t,ij}^E) \tag{3.7}$$

where

$$\begin{aligned} \gamma(x_{ti}; x_{<t,i}) &= -1 \text{ if } x_{si} = x_{ti} \text{ for all } s < t, \\ &= 1, \text{ otherwise.} \end{aligned}$$

Note that a cost function of the form in (3.7) for promoting temporal edge consistency has been used by Wright (1989) and that we have taken a similar form for the temporal consistency of pixel classes; see also Murray and Buxton (1987).

The MAP solution is thus given as that $\mathbf{X}_t$ which maximizes the posterior,

$$
\begin{aligned}
P(\mathbf{X}_t \mid \mathbf{X}_{<t}, \mathbf{Y}_{\leq t}) \;\propto\; & exp\{-\frac{1}{2}\Sigma[Y_{ti} - \mu(X_{ti}^L)]^2/\sigma^2 \\
& - \beta_1\, Cost_1(\mathbf{X}_t^L \mid \mathbf{X}_t^E) - \beta_2\, Cost_2(\mathbf{X}_t^E) \\
& - \beta_3\, Cost_3(\mathbf{X}_t^L \mid \mathbf{X}_{<t}^L) - \beta_4\, Cost_4(\mathbf{X}_t^E \mid \mathbf{X}_{<t}^E)\}\,.
\end{aligned} \tag{3.8}
$$

Reconstruction By Simulated Annealing for k Frames $(k - SA)$

It is computationally infeasible to compute the MAP solution by direct evaluation of (3.8) for all possible configurations $\mathbf{X}_t$. Instead, we shall use stochastic relaxation to obtain an approximation to the MAP solution, such as simulated annealing as popularized by Kirkpatrick et al (1983) and the Gibbs sampler of Geman and Geman(1984). We shall first describe our implementation of the Gibbs sampler, and then indicate the Metropolis approach.

Our procedure for updating $\hat{\mathbf{X}}_t$ has two phases: in the first phase we update all pixel sites and in the second phase we update all edge sites. In both phases, we perform a local update, i.e. we update the class/state of a single pixel/edge site given the provisional states of the rest. This local updating uses stochastic relaxation which allows the posterior probability to decrease as well as increase, so that it is possible to escape from local maxima and hopefully approach the global maximum. In practice, the solution obtained is only an approximation to the global maximum.

Pixel Update

From the nature of the prior model (3.2) - (3.7), we have that

$$P_T(X_{ti}^L \mid \text{everything else}) \;\propto\; exp\{-U_{ti}^L/T\} \tag{3.9}$$

where

$$
\begin{aligned}
U_{ti}^L \;=\; & \frac{1}{2}\{Y_{ti} - \mu(X_{ti}^L)\}^2/\sigma^2 + \beta_1 \sum_{j\epsilon\delta i(\leq t;k)} (1 - X_{tij}^E)[1 - 2\delta(X_{ti}^L - X_{tj}^L)] \\
& + \beta_3\, \gamma(X_{ti}^L; X_{<t,i}^L).
\end{aligned} \tag{3.10}
$$

T is the *temperature parameter* which must be varied according to an appropriate annealing schedule; we shall consider this in more detail later.

For a given temperature, T, we choose a new label X_{ti}^L by sampling from (3.9). For example, in the binary pixel class case, we choose $X_{ti}^L = 0$ with probability p_0 and $X_{ti}^L = 1$ with probability $p_1 = 1 - p_0$, where p_0 is the value of (3.9) for $X_{ti}^L = 0$. At high temperatures, $p_0 \approx p_1$, so that class assignments are made almost uniformly at random; as the temperature is lowered, the density (3.9) concentrates at the mode. Setting $T = 0$ corresponds to "instant cooling" and gives an ICM solution, which chooses the mode as the new label.

Each pixel site in the image is updated according to this strategy. Holding the temperature fixed, the edge sites are then updated.

Edge Update

We may similarly obtain the conditional posterior for an edge,

$$P(X_{tij}^E \mid \text{everything else}) \propto exp\{-U_{tij}^E/T\} \tag{3.11}$$

where the local energy

$$\begin{aligned} U_{tij}^E &= \beta_1 (1 - X_{tij}^E) [1 - 2\delta(X_{ti}^L - X_{tj}^L)] \\ &+ \beta_2\{V[c_1(X_{tij}^E)] + V[c_2(X_{tij}^E)]\} \\ &+ \beta_3 \gamma(X_{tij}^E;\ X_{<tij}^E) \end{aligned} \tag{3.12}$$

where $c_1(e)$, $c_2(e)$ are the two edge cliques (labelled arbitrarily) containing edge e as illustrated in Figure 5. The potentials $V(.)$ are given by for example Figure 4. We choose to replace X_{tij}^E with a value sampled from the distribution (3.11).

In order for convergence to the MAP solution, the temperature parameter must be reduced according to an appropriate annealing schedule; see Geman and Geman (1984). We use the schedule

$$T_r = T_1 \frac{ln\ (2)}{ln\ (1+r)}\ ,\ r = 1, \ldots, 5, \tag{3.13}$$

starting at temperature $T_1 = 2.0$. At each temperature, we perform 200 complete updates of the image (both pixels and edges) as described above. The k-frame simulated annealing ($k - SA$) algorithm is summarized in Figure 6.

The Metropolis approach to simulated annealing uses the following in place of Steps 3b(i) and 3b(ii) in the algorithm of Figure 6:

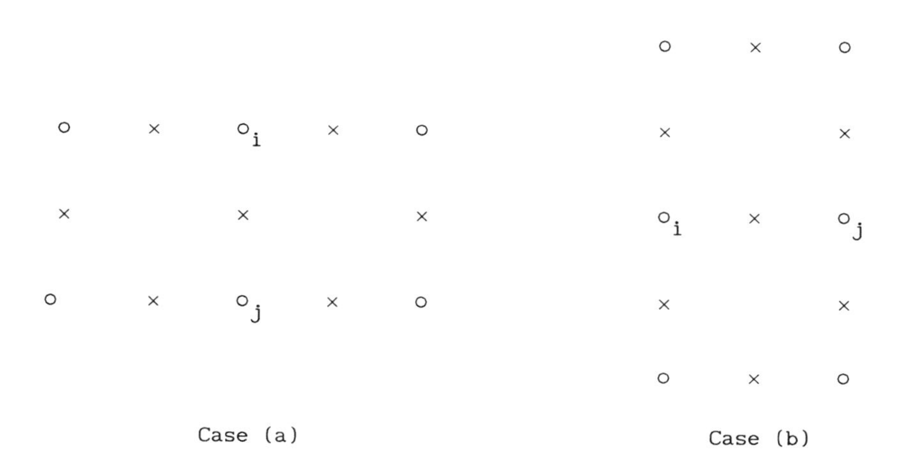

Figure 5: Edge cliques required for the update of edge X_{ij}^E when pixels i and j are (a) vertically adjacent, (b) horizontally adjacent.

Step 3b(i)* Update the label of each pixel i in frame t: Choose a candidate label $\tilde{X}_{ti}^L$, to replace X_{ti}^L, at random. Calculate the energy change

$$\Delta U = [\tilde{U}_{ti}^L - U_{ti}^L]/T ,$$

where U_{ti}^L is given in (3.10) and $\tilde{U}_{ti}^L$ is the corresponding quantity with X_{ti}^L replaced by $\tilde{X}_{ti}^L$. If $\Delta U > 0$ then replace X_{ti}^L by $\tilde{X}_{ti}^L$, otherwise replace X_{ti}^L by $\tilde{X}_{ti}^L$ with probability $exp(\Delta U)$.

Step 3b(ii)* Update X_{tij}^E for each pair of pixel neighbors $i \sim j$ in frame t: Choose a candidate, $\tilde{X}_{tij}^E$, to replace X_{tij}^E, at random. Calculate the energy change

$$\Delta U = [\tilde{U}_{tij}^E - U_{tij}^E]/T ,$$

where U_{tij}^E is given in (3.12) and $\tilde{U}_{tij}^E$ is the corresponding quantity with X_{tij}^E replaced by $\tilde{X}_{tij}^E$. If $\Delta U > 0$ then replace X_{tij}^E by $\tilde{X}_{tij}^E$, otherwise replace X_{tij}^E by $\tilde{X}_{tij}^E$ with probability $exp(\Delta U)$.

[Note that since only a single site is updated at a time, the change in total energy for the system is equivalent to the change in local energy.]

Step 1. *Initialisation step*: Assume that segmentations into p classes of frames t-k+1,...,t-1 and initial parameter estimates are available, *eg.* using a single-frame method.

Step 2. Input a new frame t and obtain an initial segmentation, *eg.* use the segmentation of frame t-1 or a single-frame segmentation of frame t.

Step 3. For r=1,...,NTEMP (=5, say) do the following:

Step 3a. Set the temperature ($T=T_r$) according to an appropriate annealing schedule, *eg.* as given by (3.13).

Step 3b. Repeat the following NITER (=200 say) times:

Step 3b(i) *Update pixel classes*: For each pixel i in frame t, choose the new class X^L_{ti} at random with probability at (3.9).

Step 3b(ii) *Update edge classes:* For each pair of adjacent pixels i~j in frame t, choose a new intermediate edge site X^E_{tij} at random with probability given by (3.11).

Step 4. Obtain new parameter estimates by maximum likelihood given the most recent segmentation.

Step 5. Increment t and goto step (2)

Figure 6: Algorithm for segmentation of a sequence of frames using annealing with Gibbs' sampler.

We have found that the two annealing approaches produce similar results in practice for binary scenes. For further details on simulated annealing and related matters, see van Laarhoven and Aarts (1987), Aarts and Korst (1989) and the references therein.

In practice, we have found that improved reconstructions of frame t are possible if information from the "future" frame $t+1$ is incorporated into the posterior. The intra-frame contributions are the same as in (3.8), but the inter-frame consistency now becomes

$$\beta_3\ Cost_3(\mathbf{X}_t^L | \mathbf{X}_{<t}^L, \mathbf{X}_{t+1}^L)\ +\ \beta_4\ Cost_4(\mathbf{X}_t^E | \mathbf{X}_{<t}^E, \mathbf{X}_{t+1}^E)\quad ,$$

throughout. The form of these cost functions is the same as before, but now utilizes classes from frame $t+1$. This approach obviously requires that we obtain a provisional estimate of frame $t+1$ before determining the reconstruction of frame t: we have simply estimated the classes of frame $t+1$ using the single frame method.

4 Iterated Conditional Modes for Sequences of k Images ($k - ICM$)

We now consider segmentation of image sequences. It is possible to define an ICM approach along the lines of Section 3, utilizing both pixels and edges, but we have found this to be too dependent on the initial estimates of $\mathbf{X}_t^L$ and $\mathbf{X}_t^E$ in practice. [Such a method may however, be useful as a final stage to "clean up" the results from simulated annealing.]

A simple k-frames version of ICM can be obtained by setting $\mathbf{X}_t^E\ \equiv 0$ in the priors of Section 3. We would then pursue a strategy of local updates, choosing the new pixel class at pixel i, X_{ti}^L, to maximize the conditional posterior (3.11).

However, here we shall consider a prior more in the spirit of the original single frame version of ICM (1-ICM). We shall view the temporal axis as a third dimension and formulate a "3-dimensional" version of ICM. In such a way, we define a prior which encompasses spatio-temporal dependences. In this section, we shall make use of only pixel labels, so that we have $\mathbf{X}_t\ \equiv\ \mathbf{X}_t^L$, with p possible labels for each pixel; *i.e.* $X_{ti}\ \epsilon\{0,\ldots,p-1\}$.

Following the single frame exposition of Besag (1986), we shall consider the following prior for $\mathbf{X}_{\leq t}$ with energy function dependent on only *pairwise* pixel interactions (based on the Ising model) given by,

$$P(\mathbf{X}_{\leq t})\ \propto\ exp[\ -\sum_{\substack{s\sim s'\\ s\leq s'\leq t}} \beta_{ss'} \sum_{i\sim j} \delta(X_{si}\ -\ X_{s'j})\ ]\ , \tag{4.1}$$

where the first summation is over all neighboring frames $s\ \sim\ s'$, $s \leq s' \leq t$, the second summation is over all neighboring pixels $si\ \sim\ s'j$ such that pixel i lies in frame s and pixel j lies in frame s', and $\delta(.)$ is the same indicator function as used in (3.4).

The $\beta_{ss'} \geq 0$ are non-negative interaction parameters controlling the relative strengths of pixel interactions between neighboring frames s and s', $s \leq s' \leq t$. The exponent in (4.1) is the sum of certain weighted potentials over all distinct neighboring pixel pairs (i, j) drawn from the composite sequence of all frames up to and including frame t. [Note that in the first summation, we have used the natural ordering of pairs of frames s, s'; however, we obtain the same result for an unordered sequence if we assume that $\beta_{ss'} = \beta_{s's}$.]

Intuitively, we might expect the magnitude of $\beta_{ss'}$ to vary in inverse relation to $|s-s'|$, since pixel classes from frames which are a large distance apart could be less compatible (because of changes in the true scene) than pixel classes from adjacent frames. In fact, we *choose* to set

$$\beta_{ss'} \quad = \quad 0 \quad \text{if} \quad |s - s'| > k - 1 \quad ,$$

i.e. there is only a k-frame "memory".

The prior (4.1) is of Gibbsian form so that the associated image sequence constitute a Markov Random Field with the specified Markov property (2.4). In particular, it is a simple matter to show that the conditional probability of class X_{ti} at pixel i within frame t given the classes, $X_{\leq t \backslash i}$, everywhere else in the sequence is simply given by,

$$\begin{aligned} P(X_{ti} |\, \mathbf{X}_{\leq t \backslash i}) &= P(X_{ti} |\, \mathbf{X}_{s,\delta i(s)},\ s = t-k+1, \ldots, t) \\ &\propto exp\{- \sum_{s=t-k+1}^{t} \beta_{st}\, u_{si}(X_{ti})\} \quad . \end{aligned} \tag{4.2}$$

where $u_{si}(x)$ is the number of neighbors of pixel i (with i in frame t) in $\delta i(s)$ which have color x.

Taking $\beta_{st} = \beta$ (some appropriate constant) for $0 \leq t - s < k$ results in simple averaging of contributions to the prior over the sequence of frames. Thus, in such a case, the *a priori* most probable color for X_{ti} given the colors everywhere else, would simply be that color most common amongst the set, $\delta i(\leq t; k)$, of neighbors of pixel i from all frames. In Section 7, we use values of $\{\beta_{st}\}$ obtained by experimentation on training data/simulations.

From (2.1) - (2.4) under the prior (4.2), we have that the posterior probability of the class of pixel i in frame t given the classes elsewhere and the data, is given by

$$P(X_{ti}|\ \mathbf{Y}_{\leq t}, \mathbf{X}_{<t}, \mathbf{X}_{t\backslash i}) \ \propto \ exp\{-\frac{1}{2\sigma^2}\{Y_{ti} - \mu(X_{ti})\}^2 - \sum_{s=t-k+1}^{t} \beta_{st}\ u_{si}(X_{ti})\} \quad . \tag{4.3}$$

The reconstruction strategy under k-ICM is simply to maximize (4.3) with respect to X_{ti} given the reconstructions $\hat{\mathbf{X}}_{<t}$ of the previous frames and provisional estimates $\tilde{\mathbf{X}}_{t\backslash i}$ at the other pixels in frame t. This is repeated for each pixel in the image at time t, and repeated iteratively for the frame until convergence is achieved. We shall assume here that the noise parameters $\mu(.)$ and σ are either known or estimates are available from a previous frame.

For example, in the $p = 2$ population case [with $X \ \epsilon\{0,1\}$], k-ICM allocates pixel i to population Π_0 $(X_{ti} = 0)$ if

$$Y_{ti} \ < \ \frac{1}{2}(\mu_0 + \mu_1) \ - \ \frac{\sigma^2}{(\mu_1 - \mu_0)} \sum_{s=t-k+1}^{t} \beta_{st}[u_{si}(1) \ - \ u_{si}(0)] \quad , \tag{4.4}$$

when $\mu_1 \equiv \mu(1) \ > \ \mu(0) \equiv \mu_0$. Note that the summation in (4.4) gives the weighted excess of neighbors of pixel i labelled as Π_1 as opposed to Π_0: so for example, if this is positive then some values of Y_{ti} smaller than the naive threshold, $\frac{1}{2}(\mu_0 + \mu_1)$, will still be accepted as belonging to Π_1.

In the special case $k = 1$, rule (4.4) gives the usual single frame ICM thresholding rule. Details of some simulation tests for the case $k = 2$ are provided in Section 7.

5 Image Registration for Fast-Varying Images

Consider now the situation of quickly varying images for which there is a non-negligible phase difference between successive frames. Our main interest is in object segmentation. We shall assume that the changes which occur between frames may be adequately represented by a rigid body motion in two-dimensions. However, the method may also be applicable for certain rigid motions in three dimensions which result in an apparent affine transformation of the object silhouette. In situations where the images contain several objects moving independently, each object is enclosed within a tracking "gate" and the registration applied to respective gates.

Suppose that we have provisional binary segmentations $\mathbf{X}_{t-1}$ and $\mathbf{X}_t$ from two frames to be registered. Let $\mathbf{Z}_{ti} \equiv (i_x, i_y)'$ denote the coordinates of the ith pixel for which $X_{ti} = 1$, this is in general a subset of locations on the pixel lattice. Suppose that

$$\mathbf{Z}_{t-1,i} \sim \mathrm{N}[\boldsymbol{\mu}_{t-1}, \boldsymbol{\Sigma}_{t-1}] \quad , \tag{5.1}$$

where $\boldsymbol{\mu}_{t-1}$ (2×1) is the average position (centroid) of $\mathbf{Z}_{t-1}$ and $\boldsymbol{\Sigma}_{t-1}$ (2×2) is the common variance matrix. [Note that this notation should not be confused with the mean and variance of the observations, for which a similar notation has been used.] Note that here we have assumed Gaussian errors, whereas in practice the $\mathbf{Z}_{ti}$ are restricted to lie on a rectangular pixel lattice. However, we shall assume that the model provides a reasonable approximation.

Now suppose that the image undergoes a rigid body motion, which may be written as the affine transformation

$$\mathbf{Z}_{ti} = \mathbf{A}_t \, \mathbf{Z}_{t-1,i} + \mathbf{b}_t \tag{5.2}$$

Where $\mathbf{A}_t$ (2×2) and $\mathbf{b}_t$ (2×1) are appropriate parameters for the transition from frame $t - 1$ to frame t. By the properties of linear transformations of multivariate normal random variables,

$$\mathbf{Z}_{ti} \mid \mathbf{Z}_{t-1,i} \sim \mathrm{N}[\mathbf{A}_t \, \boldsymbol{\mu}_{t-1} + \mathbf{b}_t, \, \mathbf{A}_t' \, \boldsymbol{\Sigma}_{t-1} \, \mathbf{A}_t] \quad . \tag{5.3}$$

The estimation of $\mathbf{A}_t$ and $\mathbf{b}_t$ can now be seen to be a multivariate regression problem. For given $\boldsymbol{\mu}_{t-1}$ and $\boldsymbol{\Sigma}_{t-1}$, it can be shown that the maximum likelihood estimates of $\mathbf{b}_t$ and $\mathbf{A}_t$ are given by

$$\mathbf{b}_t = \bar{\mathbf{Z}}_t - \hat{\mathbf{A}}_t \, \boldsymbol{\mu}_{t-1} \quad , \quad \hat{\mathbf{A}}_t = \boldsymbol{\Gamma}_{t-1}' \, \boldsymbol{\Lambda}_{t-1}^{-\frac{1}{2}} \, \hat{\boldsymbol{\Lambda}}_t^{\frac{1}{2}} \, \hat{\boldsymbol{\Gamma}}_t \quad , \tag{5.4}$$

where

$$\boldsymbol{\Sigma}_{t-1} = \boldsymbol{\Gamma}_{t-1}' \, \boldsymbol{\Lambda}_{t-1} \, \boldsymbol{\Gamma}_{t-1} \quad , \quad \mathbf{S}_t = \hat{\boldsymbol{\Sigma}}_t = \hat{\boldsymbol{\Gamma}}_{t-1}' \, \hat{\boldsymbol{\Lambda}}_{t-1} \, \hat{\boldsymbol{\Gamma}}_{t-1} \tag{5.5}$$

are spectral decompositions, and $\mathbf{S}_t$ is the sample variance matrix of the $\{\mathbf{Z}_{ti}\}$ and $\bar{\mathbf{Z}}_t$ is the average (centroid of the object in frame t) of the $\{\mathbf{Z}_{ti}\}$

Hence, from (5.2) and (5.4), we have

$$\hat{\mathbf{Z}}_{ti} = \hat{\mathbf{A}}_t \, \mathbf{Z}_{t-1,i} + \hat{\mathbf{b}}_t = [\boldsymbol{\Gamma}'_{t-1} \, \boldsymbol{\Lambda}_{t-1}^{-\frac{1}{2}} \, \hat{\boldsymbol{\Lambda}}_t^{\frac{1}{2}} \, \hat{\boldsymbol{\Gamma}}_t] \, (\mathbf{Z}_{t-1,i} - \boldsymbol{\mu}_{t-1}) + \bar{\mathbf{Z}}_t \quad . \tag{5.6}$$

In practice, however, $\boldsymbol{\mu}_{t-1}$ and $\boldsymbol{\mu}_{t-1}$ are unknown, but may be estimated by maximum likelihood conditional on $\{\mathbf{Z}_{t-1,i}\}$; *i.e.* we use,

$$\hat{\boldsymbol{\mu}}_{t-1} = \bar{\mathbf{Z}}_{t-1} \, , \quad \text{and} \quad \hat{\boldsymbol{\Sigma}}_{t-1} = \mathbf{S}_{t-1} \quad .$$

Hence, the registration transformation (5.6) becomes,

$$\hat{\mathbf{Z}}_{ti} = [\boldsymbol{\Gamma}'_{t-1} \, \boldsymbol{\Lambda}_{t-1}^{-\frac{1}{2}} \, \hat{\boldsymbol{\Lambda}}_t^{\frac{1}{2}} \, \hat{\boldsymbol{\Gamma}}_t] \, (\mathbf{Z}_{t-1,i} - \bar{\mathbf{Z}}_{t-1}) + \bar{\mathbf{Z}}_t \quad . \tag{5.7}$$

This has the geometrical interpretation of the alignment of the respective characteristic ellipses of the two frames, each having the same low-order moments as the binary segmented images.

The modified segmentation algorithm incorporating the registration stage is shown in Figure 5.1. This method has been shown to work well in practice for relatively low noise level images and has the advantages of being simple to implement and computationally inexpensive. At higher noise levels, the segmentations can become too poor to capture sufficient shape detail to perform adequate registration by this technique.

6 Other Methods

We now briefly consider extensions of some other single-frame segmentation methods to the multiple frame case. In this section, we shall consider only pixel classes; *i.e.* $\mathbf{X}_t \equiv \mathbf{X}_t^L$.

6.1 *k*-frame Alternating Mean Thresholding and Median Filtering (k-AMTMF)

Mardia and Hainsworth (1988) describe a single-frame spatial thresholding method which is fast and simple to implement. This is an iterated three-step algorithm for image segmentation, the three steps being (i) parameter estimation, (ii) choice of pixel classes to maximize a certain likelihood and (iii) smoothing of the pixel classes. This method is relatively simple to transfer to the multiple frame case.

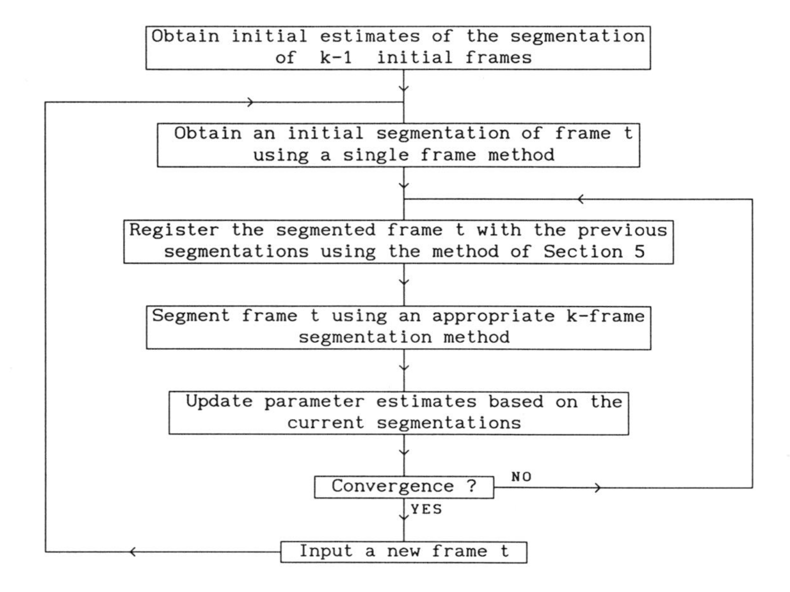

Figure 7: General segmentation algorithm for segmentation of a quickly-varying sequence of images incorporating registration.

In step (i), the parameters are estimated by maximum likelihood given the most recent estimate of the membership $\mathbf{X}_t$.

We shall assume *local spatial / temporal continuity* in the sense that

$$E[Y_{tj}] \quad = \quad \mu(X_{ti}), \quad \text{Var}(Y_{tj}) \quad = \quad \sigma^2 \quad ,$$

holds with high probability for all pixels j within a suitable small neighborhood over both spatial and temporal domains, $\delta i(\leq t; k)$, of pixel i from frame t.

As in the single frame case, in the segmentation step (ii), we choose to determine the class of pixel i based on the linear combination,

$$G_{ti} = \sum_{s=t-k+1}^{t} \sum_{j \in \delta i(s)} \gamma_{sj} Y_{sj} \quad (6.1)$$

where $\{\gamma_{sj}\}$ are appropriate weights. In practice, we have applied equal weight to each element of the neighborhood, *i.e.* $\gamma_{sj} = 1/\nu$ for all s, where ν is the number of neighbors of pixel i in $\delta i(\leq t; k)$. Under the assumption of local spatial continuity at pixel i, we have

$$G_{ti}|X_{ti} \sim N[\, \mu(X_{ti}),\, \sigma^2] \quad .$$

For each pixel i in frame t, we choose X_{ti} to maximize the conditional posterior,

$$P(X_{ti}|G_{ti}) \propto P(G_{ti}|X_{ti})P(X_{ti}) \propto P(G_{ti}|X_{ti}) \ ,$$

assuming equal prior probabilities for X_{ti}. For example, in the binary image case with $\mu(0) < \mu(1)$, we set $X_{ti} = 0$ (Black) if

$$G_{ti} < \frac{[\mu(0) + \mu(1)]}{2} \ ,$$

otherwise we set $X_{ti} = 1$.

In step (iii), we smooth the classification $\mathbf{X}_t$ of frame t using a single application of median smoothing within a 3×3 neighborhood of pixel i. [Note that here we use only neighbors from the current frame t.]

Starting with an appropriate initial classification (e.g., by naive segmentation), these three steps are repeated iteratively until convergence is achieved. Note that segmentation step (ii) utilises spatial and temporal context from all the observations, whilst the smoothing step (iii) utilises spatial context in the membership of frame t.

6.2 k-frame fuzzy reconstruction (k-FUZZY)

Our formulation follows the simplified approach of Mardia (1989). We shall consider here only the case of a two-population image (Black $\equiv$ 0, White $\equiv$ 1). Suppose now that $X_{ti} \in [0,\ 1]$ is the proportion of White in pixel i at time t. We now have,

$$E(Y_{ti}|X_{ti}) = \mu(X_{ti}) \equiv \mu(0)(1 - X_{ti}) + \mu(1)X_{ti} \quad .$$

Our extension of the single-frame method follows the lines of our k-ICM derivation in Section 4, but using the prior

$$P(\mathbf{X}_{\leq t}) \propto exp\{ - \sum_{\substack{s \sim s' \\ s \leq s \leq t}} \frac{1}{2\nu\tau^2_{ss'}} \sum_{i \sim j} |X_{si} - X_{s'j}|^{\omega}\} \tag{6.2}$$

in place of (4.1), where ω is a parameter determining the type of neighborhood averaging and ν is the neighborhood size. We shall take $\omega = 2$; in practice we have found little difference when using other values of ω. The parameters $\tau^2_{ss'}$ are spatial/temporal smoothness parameters and perform a similar role to the $\beta_{ss'}$ in the k-ICM model. We shall set

$$\tau^2_{ss'} \quad = \quad 0 \quad \text{if} \quad |s - s'| > k \quad .$$

It can be shown when $\omega = 2$, that (6.2) gives rise to the following simple conditional distribution,

$$P(X_{ti}|\ X_{<t},\ X_{t\backslash i}) \quad \propto \quad exp\{-\frac{1}{2} \sum_{s=t-k+1}^{t} \frac{(X_{si} - \tilde{X}_{si})^2}{\tau^2_{st}}\} \quad ,$$

where $\tilde{X}_{si}$ is the average value of X_{sj} for all neighbors j of i from frame s.

In this case, the value of X_{ti} maximizing the conditional posterior of X_{ti} given the classes elsewhere and the observations is given explicitly by

$$X_{ti} \quad = \quad \frac{(Y_{ti} - \mu_1)(\mu_0 - \mu_1)\sigma^{-2} \ + \ \sum_s \tau^2_{st}\tilde{X}_{si}}{(\mu_0 - \mu_1)^2\sigma^{-2} \ + \ \sum_s \tau^2_{st}} \tag{6.3}$$

Since the pixel proportions must lie in the range [0, 1], we must also clip the values obtained from (6.3); *i.e.* set

$$X_{ti} \quad = \quad \min[1,\ \max(0, X_{ti})\ ]. \tag{6.4}$$

Rule (6.3) is applied to each pixel in frame t, and the whole process repeated iteratively until convergence. In the 2-FUZZY case we have found that the values $\tau^2_{tt} = \tau^2_{t-1,t} = 0.032$ or 0.083 produce good results, although these values are somewhat image dependent. The pixel proportions are visually pleasing when viewed as gray-scale image.

At any stage, we may obtain a hardened version of the pixel proportions X_t by thresholding at an appropriate value, γ. That is, we set $X_{ti} = 0$ if $X_{ti} < \gamma$ and $X_{ti} = 0$ otherwise. In the 1-FUZZY case, the intuitive hardening threshold of $\gamma = 0.5$ is near-optimal. However, simulations in the 2-FUZZY case show that the "optimal" hardening threshold often lies in the range [0.5, 0.8] depending on the nature of the image sequence, typically around $\gamma = 0.75$ (which is supported by intuition).

7 2-ICM Examples

We first consider some applications of 2-ICM to the segmentation of slowly-varying image sequences. These examples include our "real-world" example. Throughout, we shall assume that the population (noise) parameters are unknown; these are estimated iteratively from the most recent available image segmentation. We use throughout the 8 nearest neighbors for 1-ICM and the 13 nearest neighbors for 2-ICM with $\boldsymbol{\beta} = (\beta_1, \beta_2)$, where β_s is the weight on neighbors in frame s, $s = 1, 2$.

Example 7.1: Expanding Disk Sequence

Figure 8 (a)-(e) shows a sequence of pure 64×64 images showing an expanding disc (π_1) with gray-level $\mu_1 = 120$ on a background (π_0) with gray-level $\mu_0 = 100$. The disc in the first frame has radius 17 pixels;the radius is increased by one pixel between successive frames and the center of the disk moves by one pixel in either the horizontal or vertical direction. The sequence of Figure 9(a)-(e) is obtained from Figure 8(a)-(e) by adding independent Gaussian noise with variance $\sigma^2 = 400$; this gives an expected naive error rate of 31% (SNR=1).

The 1-ICM ($\beta = 3.5$) segmentations of the sequence are shown in Figure 10(a)-(e). The corresponding 2-ICM segmentations ($\beta_1 = \beta_2 = 1.5$) are shown in Figures 11(a)-(e). [Note that Figure 11(a) is segmented using an unshown initial frame.]

It is clear that the two-frame method produces a more consistent sequence of segmentations, and viewed as a whole they present a qualitative improvement over the 1-ICM sequence of segmentations. However, if we concentrate on the pixel-by-pixel error rate of each individual frame, there is only a relatively small difference between the respective 1-ICM and 2-ICM segmentations. This arises because of the presence of registration error between each pair of successive frames. Note, however, that we have used a value of $\boldsymbol{\beta}$ in 2-ICM which promotes consistency of segmentations between frames and may reduce the error rate by a few pixels (at the expense of decreased consistency) by using $\boldsymbol{\beta} = (0.1,\ 0.6)$ for 2-ICM.

Results of 100 simulations of a two frame sequence of 32×32 images showing an expanding disk (with radius changing from 8 to 11 pixels) using the same noise model as above, indicate that good results with low average error rate ($\simeq 4.9\%$) can be obtained by taking $\boldsymbol{\beta} = (0.1,\ 0.6)$ or $\boldsymbol{\beta} = (0.1,\ 1.2)$. In fact, the error rates are relatively low over a wide range of values of β_2 when $\beta_1 = 0.1$.

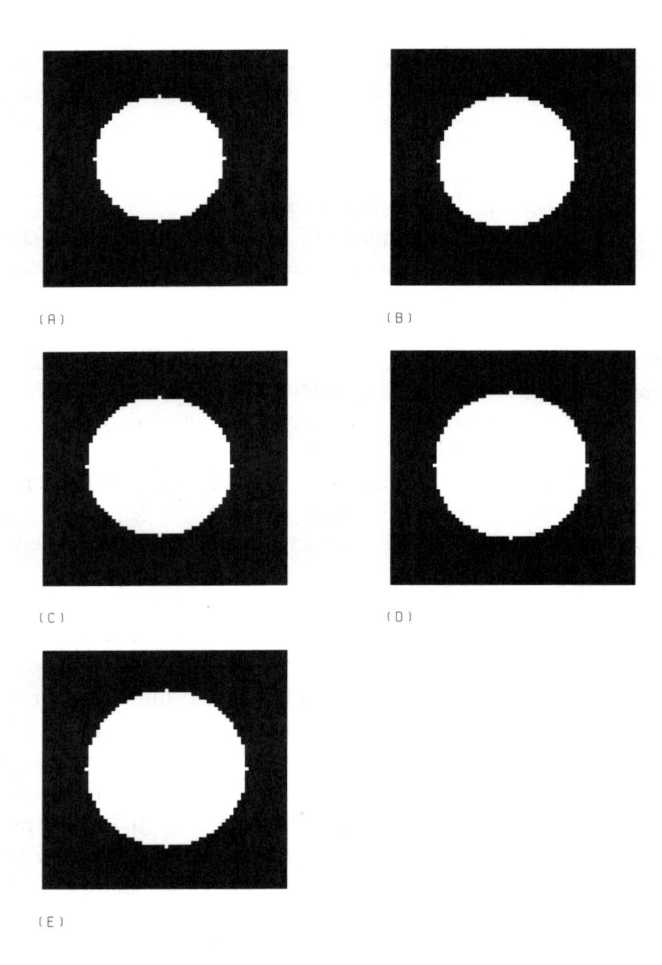

Figure 8: (a)-(e) A pure sequence of 64×64 images showing an expanding disk. The radius increases by one pixel between frames and the center of the disk moves by one pixel (either horizontally or vertically at random).

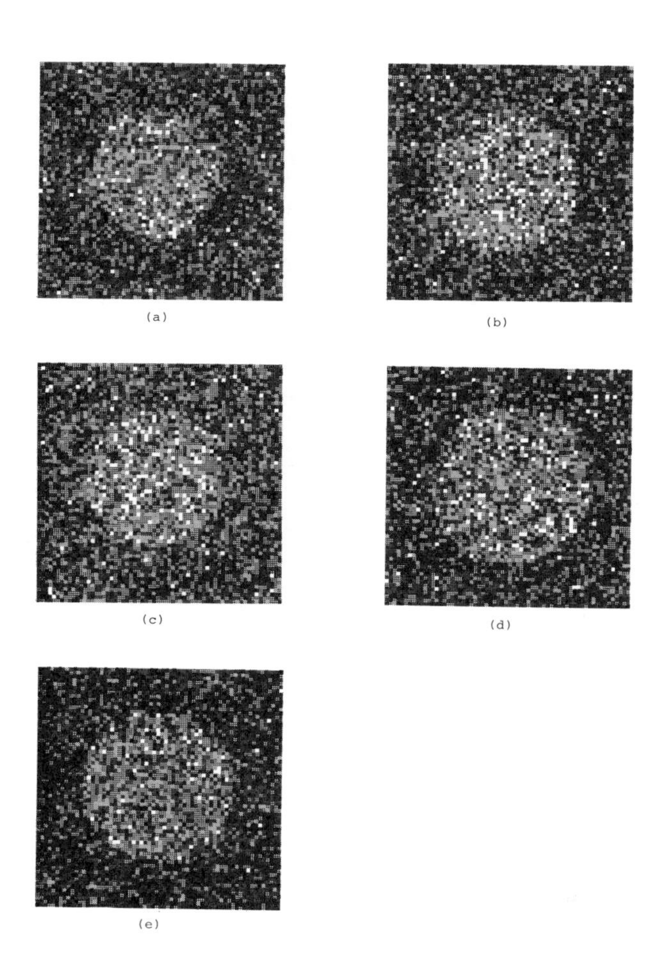

Figure 9: (a)-(e) A noisy sequence of images obtained by addition of independent Gaussian $N(0, 400)$ noise. The background mean is $\mu_0 = 100$ and the object mean is $\mu_1 = 120$.

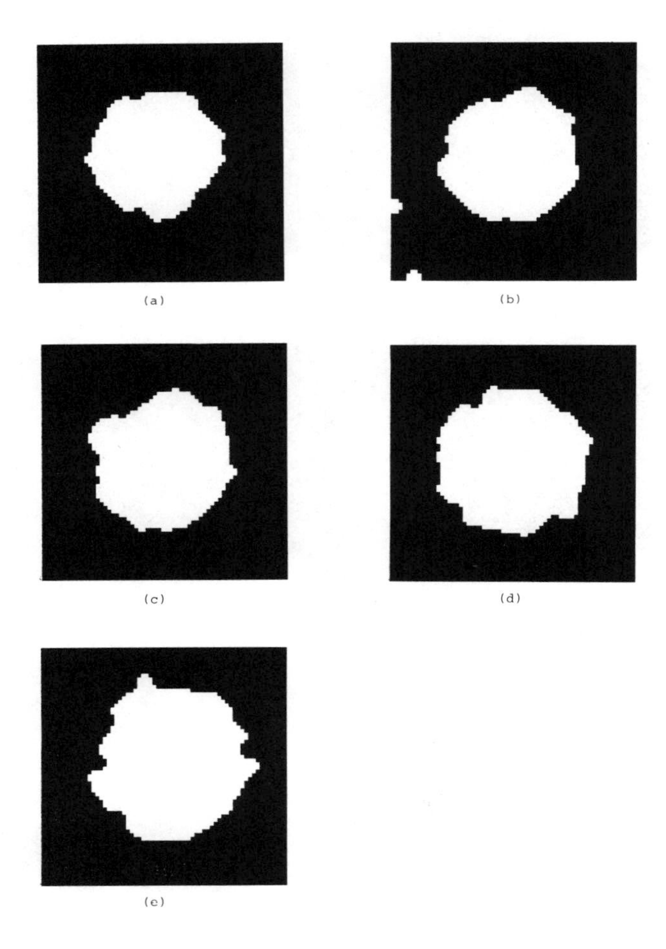

Figure 10: (a)-(e) 1-ICM ($\beta = 1.5$) segmentations of Figure 9.

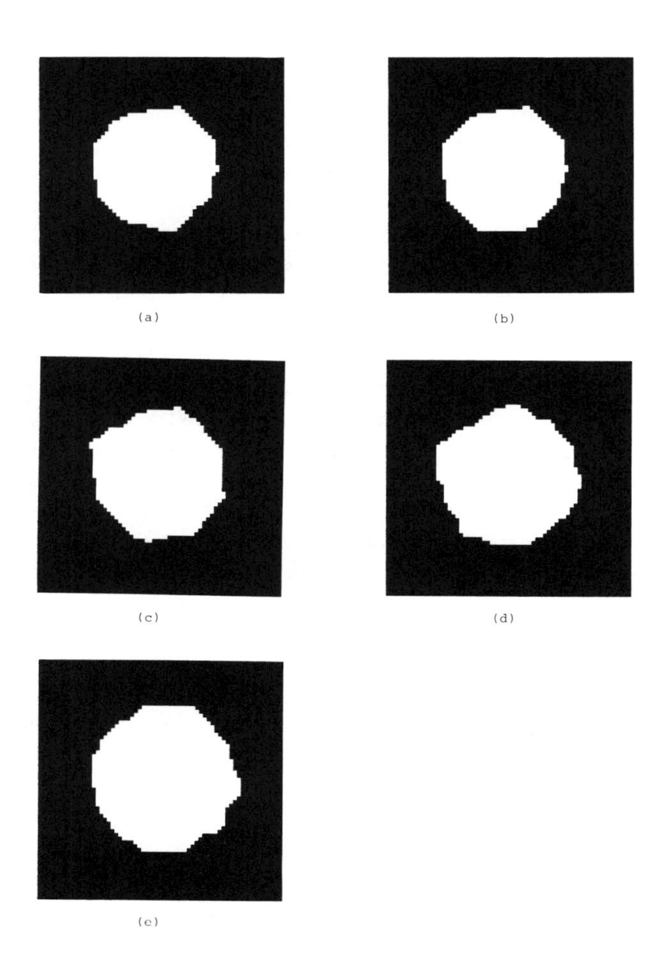

Figure 11: (a)-(e) 1-ICM ($\beta_1 = \beta_2 = 1.5$) segmentations of Figure 9(a)-(e) respectively.

As an aside, we note that if we constrain $\beta_1 = \beta_2$ (*i.e* place equal weight on all neighbors from both frames), then the minimum error rate is obtained with $\beta_1 = \beta_2 = 0.3$; however, this gives an error rate about twice that of the unconstrained values above. However, the advantage of choosing $\beta_1 \approx \beta_2$ is that, whilst the segmentation of a given frame is non-optimal in the sense of pixel-by-pixel error rate the sequence of segmented *shapes* is more consistent.

Example 7.2: FLIR Image Sequences

The remarks of the preceding example are reinforced by our real-world example. Consider the sequence of four 32×32 frames of Figure 12(a)-(d), showing a registered sequence of FLIR images of a military vehicle. These were extracted from a sequence of five 512×512 FLIR images, and registered by alignment of centroids. In this case, the ground truth is unknown, so that we must rely on *qualitative expert assessment* of the reconstructions.

The 1-ICM ($\beta = 1.5$) reconstructions are shown in Figure 13(a)-(d). The associated 2-ICM reconstructions ($\beta_1 = \beta_2 = 1.5$) are shown in Figure 14(a)-(d). [Note that reconstruction 14(a) used a previous, unshown, frame.] Here both sequences of reconstructions, viewed as individual images, are acceptable. However, the 2-ICM sequence as a whole is preferred since it presents a more consistent set of segmentations.

Note that in this "real" example, the naive (non-spatial) single frame segmentation methods produce reasonable segmentations of each individual frame since the noise level is relatively low. However, such segmentations are relatively highly inconsistent from frame to frame.

It may be argued that the images in Figure 12 contain more than two populations; *eg.* the object seems to have a hot-spot and a cool-spot, whilst the background contains clutter. Choosing the number of populations $p > 2$ to reflect this yields segmentations closer to the original gray-levels.

Example 7.3: Triangle Sequence

We now consider a synthetic example which illustrates the use of the proposed image registration method described in Section 5. Consider the sequence of two 32×32 frames shown in Figures 15(a),(b), which depict a moving triangle which undergoes a significant change in orientation. A degraded sequence is shown in Figure 15(c),(d) with expected naive error 16% (SNR=1.5). The 1-ICM reconstructions are shown in Figure 15(e),(f).

Figure 15(g) shows the result of registering Figure 15(e) with Figure 15(f) using the maximum likelihood registration method as described in Section 5. Figure 15(h) shows the (optimal) 2-ICM ($\beta_1 = 0.1,\ \beta_2 = 0.7$)

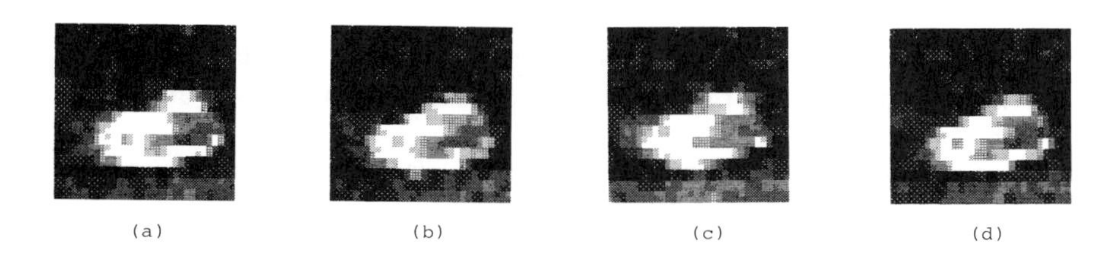

Figure 12: (a)-(d) A sequence of four registered 32×32 FLIR images showing the same vehicle, extracted from an unregistered sequence of 512×512 images.

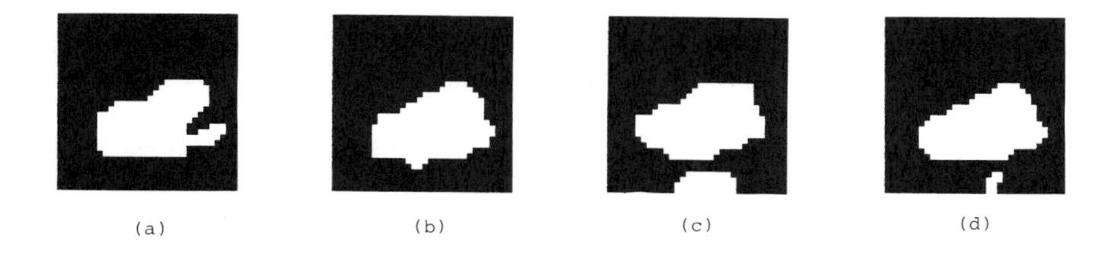

Figure 13: (a)-(d) 1-ICM ($\beta = 1.5$) reconstructions of Figure 12.

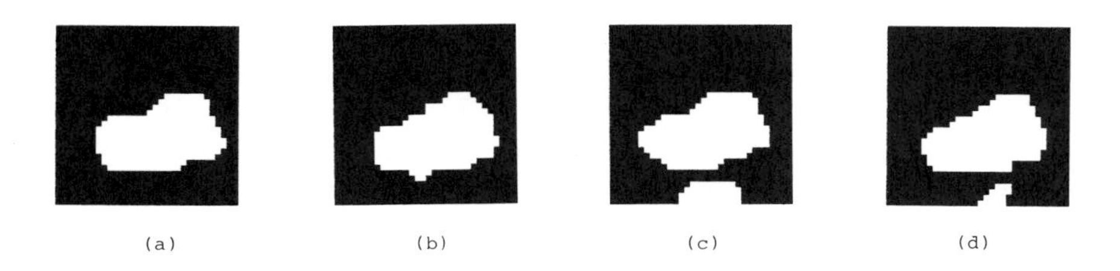

Figure 14: (a)-(d) 2-ICM ($\beta_1 = \beta_2 = 1.5$) reconstructions of Figures 12(a)-(d) respectively, assuming two populations.

reconstruction based on initial values given by the registered sequence of Figure 15(g) and Figure 15(f).

The 2-ICM reconstruction shows around a three pixel improvement over the 1-ICM reconstruction. This is a modest improvement, but it should be noted that the image contained relatively little error. The 2-ICM reconstruction could arguably be said to have captured the triangular shape better than the 1-ICM method.

Results for higher levels of noise are rather disappointing: the corners of the triangle are eroded by both 1-ICM and 2-ICM. In the latter case, this effect can be partially attributed to a failure of the registration method. It may be better to use the data to determine registration, rather than an (imperfect)1-ICM segmentation.

8 Edge / Pixel Reconstruction Examples

We now consider some simple applications of the combined edge/pixel reconstruction method using simulated annealing (SA) for 3 neighboring frames (3-SA) as described in Section 3. Here, we update frame t based on its neighboring frames $t-1$ and $t+1$ and utilize the edge potentials given in Figure 4. It is understood that in real-time applications, we are working on the penultimate frame in the sequence of images currently available.

Consider the noisy sequence of five 64×64 images in Figure 16(a)-(e) (extracted from a longer sequence). These were generated from a corresponding sequence of pure images depicting two overlapping rectangular regions of different mean gray-levels ($\mu_1 = 120$, $\mu_2 = 140$) on a uniform background ($\mu_0 = 100$), under additive independent Gaussian $N(0,\ 400)$ noise. Note that one of the rectangles is fixed, whereas the second grows as the sequence progresses. The separate $1-SA$ ($\beta_1 = 1.5$, $\beta_2 = 1.0$) pixel and edge reconstructions are shown in Figure 17(a)-(e) respectively. The $3-SA$ ($\beta_1 = 1.5$, $\beta_2 = 1.0$ $\beta_3 = 1.5$, $\beta_4 = 0$) reconstructions of frames (a) and (e) shown in Figure 17 have utilized unshown frames.

Note that here we have taken the inter-frame edge consistency weight, β_4, as zero: this reflects the observation that edge positions are variable within the image sequences, due to object/camera motion. In practice, we have obtained the best results with $\beta_4 = 0$.

We must first comment that the single frame reconstructions are remarkably good here, with 89, 70, 48, 73 and 65 misclassified pixels respectively, so that there is relatively little room for improvement. However, 3-SA produces results with 37, 21, 48, 32 and 66 misclassified pixels re-

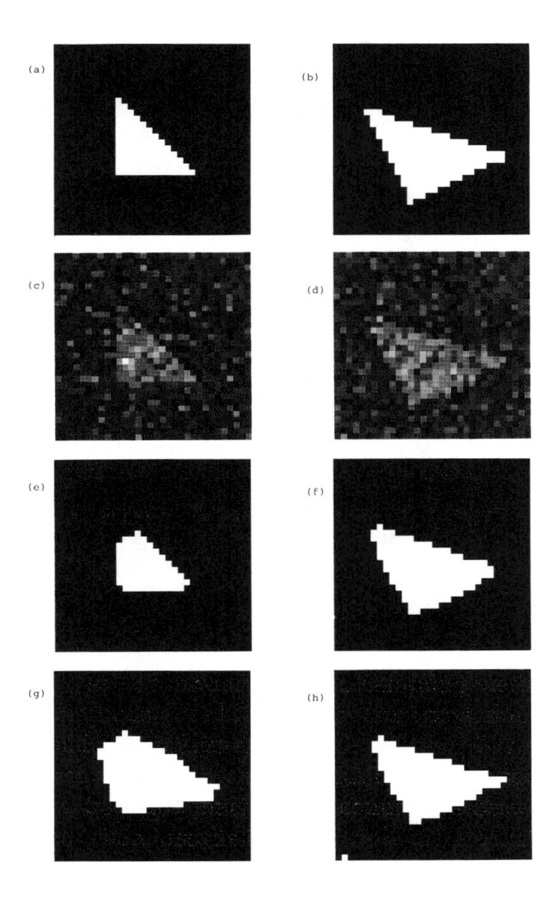

Figure 15: (a),(b) A pure sequence of two 32×32 frames showing a moving rotating and enlarging triangle. (c),(d) A noisy sequence obtained from Figure 15(a),(b) respectively by adding independent Gaussian $N(0, 100)$ noise. The background has mean $\mu_0 = 100$ and the object has mean $\mu_1 = 120$. (e),(f) 1-ICM reconstructions of Figures 15(c),(d) respectively. (g) The result of registering Figure 15(e) with Figure 15(f) using the method of Section 5. (g) 2-ICM ($\beta_1 = 0.1$, $\beta_2 = 0.7$) reconstruction of Figure 15(d) based on the registered sequence of two frames.

spectively. Thus 3-SA produces on average around half the error of 1-SA: the accuracy of the final image in the sequence was affected by poor "look-ahead" information. Further, the multiple frame method has produced a sequence of segmentations which are more consistent and arguably provide a better representation of the shapes present, for example on the visible sides of the stationary (white) rectangle. The moving rectangle is arguably more rectangular for the 3-SA reconstructions than in the 1-SA reconstructions.

We note that the 1-SA (and to a lesser extent the 3-SA) edge reconstructions contain a certain amount of noise, breaks in boundaries, etc. This is because we did not adhere to a full annealing schedule. A simple remedy may be to apply a final phase of "quenched" annealing, with the temperature parameter set to zero to rapidly freeze the edges. Although the edge reconstructions may often be of direct interest in their own right, there is some debate over what constitutes an edge in real images (see Mowforth and Gillespie, 1987), and we have used them mainly as an aid to obtaining a good pixel reconstruction, both in the sense that the misclassification error is small and that there are qualitative visual improvements.

We have found a similar relative reduction in the error rate for 3-SA compared to 1-SA at lower noise levels, although the absolute differences in the number of misclassified pixels is small. For relatively low noise levels, we have found that a value of $\beta_4 \approx 0.5$ yields slightly better results than $\beta_4 = 0$.

9 Discussion

We have considered various extensions of some popular existing single frame segmentation methods to the multiple frame case. We have found a typical reduction in pixel-by-pixel error rate of around 50% for the 3-SA method (using frames $t-1,\ t,\ t+1$) compared to the 1-SA method, although the absolute number of pixels involved is small. This arises because the single frame versions tends to produce rather good segmentations of population interiors, with errors concentrated on the object - background boundary. In the multiple-frame segmentation case further boundary confusion may result from registration errors introduced as the scene evolves. Note that $2-AMTMF$ has been found to produce significantly better results than $1-AMTMF$; the difference here is that all the data (as opposed to only estimated labels) from previous frames is used in the segmentation of the current frame.

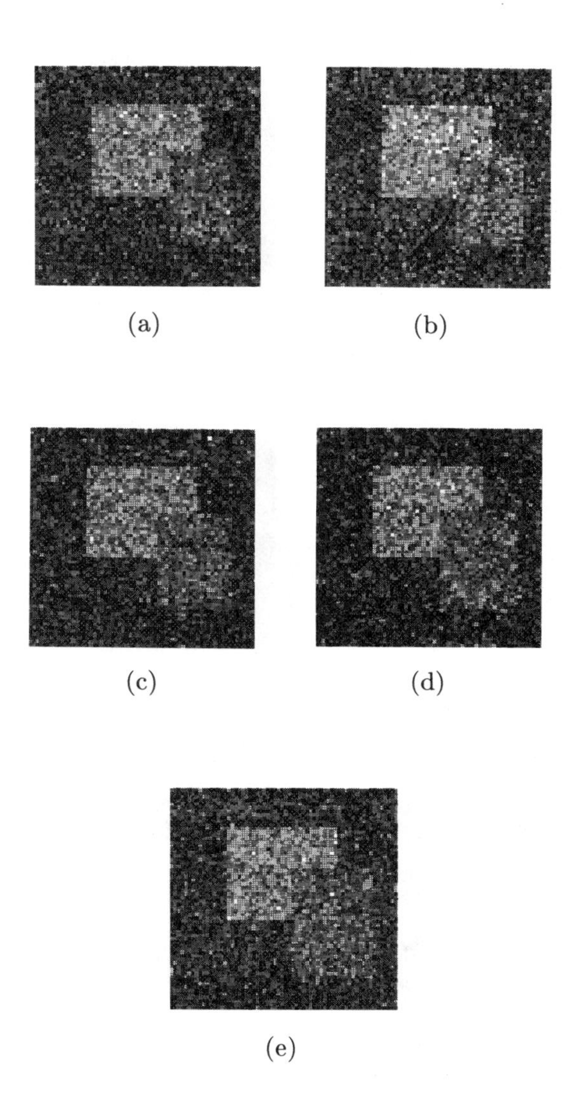

Figure 16: (a)-(e) A noisy sequence of images containing $p = 3$ populations with means $\mu_0 = 100$, $\mu_1 = 120$ and $\mu_2 = 140$ subject to independent $N(0, 400)$ noise.

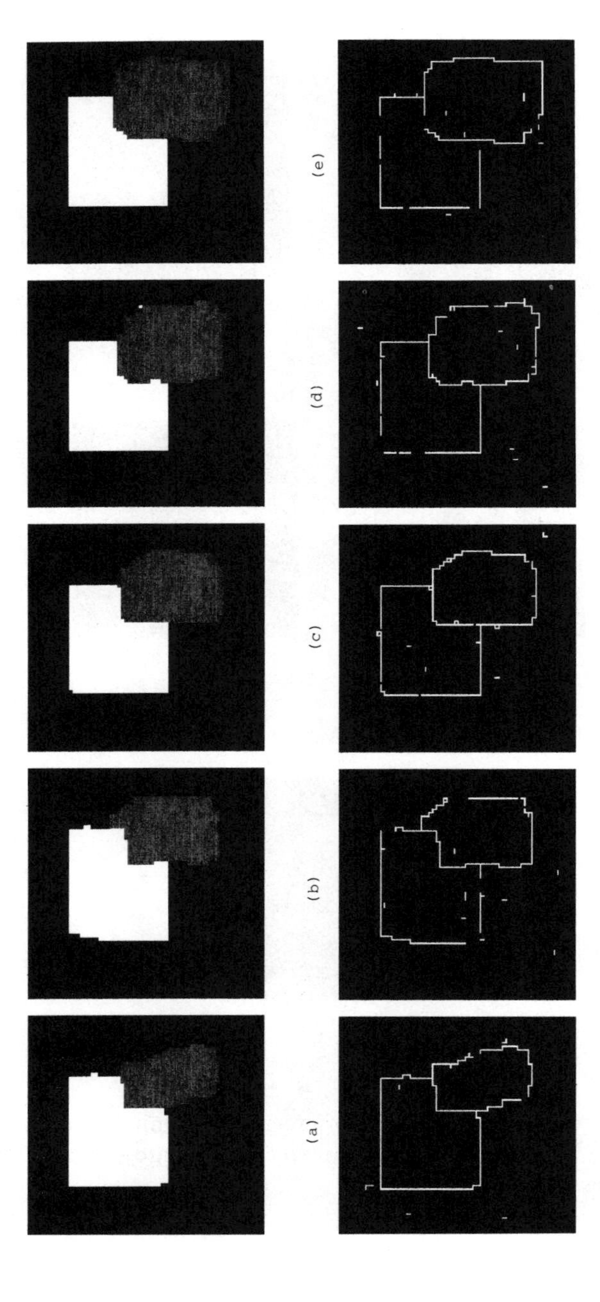

Figure 17: (a)-(e) 1-SA ($\beta_1 = 1.5$, $\beta_2 = 1.0$) pixel reconstructions of Figure 16 with the associated edge reconstructions.

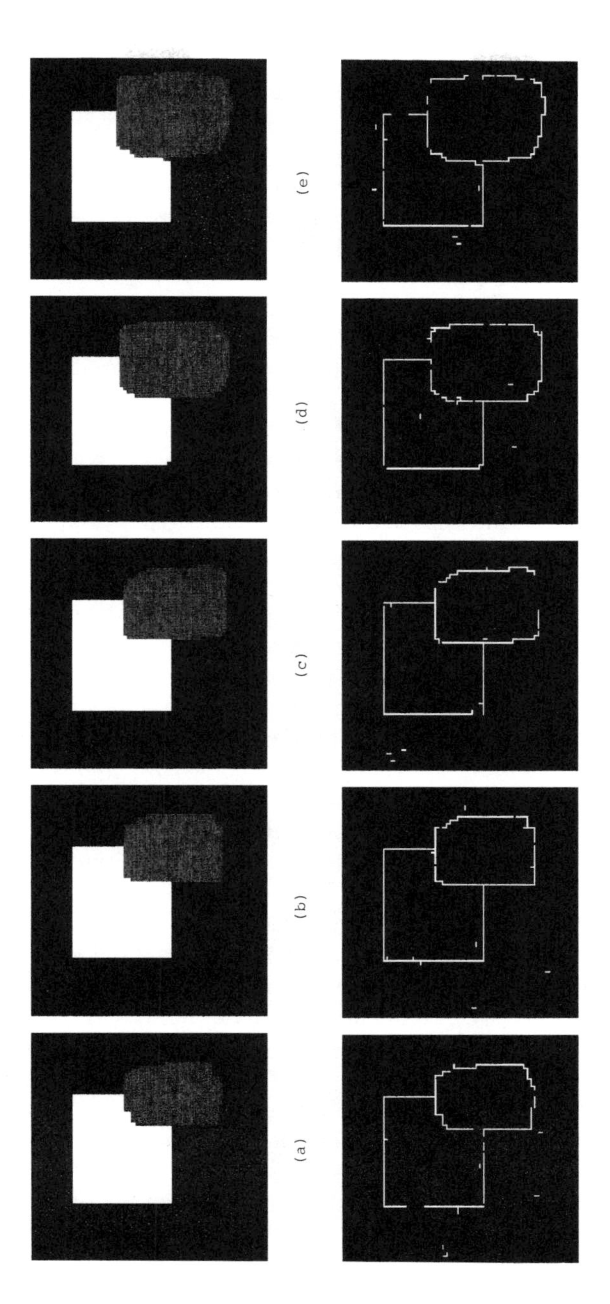

Figure 18: 3-SA ($\beta_1 = \beta_3 = 1.5,\ \beta_2 = 1.0, \beta_4 = 0$) pixel reconstruction of Figure 16 with the associated edge reconstructions.

However, we have argued that the multiple frame methods presented here can be used to produce a more consistent sequence of segmentations as illustrated in synthetic and real-world examples. We are possibly more interested in obtaining a consistent representation of the *shape* of any object present rather than achieving a minimum pixel-by-pixel error rate, which is widely regarded as a poor measure of the adequacy of a segmentation. However, we have no qualitative measure of consistency.

We have found that improved reconstructions of frames t are obtained by the 3-SA method based on frames $t-1$, $t+1$ rather than frames $t-2$, $t-1$. Thus we are working one frame behind the frame currently obtained by the camera. It is possible to similarly modify 3-ICM, 3-FUZZY and 3-AMTMF, to obtain a small improvement. When the number of populations is greater than two, k-AMTMF produces further errors on object boundaries due to averaging of gray-levels.

In our examples, we have performed unsupervised image reconstruction, estimating the noise parameters by maximum likelihood given a current reconstruction and determining the hyper-parameters of the prior by simulation study. We have found that the values of these hyper-parameters are generally image dependent, although consistency may be obtained with a wide range of values. It will be interesting to investigate the estimation of these parameters, following the work of Besag (1986) for the single frame case: this may be computationally too expensive in practice.

The inclusion of edge processes within each frame together with simulated annealing leads to an improved reconstruction. The simulated annealing ($k - SA$) method is computationally expensive compared to the other methods ($k - ICM$, $k - AMTMF$, $k - FUZZY$), but improvements in computer power and the possibility of parallel implementations (*eg.* Murray *et al*, 1986) makes the method more attractive. Further, the results for simulated annealing are less dependent on the initial values than the ICM methods.

It will be well worth considering more sophisticated priors for the edge/pixel reconstruction method of Section 3 following Geman (1989), Geman et al (1988). It may also be of benefit to make use of the observation that pixels change class relatively infrequently when viewed over time. We may thus seek to identify "change-points", t_c, for each pixel; *e.g.* in the single change-point case, the pixel has color c_1 for $t \leq t_c$ and color c_2 for $t > t_c$. We may seek to define an associated temporal change-point prior.

Since the edges are subject to variation between frames, it seems likely that an approach based on optical flow utilizing edge velocities may be useful; see Murray and Buxton (1987), Aggarwal and Nandhakumar (1988).

We have argued that we are often most interested in obtaining the *shape* of any object present in the image sequence. In poorly registered sequences, it is the object shape which remains unchanged throughout. It therefore seems plausible that a shape based segmentation method along the lines of Chow et al (1989) or Mowforth and Zhengping (1989) should be of some use here. One approach would be to seek to modify an object template from the previous frame to fit the data of the current frame by allowing deformations of the template according to an appropriate cyclic Markov chain model. In such a way, we would hope to achieve consistency of shape, but also allow innovations to be catered for in the deformation of the old template. Preliminary results show some promise.

We may also seek to include an element of cascading in to our segmentation method; see Jennison and Jubb (1989). By working on lower resolution version of the image sequence we can avoid some of the problems of image registration. For example, an object translation of a few pixels may be imperceptible after aggregating blocks of four (say) neighboring pixels into a single large pixel.

Acknowledgment

The authors are grateful to John Haddon and Mike Cooper for their helpful suggestions. This work was carried out under the support of the Procurement Executive, Ministry of Defense.

Bibliography

[1] Aarts, E.H.L. and Korst, J. (1989) *Simulated Annealing and Boltzmann Machines: A Stochastic Approach to Combinatorial Optimization and Neural Computing*, John Wiley and Sons, Chichester, UK.

[2] Aggarwal, J.K. and Nandhakumar, N. (1988) "On the Computation of Motion From Sequences of Images - A Review," *Proceedings IEEE*, 76, pp. 917-935.

[3] Besag, J. (1986) "On the Statistical Analysis of Dirty Pictures," *Journal of the Royal Statistical Society, Series B*, 36, pp.192-236.

[4] Chow, Grenander, U. and Keenan, D.M. (1989) *The Hand: A Pattern Theoretic Study of Biological Shape*, Report, Brown University, Providence, R.I., USA.

[5] Devijver, P.A. (1989) "Real-time Modeling of Image Sequences Based on Hidden Markov Mesh Random Field Models," *Report, Ecole Superiore des Telecommunications de Bretagne.*

[6] Geman, D. (1987) "Stochastic Model for Boundary Detection," *Image and Vision Computing*, 5, pp. 61-65.

[7] Geman, D. and Geman, S. (1984) "Stochastic Relaxation, Gibbs Distributions and the Bayesian Restoration of Images," *IEEE PAMI,* 6, pp. 721-741.

[8] Geman, D., Geman, S., Graffigne, C. and Dong, P. (1988) "Boundary detection by constrained Optimization," IEEE PAMI, 12, pp. 609-628.

[9] Green, P. and Titterington (1987) "Recursive Methods in Image Processing," *Bulletin of the International Statist. Institute*, pp. 51-67.

[10] Jennison, C. and Jubb, M. (1990) "Aggregation and Refinement in Binary Image Restoration," in *Spatial Statistics and Imaging*, IMS Lecture Notes series (ed. A. Possolo), Hayward, Canada.

[11] Kent, J.T. and Mardia, K.V. (1987) "Fuzzy Classification in Signal Processing," *Proceedings of the IMA Conference in Mathematics and Signal Processing*, pp. 395-407. (Oxford University Press).

[12] Kent, J.T. and Mardia, K.V. (1988) "Spatial Classification Using Fuzzy Membership Models," *IEEE PAMI*, 10, pp. 659-671.

[13] Kirkpatrick, S., Gelatt, C.D. and Vecchi, M.P. (1983) "Optimization by Simulated Annealing," *Science 220*, 13 May 1983, pp. 671-680.

[14] Mardia, K.V. (1989) "Markov Models and Bayesian Methods in Image Analysis," *Journal of Applied Statistics,* 16, pp. 125-130.

[15] Mardia, K.V. and Hainsworth, T.J. (1988) "A Spatial Thresholding Method for Image Segmentation," *IEEE PAMI*, 10, pp. 919-927.

[16] Mowforth, P. and Gillespie, L. (1987) "Edge Detection as an Ill-posed Specification Task," *Report*, TIRM-87-026, The Turing Institute, Glasgow, U.K..

[17] Mowforth, P.H. and Zhengping, J. (1989) "Model Based Tissue Differentiation in MR Brain Images," *Proceedings of the 4th Alvey Computer Vision Conference 1989*, Reading, UK, pp. 67-72.

[18] Murray, D.W. and Buxton, B.F. (1987) "Scene Segmentation from Visual Motion Using Global Optimization," *IEEE PAMI*, 9, pp. 220-228.

[19] Murray, D.W., Kashko, A. and Buxton, H. (1984) "A Parallel Approach to the Picture Restoration Algorithm of Geman and Geman", *Image and Vision Computing*, 4, pp. 133-142.

[20] Ridler, T.W., and Calvard, S. (1978) "Picture Thresholding Using an Iterative Selection Method," *IEEE Transactions on Systems, Man and Cybernetics*, SMC-8, pp. 630-632.

[21] Silverman, B.W., Jennison, C., Stander, J. and Brown, T.C. (1990) "The Specification of Edge Penalties for Regular and Irregular Pixel Images," *IEEE PAMI*, 12, pp. 1017–1024.

[22] Titterington, D.M. (1990) "Modelling and Restoration of Image Sequences," in *Mathematics in Signal Processing II*, Edited by J. A. McWhirter, Clarendon Press, Oxford.

[23] Wright, W.A. (1989) "A Markov Random Field Approach to Data Fusion and Colour Segmentation," *Image and Vision Computing*, 7, pp. 144-150.

[24] van Laarhoven, P.J.M. and Aarts, E.H.L. (1987) *Simulated Annealing: Theory and Applications.* D. Reidel Publishing Company, Dordrecht, Holland.

The MIT Vision Machine: Progress in the Integration of Vision Modules

Tomaso Poggio[‡] **and Daphna Weinshall**[†]

[‡]Artificial Intelligence Laboratory
and Center for Biological Information Processing
Massachusetts Institute of Technology
Cambridge, Massachusetts

[†] IBM Research Division
T. J. Watson Research Center
Yorktown Heights, New York

1 Introduction: The Project and Its Goals

Computer vision has developed algorithms for several early vision processes, such as edge detection, stereopsis, motion, texture, and color, which give separate cues as to the distance from the viewer of three-dimensional surfaces, their shape, and their material properties. Biological vision systems, however, greatly outperform computer vision programs. It is clear that one of the keys to the reliability, flexibility, and robustness of biological vision systems in unconstrained environments is their ability to integrate many different visual cues. For this reason, we continue the development of a *Vision Machine* system to explore the issue of integration of early vision modules. The system also serves the purpose of developing parallel vision algorithms, since its main computational engine is a parallel supercomputer, the Connection Machine.

The idea behind the Vision Machine is that the main goal of the integration stage is to compute a map of the visible discontinuities in the scene, somewhat similar to a cartoon or a line drawing. There are several reasons for this. First, experience with existing model-based recognition algorithms suggests that the critical problem in this type of recognition is to obtain a reasonably good map of the scene in terms of features such as edges and corners. The map does not need to be perfect (human recognition works with noisy and occluded line drawings) and, of course, it cannot

ISBN 0-12-170608-7

be; but it should be significantly cleaner than the typical map provided by an edge detector. Second, discontinuities of surface properties are the most important locations in a scene. Third, we have argued that discontinuities are ideal for integrating information from different visual cues.

It also is clear that there are several different approaches to the problem of how to integrate visual cues. Let us list some of the obvious possibilities:

1. There is no active integration of visual processes. Their individual outputs are "integrated" at the stage at which they are used– for example, by a navigation system. This is the approach advocated by Brooks [3]. While it makes sense for automatic, insect-like, visuo-motor tasks such as tracking a target or avoiding obstacles (e.g., the fly's visuo-motor system [19]), it seems quite unlikely for visual perception in the wide sense.

2. The visual modules are coupled so tightly that it is impossible to consider visual modules as separate, even in a first order approximation. This view is unattractive on epistemological, engineering, and psychophysical grounds.

3. The visual modules are coupled to each other and to the image data in a parallel fashion — each process represented as an array coupled to the arrays associated with the other processes. This point of view is in the tradition of Marr's $2\frac{1}{2}$-D sketch, and especially of the *intrinsic images* of Barrow and Tenenbaum [1]. Our present scheme is of this type, and exploits the machinery of Markov random field (MRF) models.

4. Integration of different vision modalities is taking place in a task-dependent way at specific locations — not over the whole image — and when it is needed — therefore not, at all times. This approach is suggested by psychophysical data on visual attention and by the idea of visual routines [21] (See also [11], [15], and [5].)

We are presently exploring the third of these approaches. We believe that the last two approaches are compatible with each other. In particular, visual routines may operate on maps of discontinuities such as those delivered by the present Vision Machine, and, therefore, be located after a parallel, automatic integration stage. In real life, of course, it may be more a matter of coexistence. We believe, in fact, that a control structure based on specific knowledge about the properties of the various modules, the specific scene, and the specific task will be needed in a later version of the Vision Machine to oversee and control the MRF integration stage itself and its parameters. It is possible that the integration stage should be much more goal-directed than what our present methods (MRF-based) allow. The main goal of our work is to find out whether this is true.

The Vision Machine project has a number of other goals. It provides a focus for developing parallel vision algorithms and for studying how to organize a real-time vision system on a massively parallel supercomputer. It attempts to alter the usual paradigm of computer vision research from past years: choose a specific problem — for example, stereo — find an algorithm, and test it in isolation. The Vision Machine allows us to develop and test an algorithm in the context of the other modules and the requirements of the overall visual task, above all visual recognition. For this reason, the project is more than an experiment in integration and parallel processing: it is a laboratory for our theories and algorithms.

Finally, the ultimate goal of the Vision Machine project is no less than the ultimate goal of vision research: to build a vision system that achieves human-level performance.

2 The Vision Machine System

The overall organization of the system is shown in Fig. 1. The images are processed in parallel through independent algorithms or modules corresponding to different visual cues. Edges are extracted using Canny's edge detector [6]. The stereo module computes disparity from the left and right images. The motion module estimates an approximation of the optical flow from pairs of images in a time sequence. The texture module computes texture attributes (such as density and orientation of textons [22]). The color algorithm provides an estimate of the spectral albedo of the surfaces, independently of the *effective illumination*, that is, illumination gradients and shading effects, as suggested by Hurlbert and Poggio [12].

The measurements provided by the early vision modules typically are noisy, and possibly sparse (for stereo and motion). They are smoothed and made dense by exploiting known constraints within each process (for instance, that disparity is smooth). This is the stage of *approximation* and *restoration* of data, performed using a Markov random field model. Simultaneously, discontinuities are found in each cue. Prior knowledge of the behavior of discontinuities is exploited; for instance, the fact that they are continuous lines, not isolated points. Detection of discontinuities is aided by the information provided by brightness edges. Thus, each cue, — disparity, optical flow, texture, and color — is coupled to the edges in brightness.

The full scheme involves finding the various types of physical discontinuities in the surfaces — *depth discontinuities* (extremal edges and blades), *orientation discontinuities, specular edges, albedo edges* (or marks), and *shadow edges* — and coupling them with each other and back to the discontinuities in the visual cues, as illustrated in Fig. 1 (See [8].) So far, we have implemented only the coupling of brightness edges to each of the cues provided by the early algorithm. As we will discuss later, the technique we use — to approximate, to simultaneously detect discontinuities, and to couple the different processes — is based on MRF models. The output

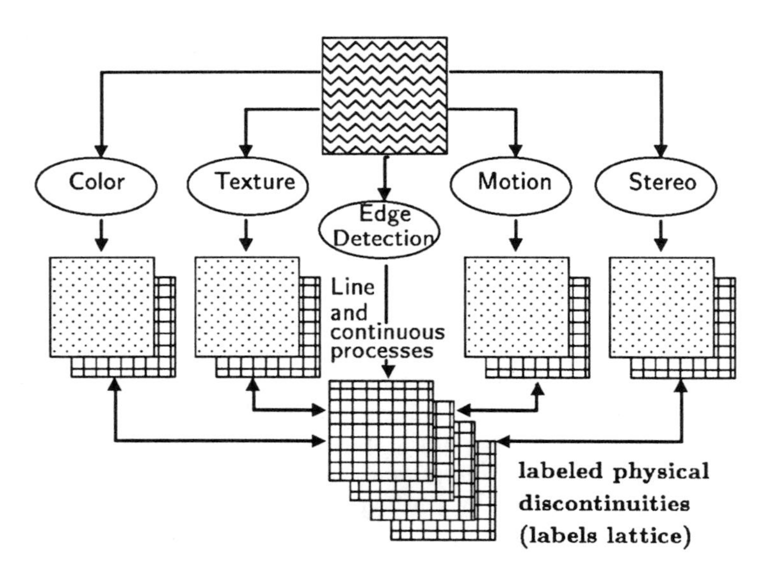

Figure 1: A sketch of the overall organization of the integration stage. The output of each of the early visual cues (or algorithms) — stereo, motion, texture, and color — are coupled to their own line processes (the crosses), i.e., their discontinuities. They also are coupled to the discontinuities in the surface properties — occluding edges (both extremal edges and blades), orientation discontinuities, specular edges, albedo discontinuities (including texture marks), and shadow edges. The image data — and, especially, the sharp changes in brightness labeled here as edges — are input to the lattices that represent the discontinuities in the physical properties of the surfaces.

of the system is a set of labeled discontinuities of the surfaces around the viewer. Thus, the scheme — an instance of inverse optics — computes *surface properties*; that is, attributes of the physical world and not any more of the images.

3 Early Vision Algorithms and their Parallel Implementation

3.1 Edge Detection

Edge detection is a key first step in correctly identifying physical changes. The apparently simple problem of measuring sharp brightness changes in the image has proven to be difficult. It is now clear that edge detection should be intended not simply as finding *edges* in the images, an ill-defined concept in general, but as measuring appropriate derivatives of the brightness data. This involves the task-dependent use of different two-dimensional derivatives. In many cases, it is appropriate to mark locations corresponding to appropriate critical points of the derivative, such as maxima or zeros. In some cases, later algorithms based on these binary features (presence or absence of edges) may be equivalent, or very similar, to algorithms that directly use the continuous value of the derivatives. A case in point is provided by our stereo and motion algorithms, to be described later. As a consequence, one should not always make a sharp distinction between edge-based and intensity-based algorithms; the distinction is more blurred, and in some cases, it is almost a matter of implementation.

In our current implementation of the Vision Machine, we are using two different kinds of edges. The first consists of zero-crossings in the Laplacian of the image filtered through an appropriate Gaussian [20]. The second consists of the edges found by Canny's edge detector [6]. Zero-crossings can be used by our stereo and motion algorithms (though we mainly have used Canny's edges at fine resolution). Canny's edges (at a coarser resolution) are input to the MRF integration scheme.

3.2 Stereo

Stereo matching is an ill-posed problem [2] that cannot be solved without taking advantage of natural constraints. The *continuity constraint* (as seen, for instance, in [16]) asserts that the world consists primarily of piecewise smooth surfaces. If the scene contains no transparent objects, then the *uniqueness constraint* applies: there can be only one match along the left or right lines of sight. If there are no narrow occluding objects, the *ordering constraint* [23] holds: any two points must be imaged in the same relative order in the left and right eyes.

The specific *a priori* assumption on which the algorithm is based is that the disparity, that is, the depth of the surface, is locally constant in a small region surrounding a pixel. It is a restrictive assumption that may be a satisfactory *local* approximation, however, in many cases. (It can be extended to more general surface assumptions in a straightforward way, but at a high computational cost.) Let $E_{\rm L}(x, y)$ and $E_{\rm R}(x, y)$ represent the left and the right images of a stereo pair, or some transformation of it, such as filtered images or a map of the zero-crossings in the two images. (More generally, they can be maps containing a feature vector at each location (x, y) in the image.)

We look for a discrete disparity $d(x, y)$ at each location x, y in the image that minimizes

$$\|E_L(x, y) - E_R(x + d(x, y), y)\|_{\text{patch}_i}$$

where the norm is a summation over a local neighborhood centered at each location (x, y); $d(x)$ is assumed constant in the neighborhood. The previous equation implies that we should look at each (x, y) for $d(x, y)$ such that

$$\int_{\text{patch}_i} (E_{\rm L}(x, y) E_{\rm R}(x + d(x, y), y))^2 dx dy \tag{3.1}$$

is maximized.

The algorithm that we have implemented on the Connection Machine actually is somewhat more complicated, since it involves geometric constraints that affect the way the maximum operation is performed [7]. The implementation currently used in the Vision Machine at the AI Laboratory uses the maps of Canny edges obtained from each image for $E_{\rm L}$ and $E_{\rm R}$.

In more detail, the algorithm is composed of the following steps:

1) Compute features for matching.

2) Compute potential matches between features.

3) Determine the degree of continuity around each potential match.

4) Choose correct matches based on the constraints of continuity, uniqueness, and ordering.

Potential matches between features are computed in the following way. Assuming that the images are registered so that the epipolar lines are horizontal, the stereo matching problem becomes one-dimensional: an edge in the left image can match any of the edges in the corresponding horizontal scanline in the right image. Sliding the right image over the left image horizontally, we compute a set of *potential match planes*, one for each horizontal disparity. Let $p(x, y, d)$ denote the value of the (x, y) entry of the potential match plane at disparity d. We set $p(x, y, d) = 1$ if there is an edge at location (x, y) in the left image and a compatible edge at location $(x - d, y)$ in the right image; otherwise, we set $p(x, y, d) = 0$. In the case of the DOG edge detector, two edges are compatible if the sign of the convolution for each edge is the same.

To determine the degree of continuity around each potential match (x, y, d), we compute a local support score $s(x, y, d) = \sum_{\text{patch}} p(x, y, d)$, where patch is a small neighborhood of (x, y, d) within the dth potential match plane. In effect, nearby points in a patch can "vote" for the disparity d. The score $s(x, y, d)$ will be high if the continuity constraint is satisfied near (x, y, d), i.e., if *patch* contains many votes. This step corresponds to the integral over the patch in the last equation.

Finally, we attempt to select the correct matches by applying the uniqueness and ordering constraints. (see earlier.) To apply the uniqueness constraint, each match suppresses all other matches along the left and right lines of sight with weaker scores. To enforce the ordering constraint, if two matches are not imaged in the same relative order in the left and right views, we discard the match with the smaller support score. In effect, each match suppresses matches with lower scores in its forbidden zone [23]. This step corresponds to choosing the disparity value that maximizes the integral of the last equation.

3.3 Motion

The motion algorithm [4] computes the optical flow field, a vector field that approximates the projected motion field. The procedure produces sparse or dense output, depending on whether it uses edge features or intensities. The algorithm assumes that image displacements are small, within a range $(\pm\delta, \pm\delta)$. It also is assumed that the optical flow is locally constant in a small region surrounding a point. This assumption is strictly only true for translational motion of 3-D planar surface patches parallel to the image plane. It is a restrictive assumption that may be a satisfactory *local* approximation, however, in many cases. Let $E_t(x, y)$ and $E_{t+\Delta t}(x, y)$ represent transformations of two discrete images separated by time interval Δt, such as filtered images, or a map of the brightness changes in the two images (more generally, they can be maps containing a feature vector at each location (x, y) in the image.)

We look for a discrete motion displacement $\underline{v} = (v_x, v_y)$ at each location x, y in the image that minimizes

$$\|E_t(x, y) - E_{t+\Delta t}(x + v_x\Delta t, y + v_y\Delta t)\|_{\text{patch}_i} ,$$

where the norm is a summation over a local neighborhood centered at each location (x, y); $\underline{v}(x, y)$ is assumed constant in the neighborhood. The previous equation implies that we should look at each (x, y) for $\underline{v} = (v_x, v_y)$ such that

$$\int_{\text{patch}_i} (E_t(x, y) - E_{t+\Delta t}(x + v_x \Delta t, y + v_y \Delta t))^2 dxdy \qquad (3.2)$$

is minimized. Alternatively, one can maximize the negative of the integrated result. The last equation represents the sum of the pointwise squared differences between a patch in the first image centered around the location (x, y) and a patch in the second image centered around the location $(x + v_x \Delta t, y + v_y \Delta t)$.

This algorithm can be translated easily into the following description. Consider a network of processors representing the result of the integrand in the previous expression. Assume for simplicity that this result is either 0 or 1. (This is the case if E_t and $E_{t+\Delta t}$ are binary feature maps.) The processors hold the result of differencing (taking the logical "exclusive or") the right and left image map for different values of (x, y) and v_x, v_y. The next stage, corresponding exactly to the integral operation over the patch, is for each processor to sum the total in an (x, y) neighborhood at the same disparity. Note that this summation operation is implemented efficiently on the Connection Machine using *scan* computations. Each processor thus collects a vote indicating support that a patch of surface exists at that displacement. The algorithm iterates over all displacements in the range $(\pm\delta, \pm\delta)$, recording the values of the integral for each displacement. The last stage is to choose $\underline{v}(x, y)$ among the displacements in the allowed range that maximizes the integral. This is done by an operation of *non-maximum suppression* across velocities out of the finite allowed set: at the given (x, y), the processor that has the maximum vote is found. The corresponding $\underline{v}(x, y)$ is the velocity of the surface patch found by the algorithm. The actual implementation of this scheme can be simplified so that the non-maximum suppression occurs during iteration over displacements, so that no actual table of summed differences over displacements need be constructed. In practice, the algorithm has been shown to be effective both for synthetic and natural images using different types of features or measurements on the brightness data, including edges (both zero-crossings of the Laplacian of Gaussian and Canny's method), which generate sparse results along brightness edges, or brightness data directly, or the Laplacian of Gaussian, or its sign, which generate dense results. Because the optical flow is computed from quantities integrated over the individual patches, the results are robust against the effects of uncorrelated noise.

The existence of discontinuities can be detected in optical flow, as in stereo, both during computation and by processing the resulting flow field. The latter field is input to the MRF integration stage. During computation, discontinuities in optical flow arising from occlusions are indicated by low normalized scores for the chosen displacement.

3.4 Color

The color algorithm that we have implemented is a very preliminary version of a module that should find the boundaries in the surface spectral reflectance function; that is, discontinuities in the surface color. The algorithm relies on the idea of *effective illumination* and on the *single-source* assumption, both introduced by Hurlbert and Poggio [12].

The single-source assumption states that the illumination may be separated into two components, one dependent only on wavelength, and one dependent only on spatial coordinates; this generally holds for illumination from a single light source. It allows us to write the image irradiance equation for a Lambertian world as

$$I^{\nu} = k^{\nu} E(x,y) \rho^{\nu}(x,y) \ ,$$

where I^{ν} is the image irradiance in the νth spectral channel (ν = red, green, blue), $\rho^{\nu}(x,y)$ is the surface spectral reflectance (or albedo), and the effective illumination $E(x,y)$ absorbs the spatial variations of the illumination and the shading due to the 3-D shape of surfaces. (k^{ν} is a constant for each channel, and depends only on the luminant.) A simple segmentation algorithm then is obtained by considering the equation,

$$H(x,y) = \frac{I^r}{I^r + I^g} = \frac{k^r \ \rho^r}{k^r \rho^r \ + \ k^g \rho^g},$$

which changes only when ρ^r or ρ^g, or both, change. Thus, H, which is piecewise constant, has discontinuities that mark changes in the surface albedo, independently of changes in the effective illumination.

The quantity $H(x,y)$ is defined almost everywhere, but typically is noisy. To counter the effect of noise, we exploit the prior information that H should be piecewise constant with discontinuities that are themselves continuous, non-intersecting lines. As we will discuss later, this restoration step is achieved by using a MRF model. This algorithm works only under the restrictive assumption that specular reflections can be neglected. Hurlbert [13] discusses in more detail the scheme outlined here and how it can be extended to more general conditions.

3.5 Texture

The *texture* algorithm is a greatly simplified parallel version of the texture algorithm developed by Voorhees and Poggio [22]. Texture is a scalar measure computed by summation of texton densities over small regions surrounding every point. Discontinuities in this measure can correspond to occlusion boundaries, or orientation discontinuities, which cause foreshortening. Textons are computed in the image by simple approximation to the methods presented in [22]. For this example, the textons are restricted to blob-like regions, without regard to orientation selection.

To compute textons, the image first is filtered by a Laplacian of Gaussian filter at several different scales. The smallest scale selects the textural elements. The Laplacian of Gaussian image then is thresholded at a nonzero value to find the regions that comprise the blobs identified by the textons. The result is a binary image with non-zero values only in the areas of the blobs. A simple summation counts the density of blobs (the portion of the summation region covered by blobs) in a small area surrounding each point. This operation effectively measures the density of blobs at the small scale, while also counting the presence of blobs caused by large occlusion edges at the boundaries of textured regions. Contrast boundaries appear as blobs in the Laplacian of Gaussian image. To remove their effect, we use the Laplacian of Gaussian image at a slightly coarser scale. Blobs caused by the texture at the fine scale do not appear at this coarser scale, while the contrast boundaries, as well as all other blobs at coarser scales, remain. This coarse blob image filters the fine blobs; blobs at the coarser scale are removed from the fine scale image. Then summation, whether with a simple scan operation or Gaussian filtering, can determine the blob density at the fine scale only. This is one example where multiple spatial scales are used in the present implementation of the Vision Machine.

4 MRF for Image Reconstruction and Integration

Whereas it is reasonable that combining the evidence provided by multiple cues — for example, edge detection, stereo, and color — should provide a more reliable map of the surfaces than any single cue alone, it is not obvious how this integration can be accomplished. The various physical processes that contribute to image formation *surface depth, surface orientation, albedo* (Lambertian and specular component), *illumination* — are coupled to the image data and therefore, to each other, through the imaging equation. The coupling, however, is difficult to exploit in a robust way, since it depends critically on the reflectance and imaging models. We argue that the coupling of the image data to the surface and illumination properties is of a more qualitative and robust sort at locations in which image brightness changes sharply and surface properties are discontinuous — in short, at edges. The intuitive reason for this is that at discontinuities, the coupling between different physical processes and the image data is robust and qualitative. For instance, a depth discontinuity usually originates a brightness edge in the image, and a motion boundary often corresponds to a depth discontinuity (and a brightness edge) in the image. This view suggests the following integration scheme for restoring the data provided by early modules. The results provided by stereo, motion, and other visual cues are typically noisy and sparse. We can improve them by exploiting the fact that they should be smooth, or even piecewise constant (as in the case of the albedo), between discontinuities. We can exploit *a priori* information about generic properties of the discontinuities themselves; for instance, that they usually are continuous and non-intersecting.

The idea then is to detect discontinuities in each cue — for instance, depth — simultaneously with the approximation of the depth data. The detection of discontinuities is helped by information on the presence and type of discontinuities in the surfaces and surface properties (See Fig. 1), which are coupled to the brightness edges in the image.

Notice that reliable detection of discontinuities is critical for a vision system, since discontinuities often are the most important locations in a scene; depth discontinuities, for example, normally correspond to the boundaries of an object or an object part. The idea thus is to couple different cues through their discontinuities and to use information from several cues simultaneously to help refine the initial estimation of discontinuities, which typically are noisy and sparse.

How can this be done? We have chosen to use the machinery of Markov random fields (MRFs), initially suggested for image processing by Geman and Geman [10]. In the following section, we will give a brief, informal outline of the technique and of our integration scheme. More detailed information about MRFs can be found in [10] and [17].

4.1 MRF Models

Consider the prototypical problem of approximating a surface given sparse and noisy data (depth data) on a regular 2-D lattice of sites. We first define the prior probability of the class of surfaces we are interested in. The probability of a certain depth at any given site in the lattice depends only upon neighboring sites (the Markov property). Because of the Clifford–

Hammersley theorem, the prior probability is guaranteed to have the Gibbs form,

$$P(f) = \frac{1}{Z} e^{-\frac{U(f)}{T}},$$

where Z is a normalization constant, T is called temperature, and $U(f) = \sum_C U_C(f)$ is an energy function that can be computed as the sum of local contributions from each neighborhood. The sum of the *potentials*, $U_C(X)$, is over the neighborhood's *cliques*. A clique is either a single lattice site or a set of lattice sites such that any two sites belonging to it are neighbors of one another. Thus, $U(f)$ can be considered as the sum over the possible configurations of each neighborhood [17]. As a simple example, when the surfaces are expected to be smooth, the prior probability can be given as sums of terms such as

$$U_c(f) = (f_i - f_j)^2,$$

where i and j are neighboring sites (belonging to the same clique).

If a model of the observation process is available (i.e., a model of the noise), then one can write the conditional probability $P(g/f)$ of the sparse observation g for any given surface f. Bayes theorem then allows one to write the posterior distribution,

$$P(f/g) = \frac{1}{Z} e^{\frac{-U(f/g)}{T}}.$$

In the simple earlier example, we have (for Gaussian noise)

$$\begin{aligned} U(f/g) &= \sum_i U_i(f/g), \\ U_i(f/g) &= \sum_j (f_i - f_j)^2 + \alpha\gamma_i(f_i - g_i)^2 , \end{aligned}$$

where $\gamma_i = 1$ only where data are available. More complicated cases can be handled in a similar manner.

The posterior distribution cannot be solved analytically, but sample distributions can be obtained using Monte Carlo techniques such as the Metropolis algorithm. These algorithms sample the space of possible surfaces according to the probability distribution $P(f/g)$ that is determined by the prior knowledge of the allowed class of surfaces, the model of noise, and the observed data. In our implementation, a highly parallel computer generates a sequence of surfaces from which, for instance, the surface corresponding to the maximum of $P(f/g)$ can be found. This corresponds to finding the global minimum of $U(f/g)$. (Simulated annealing is one of the possible techniques.) Other criteria can be used: Marroquin [18] has shown that the average surface f under the posterior distribution often is a better estimate, and one that can be obtained more efficiently by simply finding the average value of f at each lattice site.

One of the main attractions of MRFs is that the prior probability distribution can be made to embed more sophisticated assumptions about the world. Geman and Geman [10] introduced the idea of another process, the line process, located on the dual lattice, and representing explicitly the presence or absence of discontinuities that break the smoothness assumption. The associated prior energy then becomes

$$U_i(f,l) = \sum_j \left[(f_i - f_j)^2 (1 - l_{ij}) + \sum_C \beta V_C(l_{ij}) \right], \tag{4.1}$$

where l is a binary line element between site i, j. V_C is a term that reflects the fact that certain configurations of the line process are more likely than others to occur. In our world, depth discontinuities usually are themselves continuous, non-intersecting, and rarely isolated joints. These properties of physical discontinuities can be enforced locally by defining an appropriate set of energy values $V_C(l)$ for different configurations of the line process in the neighborhood of the site.

4.2 Organization of Integration

It is possible to extend the energy function of Eq. 4.1 to accommodate the interaction of more processes and their discontinuities. In particular, we have extended the energy function to couple several of the early vision modules (depth, motion, texture, and color) to brightness edges in the image. This is a central point in our integration scheme; brightness edges guide the computation of discontinuities in the physical properties of the surface, thereby coupling surface depth, motion, texture, and color, each to the image brightness data and to each other. The reason for the role of brightness edges is that changes in surface properties usually produce large brightness gradients in the image. It is exactly for this reason that edge detection is so important in both artificial and biological vision.

The coupling to brightness edges may be done by replacing the term $V_C(l_{ij})$ in Eq. (4.1) with the term,

$$V(l,e) = g(e_{ij}, V_C(l_{ij})) \ ,$$

with e_{ij} representing a measure of the presence of a brightness edge between site i, j. The term g has the effect of modifying the probability of the line process configuration depending on the brightness edge data ($V(l,e) = -\log\ p(l/e)$). This term facilitates formation of discontinuities (that is, l_{ij}) at the locations of brightness edges. Ideally, the brightness edges (and the neighboring image properties) activate, with different probabilities, the different surface discontinuities (Fig. 1), which in turn are coupled to the output of stereo, motion, color, texture, and possibly other early algorithms.

4.3 Algorithms: Deterministic and Stochastic

We have chosen to use MRF models because of their generality and theoretical attractiveness. This does not imply that stochastic algorithms must be used. For instance, in the cases in which the MRF model reduces to standard regularization [17] and the data are given on a regular grid, the MRF formulation leads not only to a purely deterministic algorithm, but also to a convolution filter. Recent work in color [12] shows that one can perform integration similar to the MRF-based scheme using a deterministic update. Geiger and Girosi [9] have shown that there is a class of deterministic schemes that are the mean-field approximations of the MRF models. These schemes have a much higher speed than the Monte Carlo schemes we used so far, while promising similar performance.

4.4 Illustrative Results

We show the results of the Vision Machine applied to the scene in Fig. 2a. The brightness edges computed by the Canny algorithm are shown later in Fig. 6. The results given by the stereo, motion, and color algorithms, after an initial smoothing to make them dense [14], are the input to the MRF machinery that integrates each of those data sets with the brightness edges. Figures 2b - d show the discontinuities found by the MRF machinery in each of the cues: b) processing of the stereo output finds depth discontinuities in the scene (mainly the outlines of the different fruit); c) motion discontinuities are found by the MRF machinery with help from brightness edges; and d) the color boundaries show regions of constant surface color, independently of shading: notice, for instance, that shadow edges do not appear as color edges.

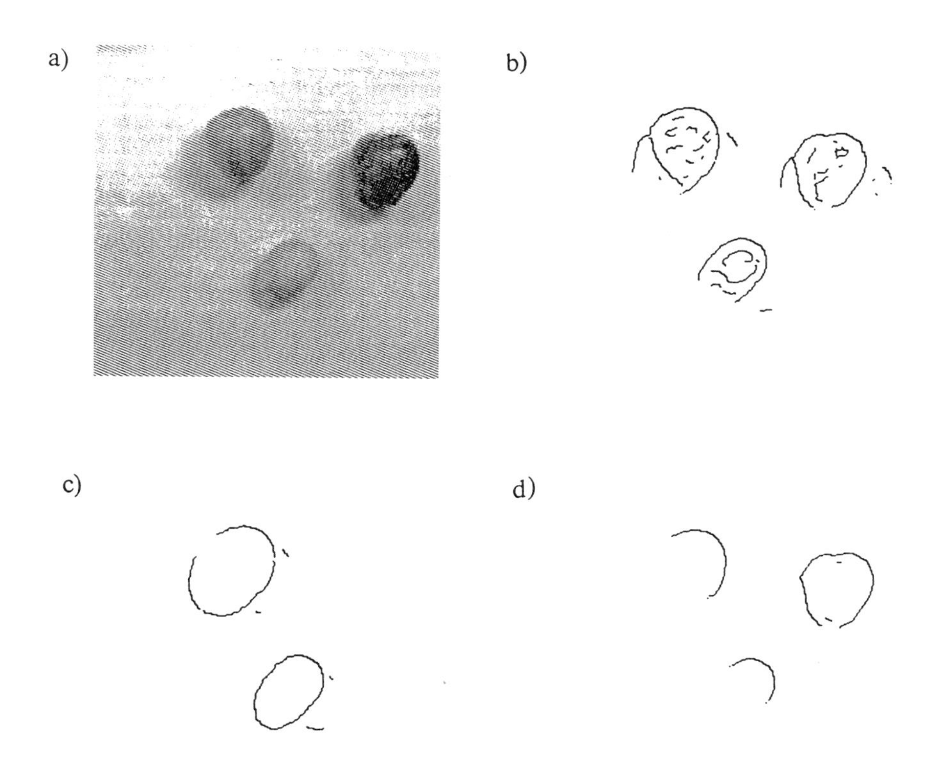

Figure 2: (a) The gray level of a color image (colors not shown here), (b) the MRF estimate of the depth discontinuities from stereo data, (c) the motion discontinuities, and (d) the discontinuities in hue. (See text.)

Figure 3: A cartoon-like representation of two objects segmented from the background using depth, motion, and texture cues by the Vision Machine.

Figure 3 shows the union of the discontinuities in depth and motion for a different scene. The integration algorithm gives a rather good "cartoon" of the original scene.

5 Integration and Labeling Discontinuities

The scheme described at the end of the preceding section provides a cartoon of discontinuities according to their origin visual process (stereo, motion, etc). We propose to label the discontinuities according to their physical origin, which would make a better representation of the three-dimensional scene for human-like tasks. We now describe an integration scheme between the MRFs of all the low-level vision modules. Preliminary results of a labeling scheme based on a linear classifier are described.

5.1 Physical Discontinuities

The most useful physical discontinuities for labeling can be understood best in relation to the reflectance equation. For simplicity, assume a Lambertian surface and a single-point source of illumination. In this case, the radiance equation takes the form

$$R = \rho \mathbf{n} \cdot \mathbf{s} \ ,$$

where ρ is the reflectance coefficient, or albedo, $\mathbf{n}$ is the vector direction of a local surface patch, and $\mathbf{s}$ is the vector direction of the illumination source.

A discontinuity in the intensity can result from a discontinuity in (at least) one of the three variables ρ, $\mathbf{n}$, or $\mathbf{s}$. We classify edges according to the following physical discontinuities: discontinuities in surface properties (ρ), called *mark* or *albedo* discontinuities (e.g., changes in the color of the surface); discontinuities in the orientation of the surface patch ($\mathbf{n}$), called *orientation* discontinuities (e.g., an edge in a polyhedron); and discontinuities in the illumination ($\mathbf{s}$), called *shadow* discontinuities. In addition, there are *occluding boundaries*, which are discontinuities in the object space (a different object), and *specular* discontinuities, which are needed since the assumption of Lambertian surface is far from being correct in most cases.

5.2 Integration via Labeling

For the purpose of labeling, we define the *labels vector* on the line processes lattice, where the value at each site is a vector with support over the space of edge labels. We first will describe the interaction between the *labels vector* $\vec{\boldsymbol{L}}_{ij}$ and the modules' line process l_{ij}^k for the k process:

Let A denote the set of labels. Generally, $\boldsymbol{L}$ is any function of the l^ks. In the simplest linear case, for each $a \in A$,

$$L_{ij}^a = \sum_k W_{ak} l_{ij}^k .$$

(We choose $L_{ij}^a(t = 0) = W_{ak} l_{ij}^k$ for k, the intensity process, as an initialization.) The weight W_{ak} should reflect the dependence of the edge label a on the discontinuities in module k. For example, depth discontinuities are very likely to be an occluding boundary, and very unlikely to be a shadow discontinuity.

The energy expression of the kth module for the ith site (as seen in Eq. (4.1) earlier) can be written as:

$$\begin{aligned} U_i^k(f, l/g, L) &= \alpha^k \gamma_i^k (f_i^k - g_i^k)^2 + \sum_j [(1 - l_{ij}^k)(f_i^k - f_j^k)^2 + \\ & \sum_C \beta^k V_C(l_{ij}^k) + \beta'^k h(L_{ij}, l_{ij}^k)] \ . \end{aligned}$$

The last term in the energy function $h(L_{ij}, l_{ij}^k)$ is the coupling between the line process of the kth module and the labels vector. We use:

$$h(L_{ij}, l_{ij}^k) = d\left(\left\{\frac{L_{ij}^a}{\sum_{a\in A} L_{ij}^a}\right\}_{a\in A}, \left\{\frac{W_{ak}}{\sum_{a\in A} W_{ak}}\right\}_{a\in A}\right) \cdot l_{ij}^k \quad , \qquad (5.2)$$

where $d()$ is some metric on probability measures, e.g., Kolmogorov - Smirnoff distance. Equation (5.2) implies that the further away the normalized vector L_{ij} is from the normalized vector $\{W_{ak}\}$ given k, where the last expression can be thought of as the probabilities for each label if there is a line in module k, the less likely it is that $l_{ij}^k = 1$.

We now can couple the Markov random fields of the different processes to a single MRF with additional cliques that are the sets of all the line processes of the different modules between pixels i, j. The energy of this MRF will be:

$$U(f, l/g) = \sum_k \sum_i U_i^k(f, l/g, L) \ .$$

Thus, $h(L_{ij}, l_{ij}^k)$ in Eq. (5.2) can be thought of as the potential of the new clique.

5.3 Implementation: Labeling with a Linear Classifier

Presently, we have implemented a part of the general scheme. More specifically, we have used a simple linear classifier to label edges at pixels where there exists an intensity discontinuity, using the output of the line process associated with each low-level vision module (see Fig. 4.). There is no mutual interaction yet between the various lines processes. We strongly use the fact that the modules' discontinuities are aligned, having been integrated with the intensity edges before, so that the nonexistence of a module discontinuity at a pixel is meaningful.

The linear classifier (Fig. 4) corresponds to a linear network where each output unit is a weighted linear combination of its inputs. (For a similar application to a problem of color vision, see [12].) The input to the network is a pixel where there exists an intensity edge and that feeds a set of qualitatively different input units. (See Figure 4.) The output is a real value vector of labels support. The final decision on the label at the pixel can be taken locally, e.g., winner takes all. We have used a more global labeling algorithm, which decides a unique label per edge based upon the output vector at each pixel along an edge segment.

Our results using this net on real images are very preliminary as yet, restricted to the domain of still-life images. The net was trained first on

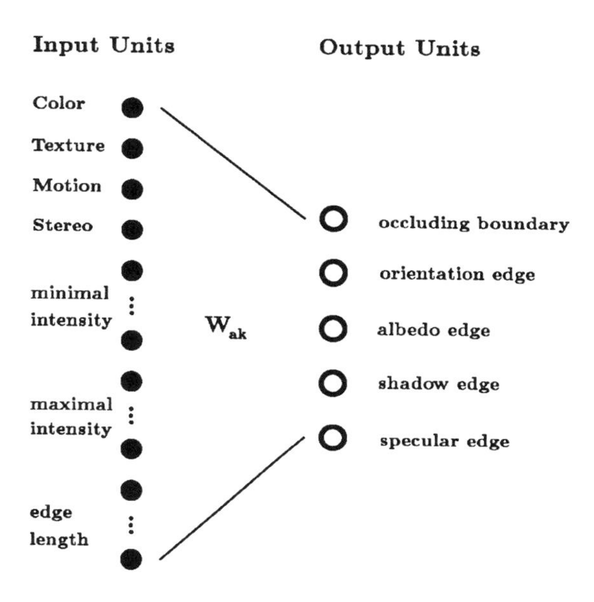

Figure 4: The linear classifier used to classify edges in our present implementation. The uppermost input units are inputs from the low-level vision modules stereo, motion, color, and texture, each feeding one unit with $+1$ if it detects a discontinuity at the pixel, -1 if there is no discontinuity, and 0 if there is no information at the pixel. The other input units give the average intensity on both sides of the edge in the immediate neighborhood of the pixel and the length of the edge, each quantized and then used to feed a number of binary cells (one per each quantized value). The output of the net is a set of units, one per label, where the output at each unit is a real number that reflects the support to the label at the current pixel.

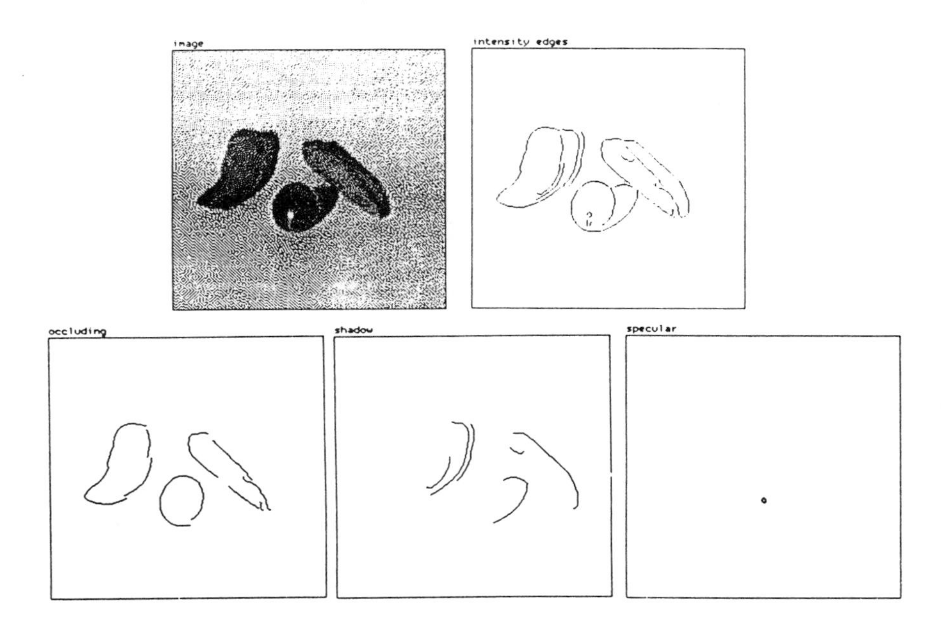

Figure 5: A still-life image and labeling by the linear classifier. Part of the input data is taken from the image and part is given manually (stereo, motion, and color discontinuities). A different still-life image was used to train the classifier.

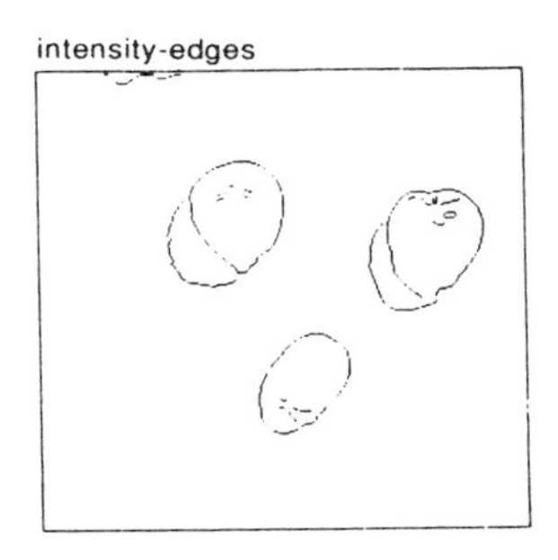

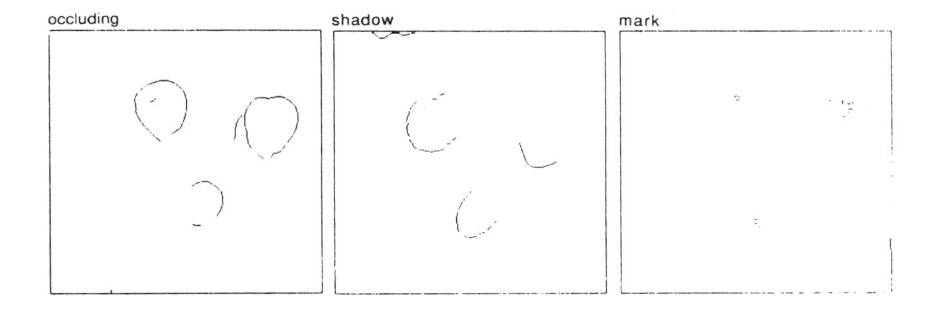

Figure 6: The labeling of the discontinuities given in Fig. 2 found by the linear classifier, trained on the same data.

images where the output of the low-level modules is given manually (and is thus almost perfect). The intensity edges, which have been used for the manual labeling of the training data, have been obtained by a Canny edge detector with the additional restriction that only edges that either are sufficiently long or sufficiently strong (the intensity changes substantially across them) are kept (a primitive saliency measure). This process helped significantly in screening *bad* edges, which reflect some local change in intensity values but not a physical discontinuity. Figure 5 shows the results of the trained net applied to a new image.

Finally, we have obtained some preliminary results on natural images using the output of the MRF machinery. Figure 6 shows the classification achieved by the linear classifier on such data (Fig. 2).

Bibliography

[1] H. G. Barrow and J. M. Tenenbaum, Recovering Intrinsic Scene Characteristics from Images, In A. R. Hanson and E. M. Riseman, editors, *Computer Vision Systems*, pp. 3–26 Academic Press, New York, 1978.

[2] M. Bertero, T. Poggio, and V. Torre, Ill-Posed Problems in Early Vision, *Proceedings of the IEEE*, vol. 76, pp. 869–889, 1988.

[3] R. Brooks, Intelligence without representation, In *Proceedings of Foundations of AI Workshop*, Massachusetts Institute of Technology, Cambridge, MA, June 1987.

[4] H. H. Bülthoff, J. J. Little, and T. Poggio, A Parallel Algorithm for Real-Time Computation of Motion, *Nature*, vol. 337, pp. 549–553, 1989.

[5] H. H. Bülthoff and H. A. Mallot, Interaction of Different Modules in Depth Perception, In *Proceedings of the 1st International Conference on Computer Vision*, pp. 295–305, June 1987.

[6] J. F. Canny, A Computational Approach to Edge Detection, *IEEE Transactions on Pattern Analysis and Machine Intelligence*, vol. 8, pp. 679–698, 1986.

[7] M. Drumheller and T. Poggio, On Parallel Stereo, In *Proceedings of IEEE Conference on Robotics and Automation*, 1986.

[8] E. Gamble, D. Geiger, T. Poggio, and D. Weinshall, Labeling Edges and the Integration of Low-Level Visual Modules, *IEEE Transactions on Systems, Man and Cybernetics*, vol. 19, pp. 1576–1581, 1989.

[9] D. Geiger and F. Girosi, Parallel and Deterministic Algorithms for MRFs: Surface Reconstruction and Integration, A.I. Memo No. 1114, Artificial Intelligence Laboratory, Massachusetts Institute of Technology, May 1989.

[10] S. Geman and D. Geman, Stochastic Relaxation, Gibbs Distributions, and the Bayesian Restoration of Images, *IEEE Transactions on Pattern Analysis and Machine Intelligence*, vol. 6, pp. 721–741, 1984.

[11] A. Hurlbert and T. Poggio, Do Computers Need Attention?, *Nature*, pp. 321(12), 1986.

[12] A. Hurlbert and T. Poggio, Synthesizing a Color Algorithm From Examples, *Science*, vol. 239, pp. 482–485, 1988.

[13] A. C. Hurlbert, *The Computation of Color*, Ph.D. Thesis, Massachusetts Institute of Technology, Cambridge, MA, 1989.

[14] J. J. Little, T. Poggio, and E. B. Gamble, Jr., Seeing in Parallel: The Vision Machine, *International Journal of Supercomputing Applications*, vol. 2, pp. 13–28, 1988.

[15] J. V. Mahoney, Image Chunking: Defining Spatial Building Blocks for Scene Analysis, Master's Thesis, Massachusetts Institute of Technology, Cambridge, MA, 1986.

[16] D. Marr and T. Poggio, Cooperative Computation of Stereo Disparity, *Science*, vol. 194, pp. 283–287, 1976.

[17] J. Marroquin, S. Mitter, and T. Poggio, Probabilistic Solution of Ill-Posed Problems in Computational Vision, *Journal of the American Statistical Association*, vol. 82, pp. 76–89, 1987.

[18] J. L. Marroquin, Optimal Bayesian Estimators for Image Segmentation and Surface Reconstruction, A.I. Memo No. 839, Artificial Intelligence Laboratory, Massachusetts Institute of Technology, Cambridge, MA, April 1985.

[19] T. Poggio and W. Reichardt, Visual Control of Orientation Behavior in The Fly (parts i and ii), *Quart. Rev. Biophys.*, vol. 3, pp. 311–439, 1976.

[20] V. Torre and T. Poggio, On Edge Detection, *IEEE Transactions on Pattern Analysis and Machine Intelligence*, vol. 8, pp. 147–163, 1986.

[21] S. Ullman, Visual Routines, *Cognition*, vol. 18, pp. 97–159, 1984.

[22] H. Voorhees and T. Poggio, Computing Texture Boundaries From Images, *Nature*, vol. 333, pp. 364–367, 1988.

[23] A. L. Yuille and T. Poggio, A Generalized Ordering Constraint for Stereo Correspondence, A.I. Memo No. 777, Artificial Intelligence Laboratory, Massachusetts Institute of Technology, Cambridge, MA, 1984.

Parameter Estimation for Gibbs Distributions from Fully Observed Data

Basilis Gidas
Division of Applied Mathematics
Brown University
Providence, Rhode Island

1 Introduction

In this chapter we study parameter estimation for Gibbs distributions—equivalently, Markov Random Fields (MRF)—over $\mathbf{Z}^d, d \geq 1$, from *fully observed (complete) data.* The distributions are parametrized by points in a finite-dimensional Euclidean space $\mathbb{R}^m, m \geq 1$; the interactions are translation invariant but not necessarily of finite range; and the single pixel random variables take values either in a finite or a compact state space. The parameters are estimated from a single realization of a MRF observed in a finite "window" (or "volume") Λ of $\mathbf{Z}^d$. We study consistency and asymptotic normality of estimators as $\Lambda \to \mathbf{Z}^d$ in the sense of van Hove (see Section 2). This framework generalizes that of time-series analysis, and it differs from the usual set up of point estimation where one studies asymptotic properties as the number of samples increases. We prove that the maximum likelihood (ML) estimators are consistent even at points of "phase transitions," irrespectively of ergodicity or stationarity. Furthermore, we establish asymptotic normality and efficiency under appropriate conditions. In [7], similar questions for ML estimators obtained from *partially observed (incomplete) data* are studied.

While consistency of ML estimators holds always, asymptotic normality fails in general if the true parameters are at phase transition points *and* the true distributions are not ergodic. In the latter case, one observes a

Partially supported by ARO DAAL03-90-G-0033 and ONR Contract N00014-88-K-0289.

ISBN 0-12-170608-7

phenomenon reminiscent to Bahadur's superefficiency. This phenomenon is studied in [8].

Gibbs distributions (to be referred also as *infinite-volume Gibbs distributions*) is a class of distributions obtained as limits of certain regular exponential families—the *finite volume Gibbs distributions* (see Section 2). They have played a fundamental role in the description of statistical mechanics systems [14,42,44] as well as of Gauge Field Theories [49]. More recently, they have been used [19,26] to provide a solid framework for studying image processing tasks, and computational vision in general. They have been used also in neural modelling and perceptual inference [1,27,28,47]. The framework of [19] has been tested successfully in a number of applications: image restoration [4,12,19,24,26,40,41], computer tomography [22], segmentation, boundary finding, and texture classification and recognition [18,19,20]. Our present work as well as the ones in [7,23] and [8] have been motivated by these applications.

The simplest example of Gibbs distributions is provided by the Ising model. This model is parametrized in terms of two parameters: the temperature T (or its inverse $\beta = 1/T$), and the external field $h \in \mathbb{R}$. The (infinite-volume) Gibbs distributions associated with a fixed pair (β, h) are probability measures on $\Omega = \{-1, 1\}^{\mathbf{Z}^d}$, obtained as (weak) limits as $\Lambda \to \mathbf{Z}^d$, of finite-volume Gibbs distributions with various "boundary conditions." The finite-volume Gibbs distributions in a finite volume $\Lambda \subset \mathbf{Z}^d$ read

$$\pi_{\Lambda,y}(x; \beta, h) = \frac{e^{-H_{\Lambda,y}(x;\beta,h)}}{Z_{\Lambda,y}(\beta, h)} \tag{1}$$

$$H_{\Lambda,y}(x; \beta, h) = -\beta \sum_{<ij>\in\Lambda} x_i x_j - h \sum_{i\in\Lambda} x_i - \beta \sum_{\substack{<i,j>:i\in\Lambda \\ j\in\Lambda^c}} x_i y_j \tag{2}$$

where $\Lambda^c = \mathbf{Z}^d \backslash \Lambda, < i, j >$ denotes nearest-neighbor pixels (i.e. $|i - j| = 1$); $Z_{\Lambda,y}(\beta, h)$ is a normalizing constant called the *partition function*; $x_i, i \in \mathbf{Z}^d$ is the single pixel random variable taking values in $\Omega_0 = \{-1, 1\}$; and $y = \{y_i : i \in \Lambda^c\}$ is a "boundary condition." The boundary condition (b.c.) y is a specified set of values of the x_i's, $i \notin \Lambda$. More generally, a b.c. is specified by specifying the distribution of the x_i's for $i \notin \Lambda$. We consider also *"free" b.c.* corresponding to $y_i = 0$ for all $i \in \Lambda^c$ and *"periodic" b.c.* corresponding to the case when Λ is a torus.

In general, Gibbs distributions are parametrized by functions (i.e., the interactions) in a certain infinite dimensional Banach space [44] (see also Section 2). In this chapter, we parametrize the distributions by an m-dimensional, $m \geq 1$, vector θ, and the finite-volume Gibbs distributions are certain regular exponential families of the form

$$d\pi_{\Lambda,y}(x;\theta) = \frac{e^{-H_{\Lambda,y}(x;\theta)}}{Z_{\Lambda,y}(x)} \Pi_{i\in\Lambda} d\mu_0(x_i), H_{\Lambda,y}(x;\theta) = \theta \cdot U_{\Lambda,y}(x), \quad (3)$$

where $x = \{x_i : i \in \Lambda\}$ with each x_i taking values in some compact space $\Omega_0; y = \{y_i : i \notin \Lambda\}$ is a b.c.; $d\mu_0(\cdot)$ is a probability measure on Ω_0 (it does not depend on θ); and $U_{\Lambda,y}(x)$ is the *energy function* made up from interactions that are assumed to be fixed. The interactions are assumed to be translation invariant but not necessarily of finite range (i.e., pixels can interact with arbitary far away pixels). The (infinite-volume) Gibbs distributions associated with a value of the parameter θ are probability measures on $\Omega = \Omega_0^{\mathbf{Z}^d}$ satisfying certain equations known as Dobrushin-Landford-Ruelle (DLR) equations [44] (see also Section 2). They can also be characterized as (weak) limits of the finite-volume Gibbs distributions (3) as $\Lambda \to \mathbf{Z}^d$ (in the sense of van Hove) with various b.c. Our consistency theorem (Section 2) does *not* hold if the Gibbs distributions are parametrized by the previously mentioned infinite dimensional Banach space. In fact in this case, consistency fails in general. Perhaps the method of sieves [21,25] could be adopted to this case to yield consistent ML estimators. This problem is mathematically interesting (nonparametric estimation) and important for applications (it amounts in estimating the interactions). We do not study it here. Also, our single pixel state space Ω_0 is assumed to be compact (or finite). Our techniques apply to noncompact Ω_0 (e.g., $\Omega_0 = \mathbb{R}^n$), but they require certain technical modifications.

The fundamental difficulty that divides the statistical inference of Gibbs distributions from that of time series is the *long-range dependence*. In time-series analysis, short-range dependence is the rule, and special models are needed to exhibit long-range dependence. In contrast, long-range dependence occurs naturally in MRF. This long-range dependence is closely related to the occurence of phase transitions ("first" or "higher order" [42]). This means that for certain values of the parameters θ there may exist more than one (infinite-volume) Gibbs distributions. In the Ising model, for example, if $d \geq 2, h = 0$ and $T < T_c$, where T_c is a *critical temperature*, then there are more than one Gibbs distributions corresponding to the pair $h = 0, T < T_c$. For $d = 2$, the set of Gibbs distributions corresponding to $h = 0, T < T_c$ is given by

$$P = \lambda P_T^+ + (1-\lambda)P_T^-, 0 \leq \lambda \leq 1, \quad (4)$$

where $P_T^{\pm}$ are two translation-invariant, ergodic, Gibbs distributions obtained from (1) and (2) with $y_i = \pm 1, i \in \Lambda^c$, as $\Lambda \to \mathbf{Z}^d$. In the general case when the distributions are parametrized by the m-dimensional vector θ, the set $G(\theta)$ of Gibbs distributions associated with a particular value of θ is known [44] to be convex, compact, and Choquet simplex. This last property implies in particular that every $P \in G(\theta)$ has a *unique* decomposition into the extremal measures of $G(\theta)$, i.e., there exists a unique probability measure $d\sigma(\cdot)$ on the set $\mathcal{E}(\theta)$ of the extremal elements of $G(\theta)$ such that

$$P = \int_{\mathcal{E}(\theta)} P_\theta(\xi) d\sigma(\xi). \tag{5}$$

For the two-dimensional Ising model, (5) reduces to (4). For translation-invariant interactions (assumed in this chapter), $G(\theta)$ always contains translation invariant-distributions, but it may also contain [13] nontranslation invariant measures. The extremal Gibbs distributions (i.e., the elements of $\mathcal{E}(\theta)$) have a trivial tail-field ("algebra at infinite" [44]). Furthermore, the translations-invariant elements of $\mathcal{E}(\theta)$ are ergodic and often have good mixing properties.

When $G(\theta)$ is not a singleton, we say that a phase transition occurs at θ. The set of parameter values where phase transitions occur form [44] a manifold of lower dimension than the parameter space. The occurence of phase transitions is closely associated with discontinuities of certain expectations, and with the nonanalytic behavior of "thermodynamic" functions such as the "pressure" (see Section 2). It is this nonuniqueness and nonanalyticity that complicates the proof of consistency of ML estimators and leads, in general, to the breakdown of asymptotic normality when the *true* distributions are not in $\mathcal{E}(\theta)$ (consistency is also complicated by the possible existence of nontranslation invariant Gibbs distributions). For the Ising model for example, consistency holds always, but if the true parameters are $h = 0, T < T_c$ *and* the true distributions is not P_T^+ or P_T^-, then the ML estimator of $h = 0$ is not asymptotically normal (although it is strongly consistent). Moreover, in this case one observes [8] a phenomenon similar to Bahadur's superefficiency. Furthermore, for the two-dimensional Ising model with no external field, we show (Theorem 3) that if the true temperature is T_c, then the ML estimator of T_c exhibits a nonstandard asymptotic behavior.

As we mentioned before, we are interested in estimating the parameters from a single realization observed in a finite volume $\Lambda \subset \mathbf{Z}^d$. More precisely, the data may arise in two ways: (i) there is an underlying infinite sample from an infinite-volume Gibbs distribution, but we observe larger and larger pieces of it, and (ii) we observe finite samples from finite-volume Gibbs distributions (with the same parameter θ) in larger volumes (the samples in different volumes are independent from each other). The two cases are slightly different. Our (strong) consistency theorem for both cases is stated in Section 2 and proven in Section 3. Our asymptotic normality and efficiency results are stated in Section 2 and proven in Section 4. They hold under appropriate condition on the Gibbs distributions. Due to the possible occurance of phase transitions and the possible existence of nontranslation invariant Gibbs distributions, the classical techniques [48,29] for proving consistency, or their extensions by Mann and Wald [35] to time series, do not directly apply to the present case. If the underlying true distribution is not translation invariant, or in case ii, the sufficient statistics for θ and the log-likelihood function may *not* converge a.s. or in probability. In Section 3 we show that (in some sense; see (32)) the limit points of the sufficient statistics for θ lie in a certain "tangent cone." This property and the strict convexity of the "pressure" (see Section 2) are basic ingredients in our proof that the ML estimator for θ is always strongly consistent.

There are at least two aspects of the statistical inference for Gibbs distributions that do not generally arise in the standard problems of point estimation:

a) In each finite volume $\Lambda \subset \mathbf{Z}^d$, the parameter θ is associated with a certain sufficient statistic (e.g., in the Ising model, the sufficient statistic for (β, h) is $\left[\sum_{<i,j>\in\Lambda} x_i x_j, \sum_{i\in\Lambda} x_i\right]$; see (1) and (2). The components of the sufficient statistic for θ, properly normalized, are themselves estimators of certain expectations of the true (infinite-volume) distribution. But these estimators may *not* be consistent if the true parameters are at points of phase transitions *and* the true distributions are *not* extremal. In the Ising model, for example, the sufficient statistic $M_\Lambda(x) = |\Lambda|^{-1}\sum_{i\in\Lambda} x_i$ for the external field h is an estimator of the average magnetization $m(T, h)$, i.e., of the mean of the true (infinite-volume) Gibbs distribution. However, if the true parameters are $h_0 = 0$ and $T_0 < T_c$, and the true distribution is not extremal (e.g., not $P^{\pm}_{T_0}$ in the two-dimensional model), then $M_\Lambda(x)$ does *not* converge a.s. or in probability to the true average magnetization $m(T_0, 0)$ (it converges a.s. to a random variable supported in $[-m^+(T_0,0), m^+(T_0,0)]$, where $m^+(T_0, 0)$ is the average magnetization of $P^+_{T_0}$). This appears to be the source of an erroneous conclusion in [38] (also repeated in [39]) that the ML estimator of $h_0 = 0$ (when $T_0 < T_c$) is not consistent.

b) In addition to estimating the parameter θ, there are two other estimation issues in the statistical inference of Gibbs distributions. First, because of the possibility of the occurence of phase transitions, the problem of estimating θ is distinct from the problem of estimating the distributions themselves. The problem of estimating the true distribution itself is equivalent to estimating the unique(!) probability measure $d\sigma(\cdot)$, which appears in the decomposition (5). For the two-dimensional Ising model this amounts to estimating the parameter λ in (4). For this model, the parameter λ is in one to one correspondence with the average magnetization $m(T, h)$. But for $T < T_c$, the average magnetization $m(T, h)$ is discontinuous in $h[m(T, h) \to m^+(T, 0)$ as $h \downarrow 0$, and $m(T, h) \to -m^+(T, 0)$ as $h \uparrow 0]$. Thus, the ML estimator of $m(T, h)$, and hence of λ, is not consistent for $T < T_c$. In general, the measure $d\sigma(\cdot)$ is related to the *order parameters* [42,46] of a model. If the true parameter θ_0 is at a "first" order phase transition, then the ML estimator of the order parameter(s) is not, in general, consistent (in constrast to the ML estimator of θ). The other estimation issue concerns the boundary condition y (see (1), (2) and (3)). Boundary conditions may be thought of as additional parameters. These parameters are unidentifiable.

From the computational point of view, ML estimation for Gibbs distributions has a drawback: the log-likelihood function involves the partition function $Z_{\Lambda,y}(\theta)$ (see (3)), which is computationally intractable. However, the log-likelihood function is convex in θ, and Geman and Geman introduced [17] a stochastic gradient algorithm that was used by Lippman [34] in the context of a probabilistic expert system. The algorithm is computationally intensive but possible. A similar algorithm has been studied more recently in [50], and a closely related algorithm has been used by Hinton and Sejnowski [27] to model learning in a theory of neuron dynamics. Besag introduced [2,3] an alternative to ML estimation—the maximum pseudo-likelihood (MPL) method. This method is computationally efficient, but in some cases it does not yield satisfactory numerical results: computer experiments for Ising type models indicate that if the true temperature T_0 is below T_c and the true external field h_0 is very small (nearly zero), then the MPL estimator of h_0 (in constrast to the MPL estimator of T_0) has a very large (experimental) bias. It would be interesting to see whether the ML method provides more accurate results in this case. The consistency of MPL estimators has been established in [20] (see [23] for an alternative proof). The MPL estimators are, under appropriate conditions, asymptotically normal but *not* efficient. Some modifications of the MPL method have been introduced in [5]. Another alternative to ML estimation is a logisticlike method introduced by Derin and Elliott [10] and analyzed extensively by Possolo [39]. References [11] and [31] also address the parameter estimation problem for Gibbs distributions.

The organization of this chapter is as follows: In Section 2, we set up our notation, summarize some properties of Gibbs distributions, and state our main results of consistency and asymptotic normality. The consistency theorem is proven in Section 3. In Section 4, we prove asymptotic normality and efficiency results. Finally, in the Appendix we provide a necessary and sufficient condition for the identifiability of the parameters of Gibbs distributions.

2 Notation and Main Results

In this section we set up our notation, summarize those properties of Gibbs distributions we will use later, and state our main results.

Let Ω_0 be a finite set or a compact space, and $\mathbf{Z}^d, d \geq 1$, be the d-dimensional lattice. With each pixel $i \in \mathbf{Z}^d$, we associate a random variable x_i taking values in Ω_0. The space Ω_0 will be referred to as the *single pixel state space.* The space $\Omega = \Omega_0^{\mathbf{Z}^d}$ is the *configuration* or *state* space. If $x \in \Omega$, then $x = \{x_i : i \in \mathbf{Z}^d\}$. If V is a subset of $\mathbf{Z}^d$, then the space $\Omega_V = \Omega_0^V$ is the configuration space in the "volume" $V \subset \mathbf{Z}^d$. We denote by $x(V) = \{x_i : i \in V\}$ a configuration in Ω_V.

Gibbs distributions are defined in terms of *interactions.* An interaction Φ is a real continuous map

$$\Phi : \cup_{\substack{V \subset \mathbf{Z}^d \\ \text{finite}}} \Omega_V \to I\!\!R.$$

Let Λ be a finite subset of $\mathbf{Z}^d$, and $\Lambda^c = \mathbf{Z}^d \backslash \Lambda$. Let y^* be a configuration in Ω, and $y = y^*(\Lambda^c) = \{y_i : i \in \Lambda^c\}$ its restriction to Ω_{Λ^c}. For an interaction Φ, the *energy* in the finite-volume Λ with *boundary condition* (b.c.) y, is

$$U_{\Lambda,y}(x(\Lambda)) = \sum_{V \subset \Lambda} \Phi(x(V)) + \sum\nolimits'_{V \subset \mathbf{Z}^d} \Phi(x(V) \vee y(V)), \tag{6}$$

where the sum $\sum'$ extends over finite $V \subset \mathbf{Z}^d$ such that $V \cap \Lambda \neq \phi, V \cap \Lambda^c \neq \phi$, and the configuration $x(V) \vee y(V)$ is defined by

$$(x(V) \vee y(V))_i = \begin{cases} x_i & \text{if} \quad i \in \Lambda \\ \\ y_i & \text{if} \quad i \in \Lambda^c \end{cases}, \quad i \in V. \tag{7}$$

In this paper the Gibbs distributions will be specified in terms of $m \geq 1$ *fixed* interactions $\Phi^{(\alpha)}$ (with corresponding energy functions $U^{(\alpha)}_{\Lambda,y}$) and will be parametrized by a vector $\theta = \{\theta^{(\alpha)}\}_{\alpha=1}^m \in I\!\!R^m$. The parameter space Θ will be taken to be $I\!\!R^m$ or a convex subset of $I\!\!R^m$. The *Hamiltonian* in the finite-volume Λ with b.c. y, is defined to be ($x \in \Omega_\Lambda$)

$$H_{\Lambda,y}(x;\theta) = -\sum_{\alpha=1}^{m} \theta^{(\alpha)} U_{\Lambda,y}^{(\alpha)}(x) \equiv -\theta \cdot U_{\Lambda,y}(x), U_{\Lambda,y} = \{U_{\Lambda,y}^{(\alpha)}(x)\}_{\alpha=1}^{m}. \quad (8)$$

We will also treat *free b.c.* corresponding to $y = 0$, and *periodic b.c.* corresponding to Λ being a torus. The interactions $\Phi^{(\alpha)}, \alpha = 1, \ldots, m$, will be assumed to be *translation invariant* and to satisfy

$$\begin{aligned} \| \Phi^{(\alpha)} \| &= \sum_{\substack{0 \in V \subset \mathbf{Z}^d \\ \text{finite}}} \sup_{x(V)} |\Phi^{(\alpha)}(x(V))| \\ &< +\infty, \alpha = 1, \ldots, m, \| U \|^2 = \sum_{\alpha=1}^{m} \| \Phi^{(\alpha)} \|^2 . \end{aligned} \quad (9)$$

The space $\mathcal{B}$ of interactions Φ satisfying Eq. (9) is a separable real Banach space. An interaction Φ is said to be of *finite range* if $\Phi(x(V)) = 0$ whenever the diameter of V is larger than some R_0—*the radius of interaction.* The set of finite range interactions equipped with the norm (9) form a real vector space $\mathcal{B}_0$, which is dense in $\mathcal{B}$. In general, Gibbs distributions are parametrized [44] by $\mathcal{B}$, but in this chapter we fix the interactions $\Phi^{(\alpha)}, \alpha = 1, \ldots, m$, in $\mathcal{B}$, and parametrize the distributions by $\theta \in I\!R^m$ as in Eq. (8). Note that the energy $U_{\Lambda,y}^{(\alpha)}$ can be written as follows:

$$U_{\Lambda,y}^{(\alpha)}(x) = \sum_{i \in \Lambda} \sum\nolimits''_{i \in V \subset \mathbf{Z}^d} \frac{\Phi^{(\alpha)}(x(V) v y(V))}{|V \cap \Lambda|}, \quad (10)$$

where the sum $\sum''$ extends over all finite $V \subset \mathbf{Z}^d$. Later we will use the functions

$$A_{U^{(\alpha)}}(x) = \sum_{\substack{0 \in V \subset \mathbf{Z}^d \\ V \text{ finite}}} \frac{\Phi^{(\alpha)}(x(V))}{|V|}, \quad A_U = \left\{ A_U^{(\alpha)} \right\}_{\alpha=1}^{m}. \quad (11)$$

Let μ_0 be a probability measure on Ω_0, and set

$$d\mu_{0,\Lambda}(x) = \Pi_{i \in \Lambda} d\mu_0(x_i). \quad (12)$$

If Ω_0 is a finite set, we take μ_0 to be the counting measure. $d\mu_0$ will be referred to as the *single pixel distribution.*

Next we define the *finite-volume Gibbs distributions* and the (infinite-volume) *Gibbs distributions.* The finite-volume Gibbs distribution in volume Λ finite $\subset \mathbf{Z}^d$ with b.c. $y \in \Omega_{\Lambda^c}$ is given by the probability measure on Ω_Λ

$$\begin{aligned} P_{\theta,y}^{(\Lambda)}(x) &= \pi_{\Lambda,y}(x;\theta)d\mu_{0,\Lambda}(x), \quad x \in \Omega_\Lambda, \\ \pi_{\Lambda,y}(x;\theta) &= \frac{e^{-H_{\Lambda,y}(x;\theta)}}{Z_{\Lambda,y}(\theta)}, \\ Z_{\Lambda,y}(\theta) &= \int_{\Omega_\Lambda} e^{-H_{\Lambda,y}(x;\theta)}d\mu_{0,\Lambda}(x). \end{aligned} \tag{13}$$

The normalizing constant $Z_{\Lambda,y}(\theta)$ is called the *partition function.* An example is given by the binary Ising model (1) and (2), which are special cases of the *general Ising model* defined by the Hamiltonian

$$H_{\Lambda,y}(x;\beta,h) = -\beta \sum_{\substack{i,j\in\Lambda \\ i\neq j}} J(|i-j|)x_i x_j - h\sum_{i\in\Lambda} x_i - \beta \sum_{\substack{i\in\Lambda \\ j\in\Lambda^c}} J(|i-j|)x_i y_j \tag{14}$$

with

$$J(k) \geq 0, \sum_{k\in\mathbf{Z}^d} J(k) < +\infty. \tag{15}$$

The single pixel distribution $d\mu_0(x_i)$ for this model is a probability measure on $\mathbb{R}$. The binary Ising model corresponds to $d\mu_0(x_i) = [\frac{1}{2}\delta(x_i+1) + \frac{1}{2}\delta(x_i-1)]dx_i$.

A (infinite-volume) Gibbs distribution associated with the interactions $\Phi^{(\alpha)}, \alpha = 1,\ldots,m$, and parametrized by $\theta = \{\theta^{(\alpha)}\}_{\alpha=1}^m \in \mathbb{R}^m$, is a probability measure P_θ on Ω whose conditional probability that $x|_{\Omega_\Lambda} = x(\Lambda)$ when it is known that $x|_{\Omega_{\Lambda^c}} = x(\Lambda^c)$ is given by

$$P_\theta^{(\Lambda)}(dx(\Lambda)|x(\Lambda^c)) = \pi_{\Lambda,x(\Lambda^c)}(x(\Lambda);\theta)d\mu_{0,\Lambda}(x) \tag{16}$$

for every finite subset of Λ of $\mathbf{Z}^d$. There is [44] an equivalent definition of Gibbs distributions: Let P be a probability measure on Ω, and $P^{(\Lambda)}$ its restriction to Ω_Λ. Then P is an infinite-volume Gibbs distribution associated with the interactions $\Phi^{(\alpha)}, \alpha = 1,\ldots,m$, and the parameter θ, if for all finite $\Lambda \subset \mathbf{Z}^d$ there exists a probability measure $\rho^{(\Lambda^c)}$ on Ω_{Λ_c} such that

$$P^{(\Lambda)}(dx(\Lambda)) = \int_{\Omega_{\Lambda^c}} \rho^{(\Lambda^c)}(d\eta)\pi_{\Lambda,\eta}(x(\Lambda);\theta)d\mu_{0,\Lambda}(x). \tag{17}$$

The measure $\rho^{(\Lambda^c)}$ can be taken [44] to be $P^{(\Lambda^c)}$—the restriction of P to Ω_{Λ^c}. Eq. (17) is known as the Dobrushin-Lanford-Ruelle (DLR) equation.

Let $G(\theta)$ be the set of all Gibbs distributions corresponding to the parameter value θ (since we fix the interactions $\Phi^{(\alpha)}, \alpha = 1, \ldots, m$, we index the set of Gibbs distributions by the parameters θ). As we mentioned in the introduction, the set $G(\theta)$ is nonempty, convex compact, and Choquet simplex. There is a concrete way to construct the Gibbs distributions and identify the set $G(\theta)$: The finite-volume Gibbs distributions (13) with prescribed b.c. y, converge weakly, along subsequences $\Lambda_n \to \mathbf{Z}^d$, to Gibbs distributions. The closed convex hull of the Gibbs distributions obtained this way is exactly the set $G(\theta)$ [44].

This definition of infinite-volume Gibbs distributions and their properties listed in the last paragraph above, do not require that the interactions be translation invariant (they should only satisfy Eq. (9)). But if the interactions are translation invariant, then there exist translation invariant Gibbs distributions. However, there may exist Gibbs distributions that are not translation invariant. This is the case [13], for example, for the three-dimensional (but not the two-dimensional) Ising model when $h = 0$ and $T = 1/\beta$ is sufficiently small. Let $I = I(\Omega)$ be the set of the translation invariant probability measures on Ω. The set $G(\theta) \cap I$ is known [44, pp. 42 and 60] to be also convex and Choquet simplex. Let $\mathcal{E}_\theta^{(0)} = \mathcal{E}(G(\theta) \cap I)$ be the set of extremal points of $G(\theta) \cap I$. The fact that $G(\theta) \cap I$ is convex and compact implies that every $P_\theta \in G(\theta) \cap I$ has a decomposition into extremal measures as follows (compare with Eq. (5)):

$$P_\theta = \int_{\mathcal{E}_\theta^{(0)}} P_\theta(\xi) d\tilde{\sigma}(\xi) \tag{18}$$

where $d\tilde{\sigma}(\xi)$ is a probability measure on $\mathcal{E}_\theta^{(0)}$. The fact that $G(\theta) \cap I$ is a Choquet simplex implies that the $d\tilde{\sigma}(\xi)$ is unique. The elements of $\mathcal{E}^{(0)}(\theta)$ are [43] ergodic and often have good mixing properties. Our consistency theorem holds for both translation and non-translation invariant Gibbs distribution (its proof is somewhat simpler for translation invariant distributions). The proof makes use of the decomposition (18).

Now we come to the parameter estimation problem. As we mentioned in the introduction, we are interested in estimating the vector-valued parameter θ from a single realization $x(\Lambda)$ in a finite volume $\Lambda \subset \mathbf{Z}^d$, and study consistency and asymptotic properties as $\Lambda \to \mathbf{Z}^d$. The sequence (or net) of observations $x(\Lambda)$ associated with an expanding sequence (net) of volumes $\Lambda \subset \mathbf{Z}^d$ may arise in two ways: *Case 1.* There is an infinite sample $x \in \Omega$ from an infinite volume Gibbs distributions, and we observe larger and larger pieces, $x(\Lambda) = x|_\Lambda$, of it or *Case 2.* The sequence of observations $\{x(\Lambda)\}$ is a sequence of samples (possible independent) from a sequence of finite-volume Gibbs distributions with the same parameter vector θ_0 (but not necessarily the same boundary condition). We introduce two maximum log-likelihood functions. The first log-likelihood function reads

$$\begin{aligned} \ell_{\Lambda,y}(x(\Lambda);\theta) &= -\frac{1}{|\Lambda|}\log \pi_{\Lambda,y}(x(\Lambda);\theta) \\ &= \frac{1}{|\Lambda|}\log Z_{\Lambda,y}(\theta) + \frac{1}{|\Lambda|}H_{\Lambda,y}(x(\Lambda);\theta). \end{aligned} \tag{19}$$

The b.c. $y \in \Omega$ is arbitrary but fixed and will be taken as known. Our second log-likelihood function is defined as follows: Let P_θ be an infinite-volume Gibbs distribution corresponding to the parameter θ. Let $P_\theta^{(\Lambda)}$ be the restriction of P_θ to Ω_Λ, and $f_\Lambda(x(\Lambda);\theta)$ the Radon-Nikodyn derivative of $P_\theta^{(\Lambda)}$ with respect to $d\mu_{0,\Lambda}$. Then the second log-likelihood function is

$$\ell'_\Lambda(x(\Lambda);\theta) = -\frac{1}{|\Lambda|}\log f_\Lambda(x(\Lambda);\theta). \tag{20}$$

From the computational point of view, the log-likelihood function (19) is more tractable than (20), but from the mathematical point of view (20) is a natural log-likelihood function. Our consistency theorem holds for both likelihood functions (the proof for (19) is slightly easier than the proof for (20)).

Before we state our results, we introduce the *pressure* (or *free energy*) function and a condition that ensures the *identifiability* of the parameters. The finite-volume pressure is defined by

$$p_{\Lambda,y}(\theta) = \frac{1}{|\Lambda|}\log Z_{\Lambda,y}(\theta) \tag{21}$$

and satisfies the following proposition [30,44]:

Proposition 1: a) $p_{\Lambda,y}(\theta)$ is a convex function of θ, b) $|p_{\Lambda,y}(\theta) - p_{\Lambda,y}(\theta')| \leq |\theta-\theta'|\|U\|$, c) $|p_{\Lambda,y}(\theta)| \leq |\theta|\|U\|$, d) the limit

$$\lim_{\Lambda\to\mathbf{Z}^d} p_{\Lambda,y}(\theta) = p(\theta) \tag{22}$$

exists and is independent of y. Furthermore, $p(\theta)$ inherits properties a), b), and c).

Remark 1. The limit $\Lambda \to \mathbf{Z}^d$ in Eq. (22) and throughout this paper will be taken in the sense of van Hove [44]. For simplicity, one may assume that Λ is a hypercube of side N, in which case $\Lambda \to \mathbf{Z}^d$ means $N \to +\infty$.

Remark 2. The infinite-volume pressure $p(\theta)$ is convex but, in general, not differentiable at phase transition points. In fact, phase transitions are closely related to the nonanalytic behavior of $p(\theta)$.

For our consistency result we will assume *identifiability* in the following sense:

Definition. We say that $\theta_0 \in I\!R^m$ (or $\theta_0 \in \Theta$) is *identifiable* if $\theta \neq \theta_0$ implies

$$G(\theta) \neq G(\theta_0). \tag{23}$$

Remark 3. Condition (23) has (see Appendix) various equivalent formulations based on the following facts: (i) $G(\theta)$ and $G(\theta')$ have one point in common iff $G(\theta) = G(\theta'), \theta, \theta' \in \Theta$, (ii) $G(\theta)$ and $G(\theta')$ have one point in common iff the pressure $p(\theta)$ is linear on the line segment connecting θ and θ', and (iii) $G(\theta)$ and $G(\theta')$ have one point in common iff the interactions $\sum_{\alpha=1}^{m} \theta^{(\alpha)} \Phi^{(\alpha)}$ and $\sum_{\alpha=1}^{m} \theta'^{(\alpha)} \Phi^{(\alpha)}$ are *physically equivalent* (see Appendix).

From (i) or (iii), we see that condition (23) is not restrictive; it simply states that for distinct θ's we have distinct physical situations. From part (ii) we see that if condition (23) holds, then the pressure is *strictly* convex.

Our consistency theorem, to be proven in Section 3, is

Theorem 1. *Let $\theta_0 = \{\theta_0^{(\alpha)}\}_{\alpha=1}^m$ be the true parameters and P_{θ_0} be any Gibbs distribution in $G(\theta_0)$. If θ_0 is identifiable, then a) for sufficiently large Λ, the minimizer $\hat{\theta}_{\Lambda,y}$ of $\ell_{\Lambda,y}(x(\Lambda);\theta)$ exists and is unique a.s., and b) independently of the b.c. y, we have $\hat{\theta}_{\Lambda,y} \to \theta$ as $\Lambda \to \mathbf{Z}^d, P_{\theta_0}$ - a.s.*

Remark 4. a) Theorem 1 holds if $\ell_{\Lambda,y}(x;\theta)$ is replaced by the log-likelihood function (19) (see Section 3), and b) the proof of Theorem 1 shows also convergence in probability, i.e., $P_{\theta_0}\{x(\Lambda) \in \Omega_\Lambda : |\hat{\theta}_{\Lambda,y} - \theta_0| \geq \epsilon\} \to 0$, as $\Lambda \to \mathbf{Z}^d$.

Remark 5. a) The b.c. y that enters in the definition of the log-likelihood function $\ell_{\Lambda,y}(x(\Lambda);\theta)$ is unrelated to the true distribution P_{θ_0}, i.e., if P_{θ_0} is obtained as a limit of a finite-volume Gibbs distribution $P_{\theta_{0,z}}^{(\Lambda)}$ with b.c. z, then y is unrelated to z. We will say that y *agrees with* P_{θ_0} if $y = z$ (a.s.), and that y *does not agree with* P_{θ_0} if $y \neq z$ (a.s.). Thus, Theorem 1 says that for consistency, the b.c. y need not agree with P_{θ_0}. But, we will see that for asymptotic normality, y must agree with P_{θ_0}. b) If we drop the identifiability condition (23), our proof of Theorem 1 shows that $\hat{\theta}_{\Lambda,y}$ eventually (P_{θ_0} - a.s.) stays inside any neighborhood of a compact set of parameter values that are *physically equivalent* to the true parameter θ_0. c) In general, Gibbs distributions are parametrized by the infinite dimensional Banach space $\mathcal{B}$ of the interactions Φ. In this case the estimation problem is a nonparametric problem, and our proof of consistency fails. In fact, in this infinite dimensional case, consistency fails in general. As we mentioned in the introduction, perhaps the method of sieves [21,25] could be adapted to obtain consistent ML estimators.

Next we state our basic theorem for the asymptotic normality and efficiency of $\hat{\theta}_{\Lambda,y}$. This theorem and another asymptotic normality result are proven in Section 4.

Theorem 2. *Let $\theta_0, \hat{\theta}_{\Lambda,y}$ be as in Theorem 1. Suppose that*

$$\lim_{\Lambda \to \mathbf{Z}^d} \frac{\partial^2 p_{\Lambda,y}(\theta)}{\partial\theta^\alpha \partial\theta^\beta} = C_{\alpha\beta}(\theta), \alpha, \beta = 1, \ldots, m \tag{24}$$

uniformly in some neighborhood D_{θ_0} of θ_0. Then

$$\sqrt{|\Lambda|}(\hat{\theta}_{\Lambda,y} - \theta_0) \underset{\Lambda \to \mathbf{Z}^d}{\overset{\mathcal{D}}{\longrightarrow}} N(0, (I(\theta_0))^{-1}), I(\theta_0) = (C_{\alpha\beta}(\theta_0)) \tag{25}$$

with respect to $P^{(\Lambda)}_{\theta_0,y}$. [Here and throughout the paper $\overset{\mathcal{D}}{\longrightarrow}$ indicates convergence in probability.]

Remark 6. a) The matrix $I(\theta_0)$ (which depends, in general, on y) is the analogue of the Fisher information matrix. Thus, Theorem 2 says that $\hat{\theta}_{\Lambda,y}$ is asymptotically efficient. b) If θ_0 is at a point of phase transition, then Eq. (24) does not in general hold; Theorem 5 of Section 4 establishes an alternative asymptotic normality result under different conditions, which may hold even if θ_0 is at a phase transition point. c) For the binary Ising model (1) and (2), (24) holds provided that the true parameter satisfies $h_0 \neq 0, \beta_0 > 0$, or $h_0 = 0, \beta_0 < \beta_c$; if $\beta > \beta_c$, then the pressure $p(\beta, h)$ has derivatives to all orders in β, and one-sided derivatives to all orders in h as $h \downarrow 0$ or $h \uparrow 0$; in two-dimensions ($d = 2$), these (one-sided) derivatives and a modification of the proof of Theorem 2 yield asymptotic normality (under $P^{\pm}_{\beta_0,0} = P^{\pm}_{1/T_0}$, see Eq. (4)) for the ML estimators $\hat{\beta}_{\Lambda,\pm}, \hat{h}_{\Lambda,\pm}$ of $\beta_0 > \beta_c$ and $h_0 = 0$.

For the binary Ising model (1) and (2) with zero external field ($h = 0$), the pressure $p(\beta) = p(\beta, 0)$ is once but not twice differentiable at $\beta = \beta_c$. The asymptotic behavior of the ML estimator for $\beta_0 = \beta_c$ appears to be complicated; in two dimensions ($d = 2$) it is given by the following theorem.

Theorem 3. *Consider the two-dimensional ($d = 2$) Ising model (1) and (2) with $h = 0$. Let $\hat{\beta}_\Lambda = \hat{\beta}_{\Lambda,p}$ be the ML estimator of $\beta_0 = \beta_c$, computed in terms of periodic b.c. (i.e., $y = p$). Then*

$$\sqrt{\frac{|\Lambda|}{\log|\Lambda|}}(\hat{\beta}_\Lambda - \beta_c) \log|\hat{\beta}_\Lambda - \beta_c| \underset{\Lambda \to \mathbf{Z}^2}{\overset{\mathcal{D}}{\longrightarrow}} N(0, \sigma^2) \tag{26}$$

with respect to the finite volume Gibbs distribution at β_c with periodic b.c; here σ^2 is a computable constant, and Λ is a hypercube of side N.

Remark 7. Since the Ising-model has a unique infinite-volume Gibbs distribution at $\beta = \beta_c$, we expect Theorem 3 to hold for any b.c.

3 Proof of Consistency

Let

$$\begin{aligned} h_{\Lambda,y}(x(\Lambda);\theta_0;\theta) &= \ell_{\Lambda,y}(x(\Lambda);\theta) - \ell_{\Lambda,y}(x(\Lambda);\theta_0) \\ &= p_{\Lambda,y}(\theta) - p_{\Lambda,y}(\theta_0) - (\theta-\theta_0)\cdot \tfrac{1}{|\Lambda|}U_{\Lambda,y}(x(\Lambda)). \end{aligned}$$

Throughout this and the next sections, $G_s(\theta)$ (resp. $G_e(\theta)$) will denote the translation invariant (resp. ergodic) Gibbs distributions, i.e., $G_s(\theta) = G(\theta)\cap I, G_e(\theta) = \mathcal{E}_\theta^0$.

Theorem 4. *a) If $P \in G_s(\theta_0)$, then P-a.s.*

$$h_{\Lambda,y}(x(\Lambda);\theta_0;\theta) \underset{\Lambda\to\mathbf{Z}^d}{\longrightarrow} h(\theta_0;\theta)(\cdot) = p(\theta) - p(\theta_0) - (\theta-\theta_0)E_P(A_U|\mathcal{S}). \quad (27)$$

Furthermore, $h(\theta_0;\theta)(\cdot) \geq 0$ with equality iff $\theta = \theta_0$; here $\mathcal{S}$ denotes the σ-field formed by the translation invariant (measurable) subsets of Ω, and A_U is given by Eq. (11). b) Let

$$\underline{h}(\theta_0;\theta)(\cdot) = \liminf_{\Lambda\to\mathbf{Z}^d} h_{\Lambda,y}(x(\Lambda);\theta_0;\theta). \quad (28)$$

Then $\underline{h}(\theta_0;\theta)(\cdot) \geq 0$ P-a.s. for every $P \in G(\theta_0)$, with equality iff $\theta = \theta_0$.

Proof: Let τ^i denote the shift operator defined by $(\tau^i x)_j = x_{i+j}$. We will show that

$$\frac{1}{|\Lambda|}\left|U_{\Lambda,y}(x(\Lambda)) - \sum_{i\in\Lambda} A_U(\tau^i x)\right| \leq \frac{c(\Lambda)}{|\Lambda|}\|U\|, x(\Lambda) = x|_\Lambda \quad (29)$$

with a constant $c(\Lambda)$ satisfying $c(\Lambda)/(\Lambda) \to 0$ as $\Lambda \to \mathbf{Z}^d$. For finite-range interactions, Eq. (29) is a straightforward consequence of Eqs.(6) and (11) (in this case $c(\Lambda) \sim |\partial\Lambda|$). For nonfinite range interactions, we use the fact that $\mathcal{B}_0$ is dense in $\mathcal{B}$, and approximate $\Phi^{(\alpha)} \in \mathcal{B}$ by a $\tilde{\Phi}^{(\alpha)} \in \mathcal{B}_0$. That is, given $\epsilon > 0$, we choose a radius of interactions R_0 so that

$$\sum_{\substack{i\in V\subset\mathbf{Z}^d \\ V \text{ finite}}} \sup_{x(V)} |\Phi^\alpha(x(V)) - \tilde{\Phi}^{(\alpha)}(x(V))| < \epsilon\|U\|, \alpha = 1,\ldots,m$$

for all $i \in \mathbf{Z}^d$ (by translation invariance). Then Eqs. (6) and (11) easily imply Eq. (29) with $c(\Lambda) \leq \epsilon|\Lambda| + 2|\partial\Lambda|$.

Now, Eq. (29) and the ergodic theorem imply

$$\frac{1}{|\Lambda|}U_{\Lambda,y}(x(\Lambda)) \underset{\Lambda\to\mathbf{Z}^d}{\longrightarrow} E_P(A_U|\mathcal{S}), P-\text{a.s.}, P \in G_s(\theta_0). \quad (30)$$

This together with Eq. (22) yield Eq. (27). If $P \in G_e(\theta_0)$, then $h(\theta_0;\theta)(\cdot)$ is a.s. a constant, and by the variational principle [41] we have $h(\theta_0;\theta) \geq 0$ with equality iff $\theta = \theta_0$. If $P \in G_s(\theta_0)$ but not ergodic, then the ergodic decomposition (18) and the result just proven for ergodic P, imply that $h(\theta_0;\theta)(\cdot) \geq 0$ with equality iff $\theta = \theta_0$.

To prove Eq. (28) we write

$$\underline{h}(\theta_0;\theta) = p(\theta) - p(\theta_0) - \sup_{Q\in G_e(\theta_0)}[(\theta-\theta_0)\cdot E_Q(A_U)] \\ + \sup_{Q\in G_e(\theta_0)}[(\theta-\theta_0)\cdot E_Q(A_U)] - \limsup_{\Lambda\to \mathbf{Z}^d}(\theta-\theta_0)\cdot \tfrac{1}{|\Lambda|}U_{\Lambda,y}(x(\Lambda)). \quad (31)$$

The second term on the right-hand side of Eq. (31) is estimated by using the following consequence of the large deviations results of [6,16]: For any $\varepsilon > 0$ and $P \in G(\theta_0)$ we have

$$\limsup_{\Lambda\to \mathbf{Z}^d} \frac{1}{|\Lambda|}\log P\{\frac{1}{|\Lambda|}U_{\Lambda,y}(x(\Lambda)) \geq \sup_{Q\in G_e(\theta_0)}[(\theta-\theta_0)\cdot E_Q(A_U)] + \varepsilon\} < 0. \quad (32)$$

This and part a) of the theorem, yield part b).

Remark 8. The use of Eq. (32) was suggested by F. Comets. Our original proof of Eq. (28) was indirect and involved the fact that the extremal elements of $G(\theta)$ have a trivial tail algerbra [44, p.19].

The following proposition is used in the proof of Theorem 1.

Proposition 2. Let $P \in G(\theta_0)$. For all sufficiently large Λ, the minimizer(s) of $\ell_{\Lambda,y}(x(\Lambda);\theta)$ exist, and P-a.s. ultimately stay in a compact subset K of $\mathbb{R}^m$.

The proof of this proposition uses the following lemma whose proof is easily obtained by following the proof of Eq. (22).

Lemma 1. Given $\varepsilon > 0$, we have for sufficiently large Λ

$$\frac{1}{|\theta|}|p_{\Lambda,y}(\theta) - p(\theta)| \leq \varepsilon, \quad \text{for all } 0 \neq \theta \in \mathbb{R}^m.$$

Proof of Proposition 2. We write

$$h_{\Lambda,y}(x(\Lambda);\theta_0;\theta) = p(\theta) - p(\theta_0) - \sup_{Q\in G_e(\theta_0)}[(\theta-\theta_0)\cdot E_Q(A_U)] \\ + p_{\Lambda,y}(\theta) - p(\theta) + p(\theta_0) - p_{\Lambda,y}(\theta_0) \\ + \sup_{Q\in G_e(\theta_0)}[(\theta-\theta_0)\cdot E_Q(A_U)] - (\theta-\theta_0)\cdot\tfrac{1}{|\Lambda|}U_{\Lambda,y}(x(\Lambda)). \quad (33)$$

We bound from below each term on the right-hand-side of Eq. (33). By Lemma 1, the second term is bounded by $-\frac{\varepsilon}{3}|\theta|$; by Proposition 1d), the third term is bounded by $-\frac{\varepsilon}{3}$; using Eq. (32), the last term is bounded by

$-\frac{\varepsilon}{3}|\theta-\theta_0|(P_{\theta_0}$ – a.s., $P_{\theta_0} \in G(\theta_0))$; the first term is nonnegative, achieves it's minimum at $\theta = \theta_0$, and by the strict convexity of $p(\theta)$ it is bounded below by $c_0|\theta|$ for all θ outside some compact subset $K \subset \mathbb{R}^m$ (here $c_0 > 0$, independent of θ). Thus,

$$\begin{aligned} \ell_{\Lambda,y}(x(\Lambda);\theta) &\geq \ell_{\Lambda,y}(x(\Lambda);\theta_0) + (c_0 - \varepsilon)|\theta| \\ &\geq \inf_{\theta' \in \mathbb{R}^m} \ell_{\Lambda,y}(x(\Lambda),\theta') + (c_0 - \varepsilon)|\theta| \end{aligned}$$

for all $\theta \notin K$ and all sufficiently large Λ. Choosing $\varepsilon > 0$ sufficiently small, we obtain the proposition.

Proof of Theorem 1: Let K be as in Proposition 2, and D an open neighborhood of θ_0. We use Eq. (33) and bound from below each term for all $\theta \in K\backslash D$. The first term is continuous in θ; hence its minimum on $K\backslash D$ is achieved, and by Theorem 4 this minimum is strictly larger than 4ε, some $\varepsilon > 0$. By Proposition 1b), the family $\{p_{\Lambda,y}(\theta)\}_\Lambda$ is uniformly equicontinuous in $K\backslash D$; this together with the pointwise convergence (22), and the compactness of $K\backslash D$, imply (by standard finite covering arguments) that $p_{\Lambda,y}(\theta) \to p(\theta)$ uniformly in $\theta \in K\backslash D$. Thus, the second term in Eq. (33) is bounded below by $-\varepsilon$ for large Λ and all $\theta \in K\backslash D$. By Eq. (22) with $\theta = \theta_0$, the second term of Eq. (33) is also bounded below by $-\varepsilon$ (large Λ). By Eq. (22) (and the linearity in $\theta - \theta_0$), the last term is also bounded below by $-\varepsilon$. Putting these bounds together we obtain

$$\begin{aligned} \ell_{\Lambda,y}(x(\Lambda);\theta) &\geq \ell_{\Lambda,y}(x(\Lambda);\theta_0) + \varepsilon \\ &\geq \inf_{\theta' \in \mathbb{R}^m} \ell_{\Lambda,y}(x(\Lambda);\theta') + \varepsilon \end{aligned}$$

for all $\theta \in K\backslash D$. This together with Proposition 2 imply that the minimizer(s) of $\ell_{\Lambda,y}(x(\Lambda);\theta)$ must eventually (a.s.) lie in D. This is true for all neighborhoods D. Taking a countable family of neighborhoods shrinking to θ_0, we obtain the theorem.

Next we prove Theorem 1 for the log-likelihood function (20). Let $\ell_\Lambda(x(\Lambda);\theta)$ be the log-likelihood function (19) with free b.c.

Proposition 3

$$|\ell'_\Lambda(x(\Lambda);\theta) - \ell_\Lambda(x(\Lambda);\theta)| \leq \frac{2c(\Lambda)}{|\Lambda|} \tag{34}$$

with a constant $c(\Lambda)$ as in Eq. (29).

Proof: By Eq. (17), it suffices to show

$$e^{-2|\theta|c(\Lambda)} \leq \frac{\pi_{\Lambda,x(\Lambda^c)}(x(\Lambda);\theta)}{\pi_\Lambda(x(\Lambda);\theta)} \leq e^{2|\theta|c(\Lambda)}. \tag{35}$$

Note that the ratio in Eq. (35) is equal to

$$\frac{Z_\Lambda(\theta)}{Z_{\Lambda,x(\Lambda^c)}(\theta)} \exp\{\theta \cdot [U_{\Lambda,x(\Lambda^c)}(x(\Lambda)) - U_\Lambda(x(\Lambda))]\}. \tag{36}$$

The difference $U_{\Lambda,x(\Lambda^c)} - U_\Lambda$ is estimated as in Eq.(29). A similar procedure may be used to derive

$$e^{-|\theta|c(\Lambda)} \leq \frac{Z_\Lambda(\theta)}{Z_{\Lambda,x(\Lambda^c)}(\theta)} \leq e^{|\theta|c(\Lambda)}.$$

These estimates easily yield Eq.(34).

Remark 9. The strict convexity of $p(\theta)$ and the convexity of $\ell_{n,y}(x(\Lambda);\theta)$ may be used to provide an alternative (and simpler) proof of Theorem 1, without using Lemma 1 (and hence, without establishing explicitely the preliminary result of Proposition 2). We prefer the proof given here, because it provides some insight into the large $|\theta|$ behavior of the log-likelihood function.

4 Asymptotic Normality and Efficiency

In this section we prove Theorems 2, and 3. We also establish (Theorem 5) another asymptotic normality result under conditions different from those of Theorem 2. Throughout this section $< \cdot >^{\Lambda,y}_{\theta}, < \cdot >^{y}_{\theta}$, denote expectations with respect to the measures $P^{(\Lambda)}_{\theta,y}$ and $P_{\theta,y}$ respectively ($P_{\theta,y}$ is the infinite-volume Gibbs distribution corresponding to b.c. y).

Proof of Theorem 2. The ML equations read

$$\frac{\partial p_{\Lambda,y}(\hat\theta_{\Lambda,y})}{\partial \theta^\alpha} - \frac{1}{|\Lambda|} U^{(\alpha)}_{\Lambda,y}(x) = 0, \alpha = 1, \ldots, m. \tag{37}$$

For finite $\Lambda \subset \mathbf{Z}^d$, the pressure $p_{\Lambda,y}(\theta)$ is an analytic function of θ. Using Taylor's theorem with remainder, we obtain

$$\sum_{\beta=1}^{m} \left[\int_0^1 \frac{\partial^2 p_{\Lambda,y}(\theta_0 + s(\hat\theta_{\Lambda,y} - \theta_0))}{\partial\theta^\alpha \partial\theta^\beta} ds \right] \cdot \sqrt{|\Lambda|}(\hat\theta^\beta_{\Lambda,y} - \theta^\beta_0)$$

$$= \frac{1}{\sqrt{|\Lambda|}} \left(U^\alpha_{\Lambda,y} - < U^\alpha_{\Lambda,y} >^{\Lambda,y}_{\theta_0} \right). \tag{38}$$

This can be written in the matrix form

$$T^{\Lambda,y}(\hat\theta_{\Lambda,y}, \theta_0) \cdot \sqrt{|\Lambda|}(\hat\theta_{\Lambda,y} - \theta_0) = \overline{U}_{\Lambda,y}(x), \tag{39}$$

where

$$\overline{U}_{\Lambda,y} = \{\overline{U}^{\alpha}_{\Lambda,y}\}_{\alpha=1}^{m}\overline{U}^{\alpha}_{\Lambda,y} = \frac{1}{\sqrt{|\Lambda|}}\left(U^{\alpha}_{\Lambda,y} - < U^{\alpha}_{\Lambda,y} >^{\Lambda,y}_{\theta_0}\right) T^{\Lambda,y}_{\alpha\beta}(\hat{\theta}_{\Lambda,y}, \theta_0)$$

$$= \int_0^1 \frac{\partial^2 p_{\Lambda,y}(\theta_0 + s(\hat{\theta}_{\Lambda,y} - \theta_0))}{\partial\theta^{\alpha}\partial\theta^{\beta}} ds. \tag{40}$$

We will prove

$$\overline{U}_{\Lambda,y} \underset{\Lambda\to\mathbf{Z}^d}{\overset{\mathcal{D}}{\longrightarrow}} N(0, I(\theta_0)) \tag{41}$$

with respect to the finite-volume Gibbs distribution $P^{(\Lambda)}_{\theta_0,y}$. Also

$$T^{\Lambda,y}_{\alpha\beta}(\hat{\theta}_{\Lambda,y}, \theta_0) \underset{\Lambda\to\mathbf{Z}^d}{\overset{\text{Prob.}}{\longrightarrow}} (I(\theta_0))_{\alpha\beta}. \tag{42}$$

The theorem then follows from Lemma 6.4.1 of [30, p.439].

Proof of Eq.(41): Let $t = \{t^{(\alpha)}\}_{\alpha=1}^{m} \in I\!R^m$. By Theorem A.8.6 of [15, p.305] (see also [36]), it suffices to prove

$$< e^{\overline{U}_{\Lambda,y}} >^{\Lambda,y}_{\theta_0} \underset{\Lambda\to\mathbf{Z}^d}{\longrightarrow} e^{\frac{1}{2}t\cdot I(\theta_0)t*} \tag{43}$$

for t in some compact subset of $I\!R^m$ containing the origin. Here t^* denote the conjugate vector to t. Consider the generating function of $U_{\Lambda,y}$

$$c_{\Lambda,y}(t) = \frac{1}{|\Lambda|}\log < e^{t\cdot U_{\Lambda,y}} >^{\Lambda,y}_{\theta_0} = p_{\Lambda,y}(\theta_0 + t) - p_{\Lambda,y}(\theta_0). \tag{44}$$

Using Taylor's theorem with remainder

$$c_{\Lambda,y}(t) = \sum_{\alpha=1}^{m} t^{\alpha}\frac{\partial p_{\Lambda,y}(\theta_0)}{\partial\theta^{\alpha}} + \sum_{\alpha,\beta=1}^{m} t^{\alpha}t^{\beta}\int_0^1 (1-s)\frac{\partial^2 p_{\Lambda,y}(\theta_0 + st)}{\partial\theta^{\alpha}\partial\theta^{\beta}} ds. \tag{45}$$

Thus,

$$< e^{t\cdot\overline{U}_{\Lambda,y}} >^{\Lambda,y}_{\theta_0} = \exp\left\{\sum_{\alpha,\beta=1}^{m} t^{\alpha}t\beta\int_0^1 (1-s)\frac{\partial^2 p_{\Lambda,y}\left(\theta_0 + \frac{st}{\sqrt{|\Lambda|}}\right)}{\partial\theta^{\alpha}\partial\theta^{\beta}} ds\right\}. \tag{46}$$

This together with Eq.(24) yields Eq.(43).

Proof of Eq.(42): By the consistency of the ML estimator $\hat{\theta}_{\Lambda,y}$, with probability tending to 1 as $\Lambda \to \mathbf{Z}^d$, we have $\theta_0 + s(\hat{\theta}_{\Lambda,y} - \theta_0) \in D_{\theta_0}$, where D_{θ_0} is a neighborhood of θ_0 so that Eq.(24) holds. Thus,

$$T^{\Lambda,y}_{\alpha\beta}(\hat{\theta}_{\Lambda,y}, \theta_0) = \frac{\partial^2 p_{\Lambda,y}(\theta_0)}{\partial\theta^{\alpha}\partial\theta^{\beta}} + o(|\hat{\theta}_{\Lambda,y} - \theta_0|). \tag{47}$$

The first term tends to $c_{\alpha\beta}$, and the second term tends to zero in probability (by the consistency of $\hat{\theta}_{\Lambda,y}$). This establishes Eq. (42) and completes the proof of the theorem.

The finite-volume Gibbs distribution $P^{(\Lambda)}_{\theta_0,y}$ in the limit (25) cannot be replaced by the infinite-volume, true Gibbs distribution $P_{\theta_0,y}$, without additional assumptions on the behavior of "boundary terms." Next, we prove a variant of Theorem 2, which involves $P_{\theta_{0,y}}$. Here we will assume that the b.c. y *agrees* (see Remark 5) with the true distribution $P_{\theta_{0,y}}$. We also assume that for θ in a neighborhood of D_{θ_0} of θ_0, $P^{(\Lambda)}_{\theta,y}$ converges (weakly) to $P_{\theta,y}$. The ML equations (37) may be written as

$$\frac{1}{|\Lambda|} < W_\Lambda >^y_{\Lambda,y} - \frac{1}{|\Lambda|} W_\Lambda(x) \quad + \quad \frac{1}{|\Lambda|}\left[< U_{\Lambda,y} >^{\Lambda,y}_{\hat{\theta}_{\Lambda,y}} - W_\Lambda >^y_{\hat{\theta}_{\Lambda,y}} \right]$$
$$+ \quad \frac{1}{|\Lambda|}\left[W_\Lambda(x) - U_{\Lambda,y}(x) \right] = 0. \tag{48}$$

where

$$W_\Lambda(x) = \sum_{i\in\Lambda} A_U(\tau^i x). \tag{49}$$

Note that the last two terms in Eq.(48) consist of "boundary" terms only. We will impose conditions that guarantee that these terms are negligible as $\Lambda \to \mathbf{Z}^d$. Furthermore, we will impose conditions that ensure that $\frac{1}{\sqrt{|\Lambda|}}[W_\Lambda - < W_\Lambda >^y_{\theta_0}]$ converges to a normal distribution (with respect to $P_{\theta_0,y}$). These latter conditions have been introduced in [37] and are stated in terms of the Fortuin-Kastelyn-Ginibre (FKG) inequalities. For completeness we state these inequalities here: A real function F on $\mathbb{R}^n$ is said to be *nondecreasing* if it is nondecreasing coordinatewise, i.e., $F(x_1, \ldots, x_n) \leq F(x_1', \ldots, x_n')$ whenever $x_\ell \leq x_\ell'$ for all $\ell = 1, \ldots, n$. Let P be a probability measure on Ω, and define D to be the $L_2(\Omega, dP)$ closure of $\{F(x_{i_1}, \ldots, x_{i_n}) \in L_2 : \ n \geq 1, x_{i_\ell} \in \Omega_0.\ i_\ell \in \mathbf{Z}^d, F$ real, and nondecreasing$\}$. Here we assume that Ω_0 is a subset of $\mathbb{R}$ (in general it must be an ordered set). We say that P satisfies the FKG inequalities if Cov $(F, G) \geq 0$, for any $F, G \in D$.

Remark 10. The FKG inequalities are related to the notion of *positive quadrant dependence* [32] of random variables. If two random variables satisfy FKG, then they are positive quadrant dependent. Like the positive quadrant dependent property, so the FKG inequalities reflect the degree of dependence of random variables. The FKG inequalities hold for the general Ising model in (14) and (15) with arbitrary single pixel distribution $d\mu_0(x_i)$.

If g is a random variable, we shall write ${}_ig = \tau^i g, i \in \mathbf{Z}^d$ for the translation of g. Also, if f, g are two random variables we will write $\mathrm{Cov}\,(f,g) = < f; g > = < fg > - < f >< g >$.

Theorem 5. *Assume that the b.c. y in Theorem 1 agrees with the true measure $P_{\theta_0} \equiv P_{\theta_0,y}$. Furthermore, assume* a)

$$\sup_{\theta \in D_{\theta_0}} \frac{1}{\sqrt{|\Lambda|}} | < U_{\Lambda,y} >_\theta^{\Lambda,y} - |\Lambda| < A_U >_\theta^y | \to 0, \quad \text{as} \quad \Lambda \to \mathbf{Z}^d \tag{50}$$

$$\frac{1}{\sqrt{|\Lambda|}} (W_\Lambda - U_{\Lambda,y}) \underset{\Lambda \to \mathbf{Z}^d}{\longrightarrow} 0, \quad P_{\theta_{0,y}} - \text{ prob.} \tag{51}$$

$$\sup_{\theta \in D_{\theta_0}} \left| \frac{\partial^2}{\partial \theta^\alpha \partial \theta^\beta} < A_U >_\theta^y \right| < +^\infty, \quad \alpha, \beta = 1, \ldots, m \tag{52}$$

$$\sum_{\substack{V \subset \mathbf{Z}^d \\ \text{finite}}} \sup_{\theta \in D_{\theta_0}} \sup_{\Lambda} \left| < \frac{1}{|\Lambda|} U_{\Lambda,y} : \Phi(x(V) \vee y(V)) >_\theta^{\Lambda,y} \right| < +\infty; \tag{53}$$

and
b) there exist functions $A'_{U^\alpha}, \alpha = 1, \ldots, m$, such that

$$A'_{U^\alpha} \pm A_{U^\alpha} \in D, \alpha = 1, \ldots, m \tag{54}$$

$$c'_{\alpha\beta} = \sum_{j \in \mathbf{Z}^d} <_0 A'_{U^\alpha} ;_j A'_{U^\beta} >_{\theta_0}^y < +^\infty, \quad \text{for all} \quad \alpha = \beta. \tag{55}$$

Then

$$\sqrt{|\Lambda|} (\hat{\theta}_{\Lambda,y} - \theta_0) \underset{\Lambda \to \mathbf{Z}^d}{\overset{\mathcal{D}}{\longrightarrow}} N(0, (I(\theta_0))^{-1}), \tag{56}$$

where $I(\theta_0) = (c_{\alpha\beta})$ and

$$c_{\alpha\beta} = c_{\alpha\beta}(\theta_0) = \sum_{j \in \mathbf{Z}^d} <_0 A_{U^\alpha} ;_j A_{U^\beta} >_{\theta_0}^y, \alpha, \beta = 1, \ldots, m. \tag{57}$$

Remark 11. a) Condition (50) holds if $\frac{1}{|\Lambda|} < U_{\Lambda,y} >_\theta^{\Lambda,y}$ approaches its limiting value $< A_U >_\theta^y$, at a certain rate. Since $W_\Lambda - U_{\Lambda,y}$ contains only "boundary" terms, condition (51) guarantees that these "boundary" terms are negligible as $\Lambda \to \mathbf{Z}^d$. b) Conditions (54) and (55) ensure the applicability of the central limit theorem of [37]. c) The finiteness of $c_{\alpha\beta}$. $\alpha, \beta = 1, \ldots, m$, is a consequence of the finiteness of $c'_{\alpha\alpha}, \alpha = 1, \ldots, m$, by Proposition 1 of [37].

Proof of Theorem 5.: First, by translation invariance

$$< \frac{1}{|\Lambda|} W_\Lambda >^y_{\theta_0} = < A_U >^y_{\theta_0} .$$

Using Taylor's theorem with remainder, we rewrite Eq.(48) as follows:

$$T^{\Lambda,y}(\hat\theta_{\Lambda,y};\theta_0) \cdot \sqrt{|\Lambda|}(\hat\theta_{\Lambda,y} - \theta_0) + \frac{1}{\sqrt{|\Lambda|}}[< U_{\Lambda,y} >^{\Lambda,y}_{\theta_{\Lambda,y}} - |\Lambda| < A_U >^y_{\hat\theta_{\Lambda,y}}]$$

$$+ \frac{1}{\sqrt{|\Lambda|}}[W_\Lambda - U_{\Lambda,y}] = \frac{1}{\sqrt{|\Lambda|}}[W_\Lambda - |\Lambda| < A_U >^y_{\theta_0}] \quad (58)$$

$$T^{\Lambda,y}_{\alpha,\beta}(\hat\theta_{\Lambda,y},\theta_0) = \frac{\partial}{\partial\theta^\beta} < A_{U^\alpha} >^y_{\theta_0} + \frac{1}{2}\sum_{\gamma=1}^{m} \frac{\partial^2}{\partial\theta^\beta\partial\theta^\gamma} < A_{U^\alpha} >^y_{\theta^*_{\Lambda,y}} (\hat\theta^\gamma_{\Lambda,y} - \theta^\gamma_0), \quad (59)$$

where $\theta^*_{\Lambda,y}$ is a point on the line segment between θ_0 and $\hat\theta_{\Lambda,y}$. Condition (53) and the "fluctuation-dissipation" theorem of [45, Proposition A.1, p.43] imply that $\frac{\partial}{\partial\theta^\beta} < A_{U^\alpha} >^y_{\theta_0}$ exists and is equal to $c_{\alpha\beta}(\theta_0)$. The second and third terms on the left-hand side of (56) are taken care of by (50) and (51), respectively. Conditions (54) and (55) ensure the applicability of the central limit theorem of [37], which yields

$$\frac{1}{\sqrt{|\Lambda|}}[W_\Lambda(x) - |\Lambda| < A_U >^y_\theta] \underset{\Lambda\to\mathbf{Z}^d}{\longrightarrow} N(0, I(\theta_0)). \quad (60)$$

Putting the above together we obtain Eq.(56) from Lemma 6.4.1 of [33]. This completes the proof of the theorem.

Proof of Theorem 3.: For the two-dimensional binary Ising model the infinite-volume pressure $p(\beta)$ is differentiable in β for all $\beta > 0$. Thus, the ML equation may be written as follows:

$$\frac{dp(\hat\beta_\Lambda)}{d\beta} - \frac{1}{|\Lambda|}U_\Lambda(x) + \left[\frac{dp_\Lambda(\hat\beta_\Lambda)}{d\beta} - \frac{dp(\hat\beta_\Lambda)}{d\beta}\right] = 0. \quad (61)$$

Let

$$\frac{dp(\beta_c)}{d\beta} = u(\beta_c).$$

The exact formula of Onsager for the pressure $p(\beta)$ yields [46, pp. 133-134]

$$\frac{dp(\hat\beta_\Lambda)}{d\beta} = u(\beta_c) + A(\hat\beta_\Lambda - \beta_c)\log|\hat\beta_\Lambda - \beta_c| + g(\hat\beta_\Lambda - \beta_c),$$

where A is a (specific) constant and

$$g(\hat\beta_\Lambda - \beta_c) = o\left(\left|(\hat\beta_\Lambda - \beta_c)\log\left|\hat\beta_\Lambda - \beta_c\right|\right|\right).$$

Thus Eq.(61) may be written as follows:

$$A\sqrt{\frac{|\Lambda|}{\log|\Lambda|}}(\hat{\beta}_\Lambda - \beta_c)\log|\hat{\beta}_\Lambda - \beta_c|(1 + G(\hat{\beta}_\Lambda - \beta_c))$$
$$+\sqrt{\frac{|\Lambda|}{\log|\Lambda|}}\left[\frac{dp_\Lambda(\hat{\beta}_\Lambda)}{d\beta} - \frac{dp(\hat{\beta}_\Lambda)}{d\beta}\right] = \frac{1}{\sqrt{|\Lambda|\log|\Lambda|}}(U_\Lambda(x) - |\Lambda|u(\beta_c)),$$

where

$$G(\beta_\Lambda - \beta_c)\overset{\text{prob.}}{\longrightarrow} 0, \quad \text{as} \quad \Lambda \to \mathbf{Z}^2.$$

By the central limit theorem of [9,38]

$$\frac{1}{\sqrt{|\Lambda|\log|\Lambda|}}(U_\Lambda(x) - |\Lambda|u(\beta_c)) \overset{\mathcal{D}}{\longrightarrow} N(0, c_0^2), \tag{62}$$

where c_0 is a (specific) positive constant. Thus Eq.(26) will be proven once we establish

$$\sqrt{\frac{|\Lambda|}{\log|\Lambda|}}\left[\frac{dp_\Lambda(\hat{\beta}_\Lambda)}{d\beta} - \frac{dp(\hat{\beta}_\Lambda)}{d\beta}\right]\overset{\text{prob.}}{\longrightarrow} 0, \quad \text{as} \quad \Lambda \to \mathbf{Z}^2.$$

Since the convergence of $\dfrac{dp_\Lambda(\beta)}{d\beta}$ to $\dfrac{dp(\beta)}{d\beta}$ is slower for $\beta = \beta_c$, it suffices to prove

$$\sqrt{\frac{|\Lambda|}{\log|\Lambda|}}\left[\frac{dp_\Lambda(\beta_c)}{d\beta} - \frac{dp(\beta_c)}{d\beta}\right] \longrightarrow 0, \quad \text{as} \quad \Lambda \to \mathbf{Z}^2. \tag{63}$$

This can again be obtained from Onsager's explicit formula. In fact Eq. (63) enters [9] the proof of Eq.(62).

5 Appendix: Identifiability of Parameters

In this appendix we elaborate on the identifiability of the parameters $\theta \in \mathbb{R}^m$, i.e., on assumption, (23).

Let $\theta, \theta', \theta \neq \theta'$ be two distinct values of the parameters, and $\{\Phi^{(\alpha)}\}_{\alpha=1}^m$ be the (fixed) interactions satisfying Eq.(9). We set

$$U = \sum_{\alpha=1}^{m} \theta^\alpha \Phi^{(\alpha)}, U' = \sum_{\alpha=1}^{m} \theta'^{(\alpha)} \Phi^{(\alpha)}. \tag{64}$$

Note that U, U' are elements of the Banach space $\mathcal{B}$. We will impose necessary and sufficient conditions on $\{\Phi^{(\alpha)}\}_{\alpha=1}^{m}$ so that the parameters are identifiable. Our presentation follows the arguments in [44,30] for physically equivalent interactions.

Let $C(\Omega)$ denote the space of bounded continuous functions on Ω. Let $V \subset \mathbf{Z}^d$ denote a finite subset. For $f \in C(\Omega)$, we denote by $E_V f$ the conditional expectation $E(f|\mathcal{S}_V)$ of f with respect to the Borel σ-algebra $\mathcal{S}_V$ of Ω_V, under the probability measure $d\mu_0(x) = \prod_{i\in\mathbf{Z}^d} d\mu_0(x_i)$. E_V may be thought as the orthogonal projection of $L_2(\Omega, d\mu_0)$ on $L_2(\Omega_V, d\mu_0)$. Now we define

$$Q_V = \sum_{W\subset V} (-1)|V\backslash W|E_W, V \subset \mathbf{Z}^d, \text{ finite.} \tag{65}$$

This is the orthogonal projection of $L_2(\Omega, d\mu_0)$ on the subspace of $L_2(\Omega_V, d\mu_0)$ orthogonal to $L_2(\Omega_W, d\mu_0)$ for all $W \subsetneq V$. As an operator on $C(\Omega), Q_V$ has a sup norm $\|Q_V\| \leq 2^{|V|}$.

Next we define an operator S on $\mathcal{B}$ by

$$(SU)(x(V)) = \sum_{W\supset V} Q_V(U(x(W)), \quad U \in \mathcal{B}. \tag{66}$$

Our necessary and sufficient condition for identifiability will be given in terms of the operator S. The interactions U and U' of Eq.(64) are said to be *physically equivalent* iff

$$S(U - U') = 0, \tag{67}$$

i.e., iff

$$\sum_{\alpha=1}^{m}(\theta^{\alpha} - \theta'^{(\alpha)}) \sum_{W\supset V} Q_V(\Phi^{(\alpha)}(x(W))) = 0 \tag{68}$$

for all finite $V \subset \mathbf{Z}^d$. We will see below that the parameters θ are identifiable iff Eq.(68) implies $\theta = \theta'$. In order to illustrate the meaning of Eq.(68), we consider the binary case, i.e., $\Omega_0 = \{-1, 1\}, \Omega = \Omega_0^{\mathbf{Z}^d}$. In this case Ω is (under pointwise multiplication) a compact abelian group with characters $x_V = \prod_{i\in V} x_i, V \subset \mathbf{Z}^d$, finite. Each $\Phi^{(\alpha)}(x(V))$ can be expanded in Fourier series

$$\Phi^{(\alpha)}(x(V)) = \sum_{W\subset V} J_W^V(\alpha) x_W \tag{69}$$

with

$$J_W^V(\alpha) = \int \Phi^{(\alpha)}(x(V)) x_W \, d\mu_0(x).$$

Then

$$(S\Phi^{(\alpha)})(x(V)) = \Big[\sum_{W \supset V} J_V^W(\alpha)\Big] x_V, V \neq \phi. \tag{70}$$

(Thus, in this case, S is the projection on x_V.) Therefore, condition (68) is equivalent to

$$\sum_{\alpha=1}^{m}(\theta^{(\alpha)} - \theta^{'(\alpha)}) \sum_{W \supset V} J_V^W(\alpha) = 0 \tag{71}$$

for all finite $V \subset \mathbf{Z}^d, V \neq \phi$. In particular, Eq.(71) implies that the parameters β and h in the Ising model (1) and (2) are identifiable. More generally, if the Hamiltonian is expressed in terms of clique parameters [17]

$$H_\Lambda(x;\theta) = -\sum_{\alpha} \theta^{(\alpha)} X_\alpha(x), \tag{72}$$

where α indexes the set of cliques, and $X_\alpha(x)$ is a sum of products of x_i's in the clique α, then the clique parameters are identifiable [we assume that there is no constant term in Eq.(72), i.e., no trivial cliques are considered].

The fact that the parameters θ are identifiable (i.e., Eq.(23) holds) iff Eq.(68) implies $\theta = \theta^{'}$ is a consequence of the following theorem.

Theorem 6. *Suppose that $\Phi^{(\alpha)}, \alpha = 1, \ldots, m$, are translation invariant and satisfy (11). Then the following statements are equivalent: a) U and $U^{'}$ are pysically equivalent, i.e. Eq.(68) holds; (b) the infinite-volume pressure $p(\theta)$ is linear on the line segment between θ and $\theta^{'}$; (c) U and $U^{'}$ have in common some infinite-volume Gibbs distribution (i.e., $G(\theta), G(\theta^{'})$ have in common at least one point); and (d) every Gibbs distribution for U (i.e., for θ) is also a Gibbs distribution for $U^{'}$ (i.e., for $\theta^{'}$), and vice versa.*

For a proof of this theorem in a more general setup, see [44, pp.63,64] and [30, p.75].

Acknowledgment: I would like to acknowledge my debt to D. Geman, S. Geman, U. Grenander, and D.E. McClure for a flow of ideas. I also wish to thank S.R.S. Varadhan for inspiring conversations on the statistical inference for Gibbs distributions, and A. Sokal for his feedback.

Bibliography

[1] Ackley, D.H., G.E. Hinton, and T.J. Sejnowski: "A Learning Algorithm for Boltzmann Machines," *Cognitive Sciences* 9 (1985) 147–169.

[2] Besag, J.: "Spatial Interaction and Statistical Analysis of Lattice Systems" (with discussion), *J. Roy. Stat. Soc.*, Series B, 36 (1974) 192–236.

[3] Besag, J.: "On the Statistical Analysis of Dirty Pictures" (with discussion), *J. Roy. Stat. Soc.*, Series B, 48 (1986).

[4] Besag, J.: "Efficiency of Pseudo-Likelihood Estimation for Simple Gaussian fields," *Biometrika* 64 (1977) 616–618.

[5] Chalmond, B.: "Image Restoration Using an Estimated Markov Model," preprint, Mathematics Dept., University of Paris, Orsay (1986).

[6] Comets, F.: "Large Deviation Estimates for a Conditional Probability Distribution. Applications to Random Interactions Gibbs Measures," *Prob. Th. and Rel. Fields* 80 (1989) 407–432.

[7] Comets, F., and B. Gidas: "Parameter Estimation for Gibbs Distributions from Partially Observed Data," *Ann. Applied Prob.* 2 (1992) 142–170.

[8] Comets, F., and B. Gidas: "Asymptotics of Maximum Likelihood Estimators for the Curie-Weiss Model," *Ann. Stat.* 19 (1991) 557–578.

[9] DeConinck, J.: "Scaling Limit of the Energy Variable for the Two-Dimensional Ising Ferromagnet," *Comm. Math. Phys.* 95 (1984) 53–59.

[10] Derin, H., and H. Elliott: "Modelling and Segmentation of Noisy and Textured Images Using Gibbs Random Fields," *IEEE Transactions*, PAMI-9 (1987) 39–55.

[11] Derin, H., H. Elliott, R. Cristi, and D. Geman: "Bayes Smoothing Algorithms for Segmentation of Binary Images Modelled by Markov Random Fields," *IEEE Transactions* PAMI-6 (1984) 707–720.

[12] Derin, H., and C.S. Won: "Estimating the Parameters of a Class of Gibbs Distributions," Conference on Information Sciences and Systems (1988) 222–228.

[13] Dobrushin, R.L.: "Gibbs State Describing Co-existence of Phases for a Three-Dimensional Ising Model," *Theory of Probability Appl.* 17 (1972) 582–600.

[14] Domb, C., and M.S. Green: *Phase Transitions and Critical Phenomena*, Vols. 1–6, Academic Press (1972–1978).

[15] Ellis, R.S.: *Entropy, Large Deviations, and Statistical Mechanics*, Springer-Verlag (1985).

[16] Föllmer, H., and S. Orey: "Large Deviations for the Empirical Field of Gibbs measures," *Ann. Prob.* 16 (1988) 961–977.

[17] Geman, D., and S. Geman: "Stochastic Relaxation, Gibbs Distributions, and the Bayesian Restoration of Images," in *Maximum Entropy and Bayesian Methods in Science and Engineering*, Dordrecht, Hollard (1988), eds.: C.R. Smith and G.J. Erickson.

[18] Geman, D., S. Geman, and C. Graffigne: "Locating Texture and Object Boundaries," in *Pattern Recognition Theory and Applications*, NATO ASI Series, Springer-Verlag (1986), ed.: P. Derijer.

[19] Geman, S., and D. Geman: "Stochastic Relaxation, Gibbs Distributions, and the Bayesian Restoration of Images," *IEEE Trans.* PAMI-6 (1984) 721–741.

[20] Geman, S., and C. Graffigne: "Markov Random Field Image Models and Their Applications to Computer Vision," *Proceedings of the International Congress of Mathematics 1986* (1987) ed.: A.M. Gleason, American Mathematical Society.

[21] Geman, S., and C.-R. Hwang: "Nonparametric Maximum Likelihood Estimation by the Method of Sieves," *Annals of Statistics* 10 (1982) 401–414.

[22] Geman, S., and D.E. McClure: "Bayesian Image Analysis: An Application to single Photon Emission Tomography," *1985 Proceedings of the American Statistical Association*, Statistical Computing Section, (1985).

[23] Gidas, B.: "Consistency of Maximum Likelihood and Pseudo-likelihood Estimators for Gibbs Distributions," Proceedings of the Workshop on *Stochastic Differential Systems with Applications in Electrical/Computer Engineering, Control Theory, and Operations Research*, IMA, University of Minnesota, 1986.

[24] Gidas, B.: "A Renormalization Group Approach to Image Processing Problems," *IEEE Trans.* PAMI-11(1989) 164–180.

[25] Grenander, U.: *Abstract Inference,* Willey (1981).

[26] Grenander, U.: *Tutorial in Pattern Theory,* Div. Appl. Math., Brown University (1983).

[27] Hinton, G.E., and T.J. Sejnowski: "Optimal Perceptual Inference", in *Proc. IEEE Conf. Comp. Vision Pattern Recognition* (1983).

[28] Hopfield, J.J.: "Neural Networks and Physical Systems with Emergent Collective Computational Abilities," *Proc. Natl. Acad. Sci. USA*, 79 (1982) 2554–2558.

[29] Huber, P.J.: "The Behavior of Maximum Likelihood Estimates Under Nonstandard Conditions," in *Proc. Fifth Berkeley Symposium on Mathematical Statistics and Probability,* vol. 1, Univ. of California Press, Berkeley.

[30] Israel, R.B.: *Convexity in the Theory of Gases*, Princeton University Press, 1979.

[31] Kashyap, R. L., and R. Chellappa: "Estimation and Choice of Neighbors in Spatial-Interaction Models of Images", *IEEE Trans. on Information Theory,* Vol. IT- 29 (1983) 60-72.

[32] Lehmann, E.L.: "Some Concepts of Dependence," *Ann. Math. Stat.* 37 (1966) 1137–1153.

[33] Lehmann, E.L.: *Theory of Point Estimation,* Willey (1983).

[34] Lippmann, A.: "A Maximum Entropy Method for Expert System Construction," Ph.D. Thesis, Div. of Appl. Math., Brown University, 1986.

[35] Mann, H.B., and A. Wald: "On the Statistical Treatment of Linear Stochastic Difference Equations," *Econometrica* 11 (1943) 173–200.

[36] Martin-Löf, A.: "Mixing Properties, Differentiability of the Free Energy and the Central Limit Theorem for a Pure Phase in the Ising Model at Low Temperature," *Comm. Math. Phys.* 32 (1973) 75–92.

[37] Newman, C.: "A General Central Limit Theorem for FKG Systems," *Comm. Math. Phys.* 91 (1983) 75–80.

[38] Pickard, D.K.: "Asymptotic Inference for Ising Lattice III. Non-zero Field and Ferromagnetic States," *J. Appl. Prob.* 16 (1979) 12–24.

[39] Possolo, A.: "Estimation of Binary Markov Random Fields," Univ. of Washington, Department of Statistics, preprint (1986).

[40] Ripley, B.D.: "Statistics, Images, and Pattern Recognition," *Canadian J. of Statistics* 14 (1986) 83–111.

[41] Ripley, B.D.: "Statistical Inference for Spatial Processes," preprint (1987), Dept. of Mathemtics, University of Strathclyde, Glasgow.

[42] Ruelle, D.: *Statistical Mechanics*, W.A. Benjamin, Inc. (1969).

[43] Ruelle, D.: "A Heuristic Theory of Phase Transition," *Commun. Math. Physics* 53 (1977) 195–208.

[44] Ruelle, D.: *Thermodynamic Formalism*, Addison-Wesley (1978).

[45] Sokal, A.D.: "More Inequalities for Critical Exponents," *Jour. Stat. Phys.* 25 (1981) 25–50.

[46] Thomson, C.J.: *Mathematical Statistical Mechanics*, Macmillan Company, 1972.

[47] von der Marlsburg, C., and E. Bienenstock: "A Neural Network for the Retrieval of Superimposed Connection," Europhys. Lett (1987), 121–126.

[48] Wald, A.: "Note on the consistency of the Maximum Likelihood Estimate," *Ann. Math. Stat.* 20 (1949) 595–601.

[49] Wilson, K.: "Monte Carlo Calculations for Lattice Gauge Theory," in *Recent Developments in Gauge Theories*, Plenum, N.Y. (1980), eds.: G. 't Hooft, *et al.*

[50] Younes, L.: "Estimation and Annealing for Gibbs Fields," Annales de l'Institut Henri Poincare, 1988, Vol. 3.

On Sampling Methods and Annealing Algorithms

Saul B. Gelfand[†] **and Sanjoy K. Mitter**[‡]

[†]School of Electrical Engineering
Purdue University
West Lafayette, Indiana

[‡]Department of Electrical Engineering and Computer Science
and Laboratory for Information and Decision Systems
Massachusetts Institute of Technology
Cambridge, Massachusetts

1 Introduction

Discrete Markov random fields (MRFs) defined on a finite lattice have seen significant application as stochastic models for images [1, 2]. There are two fundamental problems associated with image processing based on such random field models. First, we want to generate realizations of the random fields to determine their suitability as models of our prior knowledge. Second, we want to collect statistics and perform optimizations associated with the random fields to solve model-based estimation problems, e.g., image restoration and segmentation.

According to the Hammersley–Clifford Theorem [3], MRFs which are defined on a lattice are in one-to-one correspondence with Gibbs distributions. Starting with [4], there have been various constructions of Markov chains that possess a Gibbs invariant distribution, and whose common characteristic is that their transition probabilities depend only on the ratio of the Gibbs probabilities (and not on the normalization constant). These chains can be used via Monte Carlo simulation for sampling from Gibbs distributions at a fixed temperature, and for finding globally minimum energy states by slowly decreasing the temperature as in the simulated annealing (or stochastic relaxation) method [5, 6]. Certain types of diffusion processes that also have a Gibbs invariant distribution can be used for the same purposes when the random fields are continuous-valued [7, 8].

ISBN 0-12-170608-7

Many of the fundamental ideas on MRF-based image processing stem from [6], which introduced the idea of modeling an image with a compound random field for both the intensity and boundary processes. This prior random field is an MRF characterized by a Gibbs distribution. A measurement model is specified for the observed image, and the resulting posterior random field also is an MRF characterized by a Gibbs distribution. A maximum *a posteriori* probability (MAP) estimate of the image based on the noisy observations then is found by minimizing the posterior Gibbs energy via simulated annealing.

There have been numerous variations and extensions of the ideas in [6], including different estimation criteria, different methods to perform the annealing, and different methods to determine the random field parameters [9–12]. We note that some of the alternative estimators that have been proposed do not use annealing but rather collect statistics at a fixed temperature, e.g., the maximizer of the posterior marginals (MPM) and the thresholded posterior mean (TPM) estimators [9]. The scope of the MRF image models also has been enlarged over time. Most of the early work on Monte Carlo sampling methods and annealing algorithms as applied to MRF-based image processing considered finite-valued MRFs (e.g., generalized Ising models) to model discrete gray-level distributions [6]. Some more recent work has dealt with continuous-valued MRFs (e.g. Gauss-Markov models) to model continuous gray level distributions [13, 14]. In certain applications, it may be advantageous to use a continuous Gauss-Markov random field model for computational and modeling considerations even when the image pixels actually can take only a finite (but large) number of gray-level values. Both Markov chain sampling methods and annealing algorithms, and diffusion-type sampling methods and annealing algorithms, have been used in continuous-valued MRF-based image processing.

It also should be noted that the annealing algorithm has been used in image processing applications to minimize cost functions not derived from an MRF model (*cf.* [15] for an application to edge detection), and many other non-image processing applications as well. There has been a lot of research on the convergence of discrete-state Markov chain annealing algorithms and diffusion annealing algorithms, but very few results are known about continuous-state Markov chain annealing algorithms.

Our research, described in detail in [16–19], addresses the following questions:

1. What is the relationship between the Markov chain sampling methods/annealing algorithms and the diffusion sampling methods/annealing algorithms?

2. What types of convergence results can be shown for discrete-time approximations of the diffusion annealing algorithms?

3. What types of convergence results can be shown for continuous-state Markov chain annealing algorithms?

In this chapter, we summarize some of our results. In Section 2, we show that continuous-time interpolations of certain continuous-state Markov chain sampling methods and annealing algorithms converge weakly to diffusions. In Section 3, we establish the convergence of a large class of discrete time modified stochastic gradient algorithms related to the diffusion annealing algorithm. Also in Section 3, we establish the convergence of certain continuous-state Markov chain annealing algorithms, essentially by showing that they can be expressed in the form of modified stochastic gradient algorithms. This last result gives a unifying view of the Markov chain and diffusion versions of simulated annealing algorithms. In Section 4, we briefly examine some directions for further work.

2 Convergence of Markov Chain Sampling Methods and Annealing Algorithms to Diffusion

In this section, we analyze the dynamics of a class of continuous-state Markov chains, which arise from a particular implementation of the Metropolis and the related Heat Bath Markov chain sampling methods [20]. Other related sampling methods (*cf.* [21]) can be analyzed similarly. We show that certain continuous-time interpolations of the Metropolis and Heat Bath chains converge weakly (i.e., in distribution on path space) to Langevin diffusions. This establishes a much closer connection between the Markov chains and diffusions than just the fact that both are Markov processes that possess an invariant Gibbs distribution. We actually show that the interpolated Metropolis and Heat Bath chains converge to the same Langevin diffusion running at different time scales. This establishes a connection between the two Markov chain sampling methods that in general, is not well understood. Our results apply to both (fixed temperature) sampling methods and (decreasing temperature) annealing algorithms.

We start by reviewing the discrete-state Metropolis and Heat Bath Markov chain sampling methods. Assume that the state space Σ is countable. Let $U(\cdot)$ be the real-valued energy function on Σ for the system. Also, let T be the (positive) temperature of the system. Let $q(i,j)$ be a stationary transition probability from i to j for $i,j \in \Sigma$. The general form of the one-step transition probability from i to j for the discrete-state Markov chain $\{X_k\}$ we consider is given by

$$p(i,j) = q(i,j)s(i,j) + m(i)1(j = i), \tag{1}$$

where

$$m(i) = 1 - \sum_j q(i,j)s(i,j), \tag{2}$$

$s(i,j)$ is a weighting factor ($0 \leq s(i,j) \leq 1$), and $1(\cdot)$ is an indicator function. Let $[a]_+$ denote the positive part of a, i.e., $[a]_+ = \max\{a, 0\}$. The weighting factor $s(i,j)$ is given by

$$s_M(i,j) = \exp(-[U(j) - U(i)]_+/T) \tag{3}$$

for the Metropolis Markov chain, and by

$$s_H(i,j) = \frac{\exp(-(U(j) - U(i))/T)}{1 + \exp(-(U(j) - U(i))/T)} \tag{4}$$

for the Heat Bath Markov chain.

Let

$$\pi(i) = \frac{1}{Z}\exp(-U(i)/T), \quad i \in \Sigma; \quad Z = \sum_i \exp(-U(i)/T)$$

(assuming $Z < \infty$). If the stochastic matrix $Q = [q(i,j)]$ is symmetric and irreducible, then the detailed balance equation,

$$\pi(i)p(i,j) = \pi(j)p(j,i), \quad i,j \in \Sigma,$$

is satisfied, and it follows easily that $\pi(i)$, $i \in \Sigma$, are the unique stationary probabilities for both the Metropolis and Heat Bath Markov chains. Hence these chains may be used to sample from and to compute mean values of functionals with respect to a Gibbs distribution with energy $U(\cdot)$ and temperature T [22]. The Metropolis and Heat Bath chains can be interpreted (and simulated) in the following manner. Given the current state $X_k = i$, generate a candidate state $\tilde{X}_k = j$ with probability $q(i,j)$. Set the next state $X_{k+1} = j$ if $s(i,j) > \Theta_k$, where Θ_k is an independent random variable uniformly distributed on the interval $[0,1]$; otherwise set $X_{k+1} = i$.

We can generalize the discrete state Markov chain sampling methods described previously to a continuous d-dimensional Euclidean state space as follows. Let $U(\cdot)$ be a smooth real-valued energy function on $\Sigma = \mathbf{R}^d$, and let T be the (positive) temperature. Let $q(x,y)$ be a stationary transition density from x to y for $x, y \in \mathbf{R}^d$. The general form of the one-step transition probability density for the continuous-state Markov chain $\{X_k\}$ we consider is given by

$$p(x,y) = q(x,y)s(x,y) + m(x)\delta(y-x), \tag{5}$$

where

$$m(x) = 1 - \int q(x,y)s(x,y)dy, \tag{6}$$

$s(x,y)$ is a weighting factor ($0 \le s(x,y) \le 1$), and $\delta(\cdot)$ is a Dirac-delta function. Here, $s(\cdot,\cdot) = s_M(\cdot,\cdot)$ or $s(\cdot,\cdot) = s_H(\cdot,\cdot)$ for the generalized Metropolis and Heat Bath chains, respectively. (See Eqs. (3) and (4).)

The continuous state Metropolis and Heat Bath Markov chains can be interpreted (and simulated) analogously to the discrete-state versions. In particular, $q(x,y)$ is a conditional probability density for generating a candidate state $\tilde{X}_k = y$, given the current state $X_k = x$. For our analysis, we shall consider the case where only a single component of the current state is changed to generate the candidate state, and the component is selected at random with all components equally likely. Furthermore, we shall require that the candidate value of the selected component depend only on the current value of the selected component. Let x_i denote the ith component of the vector $x \in \mathbf{R}^d$. Let $r(x_i, y_i)$ be a transition density from x_i to y_i for $x_i, y_i \in \mathbf{R}$. Hence we set

$$q(x,y) = \frac{1}{d}\sum_{i=1}^{d} r(x_i,y_i)\prod_{j\neq i}\delta(y_j - x_j) \tag{7}$$

Suppose we take

$$r(x_i,y_i) = 1(x_i = -1)\delta(y_i - 1) + 1(x_i = 1)\delta(y_i + 1) \tag{8}$$

In this case, if the ith coordinate of the current state X_k is selected (at random) to be changed in generating the candidate state $\tilde{X}_k$, then $\tilde{X}_{k,i}$ is ± 1 when $X_{k,i}$ is ∓ 1. If, in addition,

$$U(x) = -\sum_{j\neq i} J_{ij}x_i x_j, \quad x \in \mathbf{R}^d,$$

then $\{X_k\}$ corresponds to a discrete-time kinetic Ising model with interaction energies J_{ij} [20].

Suppose we instead take

$$r(x_i,y_i) = \frac{1}{\sqrt{2\pi\sigma^2}}\exp\left[\frac{-(y_i - x_i)^2}{2\sigma^2}\right]. \tag{9}$$

In this case, if the ith coordinate of the current state X_k is selected (at random) to be changed in generating the candidate state $\tilde{X}_k$, then $\tilde{X}_{k,i}$ is conditionally Gaussian with mean $X_{k,i}$ and variance σ^2. In the sequel, we shall show that a family of interpolated Markov chains of this type converges (weakly) to a Langevin diffusion.

For each $\varepsilon > 0$, let $r_\varepsilon(\cdot,\cdot)$ denote the transition density in Eq. (9) with $\sigma^2 = \varepsilon$, and let $p_\varepsilon(\cdot,\cdot)$ denote the corresponding transition density in Eqs. (5)–(7). Let $\{X_k^\varepsilon\}$ denote the Markov chain with transition density $p_\varepsilon(\cdot,\cdot)$ and initial condition $X_0^\varepsilon = X_0$. Interpolate $\{X_k^\varepsilon\}$ into a continuous-time process $\{X^\varepsilon(t), t \geq 0\}$ by setting

$$X^\varepsilon(t) = X^\varepsilon_{[t/\varepsilon]}, \quad t \geq 0,$$

where $[a]$ is the largest integer less than or equal to a. Now the precise definition of the weak convergence of the process $X^\varepsilon(\cdot)$ to a process $X(\cdot)$ (as $\varepsilon \to 0$) is given in [23]. The significance of the weak convergence is that it implies not only the convergence of the multivariate distributions but also the convergence of the distributions of many interesting path functionals, such as maxima, minima, and passage times. (See [23] for a full discussion.) To establish weak convergence here, we require the following condition on $U(\cdot)$:

(A) $U(\cdot)$ is continuously differentiable, and $\nabla U(\cdot)$ is bounded and Lipshitz-continuous.

Theorem1: Assume (A). Then there is a standard d-dimensional Wiener process $W(\cdot)$ and a process $X(\cdot)$ (with $X(0) = X_0$ in distribution), nonanticipative with respect to $W(\cdot)$, such that $X^\varepsilon(\cdot) \to X(\cdot)$ weakly as $\varepsilon \to 0$, and

a) for the Metropolis method,

$$dX(t) = -\frac{\nabla U(X(t))}{2Td}dt + \frac{1}{\sqrt{d}}dW(t), \tag{10}$$

b) for the Heat Bath method,

$$dX(t) = -\frac{\nabla U(X(t))}{4Td}dt + \frac{1}{\sqrt{2d}}dW(t). \tag{11}$$

Proof: See [16].

Note that Theorem 1 justifies our claim that the interpolated Metropolis and Heat Bath chains converge to Langevin diffusions running at different time scales. Indeed, suppose $Y(\cdot)$ is a solution of the Langevin equation,

$$dY(t) = -\nabla U(Y(t))dt + \sqrt{2T}dW(t), \tag{12}$$

with $Y(0) = X_0$ in distribution. Then for $\tau(t) = t/2Td$, $Y(\tau(\cdot))$ has the same multivariate distributions as $X(\cdot)$ satisfying Eq. (10), while for $\tau(t) = t/4Td$, $Y(\tau(\cdot))$ has the same multivariate distributions as $X(\cdot)$ satisfying Eq. (11). Observe that the limit diffusion Eq. (10) for the Metropolis chain runs at twice the rate of the limit diffusion Eq. (11) for the Heat Bath chain, independent of the temperature.

To obtain Markov chain annealing algorithms, we simply replace the fixed temperature T in the preceding Markov chain sampling methods by a temperature schedule $\{T_k\}$ (where typically $T_k \to 0$). We can establish a weak convergence result for a nonstationary continuous-state Markov chain of this type as follows. Suppose $T(\cdot)$ is a positive continuous function on $[0, \infty)$. For $\varepsilon > 0$, let

$$T_k^\varepsilon = T(k\varepsilon), \quad k = 0, 1, \ldots,$$

and let $\{X_k^\varepsilon\}$ be as in the preceding but with temperature schedules $\{T_k^\varepsilon\}$. It can be shown that Theorem 1 is valid with T replaced by $T(t)$ in Eqs. (10) and (11). Hence the Markov chain annealing algorithms converge weakly to time-scaled versions of the Markov diffusion annealing algorithm,

$$dY(t) = -\nabla U(Y(t))dt + \sqrt{2T(t)}dW(t). \tag{13}$$

We remark that there has been a lot of work establishing convergence results for discrete state Markov chain annealing algorithms [6, 24–27], and also for the Markov diffusion annealing algorithm [7, 28, 29]. However, there are very few convergence results for continuous-state Markov chain algorithms. We note that the weak convergence of a continuous-state chain to a diffusion together with the convergence of the diffusion to the global minima of $U(\cdot)$ does not directly imply the convergence of the chain to the global minima of $U(\cdot)$; see [30] for a discussion of related issues. However, establishing weak convergence is an important first step in this regard. Indeed, a standard method for establishing the asymptotic (large-time) behavior of a large class of discrete-time recursive stochastic algorithms involves first proving weak convergence to an ODE limit. The standard method does not quite apply here because we have a discrete-time algorithm converging weakly to a nonstationary SDE limit; but calculations similar to those used to establish the weak convergence, in fact, do prove useful in ultimately establishing the convergence of continuous-state Markov chain annealing algorithms, as discussed in Section 3.2.

3 Recursive Stochastic Algorithms for Global Optimization in $\mathbf{R}^d$

3.1 Modified Stochastic Gradient Algorithms

In this section, we consider a class of algorithms for finding a global minimum of a smooth function $U(x)$, $x \in \mathbf{R}^d$. Specifically, we analyze the convergence of a modied stochastic gradient algorithm,

$$X_{k+1} = X_k - a_k(\nabla U(X_k) + \xi_k) + b_k W_k, \tag{14}$$

where $\{\xi_k\}$ is a sequence of $\mathbf{R}^d$-valued random variables, $\{W_k\}$ is a sequence of standard d-dimensional independent Gaussian random variables, and $\{a_k\}$, $\{b_k\}$ are sequences of positive numbers with a_k, $b_k \to 0$. An algorithm of this type arises by artificially adding the $b_k W_k$ term (via a Monte Carlo simulation) to a standard stochastic gradient algorithm,

$$Z_{k+1} = Z_k - a_k(\nabla U(Z_k) + \xi_k). \tag{15}$$

Algorithms like Eq. (15) arise in a variety of optimization problems, including adaptive filtering, identification, and control; the sequence $\{\xi_k\}$ is due to noisy or imprecise measurements of $\nabla U(\cdot)$ (*cf.* [31]). The asymptotic behavior of $\{Z_k\}$ has been much studied. Let S and S^* be the set of local and global minima of $U(\cdot)$, respectively. It can be shown, for example, that if $U(\cdot)$ and $\{\xi_k\}$ are suitably behaved, $a_k = A/k$ for k large, and $\{Z_k\}$ is bounded, then $Z_k \to S$ as $k \to \infty$ w.p.1. However, in general, $Z_k \not\to S^*$ (unless, of course, $S = S^*$). The idea behind adding the additional $b_k W_k$ term in Eq. (14), compared with Eq. (15), is that if b_k tends to zero slowly enough, then possibly $\{X_k\}$ (unlike $\{Z_k\}$) will avoid getting trapped in a strictly local minimum of $U(\cdot)$. (This is the usual reasoning behind simulated annealing-type algorithms.) We shall show in fact that if $U(\cdot)$ and $\{\xi_k\}$ are suitably behaved, $a_k = A/k$ and $b_k^2 = B/kloglogk$ for k large with $B/A > C_0$ (where C_0 is a positive constant that depends only on $U(\cdot)$), and $\{X_k\}$ is tight, then $X_k \to S^*$ as $k \to \infty$ in probability. We also give a condition for the tightness of $\{X_k\}$. We note that the convergence of Z_k to S can be established under very weak conditions on $\{\xi_k\}$, assuming $\{Z_k\}$ is bounded. Here, the convergence of X_k to S^* is established under somewhat stronger conditions on $\{\xi_k\}$, assuming that $\{X_k\}$ is tight (which is weaker than boundedness).

The analysis of the convergence of $\{Z_k\}$ usually is based on the asymptotic behavior of the associated ordinary differential equation (ODE),

$$\dot{z}(t) = -\nabla U(z(t)) \tag{16}$$

(*cf.* [31, 32]). This motivates our analysis of the convergence of $\{X_k\}$ based on the asymptotic behavior of the associated stochastic differential equation (SDE),

$$dY(t) = -\nabla U(Y(t))dt + c(t)dW(t), \tag{17}$$

where $W(\cdot)$ is a standard d-dimensional Wiener process and $c(\cdot)$ is a positive function with $c(t) \to 0$ as $t \to \infty$. This is just the diffusion annealing algorithm discussed in Section 2 (Eq. (13)), with $T(t) = c^2(t)/2$. The asymptotic behavior of $Y(t)$ as $t \to \infty$ has been studied intensively by a number of researchers. In [7, 29], convergence results were obtained by considering a version of Eq. (17) with a reflecting boundary; in [28], the reflecting boundary was removed. Our analysis of $\{X_k\}$ is based on the analysis of $Y(t)$ developed in [28], where the following result is proved: if $U(\cdot)$ is well-behaved, and $c^2(t) = C/logt$ for t large with $C > C_0$ (the same constant C_0 as before), then $Y(t) \to S^*$ as $t \to \infty$ in probability. To see intuitively how $\{X_k\}$ and $Y(\cdot)$ are related, let $t_k = \sum_{n=0}^{k-1} a_n$, $a_k = A/k$, $b_k^2 = B/kloglogk$, $c^2(t) = C/logt$, and $B/A = C$. Note that $b_k \sim c(t_k)\sqrt{a_k}$. Then we should have

$$\begin{aligned} Y(t_{k+1}) &\simeq Y(t_k) - (t_{k+1} - t_k)\nabla U(Y(t_k)) + c(t_k)(W(t_{k+1}) - W(t_k)) \\ &= Y(t_k) - a_k \nabla U(Y(t_k)) + c(t_k)\sqrt{a_k} V_k \\ &\simeq Y(t_k) - a_k \nabla U(Y(t_k)) + b_k V_k, \end{aligned}$$

where $\{V_k\}$ is a sequence of standard d-dimensional independent Gaussian random variables. Hence (for $\{\xi_k\}$ small enough) $\{X_k\}$ and $\{Y(t_k)\}$ should have approximately the same distributions. Of course, this is a heuristic; there are significant technical difficulties in using $Y(\cdot)$ to analyze $\{X_k\}$ because we must deal with long time intervals and slowly decreasing (unbounded) Gaussian random variables.

An algorithm like Eq. (14) was first proposed and analyzed in [29]. However, the analysis required that the trajectories of $\{X_k\}$ lie within a fixed ball (which was achieved by modifying Eq. (14) near the boundary of the ball). Hence such a version of Eq. (14) is suitable only for optimizing $U(\cdot)$ over a compact set. Furthermore, the analysis also required ξ_k to be zero to obtain convergence. In our first analysis of Eq. (14) in [17], we also required that the trajectories of $\{X_k\}$ lie in a compact set. However, our analysis did not require ξ_k to be zero, which has important implications when $\nabla U(\cdot)$ is not measured exactly. In our later analysis of Eq. (14) in [18], we removed the requirement that the trajectories of $\{X_k\}$ lie in a compact set. From our point of view, this is the most significant difference between our work in [18] and what is done in [29, 17] (and more generally in other work on global optimization, such as [33]): we deal with unbounded processes and establish the convergence of an algorithm that finds a global minimum of a function when it is not specified *a priori* what bounded region contains such a point.

We now state the simplest result from [18] concerning the convergence of the modified stochastic gradient algorithm Eq. (14). We will require

$$a_k = \frac{A}{k}, \quad b_k = \frac{\sqrt{B}}{\sqrt{kloglogk}}, \quad k \text{ large}, \tag{18}$$

and the following conditions:

(A1) $U(\cdot)$ is a C^2 function from $\mathbf{R}^d$ to $[0,\infty)$ such that the $S^* = \{x : U(x) \leq U(y) \forall y\} \neq \oslash$. (We also require some mild regularity conditions on $U(\cdot)$; see [18]).

(A2) $\underline{\lim}_{x\to\infty} \frac{|\nabla U(x)|}{|x|} > 0$, $\overline{\lim}_{x\to\infty} \frac{|\nabla U(x)|}{|x|} < \infty$.

(A3) $\lim_{x\to\infty} \left\langle \frac{\nabla U(x)}{|\nabla U(x)|}, \frac{x}{|x|} \right\rangle = 1$.

(A4) For $k = 0, 1, \ldots$, let $\mathcal{F}_k$ be the σ-field generated by $X_0, W_0, \ldots, W_{k-1}, \xi_0, \ldots, \xi_{k-1}$. There exists an $L \geq 0$, $\alpha > -1$, and $\beta > 0$ such that

$$E\{|\xi_k|^2|\mathcal{F}_k\} \leq La_k^\alpha \quad |E\{\xi_k|\mathcal{F}_k\}| \leq La_k^\beta \quad w.p.1,$$

and W_k is independent of $\mathcal{F}_k$.

Theorem 2: Assume (A1)–(A4) hold. Let $\{X_k\}$ be given by Eq. (14). Then there exists a constant C_0, such that for $B/A > C_0$,

$$X_k \to S^* \text{ as } k \to \infty$$

in probability.

Proof: See [18]

Remarks:

1. The constant C_0 plays a critical role in the convergence of x_k as $k \to \infty$ and also $Y(t)$ as $t \to \infty$. In [28], it is shown that the constant C_0 (denoted there by c_0) has an interpretation in terms of the action functional for a family of perturbed dynamical systems; see [28] for a further discussion of C_0, including some examples.

2. It is possible to modify Eq. (14) in such a way that only the lower bound and not the upper bound on $|\nabla U(\cdot)|$ in (A2) is needed. (See [18].)

3. In [18], we actually separate the problem of convergence of $\{X_k\}$ into two parts: one to establish tightness and another to establish convergence given tightness. This is analogous to separating the problem of convergence of $\{Z_k\}$ into two parts: one to establish boundedness and another to establish convergence given boundedness (*cf.* [31]). Now in [18], the conditions given for tightness are much stronger than the conditions given for convergence assuming tightness. For a particular algorithm it often is possible to prove tightness directly, resulting in somewhat weaker conditions than those given in Theorem 2.

3.2 Continuous-State Markov Chain Algorithm

In this section, we examine the convergence of a class of continuous-state Markov chain annealing algorithms similar to those described in Section 2. Our approach is to write such an algorithm in the form of a modified stochastic gradient algorithm of the type considered in Section 3.1. A convergence result is obtained for global optimization over all of $\mathbf{R}^d$. Some care is necessary to formulate a Markov chain with appropriate scaling. It turns out that writing the Markov chain annealing algorithm in the form of Eq. (14) is rather more complicated than writing standard variations of gradient algorithms that use some type of (possibly noisy) finite difference estimate of $\nabla U(\cdot)$ in the form of Eq. (15) (*cf.* [31]). Indeed, to the extent that the Markov chain annealing algorithm uses an estimate of $\nabla U(\cdot)$, it does so in a much more subtle manner than a finite difference approximation.

Although some numerical work has been performed with continuous-state Markov chain annealing algorithms [13, 14], there has been very little theoretical analysis, and furthermore, the analysis of the continuous-state case does not follow from the finite state case in a straightforward way (especially for an unbounded state space). The only analysis we are aware of is in [13], where a certain asymptotic stability property is established. Since our convergence results for the continuous-state Markov chain annealing algorithm are based ultimately on the asymptotic behavior of the diffusion annealing algorithm, our work demonstrates and exploits the close relationship between the Markov chain and diffusion versions of simulated annealing.

We shall perform our analysis of continuous-state Markov chain annealing algorithms for a Metropolis-type chain. We remark that convergence results for other continuous-state Markov chain sampling method-based annealing algorithms (such as the Heat Bath method) can be obtained by a similar procedure. Recall that the one-step transition probability density for a continuous-state Metropolis-type (fixed temperature) Markov chain is given (as indicated in Eqs. (3),(5), and (6)) by

$$p(x,y) = q(x,y)s(x,y) + m(x)\delta(y-x),$$

where

$$m(x) = 1 - \int q(x,y)s(x,y)dy$$

and

$$s(x,y) = \exp(\frac{-[U(y)-U(x)]_+}{T}).$$

Here, we have dropped the subscript on the weighting factor $s(x,y)$. If we replace the fixed temperature T by a temperature sequence $\{T_k\}$, we get a Metropolis-type annealing algorithm.

Our goal is to express the Metropolis-type annealing algorithm as a modified stochastic gradient algorithm like Eq. (14) to establish its convergence. This leads us to choosing a nonstationary Gaussian transition density,

$$q_k(x,y) = \frac{1}{(2\pi b_k^2 \sigma_k^2(x))^{d/2}} \exp\left(-\frac{|y-x|^2}{2b_k^2\sigma_k^2(x)}\right), \tag{19}$$

and a state-dependent temperature sequence,

$$T_k(x) = \frac{b_k^2\sigma_k^2(x)}{2a_k}, \tag{20}$$

where

$$\sigma_k(x) = \max\{a_k^\gamma |x|, 1\} \tag{21}$$

and γ is a positive number. With these choices, the Metropolis-type annealing algorithm can be expressed as

$$X_{k+1} = X_k - a_k(\nabla U(X_k) + \xi_k) + b_k W_k$$

for appropriately behaved $\{\xi_k\}$. We remark that the state-dependent term $\sigma_k(x)$ in Eqs. (19) and (20) produces a drift toward the origin proportional to $|x|$, which is needed to establish tightness of the annealing chain.

This discussion leads us to the following continuous-state Metropolis-type annealing algorithm. Let $N(m, \Lambda)$ denote d-dimensional normal measure with mean m and covariance matrix Λ.

Continuous-State Metropolis-Type Annealing Algorithm

Let $\{X_k\}$ be a Markov chain with one-step transition probability at time k given by

$$P\{X_{k+1} \in A | X_k = x\} = \int_A s_k(x,y) dN(x, b_k^2 \sigma_k^2(x) I)(y) + m_k(x) 1_A(x), \tag{22}$$

where

$$m_k(x) = 1 - \int s_k(x,y) dN(x, b_k^2 \sigma_k^2(x) I)(y) \tag{23}$$

and

$$s_k(x,y) = \exp\left(-\frac{2a_k[U(y) - U(x)]_+}{b_k^2 \quad \sigma_k^2(x)}\right). \tag{24}$$

We now state a convergence result from [19] concerning the convergence of the continuous-state Metropolis-type annealing algorithm. Let the sequences $\{a_k\}$ and $\{b_k\}$ be given by Eq. (18).

Theorem 3: Assume (A1)–(A3) hold. Let $\{X_k\}$ be the Markov chain with transition probability given by Eqs. (22)–(24) and with $0 < \gamma < \frac{1}{4}$. Then there exists a constant C_0, such that for $B/A > C_0$,

$$X_k \to S^* \text{ as } k \to \infty$$

in probability.

Proof: See [19].

Remarks:

1. The constant C_0 is the same constant described in Remark 1 following Theorem 2.

2. It is possible to modify Eqs. (22)–(24) in such a way that only the lower bound and not the upper bound on $|\nabla U(\cdot)|$ from (A2) is needed. (See [19].)

4 Conclusions

Monte Carlo sampling methods and annealing algorithms have found significant application to MRF-based image processing. These algorithms fall broadly into two groups: Markov chain and diffusion methods. The discrete-state Markov chain algorithms have been used with finite-range

MRF models, while both continuous-state Markov chain and diffusion algorithms have been used with continuous-range MRF models. We note that there are some very interesting questions related to the parallel implementation of these Monte Carlo procedures that we have not discussed here; see [34].

In this chapter, we summarized some of our research, which has investigated the relationship between the various Markov chain and diffusion sampling methods and annealing algorithms. We demonstrated the weak convergence of certain interpolated continuous-state Markov chain sampling methods and annealing algorithms to diffusions. We also established the large-time convergence of a class of discrete-time modified stochastic gradient algorithms based on the asymptotic behavior of the associated diffusion annealing algorithm. We further established the large-time convergence of a continuous-state Markov chain annealing algorithm by writing it in the form of such a modified stochastic gradient algorithm. The convergence here is to the global minima of an energy cost function defined on the entire d-dimensional Euclidean space.

It seems to us that some experimental comparisons of continuous-state Markov chain and diffusion-type annealing algorithms (practically implemented by the modified stochastic gradient algorithms described earlier) on image segmentation and restoration problems would be of some interest. We are not aware of any explicit comparisons of this type in the literature. It also might be useful to examine the application of the modified stochastic gradient algorithms to adaptive pattern recognition, filtering, and identification, where stochastic gradient algorithms frequently are employed. Because of the slow convergence of the modified stochastic gradient algorithms, off-line applications probably will be required. One particular application that might prove fruitful is training multilayer feedforward *neural nets*, which is a nonconvex optimization problem often plagued with local minima [35].

5 Acknowledgment

The research of the first author has been supported by the National Science Foundation under contract ECS-8910073. The research of the second author has been supported by the Air Force Office of Scientific Research under contract 89-0279B and by the Army Research Office contract DAAL03-86-K-0171 (Center for Intelligent Control Systems).

Bibliography

[1] R. L. Kashyap, R. Chellappa, Estimation and Choice of Neighbors in Spatial Interaction Models of Images, *IEEE Trans. Info. Theory*, Vol. 29, 1983, pp. 60–72.

[2] J. W. Woods, Two-Dimensional Discrete Markovian Fields, *IEEE Trans. Info. Theory*, Vol. 18, 1972, pp. 232–240.

[3] J. Besag, Spatial Interaction and the Statistical Analysis of Lattice Systems, *J. Royal Stat. Soc.*, Vol. 34, 1972, pp. 75–83.

[4] N. Metropolis, A. W. Rosenbluth, M. N. Rosenbluth, A. H. Teller, and E. Teller, Equation of State Calculations by Fast Computing Machines, *J. Phys. Chem.*, Vol. 21, No. 6, 1953, pp. 1087.

[5] S. Kirkpatrick, C. D. Gelatt, and M. P. Vecchi, Optimization by Simulated Annealing, *Science*, Vol. 220, 1983, pp. 671–680.

[6] S. Geman and D. Geman, Stochastic Relaxation, Gibbs Distribution, and the Bayesian Restoration of Images, *IEEE Trans. Pattern Anal. and Machine Intell.*, Vol. 6, 1984, pp. 721–741.

[7] S. Geman and C. R. Hwang, Diffusions for Global Optimization, *SIAM Journal Control and Optimization*, Vol. 24, 1986, pp. 1031–1043.

[8] U. Grenander, *Tutorial in Pattern Theory*, Div. of Applied Math, Brown University, Providence, RI, 1984.

[9] J. L. Marroquin, S. Mitter, and T. Poggio, Probabilistic Solution of Ill-Posed Problems in Computational Vision, *J. Amer. Statist. Assoc.*, Vol. 82, 1987, pp. 76–89.

[10] B. Gidas, A Renormalization Group Approach to Image Processing Problems, *IEEE Trans. Pattern Anal. and Machine Intell.*, Vol. 11, 1989, pp. 164–180.

[11] S. Lakshmanan, and H. Derin, Simultaneous Parameter Estimation and Segmentation of Gibbs Random Fields Using Simulated Annealing, *IEEE Trans. Pattern Anal. and Machine Intell.*, Vol. 11, 1989, pp. 799–813.

[12] D. Geman, S. Geman, C. Graffigne, and P. Dong, Boundary Detection by Constrained Optimization, *IEEE Trans. Pattern Anal. and Machine Intell.*, Vol. 12, 1990, pp. 609–628.

[13] F. J. Jeng and J. W. Woods, Simulated Annealing in Compound Gaussian Random Fields, *IEEE Trans. Info. Theory*, Vol. 36, 1990, pp. 94–107.

[14] T. Simchony, R. Chellappa, and Z. Lichtenstein, Relaxation Algorithms for MAP Estimation of Grey-Level Images with Multiplicative Noise, *IEEE Trans. Info. Theory*, Vol. 36, 1990, pp. 608–613.

[15] H. L. Tan, S. B. Gelfand, and E. J. Delp, A Cost Minimization Approach to Edge Detection Using Simulated Annealing, *IEEE Trans. Pattern Anal. and Machine Intell.*, Vol. 14, 1992, pp. 3–18.

[16] S. B. Gelfand, and S. K. Mitter, Weak Convergence of Markov Chain Sampling Methods and Annealing Algorithms to Diffusions, *J. Optimization Theory and Applications*, Vol. 68, 1991, pp. 483–498.

[17] S. B. Gelfand, and S. K. Mitter, Simulated Annealing-Type Algorithms for Multivariate Optimization, *Algorithmica*, Vol. 69, 1991, pp. 419–436.

[18] S. B. Gelfand and S. K. Mitter, Recursive Stochastic Algorithms for Global Optimization in $\mathbf{R}^d$, *SIAM Journal Control and Optimization*, Vol. 29, 1991, pp. 999–1018.

[19] S. B. Gelfand and S. K. Mitter, Metropolis-Type Annealing Algorithms for Global Optimization in $\mathbf{R}^d$, to appear in *SIAM Journal Control and Optimization.*

[20] K. Binder, *Monte Carlo Methods in Statistical Physics,* Springer-Verlag, Berlin, 1978.

[21] W. K. Hastings, Monte Carlo Sampling Methods Using Markov Chains and Their Applications, *Biometrika*, Vol. 57, 1970, pp. 97–109.

[22] K. L. Chung, *Markov Processes with Stationary Transition Probabilities*, Springer-Verlag, Heidelberg, Germany, 1960.

[23] P. Billingsley, *Convergence of Probability Measures*, John Wiley & Sons, New York, 1968.

[24] B. Gidas, Nonstationary Markov Chains and Convergence of the Annealing Algorithm, *J. Statistical Physics*, Vol. 39, 1985, pp. 73–131.

[25] B. Hajek, Cooling Schedules for Optimal Annealing, *Mathematics of Operations Research*, Vol. 13, 1988, pp. 311–329.

[26] D. Mitra, F. Romeo, and A. Sangiovanni-Vincentelli, Convergence and Finite-Time Behavior of Simulated Annealing, *Advances in Applied Probability*, Vol. 18, 1986, pp. 747–771.

[27] J. Tsitsiklis, Markov Chains with Rare Transitions and Simulated Annealing, *Mathematics of Operations Research*, Vol. 14, 1989, pp. 70–90.

[28] T. S. Chiang, C. R. Hwang, and S. J. Sheu, Diffusion for Global Optimization in $\mathbf{R}^n$, *SIAM Journal Control and Optimization*, Vol. 25, 1987, pp. 737–752.

[29] H. J. Kushner, Asymptotic Global Behavior for Stochastic Approximation and Diffusions with Slowly Decreasing Noise Effects: Global Minimization via Monte Carlo, *SIAM Journal Applied Mathematics*, Vol. 47, 1987, pp. 169–185.

[30] H. J. Kushner, *Approximation and Weak Convergence Methods for Random Processes*, M.I.T. Press, Cambridge, MA, 1984.

[31] H. J. Kushner and D. Clark, *Stochastic Approximation Methods for Constrained and Unconstrained Systems*, Springer-Verlag, Berlin, Germany, 1978.

[32] L. Ljung, Analysis of Recursive Stochastic Algorithms, *IEEE Trans. on Automatic Control*, Vol. AC-22, 1977, pp. 551–575.

[33] L. C. W. Dixon and G. P. Szego, *Towards Global Optimization,* North-Holland, 1978.

[34] J. L. Marroquin, *Probabilistic Solution of Inverse Problems,* Ph.D. Thesis, LIDS-TH-1500, Laboratory for Information and Decision Systems, MIT, Cambridge, MA, 1985.

[35] T. Khanna, *Foundations of Neural Networks,* Addison-Wesley, Reading, MA, 1990.

Adaptive Gibbsian Automata

Jose Luis Marroquin and Arturo Ramirez
Centro de Investigacion en Matematicas
36000-Guanajuato, Gto.
Mexico

1 Introduction

The well known power of Bayesian estimation methods, based on Markov random field (MRF) models, for solving image reconstruction problems depends, to a large extent, on the availability of certain engines that simulate these fields. These algorithms (the most widely used being the Metropolis [9] and the Gibbs sampler or stochastic relaxation [2] algorithms) are particular instances of a class of dynamical systems known as stochastic cellular automata (SCA, for short [11] and, technically, are regular Markov chains whose invariant distribution is the Gibbs measure of the corresponding MRF. Thus, they may be used to generate typical configurations (images) of the fields in question to check the validity of the assumptions that are embodied in them. They are also used to compute optimal Bayesian estimators that solve the corresponding image reconstruction problems [2] [8].

The asymptotic behavior of these algorithms is relatively well understood [2] [10] . The transient dynamical behavior (and thus, the problem of the convergence rate of optimal estimators), however, has received little attention. This behavior is important, because in a perceptual system (either natural or artificial) it is often desirable to obtain "quick and dirty" solutions that are progressively refined. Also, it is desirable to construct algorithms that take as much advantage as possible of certain information about the solution that may not be encoded in the parameters of the MRF model, but rather, in the initial state of the stochastic automaton.

This is the problem we address in his chapter; we will present design methods for SCA that have the desired asymptotic behavior (i. e., the appropriate invariant Gibbs measure) but that have better transient behavior (faster decreasing rates for the estimation error) than the "classical" designs.

MARKOV RANDOM FIELDS
Theory and Applications

ISBN 0-12-170608-7

To construct such methods, one has to understand first which are the available degrees of freedom for the design of a SCA, once the MRF model is specified. The first part of this chapter (Sections 2, 3 and 4), thus, is devoted to the study of a large class of Gibbsian automata whose general structure and properties were introduced in [7]. This class includes the classical designs and also a new subclass that has the property of having nonreversible dynamic behavior. The automata that belong to it are characterized by the fact that the overall automaton may be formed by a composition of designs that are chosen for each local configuration in an independent way (they are thus called *uncoupled Gibbsian automata*) and are in this sense "adaptive." Of particular interest is the construction of nonreversible, adaptive SCA that are able to move certain configurations around a lattice in a specified way, while retaining the appropriate asymptotic behavior; in this way, it is possible to construct networks whose state converges very fast to an acceptable solution, if one knows an approximate one, and that asymptotically compute the optimal estimator. The design of these automata is discussed in Section 5.

2 Definitions and Notation

In this section we will define the notation and main concepts that will be used throughout this chapter.

We will deal with SCA that generate MRFs on a finite lattice L with N sites. We define, as usual, a neighborhood system $V = \{V_i, i \in L\}$ as a collection of nonempty subsets of sites of L such that

i) $i \notin V_i$, for all $i \in L$,

ii) $i \in V_j$ iff $j \in V_i$, for all $i, j \in L$,

where V_i is the *neighborhood* of site i. A *clique* C of V is either a single site, or a nonempty set of sites of L, such that all sites that belong to C are neighbors of each other.

We will use a finite *state space* Ω, which is defined as a set of functions $\{x : L \mapsto Q\} = Q^N$, where $Q = \{q_1, \dots q_M\}$ is a finite set with M elements (the functions $x(i) = x_i$ are also called *state variables*).

For every subset $S \subseteq L$ we define a *configuration* on S as the restriction $x \mid S$, for $x \in \Omega$, and denote it by $\phi(x, S)$ (thus, $\phi(x, S)$ may be considered as a vector whose components correspond to the values of the state variables $\{x_i, i \in S\}$). The set of all possible configurations on S is denoted by $\mathcal{C}(S) = \{y : S \mapsto Q\}$.

A stochastic cellular automaton ξ on $\{L, V, \Omega\}$ is a discrete-time dynamical system that takes values in Ω :

$$\begin{aligned} \xi: \quad & \mathbf{N} \mapsto \Omega \\ & t \mapsto \xi^{(t)}. \end{aligned}$$

This system is such that $\xi^{(t)}$ changes to $\xi^{(t+1)}$ according to some local probabilistic rule: $\xi_i^{(t+1)}$ depends, probabilistically, on the values of $\xi_i^{(t)}$ at the sites $\{j \in V_i \cup \{i\}\}$. The transition functions are defined by a finite set of parameters (the transition probabilities) that depend on the local configuration in the neighborhood of each site. These parameters completely characterize a SCA; if their value does not depend on t, the automaton is called *time invariant*. In this case we will write the transition probabilities as

$$P_{i,\alpha,r}(q) = \Pr(\xi_i^{(t+1)} = q \mid \xi_i^{(t)} = r \text{ and } \phi(\xi^{(t)}, V_i) = \alpha)),$$

with $0 \leq P_{i,\alpha,r}(q) \leq 1$ for all $i \in L$, all $\alpha \in \mathcal{C}(V_i)$ and all $q, r \in Q$. These probabilities satisfy

$$\sum_{q \in Q} P_{i,\alpha,r}(q) = 1. \tag{2.1}$$

A SCA will be called *serial* if the state variables change value one at a time in a fixed specified sequence, called the *sweeping strategy*, and it will be called *asynchronous* if at a given time only one variable, selected at random with uniform probability, updates its state.

A SCA is technically a Markov chain, whose states correspond to global configurations $\phi(\xi^{(t)}, L)$. A time invariant SCA corresponds to a homogeneous chain, and will be called *ergodic* if the chain is regular [5]. In this case, we will refer to the invariant probability measure of the chain as the *invariant measure* of the automaton. An asynchronous, ergodic SCA will be called *serializable* if its invariant measure remains the same when any serial sweeping strategy is adopted for the update, without changing the parameters of the automaton.

An ergodic SCA will be called *Gibbsian* if its invariant measure corresponds to the probability distribution of a MRF on L with the same state space and neighborhood system as the automaton, i.e., if

$$\pi(x) = \lim_{t \to \infty} \Pr(\xi^{(t)} = x) = \frac{1}{Z} e^{-U(x)}$$

with $\pi(x) > 0$ for all $x \in \Omega$, and

$$U(x) = \sum_C v_C(x),$$

where C ranges over the cliques of the neighborhood system V and the functions $v_C(\cdot)$ (called *potential functions*) depend only on the states of the variables indexed by the sites in C.

$X(i,q)$ will be defined as the global configuration that is identical to the configuration $x \in \Omega$, except at site i, where it is equal to $q \in Q$. If π is the Gibbsian measure of a MRF on $\{L, V, \Omega\}$, the relative likelihood of state q to state r in site i is defined as

$$\begin{aligned}
\psi_{i,\phi(x,V_i)}(q,r) &= \frac{\pi\left(X(i,q)\right)}{\pi\left(X(i,r)\right)} \qquad (2.2)\\
&= e^{U(X(i,r))-U(X(i,q))} = e^{\sum_{C:i\in C}(v_C(X(i,r))-v_C(X(i,q)))}.
\end{aligned}$$

These relative likelihoods have the property that, for all $r, s, q \in Q$, all $i \in L$ and all $\alpha \in \mathcal{C}(V_i)$:

$$\psi_{i,\alpha}(r,q) = \psi_{i,\alpha}(r,s)\psi_{i,\alpha}(s,q), \qquad (2.3)$$

so that $\psi_{i,\alpha}(q,q) = 1$. This implies that the quantities

$$\{\psi_{i,\alpha}(q_1,q_2), \ldots, \psi_{i,\alpha}(q_{M-1},q_M)\}$$

completely characterize a given MRF model. The conditional probabilities may be obtained from them using the relations:

$$\Pr(x_i = q \mid \phi(x,V_i) = \alpha) = \frac{1}{\sum_{r\in Q}\psi_{i,\alpha}(r,q)}.$$

This representation of a MRF may be considered "canonical" in the sense that it gives the precise number of effective degrees of freedom available for the specification of the field. Note that this is not the case of the potential functions, because in general it is possible to specify different sets of potentials (with different numbers of elements) that in fact give the same Gibbsian distribution.

An ergodic SCA is called *reversible* if the corresponding chain is regular and reversible. It is well known [5] that reversibility of a regular chain is equivalent to the *detailed balance* condition

$$\pi(x)P(x,y) = \pi(y)P(y,x),$$

where $\pi(\cdot)$ is the invariant measure of the chain and $P(x,y)$ is the transition probability between global configurations x and y (this condition was first established by Kolmogorov [6]). For asynchronous Gibbsian automata, it takes the form

$$P_{i,\alpha,r}(q) = \psi_{i,\alpha}(q,r)P_{i,\alpha,q}(r). \qquad (2.4)$$

3 Uncoupled Gibbsian Automata

In this section, we will introduce the general family of automata that is the subject of this chapter and discuss its structure and general properties. We will start with the case of asynchronous automata and then extend the results to serial ones.

Let ξ be an asynchronous, time invariant SCA on $\{L, V, \Omega\}$. The necessary and sufficient conditions that a given probability measure π must satisfy to be an invariant measure of ξ are that, for all $x \in \Omega$,

$$\pi(x) = \sum_{y \in \Omega} P(y, x)\pi(y), \tag{3.5}$$

where $P(y, x)$ is the one-step transition probability between global configurations y and x [5]. Since ξ is asynchronous, only one site (selected at random) is updated at a time, so,

$$\begin{aligned} P(y,x) &= \tfrac{1}{N}\textstyle\sum_{i\in L} \Pr(\xi_i^{(t+1)} = x_i \mid \xi^{(t)} = x), && \text{if } y = x; \\ &= \tfrac{1}{N}\Pr(\xi_i^{(t+1)} = x_i \mid \xi^{(t)} = X(i,q)), && \text{if } y = X(i,q) \neq x; \\ &= 0, && \text{otherwise.} \end{aligned}$$

Using these relations, we may rewrite Eq. (3.5) in terms of the parameters of the automaton as

$$\pi(x) = \frac{1}{N} \sum_{i \in L} \sum_{q \in Q} P_{i,\phi(x,V_i),q}(x_i)\pi(X(i,q)) \tag{3.6}$$

for all $x \in \Omega$.

Every SCA whose parameters $\{P_{i,\alpha,r}(q)\}$ satisfy this system will have π as its invariant measure. A particular class of SCA is obtained when the set of solutions of Eq. (3.6) is replaced by the set of solutions to the systems

$$\pi(x) = \sum_{q \in Q} P_{i,\phi(x,V_i),q}(x_i)\pi(X(i,q)) \tag{3.7}$$

for each site $i \in L$ and each $x \in \Omega$, which are the discrete–time equivalent of the *partial balance* conditions that appear in the theory of continuous–time Markov chains [4].

Every automaton that satisfies Eq. (3.7) will also satisfy Eq. (3.6); system (3.7), however, is much simpler, since one may solve for the SCA parameters *for each site and each configuration separately*, because the equations are completely decoupled. For this reason, we will call the automata that belong to this class, *uncoupled automata*.

In geometric terms, the set of all uncoupled SCA with a given invariant measure π is the cartesian product (for each site i and for each configuration $\alpha \in \mathcal{C}(V_i)$) of a family of sets $\mathcal{G}_{i,\alpha}$, each one of which is the intersection of three families of sets:

- The hypercube $\{P_{i,\alpha,q}(r) : 0 \leq P_{i,\alpha,q}(r) \leq 1\} \subset \mathbf{R}^{M^2}$ (where M is the cardinality of Q).
- The hyperplanes defined by Eq. (2.1):

$$\sum_{r \in Q} P_{i,\alpha,q}(r) = 1.$$

- The hyperplanes defined by the solutions of system (3.7).

The uncoupled automata have other important properties:

Theorem 1 *Every uncoupled SCA with positive parameters is serializable and Gibbsian.*

Proof:
Let ξ be an uncoupled SCA, and let π be its invariant measure. If all the parameters $\{P_{i,\alpha,r}(q)\}$ are strictly positive, the transition matrix for the full sweep chain (with any sweeping strategy) will have no zero entries, and hence, it will be regular. [5]

Consider now the (nonergodic) one-step chain that updates only site i: since the automaton is uncoupled, it satisfies Eq. (3.7) and hence, π is an invariant measure for this chain. For any sweeping strategy, the full sweep transition probability matrix is the product of the one-step transition matrices (taken in the order of the sweep), and hence, π will also be an invariant measure for this chain.

To see that ξ is Gibbsian, we note that the invariant measure of the one–step chain that updates site i, for a given, fixed x, is precisely the conditional probability distribution $\pi(x_i \mid x_j, j \neq i)$. Since this distribution is completely determined by the transition probabilities $\{P_{i,\phi(x,V_i),r}(q)\}$, it depends only on the local configuration $\phi(x, V_i)$, which means that ξ is Gibbsian [4]. □

Remarks:

1. *By the same reasoning, one can see that if one has several uncoupled automata with positive parameters and the same invariant measure π, the composite (time-variant) automaton that alternates full sweeps of these automata, with any sweeping strategies, will also have π as its invariant measure.*

2. *Note that the condition that the parameters are positive is sufficient but not necessary for the regularity of the full sweep chain [5]. If this condition does not hold, one may still have regularity for particular sweeping strategies, but this is, in general, difficult to verify.*

Consider now the Gibbsian invariant measure π of an uncoupled automaton with neighborhood system V. In this case, we can divide both sides of Eq. (3.7) by $\pi(x)$ and use Eq. (2.2) to get

$$1 = \sum_{r \in Q} P_{i,\phi(x,V_i),r}(x_i)\psi_{i,\phi(x,V_i)}(r, x_i)$$

for each $x \in \Omega$ and each $i \in L$.

Eliminating the redundant equations, we may rewrite this system as

$$\sum_{r \in Q} P_{i,\alpha,r}(q)\psi_{i,\alpha}(r, q) = 1$$

for every $i \in L$, $\alpha \in \mathcal{C}(V_i)$ and $q \in Q$.

To summarize, then, given a specific MRF model on $\{L, V, \Omega\}$, i.e., given the quantities $\{\psi_{i,\alpha}(q_1, q_2), \ldots, \psi_{i,\alpha}(q_{M-1}, q_M)\}$, every automaton that satisfies, for each i and α the conditions

$$\sum_{r \in Q} P_{i,\alpha,r}(q)\psi_{i,\alpha}(r, q) = 1, \quad \text{for every } q \in Q \tag{3.8}$$

$$\sum_{q \in Q} P_{i,\alpha,r}(q) = 1, \quad \text{for every } r \in Q \tag{3.9}$$

$$0 < P_{i,\alpha,r}(q) < 1, \quad \text{for all } r, q \in Q \tag{3.10}$$

is serializable and will have the given Gibbsian measure as its long run distribution.

The set of all SCA that satisfy these conditions will be called the set of *uncoupled Gibbsian automata*, and will be denoted by $\mathcal{U} = \{\mathcal{G}_{i,\alpha}, i \in L, \alpha \in \mathcal{C}(V_i)\}$. Note that all reversible Gibbsian automata are uncoupled: to see this, recall that a reversible SCA satisfies Eqs. (2.1) and (2.4), and therefore, Eq. (3.8). We will denote the set of all reversible Gibbsian automata by $\mathcal{R} = \{\mathcal{R}_{i,\alpha} : i \in L, \alpha \in \mathcal{C}(V_i)\}$.

The fundamental properties of the uncoupled Gibbsian automata are given by the following theorem:

Theorem 2 *For every $i \in L$ and $\alpha \in \mathcal{C}(V_i)$, $\mathcal{G}_{i,\alpha}$ and $\mathcal{R}_{i,\alpha}$ are subsets of the vector space of dimension M^2: $\{P_{i,\alpha,r}(q) \in \mathbf{R} : r, q \in Q\}$ that satisfy* *(i)* $\mathcal{R}_{i,\alpha} \subset \mathcal{G}_{i,\alpha}$, *and both are bounded convex sets;*

(ii) The dimension of $\mathcal{R}_{i,\alpha}$ *is* $M(M-1)/2$, *and the dimension of* $\mathcal{G}_{i,\alpha}$ *is* $(M-1)^2$;

(iii) $\mathcal{R}_{i,\alpha} = \mathcal{G}_{i,\alpha}$ *iff* $M = 2$; *and*

(iv) $\bar{\mathcal{G}}_{i,\alpha}$ *and* $\bar{\mathcal{R}}_{i,\alpha}$ *are compact, convex polytopes (polyhedra of higher dimension) with a finite number of vertices.*

The proof of this theorem may be found in [7], as well as a discussion about Gibbsian automata outside $\mathcal{U}$.

4 Design Procedures

The general procedures for selecting values for finding points in each $\mathcal{G}_{i,\alpha}$ and $\mathcal{R}_{i,\alpha}$ are based on the fact that the intersection of the convex sets $\mathcal{G}_{i,\alpha}$ or $\mathcal{R}_{i,\alpha}$ with the hyperplanes that correspond to the values of the elements $P_{i,\alpha,s}(t)$ that have already been specified remains convex and the projection of these sets on each of the remaining $P_{i,\alpha,r}(q)$ axes is also convex (i.e., an interval). The general design algorithm, therefore, consists of the specification of the update mechanism for the valid intervals (L_{kl}, U_{kl}), for each element $\{P_{i,\alpha,q_k}(q_l)\}$ whose value remains to be selected [7].

Another way of designing an automaton is based on the fact that any point in the interior of a convex set may be specified as a convex combination of elements in its boundary (note that these boundary elements are not necessarily ergodic nor serializable). These extreme SCA may also be combined with interior points, to obtain automata that combine extreme properties with other, more stable ones (see Section 5.4). This is the approach that we will follow in this chapter. To simplify the presentation, we will use the following notation: since the SCA are uncoupled, the design procedures will be given for a fixed $i \in L$ and $\alpha \in \mathcal{C}(V_i)$; therefore, the indices (i, α) may be omitted. Thus, we will denote $\psi_{i,\alpha}(q_k, q_l)$ by ψ_{kl}, and the parameters $\{P_{i,\alpha,q_k}(q_l)\}$ by P_{kl}, for $1 \leq k, l \leq M$. In this way, a SCA design, for each (i, α) is represented by the $M \times M$ matrix P, which is sometimes called the *one–step transition matrix.*

One strategy to obtain boundary points is based in the fact that if we partition the state space Q into n disjoint subsets and require that $P_{ij} \neq 0$ only if i and j correspond to states in the same subset, we will obtain SCA with parameter matrices of the form

$$\begin{bmatrix} P_1 & 0 & \dots & 0 \\ 0 & P_2 & 0 & \dots \\ 0 & \dots & 0 & P_n \end{bmatrix},$$

which are in the boundary. The elements of every submatrix P_i satisfy equations and constraints of the same form as the original matrix P, and hence their value may be determined by the same methods. Note that if a submatrix consists of a single element, the only valid value for this element is 1.

4.1 Vertices

Vertices are particularly important extreme points. They may be obtained using the general procedures described in [7], selecting, at each stage, an extreme value (either L_{ij} or U_{ij}) for every element. To study the vertices in a systematic way, it is useful to introduce the concept of *canonical form*:

Let $\gamma : \{1, \dots M\} \mapsto \{1, \dots M\}$ be a permutation that reorders the states of Q (for a fixed i, α) in such a way that the conditional probabilities

$$\hat{p}_k = \Pr(x_i = q_{\gamma(k)} \mid \phi(x, V_i) = \alpha)$$

satisfy $\hat{p}_1 \leq \hat{p}_2 \leq \dots \leq \hat{p}_M$. The order $\{\gamma(1), \dots, \gamma(M)\}$ is called the *canonical order* for Q. Note that by Eq. (2.2), it implies that

$$\hat{\psi}_{ij} = \psi_{\gamma(i)\gamma(j)} \leq 1$$

for $i < j$. The automaton with parameter matrix $\hat{P}$, obtained using $\hat{\psi}$ is said to be in *canonical form*. The matrix P that corresponds to the original order may be recovered from $\hat{P}$ using

$$P_{ij} = \hat{P}_{\gamma^{-1}(i)\gamma^{-1}(j)}.$$

Using the canonical form, one may give general families of vertices: for $M = 2$, $\mathcal{G}_{i,\alpha} = \mathcal{R}_{i,\alpha}$ has only two vertices (see Fig. 1–a):

$$\begin{bmatrix} 1 & 0 \\ 0 & 1 \end{bmatrix} \text{ and } \begin{bmatrix} 0 & 1 \\ \hat{\psi}_{12} & 1 - \hat{\psi}_{12} \end{bmatrix}.$$

Figure 1-b shows $\mathcal{R}_{i,\alpha}$ for $M = 3$ (note that it is in the space spanned by $(\hat{P}_{12}, \hat{P}_{13}, \hat{P}_{23})$, since these are the only degrees of freedom); there are two different cases:

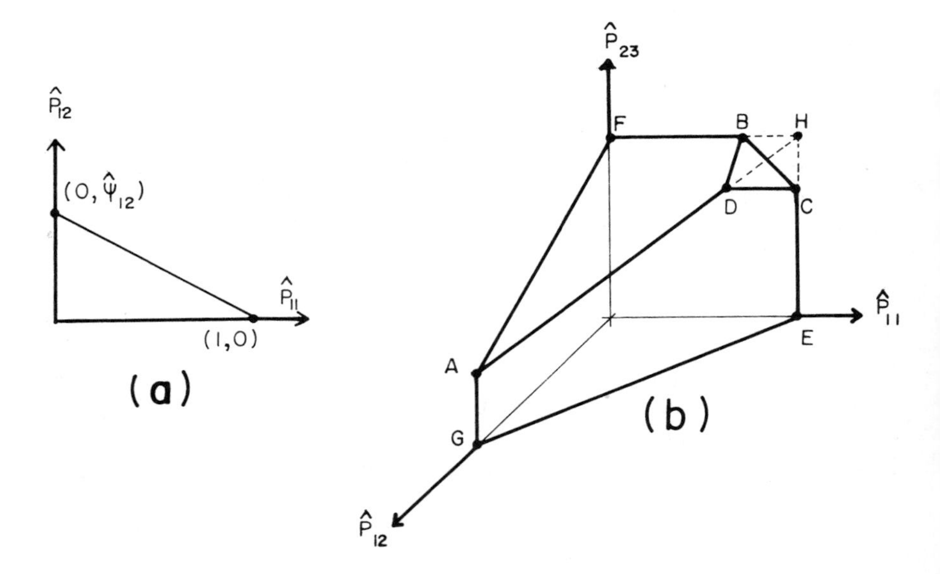

Figure 1: (a) Canonical form of $\mathcal{G}_{i,\alpha} = \mathcal{R}_{i,\alpha}$ for $M = 2$. (b) Canonical form of $\mathcal{R}_{i,\alpha}$ for $M = 3$.

1. When $\hat{\psi}_{21}(\hat{\psi}_{32} - 1) < 1$, in which case there are eight vertices, that correspond to the points

$$\begin{aligned} O &= (0,0,0) \\ A &= (1,0,1-\hat{\psi}_{12}) \\ B &= (0,\hat{\psi}_{31}-\hat{\psi}_{21},1) \\ C &= (0,1,\hat{\psi}_{32}-\hat{\psi}_{12}) \\ D &= \left(\frac{1+\hat{\psi}_{21}-\hat{\psi}_{31}}{2}, \frac{1+\hat{\psi}_{31}-\hat{\psi}_{21}}{2}, \frac{1+\hat{\psi}_{32}-\hat{\psi}_{12}}{2}\right) \\ E &= (0,1,0) \\ F &= (0,0,1) \\ G &= (1,0,0). \end{aligned}$$

2. When $\hat{\psi}_{21}(\hat{\psi}_{32}-1) \geq 1$, in which case the vertices B, C and D collapse

into the vertex $H = (0,1,1)$.

One can show that D is the only vertex of $\mathcal{R}_{i,\alpha}$ that is *not* a vertex of $\mathcal{G}_{i,\alpha}$ [7].

A table of all the vertices of $\mathcal{G}_{i,\alpha}$ for M = 3 is presented in [7]. We now give some examples of families of vertices that always exist for $M \geq 3$. For all these families, $\hat{P}_{ij} = 0$ for all indices i, j, except for those indicated:

$$\text{i)} \quad \begin{array}{lcll} \hat{P}_{1,M} & = & 1; & \\ \hat{P}_{i+1,i} & = & \hat{\psi}_{i,i+1}, & \text{for } i = 1, \ldots M-1; \\ \hat{P}_{i,M} & = & 1 - \hat{\psi}_{i-1,i}, & \text{for } i = 2, \ldots M \end{array} \tag{4.11}$$

$$\text{ii)} \quad \begin{array}{lcll} \hat{P}_{M,1} & = & \hat{\psi}_{1,M}; & \\ \hat{P}_{i,i+1} & = & 1, & \text{for } i = 1, \ldots M-1; \\ \hat{P}_{M,i} & = & \hat{\psi}_{i,M} - \hat{\psi}_{i-1,M}, & \text{for } i = 2, \ldots M \end{array} \tag{4.12}$$

$$\text{iii)} \quad \begin{array}{lcll} \hat{P}_{M,1} & = & \hat{\psi}_{1,M}; & \\ \hat{P}_{i,i+1} & = & \hat{\psi}_{1,i}, & \text{for } i = 1, \ldots M-1; \\ \hat{P}_{i,i} & = & 1 - \hat{\psi}_{1,i}, & \text{for } i = 1, \ldots M \end{array} \tag{4.13}$$

$$\text{iv)} \quad \begin{array}{lcll} \hat{P}_{1,M} & = & 1; & \\ \hat{P}_{i+1,i} & = & \hat{\psi}_{i,i+1}, & \text{for } i = 1, \ldots M-1; \\ \hat{P}_{i,i} & = & 1 - \hat{\psi}_{i-1,i}, & \text{for } i = 2, \ldots M \end{array} \tag{4.14}$$

$$\text{v)} \quad \begin{array}{lcll} \hat{P}_{i+1,i} & = & \sum_{j=0}^{i-1} \psi_{i-j,i}, & \text{for } i = 1, \ldots M-1 \\ \hat{P}_{i-1,i} & = & \hat{\psi}_{i,i-1} \hat{P}_{i,i-1}, & \text{for } i = 2, \ldots M \\ \hat{P}_{M,M} & = & 1 - \hat{P}_{M,M-1} & \end{array} \tag{4.15}$$

Families (i) through (iv) correspond to nonreversible vertices, and family (v) to reversible ones.

5 Adaptive Gibbsian Automata

In the last two sections we saw how it is possible to design Gibbsian automata that simulate a given MRF model for each site $i \in L$ and each local configuration $\alpha \in \mathcal{C}(V_i)$ in an independent way. We will now show how these principles may be used to construct automata that are adaptive, in the sense that one chooses different designs for different classes of local configurations. This adaptive behavior allows one to construct high performance Bayesian estimators for specific reconstruction problems. Specifically, these automata will be able to take advantage of additional information about the solution that is encoded in the initial state of the SCA to improve significantly the convergence rate of the estimator.

To simplify the following discussion, we will consider only MRFs that are *homogeneous* in the sense that

$$\psi_{i,\alpha}(r,q) = \psi_{j,\alpha}(r,q) = \psi_{\alpha}(r,q)$$

for all $i, j \in L$. To include the important case of posterior Gibbsian distributions obtained by coupling, to a given MRF, observations corrupted by noise that is independent for each site, we will consider an enlarged neighborhood $\bar{V}_i$, which is obtained as the union of V_i and the corresponding site of the coupled observations lattice (we also assume that the observations take values in Q). This homogeneity assumption also implies that $\mathcal{C}(\bar{V}_i) = \mathcal{C}(\bar{V}_j) = \mathcal{C}(\bar{V})$ for all $i, j \in L$. The resulting SCA will also be homogeneous, and we will denote their parameters by $\{P_{\alpha,r}(q)\}$, for $\alpha \in \mathcal{C}(\bar{V})$, and $r, q \in Q$.

To construct an adaptive automaton, one partitions the set $\mathcal{C}(\bar{V})$ into nonoverlapping classes $\{S_1 \ldots S_K\}$ and then selects an appropriate design for each class.

Once a specific design is found for a particular class S_k, it is often desirable to find an "equivalent" design for other classes that are related to S_k by some symmetry operation: of particular interest are permutations of the labeling of the states and of the sites of the neighborhood. We will now study them in detail.

5.1 State Symmetric Automata

Let $S_k = \{\alpha_1, \ldots, \alpha_n\}$, and $S_l = \{\beta_1, \ldots, \beta_n\}$ be two subsets of $\mathcal{C}(\bar{V})$, and let $\sigma : Q \mapsto Q$ be a permutation on the labeling of the states. We will denote, by $\sigma(r)$, the image of state r under the permutation and, by $\sigma(\alpha)$, the configuration that is obtained from α by permuting the states of all the sites of $\bar{V}$. Suppose that

1. if $\alpha = \phi(x, \bar{V}_i) \in S_k$, then $\sigma(\alpha) = \phi(\sigma(x), \bar{V}_i) \in S_l$;

2. $\psi_{\sigma(\alpha)}(\sigma(r), \sigma(q)) = \psi_{\alpha}(r, q)$, for all $\alpha \in S_k$, and $r, q \in Q$.

If ξ is the SCA design chosen for class S_k, then the *state symmetric automaton* $\hat{\xi}$ is obtained from the parameters of ξ by the relations

$$\hat{P}_{\sigma(\alpha),r}(q) = P_{\alpha,\sigma^{-1}(r)}(\sigma^{-1}(q)) \tag{5.16}$$

for all $\alpha \in S_k$ and $r, q \in Q$.

The automaton $\hat{\xi}$ will have the same qualitative dynamic behavior as ξ when the labeling of the states is permuted using σ (see the example of Section 5.3). Note that this symmetric design is possible only if the potentials of the underlying MRF are symmetric in the same sense, so that condition 2 is satisfied. In this case, it is clear that if ξ satisfies conditions (3.8), (3.9) and (3.10), $\hat{\xi}$ will satisfy them also, since the canonical forms are the same for both automata.

5.2 Site Symmetric Automata

Consider now the two sets of configurations $T_k = \{\alpha_1, \ldots \alpha_n\}$, $T_l = \{\beta_1, \ldots, \beta_n\}$, and let $\tau : \bar{V} \mapsto \bar{V}$ be a permutation of the sites of the neighborhood of each site, such that

1. if $\alpha = \phi(x, \bar{V}_i) \in T_k$, then $\tau(\alpha) = \phi(x, \tau(\bar{V}_i)) \in T_l$, for all $i \in L$;
2. $\psi_{\tau(\alpha)}(r, q) = \psi_\alpha(r, q)$ for all $\alpha \in T_k$ and all $r, q \in Q$.

If ξ is the SCA design chosen for class T_k, then the *site symmetric automaton* $\tilde{\xi}$ is obtained from the parameters of ξ by

$$\tilde{P}_{\tau(\alpha),r}(q) = P_{\alpha,r}(q) \tag{5.17}$$

for all $\alpha \in T_k$ and $r, q \in Q$.

As before, if the potentials of the MRF are such that condition 2 is satisfied, $\tilde{\xi}$ will have the same qualitative behavior as ξ when the permutation τ is applied to the sites of the neighborhood of each site. Note that in this case, conditions (3.8) and (3.9) give the same set of equations for both automata.

We will now give an example of the application of these concepts to the design of optimal estimators for a specific image reconstruction problem.

5.3 An Adaptive Nonreversible Automaton

In image reconstruction, there are situations in which one knows *a priori* that a certain configuration (e.g., a uniform region of a given shape, a line, etc.) should be present at some approximate location, but the precise location and size are unknown. Thus, for example, in multiscale processing, one may wish to find the precise location (on a fine grid) of a boundary found on a coarser one; one may wish to couple observations from different sources that are obtained with different precision and resolution, etc.

If a Gibbsian SCA is used for the reconstruction task, it would be nice to take advantage of this additional information to improve its convergence rate. We will now present an example in which this improvement is achieved by means of a nonreversible adaptive automaton.

Since for binary fields all Gibbsian SCA are reversible, the simplest example of nonreversible automata must use a ternary MRF. We will construct it in the following way:

Consider an Ising field where the uniform regions (with states 0 or 1, respectively) are separated by boundaries (pixels with state 2) that tend to be predominantly vertical. To implement this behavior, we will assume ferromagnetic interactions of strength β between "uniform" pixels and of strength $\eta < \beta$ between a "uniform" and a "boundary" pixel. Since the boundaries tend to be vertical, we will assume vertical ferromagnetic and horizontal antiferromagnetic interactions between boundary pixels. The potentials for nearest neighbor pairs of sites $[i, j]$ are thus:

$$\begin{aligned} V_2(f_i, f_j) &= V_2(f_j, f_i) \\ &= -\beta, \text{ if } f_i = f_j \text{ and } f_i, f_j \in \{0,1\}; \\ &= \beta, \text{ if } f_i \neq f_j \text{ and } f_i, f_j \in \{0,1\}; \\ &= -\eta, \text{ if } f_i = 2 \text{ and } f_j \in \{0,1\}; \\ &= \eta, \text{ if } f_i = f_j = 2 \text{and} (i,j) \text{is a pair of horizontal} \\ &\qquad\qquad\qquad\qquad\qquad\qquad \text{neighbors;} \\ &= -\mu, \text{ if } f_i = f_j = 2 \text{ and} (i,j) \text{is a pair of vertical} \\ &\qquad\qquad\qquad\qquad\qquad\qquad \text{neighbors;} \\ &= 0, \text{otherwise.} \end{aligned}$$

To prevent clustering of boundary pixels, we include a potential V_3 that penalizes clusters of three sites with state = 2 (see Fig. 2) by making $V_3(f_i, f_j, f_k) = \nu$ for those clusters and $V_3(f_i, f_j, f_k) = 0$ otherwise. [t] Figure 3 shows a typical configuration of the field (on a toroidal lattice).

5.3.1 Observations

Suppose that we have observations g obtained at sparse locations and with a probability of error equal to ϵ, i.e.,

$$\Pr(g_i \neq f_i \mid f) = \epsilon.$$

We will assume that we have available observations with density ρ_1 in uniform regions (i.e., regions with $f_i \in \{0, 1\}$) and with density ρ_2 at the

Figure 2: Clusters of three sites i, j, k with state = 2 (modulo rotations) penalized by setting $V_3(f_i, f_j, f_k) = \nu$ (see text).

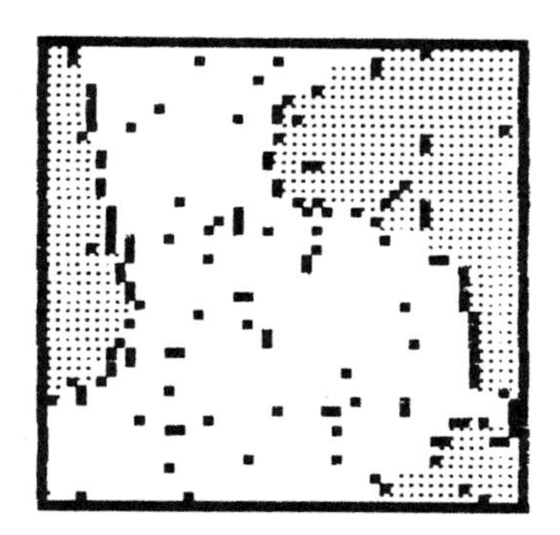

Figure 3: Typical equilibrium configuration of the ternary field described in the text. The values of the parameters are $\beta = \mu = 1$, $\eta = 0.1$ and $\nu = 8.0$. State 0 is represented by a white pixel, state 1 by a grey one, and state 2 by a black one.

boundaries. Calling S the subset of sites where observations are available, the posterior probability function is a Gibbsian measure with energy given by

$$U_P = \sum_{[i,j]} V_2(f_i, f_j) + \sum_{[i,j,k]} V_3(f_i, f_j, f_k) + \alpha \sum_{i \in S} (1 - \delta(f_i - g_i))$$

$$\text{with } \alpha = \ln\left(\frac{1-\epsilon}{\epsilon}\right)$$

where $[i, j]$ and $[i, j, k]$ denote nearest neighbor pairs and configurations

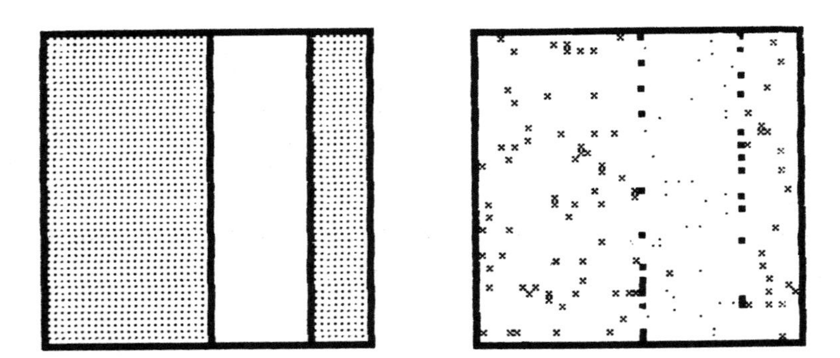

Figure 4: Original image (left) and sparse observations (right) for the estimation problem described in the text. The observation is denoted by an x for a grey pixel (state = 1) and by a dot for a white pixel (state = 0).

like the ones of Fig. 2, respectively, and $\delta(x) = 1$, if $x = 0$ and it equals 0, otherwise.

5.3.2 Construction of the Adaptive Automaton

Consider now the following estimation problem:

Suppose we have an image that consists of a vertical bar of pixels at state 0, separated by boundaries of pixels at state 2 from a uniform background of pixels at state 1 (see Fig. 4), and that one knows the approximate width and location of this bar.

To find the exact location and width from a set of sparse observations, one would like to have an automaton that, when initialized with the approximate solution, moves it around the lattice in a controlled fashion and "locks" it in the exact position. To design this automaton, we explored the behavior of the extreme SCA corresponding to the prior measure, in particular those specified by the nonreversible vertices, for configurations that may correspond to a boundary between uniform regions at states 1 and 0. Thus, we explored the adaptive designs

$$P_{\alpha,r}(q) = V_{\alpha,r}(q)$$

for $\alpha \in \mathcal{C}(\bar{V}_i)$, such that $\xi_{i_x+1,i_y} = 0$ and $\xi_{i_x-1,i_y} = 1$,

$$P_{\alpha,r}(q) = P^*_\alpha(q), \text{ elsewhere,}$$

where the site index $i = (i_x, i_y)$, and $V_{\alpha,r}(q)$ and $P^*_\alpha(q)$ are the corresponding transition probabilities of the examined vertex and the Gibbs Sampler automaton, respectively. We used a semisynchronous updating strategy, which is obtained by coloring the sites of the lattice in such a way that two neighbors are never of the same color (in this case, four colors suffice) and then updating the sites of each color in turn (note that the sweeping order for the sites of a given color is irrelevant, and in fact, they may be updated in parallel [2]).

In this way, we found that vertex (4.12), whose canonical form for $M = 3$ is

$$\begin{bmatrix} 0 & 1 & 0 \\ 0 & 0 & 1 \\ \hat{\psi}_{i,\alpha}(0,2) & \hat{\psi}_{i,\alpha}(1,2) - \hat{\psi}_{i,\alpha}(0,2) & 1 - \hat{\psi}_{i,\alpha}(1,2) \end{bmatrix}, \tag{5.18}$$

in fact moves this boundary from left to right. The average speed of this motion may be controlled by selecting, instead of V_α, an automaton at a relative distance λ from the Gibbs Sampler automaton along the line that joins these two SCA, i.e.,

$$P_{\alpha,r}(q) = \lambda P^*_\alpha(q) + (1-\lambda)V_{\alpha,r}(q) \tag{5.19}$$

for $\alpha \in \mathcal{C}(\bar{V}_i)$, such that $\xi_{i_x+1,i_y} = 0$ and $\xi_{i_x-1,i_y} = 1$, and

$$P_{\alpha,r}(q) = P^*_\alpha(q) \text{ , elsewhere,}$$

for $\lambda \in (0,1)$.

Once this automaton is found, the corresponding SCA that moves the right boundary of the bar with the same direction and speed is just the state-symmetric automaton, obtained with the permutation σ that interchanges states 0 and 1 (see Fig. 5).

To design an automaton that moves the bar in the opposite direction (from right to left), one uses the corresponding site-symmetric automaton, obtained by a permutation that reflects the sites of the neighborhood about a vertical line that passes through the central site (see Fig. 6).

These two designs may now be combined—alternating between them every fixed number of iterations—to obtain an "oscillating" bar. If one uses the posterior measure for the design, this bar will search for the best position to lock up.

The motion speed (i.e., the value of λ) determines, of course, the rate of convergence; if this value is too high, however, it may prevent the bar from locking up at the correct position (even though its speed will decrease as it passes over it) and thus have a negative effect on the convergence rate. The highest value of λ that still allows for a correct lock up depends on the signal to noise ratio and, at present, must be determined experimentally.

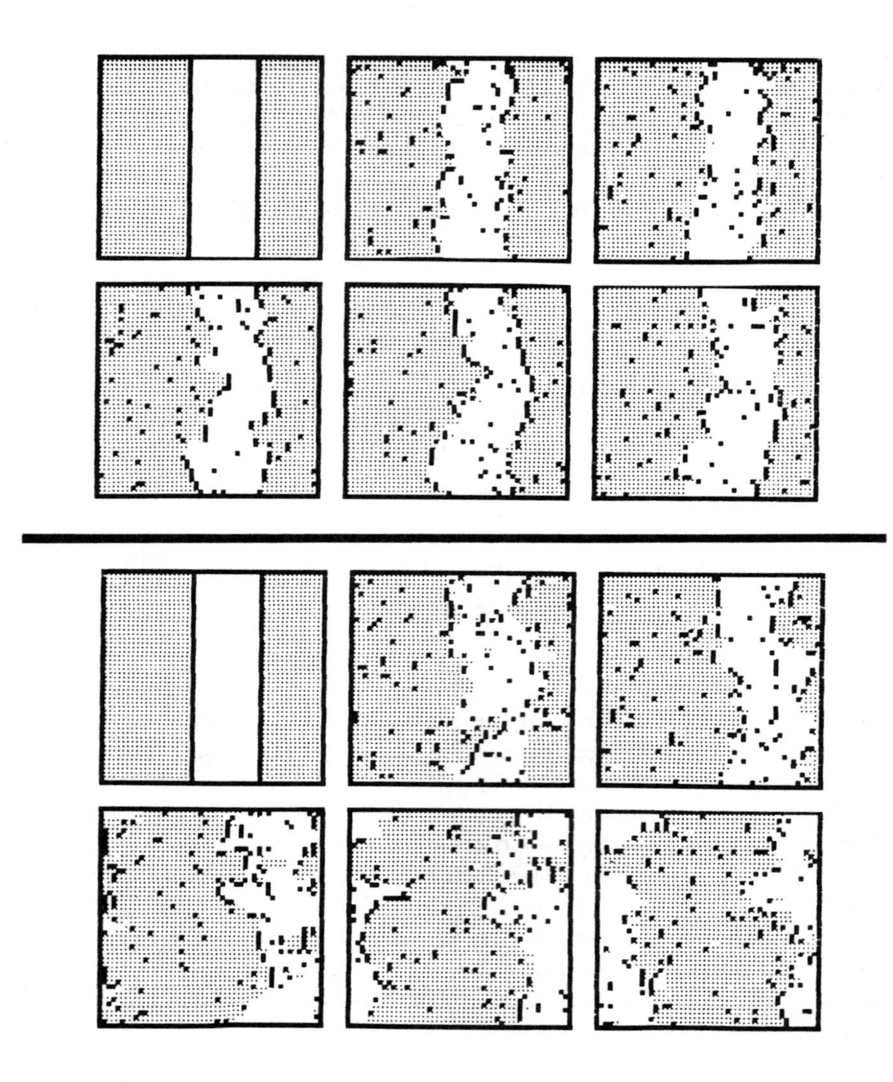

Figure 5: Horizontal motion of a vertical bar on a toroidal lattice at different average speeds: the upper panel shows the instantaneous state of the adaptive automaton described in the text, with $\lambda = 0.3$, taken every 10 iterations. The lower panel shows the same automaton with $\lambda = 0.8$.

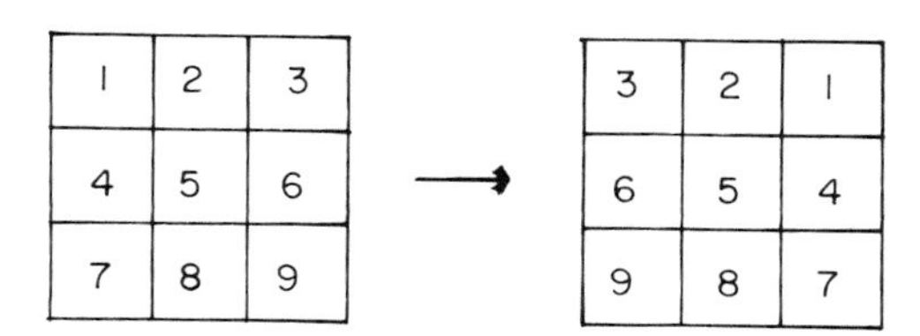

Figure 6: Permutation τ for the site symmetric automaton that reverses the horizontal direction of motion.

5.3.3 Experimental Performance

Figure 7 shows a plot of the MPM estimation error (averaged over 50 independent runs) against the iteration number for the following cases:

1. The Gibbs Sampler automaton initialized with a random configuration;
2. The Gibbs Sampler automaton initialized with an approximate solution;
3. The "oscillating" adaptive automaton described previously initialized with the same approximate solution; and
4. Same as 3, but with the adaptive design replaced by the Gibbs Sampler after 30 iterations.

In all cases, we used a semi-synchronous sweeping strategy. The parameters of the field and the observations are: $\mu = \beta = 1.0$; $\eta = 0.1$; $\rho_1 = 0.05$; $\rho_2 = 0.3$; and $\epsilon = 0.01$. The original image is a vertical bar (of pixels with state = 0) 15 pixels wide that starts at column 25 of a toroidal lattice of size 50×50. The state of the rest of the pixels is equal to 1, except at the boundaries, where it equals 2 (see Fig. 4).

The approximate solution used to initialize cases 2, 3 and 4 is a vertical bar 15 pixels wide, starting at column 20 (see Fig. 8). For the adaptive design, we used Eq. (5.19) with $\lambda = 0.3$ and changed to the site-symmetric design to reverse the direction of motion every 25 iterations. This automaton, therefore repeatedly sweeps the area where the correct solution is supposed to be, although in fact, it locks up in the correct position at the first sweep. Figure 8 shows the instantaneous state of each of these SCA, taken every 15 iterations, in a typical run.

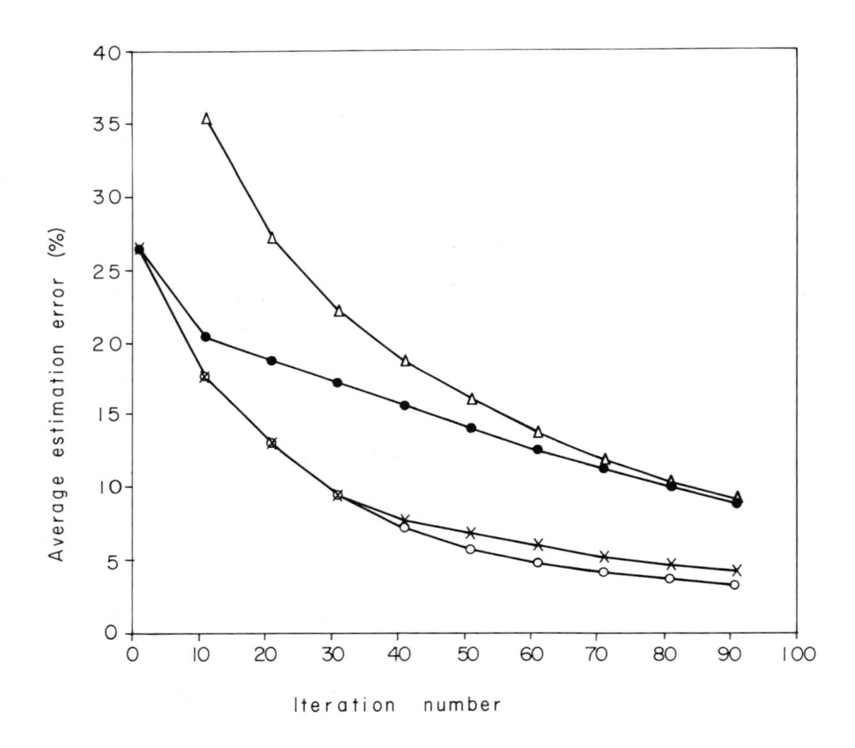

Figure 7: Average MPM estimation error in the reconstruction of a ternary image of a vertical bar from sparse samples for the Gibbs Sampler automaton with a random initial state (triangles); the Gibbs Sampler initialized with an approximate solution (black circles); and the "oscillating" adaptive automaton described in the text, initialized with an approximate solution (Xs) and the same adaptive design, replaced by the Gibbs Sampler after iteration 30 (white circles).

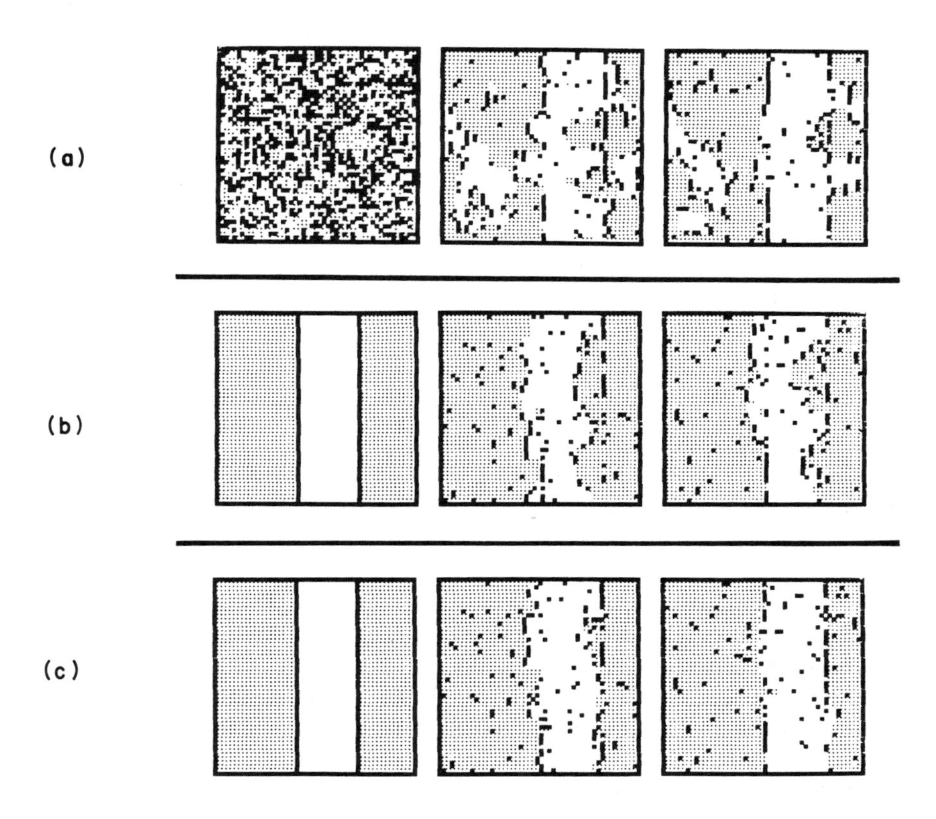

Figure 8: Instantaneous state (taken every 15 iterations) of a typical run of the posterior SCA for the estimation of a vertical bar for (a) the Gibbs Sampler, initialized with a random configuration; (b) the Gibbs Sampler, initialized with an approximate solution; and (c) the "oscillating" automaton described in the text (the direction of motion is reversed every 25 iterations).

In practical applications, one usually waits for a certain number of iterations before starting to gather statistics to compute the MPM estimator; in other words, one approximates the marginal probability for state q at site i and at time t by

$$P_i^{(t)}(q) = \frac{1}{t - t_0} \sum_{\tau = t_0}^{t} \delta(\xi_i^{(t)} - q)$$

and computes the optimal estimator using

$$f_i^{*(t)} = \arg\max_{q \in Q} P_i^{(t)}(q).$$

Figure 9 shows, for a typical run, the evolution of the optimal estimator, computed using the Gibbs Sampler (1 and 2) and the adaptive scheme 4, for $t_0 = 15$ and $t = 20$, 25, 30 and 115. As one can see, the adaptive algorithm converges very fast to an acceptable solution; in the long run, all estimators converge to the exact solution.

5.4 Discussion

The example that we have presented indicates that there are situations in which one may obtain significant improvements on the convergence rate of Bayesian estimators if one selects adaptive SCA instead of the conventional designs.

The Gibbs Sampler automaton maximizes, over $\mathcal{G}_{i,\alpha}$, the minimum transition probability of the system [7], which means that it "mixes up" the states at a high rate. For this reason, its asymptotic behavior is usually very good. Therefore, the best results are usually obtained by shifting to the Gibbs Sampler once the adaptive algorithms have reached a good instantaneous configuration. It should be possible to detect this optimal shifting time in an automatic way and, thus, to obtain automata that adapt in time as well as with respect to the local configurations. At present, however, we are using a fixed number of iterations.

Vertices of the form (5.18) are appropriate for moving certain configurations in a lattice. Vertical motion may be obtained by rotating the neighborhood and finding the corresponding site-symmetric automaton, provided, of course, that the potentials of the MRF model are modified to make them symmetric with respect to this operation, e.g., provided that preference for vertical boundaries is eliminated. Motion of arbitrary direction may be obtained by combinations of these automata.

It is certainly possible that other nonreversible automata exhibit better motion behavior for these or other configurations; therefore, the adaptive

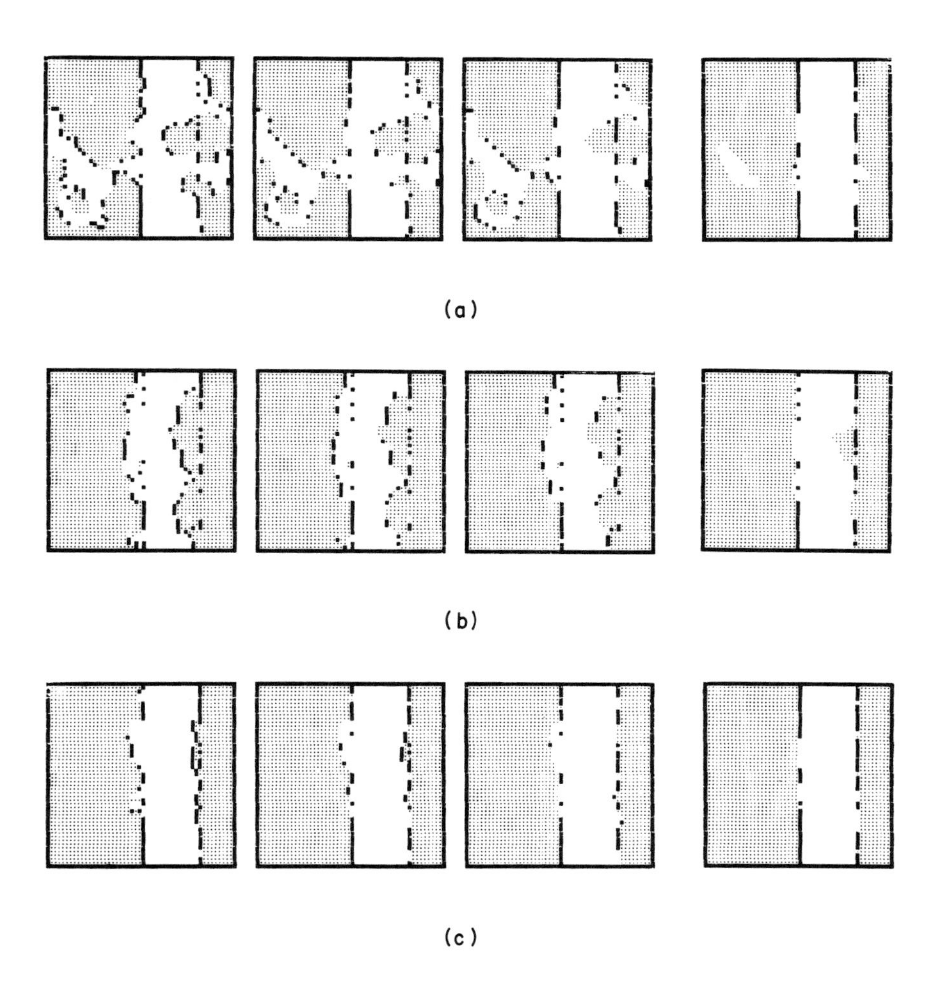

Figure 9: MPM estimator computed using (a) the Gibbs Sampler with random starting point; (b) the Gibbs Sampler initialized with an approximate solution; and (c) the adaptive scheme described in the text replaced with the Gibbs Sampler after 30 iterations. The first three panels of each group correspond to iterations 20, 25 and 30, respectively, and the fourth panel to iteration 115. The gathering of statistics for the computation of the estimator started in all cases at iteration 15.

SCA presented in this section should not be considered "optimal" in any sense but just a simple example of how this kind of design may be used to improve the rate of convergence.

6 Summary and Conclusions

It has been our purpose in this chapter to present, in a systematic way, the structure of a large class of SCA that have a prescribed Gibbsian measure. We think that it is important to understand this structure to be able to use the degrees of freedom available for the design of automata that are optimal in some sense.

We showed how this class, the class of uncoupled Gibbsian automata, is structured as the Cartesian product—taken over all the sites of the lattice and all valid configurations in the neighborhood of each site—of the interior of convex polytopes: $\{\mathcal{G}_{i,\alpha}\}$, of dimension $(M-1)^2$, where M is the size of the state space for each site. A specific Gibbsian SCA may be obtained by selecting a point (that is, a set of transition probabilities) from each $\{\mathcal{G}_{i,\alpha}\}$ in an independent way.

These automata include, as particular cases, the set of all reversible Gibbsian SCA and, thus, all classical designs, such as the Metropolis, Gibbs Sampler and Heat Bath algorithms [8]. They also include an even larger set that has not been explored so far: the nonreversible Gibbsian automata, which may be used, among other things, to move around specific configurations (approximate solutions) in a lattice in such a way that they "lock up" in the correct position.

The uncoupled Gibbsian automata have also the property of being serializable, which means not only that one may use any sweeping strategy for the update (including semi-synchronous modes in parallel engines), but also that one may alternate between different designs (to make, for example, certain configurations to "oscillate" in the lattice) or switch to a design with good asymptotic behavior—such as the Gibbs Sampler—once the error has been reduced by an adaptive strategy.

There are three basic tools for the design of adaptive automata:

1. The general design algorithm presented in [7]: conceptually, one may think of this algorithm in terms of a user interface (which in fact is not difficult to implement) that presents, for a given site i and local configuration α, the valid range for each one of the parameters $P_{i,\alpha,r}(q)$ whose value has not been selected yet; these limits are then updated every time a selection is made.

2. Vertices of $\{\mathcal{G}_{i,\alpha}\}$ and $\{\mathcal{R}_{i,\alpha}\}$: these vertices often exhibit some interesting "extreme" behavior and may be useful for obtaining designs as convex combinations that involve other vertices or interior points. They may be found using the general design procedure, selecting an extreme value (either the upper or lower limit) for each one of the parameters. It is also possible to use one of the existing general methods for finding vertices of polytopes [1]. We presented in Section 4.3 the general form of some vertices that always exist (for $M \geq 3$); they may be used to perform a partial exploration of the design space.

3. Symmetric automata: once a design for a class of local configurations is found, one may exploit the symmetries of the MRF model to find designs for corresponding symmetric classes (see Sections 5.1 and 5.2).

We presented an example of how to use these tools for the design of adaptive automata that take advantage of additional information about the solution to achieve high convergence rates (another example may be found in [7]). It may also be possible to use mathematical programming techniques, such as convex programming, to find designs that are optimal in some other sense; thus, for example, the Gibbs Sampler automaton may be found as the maximizer of the ergodic coefficient [3] of the one-step uncoupled transition matrices: for these matrices, the ergodic coefficient $\alpha(P_{i,\beta})$ is

$$\alpha(P_{i,\beta}) = 1 - \sup_{q,s} \sum_{r \in Q} \max\left(0, P_{i,\beta,q}(r) - P_{i,\beta,s}(r)\right),$$

which is obviously maximized when $P_{i,\beta,q}(r) = P_{i,\beta,s}(r)$, for all $q, s, r \in Q$. Similar properties may be useful for theoretical purposes; however, we have not explored these possibilities yet.

The main question that remains open at this point is, in our opinion, to clarify the relation between specific designs and the corresponding qualitative behavior: in other words, the construction of an appropriate "user interface" for the design of adaptive automata that implement high performance estimators for specific reconstruction tasks.

Acknowledgments

We wish to thank Diego Bricio Hernandez for his helpful comments.

Bibliography

[1] M.E. Dyer and L.G. Proll, *An Algorithm for Finding all Extreme*

Points of a Convex Polytope, Mathematical Programming, 12 (1977), p. 81–96.

[2] S. Geman and D. Geman, *Stochastic Relaxation, Gibbs Distributions and the Bayesian Restoration of Images*, IEEE Transactions on Pattern Analysis and Machine Intelligence, 6 (1984), p. 721–741.

[3] D.L. Isaacson and R.W. Madsen, *Markov Chains Theory and Applications*, John Wiley, New York (1976).

[4] F.P. Kelly, *Reversibility and Stochastic Networks*, John Wiley, New York (1979).

[5] J.G. Kemeny and J.L. Snell, *Finite Markov Chains and their Applications*, Van Nostrand, New York (1960).

[6] A. Kolmogorov, *Zur Theorie der Markoffschen Ketten*, Math. Ann., 12 (1936), p. 155–160.

[7] J.L. Marroquin and A. Ramirez, *Stochastic Cellular Automata with Invariant Gibbsian Measures*, IEEE Transactions in Information Theory 37, 3 (1991), p. 541–551.

[8] J.L. Marroquin, S. Mitter and T. Poggio, *Probabilistic Solution of Ill–Posed Problems in Computational Vision*, Journal of the American Statistical Association, 82, 397 (1987), p. 76–89.

[9] N. Metropolis, A. Rosenbluth, M. Rosebluth, A. Teller and E. Teller, *Equation of State Calculations by Fast Computing Machines,* Journal of Physical Chemistry, 21 (1953), p. 1087–1092.

[10] D. Mitra, F. Romeo and A.L. Sangiovanni-Vincentelli, *Convergence and Finite Time Behavior of Simulated Annealing*, Advances in Applied Probability, 18 (1986) p. 747–771.

[11] J. Von Neumann, *Theory of Self–Reproducing Automata*, (edited by A. Burks) University of Illinois Press, Urbana, Ill. (1966).

Range Image Segmentation Using MRF Models

Anil K. Jain and Sateesha G. Nadabar
Department of Computer Science
Michigan State University
East Lansing, MI

1 Introduction

Considerable research in computer vision has been devoted to deriving three-dimensional shape or depth information from intensity images using various cues such as shading, texture, motion, and stereo. Recently, another approach which is motivated by engineering considerations is gaining interest within the computer vision research community. In this approach, three-dimensional geometric shape is represented directly in terms of depth measurements. Typically, these measurements are represented in a two-dimensional array, where the value at each pixel represents the distance (range) of the corresponding surface point from the sensor. Such images are known as range images or depth maps [15]. Since range images represent geometric measurements of the surface shape directly and accurately, several vision applications have benefited by extracting more reliable features from the range data.

Many high level computer vision tasks such as object recognition and image interpretation, require a reliable and accurate description of image segments. Image segmentation techniques partition an image into regions which are homogeneous with respect to some image property such as gray level, color, texture or surface type. Segmentation is necessary primarily for extracting 'invariant' features useful for the recognition and interpretation task. Because of our interest in 3-D object recognition, we are interested here in segmenting range images into surface patches [3, 12, 13] which match physical object surfaces. Figure 1(a) shows a range image of a scene containing two wooden blocks, and Figure 1(c) shows its ideal or desired segmentation. An unregistered intensity image of the same scene is shown in Figure 1(b) which highlights the crease and curvature edges in the scene.

Research supported by NSF grant, IRI 89-01513

ISBN 0-12-170608-7

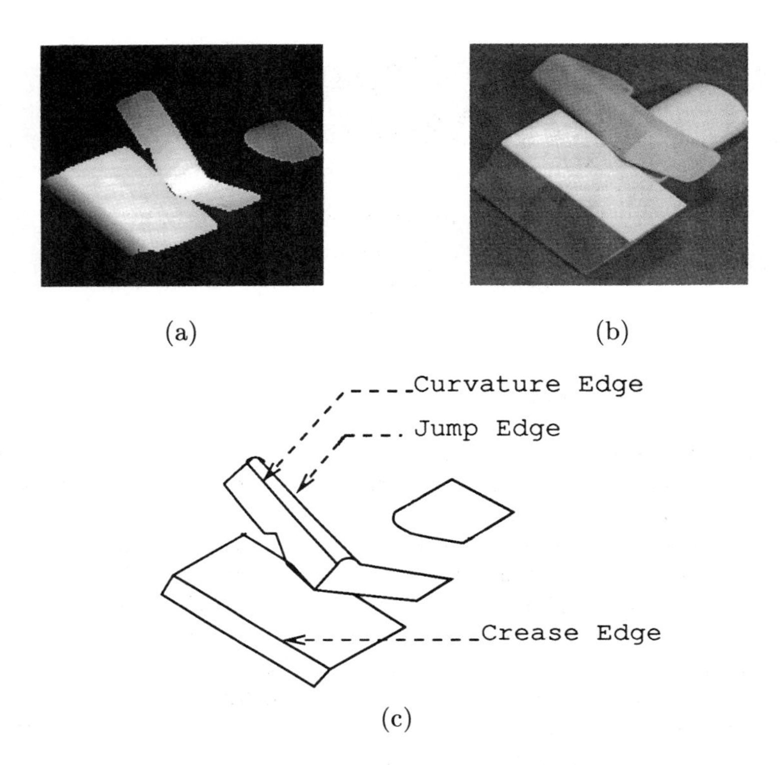

Figure 1: A typical range image (a) and its ideal segmentation (c). An intensity image of the scene (not registered) is shown in (b) to highlight the crease and curvature edges.

There are two common approaches to the problem of segmenting range images into surface patches: region-based [3, 13] and edge-based [12, 17]. Region-based segmentation schemes group 'similar' pixels to form surface patches. Rimey and Cohen [19] propose a hierarchical, maximum likelihood approach to segment range data. Their method tests the hypothesis that the data within a window is homogeneous against the hypothesis that it comes from a mixture of distributions. Hoffman and Jain [13] have used a clustering algorithm for surface segmentation. The (x, y, z) coordinates of pixels and the three components of surface normals at each of these pixels together form the six features used for clustering. The typically oversegmented solution produced during clustering is refined by merging adjacent patches based on their surface types and the strength of edges between them. Besl and Jain [3] also report good surface segmentation results on noisy images using variable-order surface fitting.

Edge-based segmentation schemes utilize the dissimilarity in properties of adjacent surfaces to compute the surface boundary map of the image. Closed boundaries define a surface patch. Edge detection in range images is quite different from edge detection in intensity images. While it is, in general, not possible to associate edges in an intensity image to physical boundaries in the scene because of the effect of shadows, direction of light source, ambient light, etc., edges in range images almost always correspond to the boundaries between surfaces. Therefore, an edge in a range image is an intrinsic property that is very useful in 3-D object recognition [15]. In intensity images, an edge is detected due to discontinuity in intensity. The equivalent type of edge in range images is called a jump or depth edge. In addition to this, two other types of edges are commonly identified in range images [13]: crease edges and curvature edges. A crease edge is formed by the discontinuity of surface normals, whereas curvature edges are formed by discontinuities in curvature [13]. Figure 1(c) shows examples of jump, crease and curvature edges. Curvature edges are the most difficult to detect. As a result, only jump and crease edges have been widely used in the vision literature [13, 17]. Inokuchi *et al.* [14], Mitiche and Aggarwal [17], and Fan *et al.* [12] have proposed different methods for detecting edges in range images.

Both region-based and edge-based segmentation methods have their own advantages and drawbacks. For example, region-based schemes usually assume that regions or surfaces in range images can be globally approximated by an analytic function [3], whereas edge-based schemes need only approximate the surfaces near the discontinuities. Region-based methods guarantee closed boundaries for regions, a feature that is missing in edge-based methods. Recently, Yokoya and Levine [20] proposed a hybrid method for range image segmentation which uses both region-based and edge-based segmentations. However, several thresholds need to be specified a priori in their approach.

This chapter deals with the problem of identifying jump and crease edges in range images. We also propose a procedure for merging edge-based and region-based segmentations. The main contribution of this chapter, is to model the a priori knowledge of edge labels in range images by a Markov random field (MRF) and to derive the a posteriori probability of edge labels in a Bayesian framework. The success of the method is demonstrated on several real range images. A simple hybrid segmentation method which combines the initial region-based segmentation of Hoffman and Jain [13] and our MRF model-based boundary detection method has been developed. The edge map obtained from the edge detection method is used in determining whether two adjacent surface patches can be merged. The hybrid method is shown to result in better segmentations than the individual approaches.

The chapter is organized as follows. In Section 2 we give a brief description of Markov random field models. Section 3 develops the methodology for edge detection in range images in the Bayesian framework using Markov random field models. Experimental results are presented in Section 4. The hybrid method for segmentation is presented in Section 5. Section 6 gives conclusions based on our experimental results.

2 Markov Random Field Models

Markov random field models have been successfully used to represent contextual information in many 'site' labeling problems [2, 5, 7, 8, 9, 11, 16]. A site labeling problem involves classification of each site (pixel, edge element, region) into a small number of classes based on an observed value (or vector) at each site. For example, in image segmentation problems, pixels are labeled based on observed value(s) at each pixel, where the number of labels correspond to the number of true segments or regions. We assume that the desired labeling is the true image. The goal of a site labeling problem, then, is to recover the true image from the noisy observed image. The site labeling problem, therefore, is an inverse problem with many possible solutions and is ill-posed [16]. Contextual information plays an important role here because in many site labeling problems, the true label of a site is compatible in some sense with the labels of the neighboring sites. Context represents our a priori assumptions about the physical world such as continuity and smoothness. More specifically, vision researchers have tried to incorporate the following constraints in segmentation and grouping modules: (i) neighboring pixels are likely to belong to the same region; (ii) close parallel edges are unlikely; (iii) isolated edge elements are improbable; (iv) surfaces are smooth. In computer vision, constraints are used to convert an ill-posed problem into a well posed one. Such methods have been called regularization methods [18].

There are two main approaches to regularization proposed in the literature: (i) mechanical models [4] and (ii) probabilistic models [2, 5, 11, 16]. In the mechanical model approach, variational principles are used to reduce the class of solutions that are acceptable. This transforms the original problem into an optimization problem, where the cost function consists of two terms, one of which regularizes the solution (a priori knowledge) and the other which binds the solution to the data. In standard regularization methods, discontinuities are not taken into account [18]. Blake and Zisserman [4] use weak constraints to incorporate discontinuities into the cost function.

In the probabilistic approach, a priori knowledge is modeled by a Markov random field instead of restricting the class of admissible solutions, as in the mechanical model approach [16], and Bayesian estimation is used to obtain the solution. The probabilistic approach is flexible and allows fusion of data from multiple sources [6, 16]. In this chapter, we are interested in the probabilistic approach. The following is the procedure for deriving an optimal labeling in the Bayesian framework by modeling the a priori knowledge about site labels by a Markov random field.

Let $\mathbf{X} = \{X_1, X_2, ..., X_M\}$ be the M-tuple random vector representing the 'true' labels. Each X_t takes a value from the set of labels, $A = \{1,...,G\}$. For notational convenience, a linear ordering of the M sites is assumed resulting in the vector $\mathbf{X}$. Let vector $\mathbf{Y} = \{Y_1, Y_2, ..., Y_M\}$ represent the observations at the M sites. For computational reasons, it is assumed that the observations are conditionally independent [2]. Therefore, the density for $\mathbf{Y}$, given the true labeling, is

$$f(\mathbf{y}|\mathbf{X} = \mathbf{x}) = \Pi_{t=1}^{M} f(y_t|x_t). \tag{1}$$

Note that $f(y_t|x_t)$ is the conditional density function for the observation Y_t, given the true label x_t at site t. Contextual information is incorporated through a Markov random field (MRF) model of the statistical dependence among the labels on neighboring sites in $\mathbf{X}$ which is equivalent to the Gibbs process [1]. The (a priori) probability density function for $\mathbf{X}$ is a Gibbs random field [1, 11]

$$P(\mathbf{X} = \mathbf{x}) = e^{-U(\mathbf{x})}/Z, \tag{2}$$

where Z is the partition function or the sum of the numerator over all possible labelings and $U(.)$ is the energy function

$$U(\mathbf{x}) = \sum_{c \in C} V_c(\mathbf{x}), \tag{3}$$

where C is the set of all cliques with respect to the neighborhood system and $V_c(.)$ are the potential functions which map the local interactions of the elements of clique c to energy contributed by the clique towards the total energy. The functions $V_c(.)$ encode our a priori knowledge about the spatial dependence of labels at neighboring sites. For site labeling problems, natural constraints such as smoothness and continuity are enforced using $V_c(.)$. We emphasize that the random field model is used only to incorporate context into the site labeling problem and is not expected to be an accurate model of the true image.

Under assumptions (1) and (2), the a posteriori probability mass function for the site labels $\mathbf{X}$, given the observations $\mathbf{Y} = \mathbf{y}$, also has the form of a Gibbs random field

$$P(\mathbf{X} = \mathbf{x}|\mathbf{Y} = \mathbf{y}) = e^{-U(\mathbf{x}|\mathbf{y})}/Z_{\mathbf{y}}, \quad (4)$$

where $Z_{\mathbf{y}}$ is a normalizing constant, and the corresponding energy function is

$$U(\mathbf{x}|\mathbf{y}) = \sum_{t=1}^{M}[-ln(f(y_t|x_t))] + \sum_{c \in C} V_c(\mathbf{x}). \quad (5)$$

The local properties of a Markov random field (MRF) can be derived from the Gibbs random field. Let $\mathbf{X}_{\partial \mathbf{t}}$ be a (vector) random variable representing the labels of neighbors of pixel t. The conditional probability of X_t can be written as follows [1]:

$$P(X_t = x_t|\mathbf{X}_{\partial \mathbf{t}} = \mathbf{x}_{\partial \mathbf{t}}, \mathbf{Y} = \mathbf{y}) = e^{-U_t(x_t, \mathbf{x}_{\partial \mathbf{t}}, \mathbf{y})}/Z_t, \quad (6)$$

where Z_t is a normalizing constant, and

$$U_t(x_t, \mathbf{x}_{\partial \mathbf{t}}, \mathbf{y}) = -ln(f(y_t|x_t)) + \sum_{c:t \in c} V_c(\mathbf{x}_{\partial \mathbf{t}}), \quad (7)$$

where the summation is over all the cliques to which the site t belongs.

The site labeling problem can now be stated as follows: given the observation vector $\mathbf{y}$, estimate the labels in the true image. Contextual information is represented by the rightmost terms ($\sum_{c \in C} V_c(\mathbf{x})$ and $\sum_{c:t \in c} V_c(\mathbf{x}_{\partial \mathbf{t}})$) in Eqs. (5) and (7). If these terms were removed, energy minimization algorithms would assign labels independently to each site. The MAP (maximum a posteriori) estimate is the vector $\hat{\mathbf{x}}$ which maximizes $P(\mathbf{X} = \mathbf{x}|\mathbf{Y} = \mathbf{y})$ with respect to $\mathbf{x}$. Maximizing this a posteriori probability mass function, which is a function of M variables, is a formidable task since for a 256×256 image $M = 65,536$. Geman and Geman [11] have used simulated annealing to obtain an approximation to the MAP estimate in the context of image restoration, but their procedure is also computationally expensive.

Several alternative estimates have been suggested for obtaining the a posteriori labeling. Marroquin *et al.* [16] have suggested that minimizing the expected value of an error functional is a better criterion for some problems. For example, for the segmentation problem an appropriate criterion is to minimize the expected value of the percentage of pixels misclassified. For this criterion, the optimal labeling is also the labeling that maximizes the a posteriori marginal probability (MPM). They propose a method of computing the MPM estimate using a Monte Carlo method. Besag [2] proposed a deterministic iterative scheme called iterated conditional modes (ICM), in which the labeling of a site is iteratively updated so as to maximize the local conditional probabilities $P(X_t = x_t | \mathbf{X}_{\partial \mathbf{t}} = \mathbf{x}_{\partial \mathbf{t}}, \mathbf{Y} = \mathbf{y})$. This algorithm finds a local minimum of the energy function given in Eq. (5). Two important advantages of the ICM algorithm are (i) it is a computationally attractive scheme, and (ii) it avoids the undesirable large scale properties of the MRF prior (see [2] for details). For an experimental comparison of the above three methods, in the context of intensity image segmentation, see Dubes *et al.* [9].

In this chapter, we have used a method of optimization proposed by Chou *et al.* [5], known as the highest confidence first (HCF) algorithm, because, it is deterministic, computationally attractive and its performance in this application is better than the ICM algorithm. The HCF algorithm is similar to the ICM algorithm except for the order in which sites are visited. While the ICM algorithm does not impose any single order for visiting the sites, the HCF algorithm requires that the site that is visited next be the one which causes the largest energy reduction. Additionally, HCF starts by assigning a NULL (non-committal) label to each site initially. The site commits to a label only if it gains enough confidence in its decision. Cliques with NULL labels contribute zero potential. The HCF algorithm can be summarized as follows:

1. Initialize all sites to NULL label.
2. Compute a stability measure for each site t as follows:

   - If the site t has a non-NULL label x_t, then

   $$\text{stability} = U_t(h, \mathbf{x}_{\partial \mathbf{t}}, \mathbf{y}) - U_t(x_t, \mathbf{x}_{\partial \mathbf{t}}, \mathbf{y})$$

   - If the site has a NULL label, then

   $$\text{stability} = U_t(g, \mathbf{x}_{\partial \mathbf{t}}, \mathbf{y}) - U_t(i, \mathbf{x}_{\partial \mathbf{t}}, \mathbf{y})$$

   where

   $$g = \underset{l \in A}{argmin} \; U_t(l, \mathbf{x}_{\partial \mathbf{t}}, \mathbf{y})$$

   $$h = \underset{l \in A, l \neq x_t}{argmin} \; U_t(l, \mathbf{x}_{\partial \mathbf{t}}, \mathbf{y})$$

   $$i = \underset{l \in A, l \neq g}{argmin} \; U_t(l, \mathbf{x}_{\partial \mathbf{t}}, \mathbf{y})$$

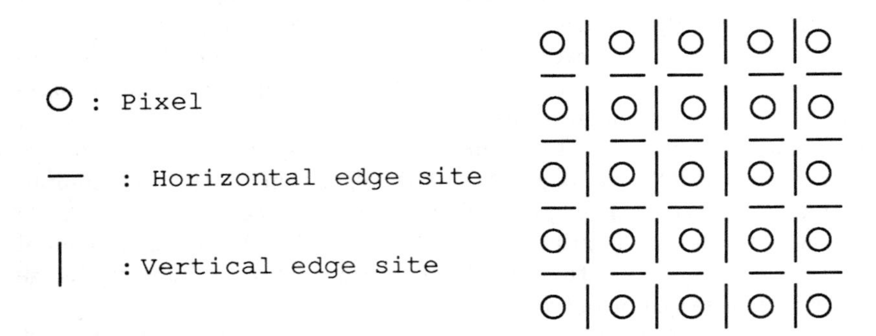

Figure 2: Pixel and Edge sites

3. Repeat
   (a) Select the site with minimum stability
   (b) Change its label to the one which corresponds to the lowest energy.
   (c) Recompute stability for the site and its neighbors.

   until (stability $\geq$ 0) for all sites.

3 Edge Detection as Bayesian Estimation

We view the problem of edge detection in range images as a site labeling problem. Our approach is similar to the approach used by Chou *et al.* [5] for edge detection in intensity images. The set of site labels are {EDGE, NON-EDGE}. The edge or line sites are the imaginary lines between adjacent pixels as shown in Figure 2. Figure 3 shows the 8-neighborhoods (second-order neighborhood) for horizontal and vertical edge sites. We will use Markov random fields to represent the a priori knowledge about the formation of edges in range images—such as continuity and smoothness of edges—using appropriate clique potentials. The HCF algorithm will be used to obtain the final edge labeling.

A complete specification of the MRF model requires that the observation vector $\mathbf{Y}$, the conditional distribution $f(y_t|x_t)$, and clique potential functions $V_c(.)$ be specified. For the case of binary edge labels {EDGE, NON-EDGE}, it is easy to see that we only need the likelihood ratio $\frac{f(y_t|\text{EDGE})}{f(y_t|\text{NON-EDGE})}$ at each line site and the clique potential functions $V_c(.)$ [5].

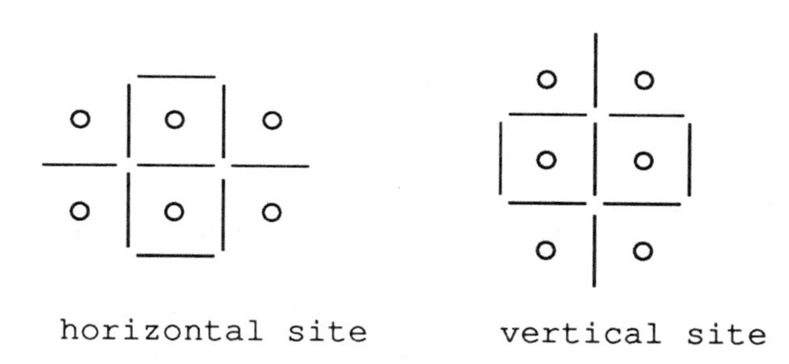

Figure 3: Edge site (second-order) neighborhoods

3.1 Computation of Edge Likelihoods

In the case of edge labeling, the edge sites are defined to lie between two adjacent pixels and there is no *direct* observation representing the edge strength at an edge site. Therefore, the edge strengths have to be derived from the range values observed at pixels in the neighborhood of the edge site. The size of the neighborhood must be large enough to differentiate true edges from noisy spikes, and it should be small enough not to involve multiple edges. It is imperative that different neighborhoods be used for vertical and horizontal edge sites.

We define a statistic which indicates the edge strength at an edge site, and we derive its distribution. Obviously, two such statistics will be required, one each for jump edges and crease edges.

3.1.1 Jump Edge Statistic

We want to define a statistic for testing the hypothesis that there is a jump edge at an edge site. This statistic should take high values when there is a large depth difference at an edge site and low values otherwise. Other important considerations in selecting a jump edge statistic are: (i) its distribution should be known under both the null and the alternative hypothesis, and (ii) it should be simple to compute. These considerations lead us to use the absolute value of the difference in the average depth values on the two sides of the edge site as the test statistic. If the N depth values in the neighborhood of an edge site are numbered as in Figure 4, then the jump edge statistic t_d is given by

$$t_d = |\frac{2}{N}(\sum_{i=1}^{N/2} d_i - \sum_{i=\frac{N}{2}+1}^{N} d_i)|, \tag{8}$$

where d_i is the range or depth value at the ith pixel.

The null hypothesis is that there exists no jump edge at a site, and the alternative hypothesis states that there exists a jump edge at that site. Note that the alternative hypothesis here is a composite hypothesis because there are several ways in which a jump edge can be formed:

1. *Distribution under the null hypothesis (f_{JO}):* It is easy to determine that the distribution of t_d under the null hypothesis is a t-distribution with $\frac{N}{2}$ degrees of freedom, modified such that negative values are not allowed. For computational reasons, we have approximated the t-distribution by a standard normal distribution without any significant loss in accuracy. The observed z statistic is computed from the observed t_d statistic, $t_{d_{obs}}$ as

$$z_{obs} = \frac{t_{d_{obs}}}{\sqrt{(4/N)variance(t_d)}}. \tag{9}$$

   The variance of t_d can be estimated either from a sample range map from the sensor, or from the given range image itself.

2. *Distribution under the alternative hypothesis (f_{JA}):* The computation of distribution of the statistic under the alternative hypothesis is complicated by the composite nature of the alternative hypothesis. The depth differences at true jump edges are not identical; depth differences for true jump edges vary from one jump edge to another and from one image to another. This prevents us from using the popular normal shift alternative hypothesis. We assume that any depth difference greater than a minimum depth difference (taken here to be the standard deviation of t_d, sd(t_d)) is equally likely. Figure 5 shows the assumed distribution of the statistic given that a jump edge exists at the edge site. For reasons described in section 4, we have scaled the depth values to lie between 0 and 255 so the value of max t_d is 255 for the jump edge statistic.

The jump edge likelihood ratio $L_J(t)$, at edge site t, can then be computed by using the distribution of the jump edge statistic under the null

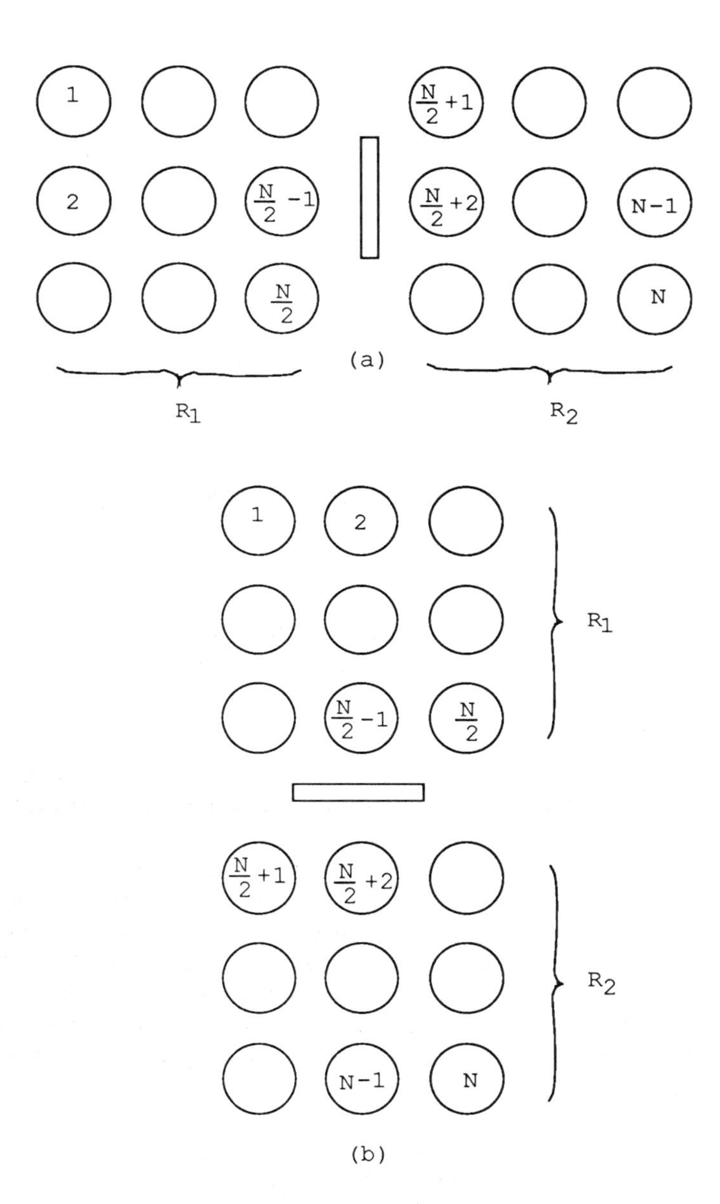

Figure 4: Pixel numbering in the neighborhood of (a) a vertical edge site and (b) a horizontal edge site.

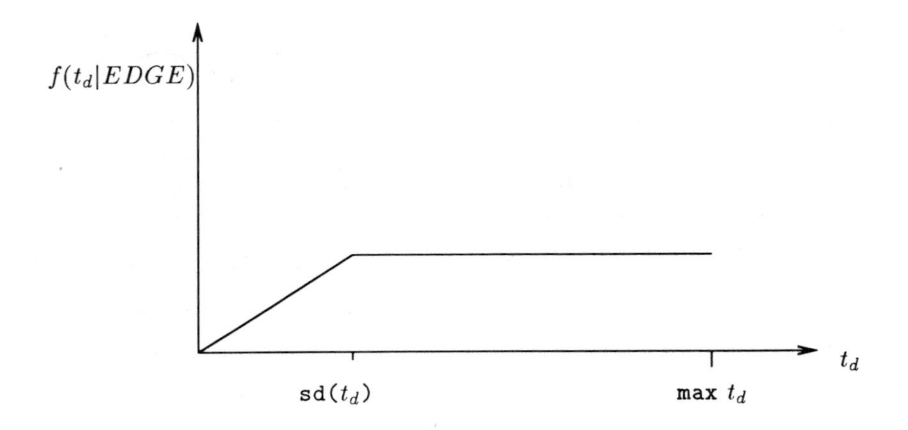

Figure 5: Distribution of t_d under the alternative hypothesis

and the alternative hypotheses as

$$L_J(t) = \frac{f_{JA}(t_{d_{obs}}|\text{jump edge})}{f_{J0}(t_{d_{obs}}|\text{non} - \text{jump edge})}.$$

3.1.2 Crease Edge Statistic

A statistic for crease edges can not be computed directly from the depth values in the neighborhood of an edge site, since average depth difference could be zero at crease edges. One can use the change in direction of the surface normal or curvature values to determine the likelihood of a crease edge at an edge site. Since surface normal estimation is more accurate than curvature estimation in a small neighborhood [10], we have used surface normals. Surface normals were computed by using only four of the 8 neighbors closest in depth value to the center pixel along with the center pixel. This reduces the undesired effect of using pixels from across edges, which introduces errors in the computed surface normal. We define the crease edge statistic t_n to be

$$t_n = ||\frac{2}{N}(\sum_{i=1}^{N/2} \mathbf{n}_i - \sum_{i=\frac{N}{2}+1}^{N} \mathbf{n}_i)||, \tag{10}$$

where $\mathbf{n}_i$ is the unit vector in the direction of the surface normal at pixel i.

As in the case of the jump edge hypotheses, the null hypothesis for the detection of crease edges is that there exists no crease edge at a site, and the alternative hypothesis states that there exists a crease edge at that site. To compute the crease edge likelihood ratio, we need the following distributions of t_n under the two hypotheses:

1. *Distribution under the null hypothesis (f_{CO}):* We again use the t-distribution as an approximation which is justified since t_n is similar to the classical 't-test' except that the observations here are vectors. As in the case of jump edge statistic, we use standard normal approximation to the t-distribution for computational reasons. The observed z statistic for crease edges is computed from the observed t_n statistic, $t_{n_{obs}}$ as
$$z_{obs} = \frac{t_{n_{obs}}}{\sqrt{(4/N)variance(t_n)}}. \tag{11}$$
The variance of t_n can be estimated similar to that of t_d.

2. *Distribution under the alternative hypothesis (f_{CA}):* The distribution of t_n under the alternative hypothesis is taken to be identical in form to that of t_d for the equivalent case; only the maximum and knee values in Figure 5 are different and are `max` t_n and `sd(`t_n`)`, respectively. The value of `max` t_n is 2.0 for the crease edge statistic.

The crease edge likelihood ratio $L_C(t)$, at edge site t, can then be computed as
$$L_C(t) = \frac{f_{CA}(t_{n_{obs}}|\text{crease edge})}{f_{CO}(t_{n_{obs}}|\text{non} - \text{crease edge})}.$$

3.1.3 Difficulties with Gradients and Discontinuities

The jump and crease edge statistics described above have been designed under the assumption that, in the neighborhoods of an edge site (R_1 and R_2 in Figure 4), the observations are i.i.d. Gaussian. In practice this assumption is violated in two ways: (i) there could be discontinuities within R_1 and R_2 as in the case of an edge site close to a boundary between two surfaces; (ii) the true depth values (or surface normal directions) may not be constant in R_1 and R_2 (e.g., slanted surfaces for jump edges and curved surfaces for crease edges). These violations lead to undesirable effects such as classifying a thick strip of edge sites near a boundary as edges, and edge sites within regions with high depth or surface normal gradients being identified as edges. Similar problems are encountered in the method of mechanical regularization using weak constraints [4].

We incorporate two modifications to each of the two statistics so that the modified statistics can be used in a hypothesis test derived under the i.i.d. Gaussian assumption. We first replace the statistics t_d and t_n by deviates from local averages computed in a 1×5 (or 5×1 for vertical edge sites) window of edge sites parallel to the center site. For edge sites away from the boundary, the local average and the individual values are close so that the modified statistic will take small values for these sites. For edge sites on a boundary, individual values are higher than the local averages and the modified statistics take larger values. The second modification performs a nonmaxima suppression in the same window by suppressing the statistic value at the center site if it is not either the largest or the next largest value in the window. The second largest value in the window is not suppressed to account for parallel edges that are very often formed near corners.

3.1.4 Combining Jump and Crease Edge Likelihoods

We have so far described the computation of jump and crease edge likelihood ratios. As a result, each edge site will have two likelihood ratios. How should we combine this information to obtain the edge labeling? At one extreme, we could have two separate labeling processes, one each for jump and crease edges, that make use of the corresponding likelihood ratios only. But this approach does not account for the interaction between the two processes, which makes it difficult to enforce continuity constraints. At the other extreme, one could use the two likelihood ratios as components of an observation vector, specify appropriate cliques and potential functions, and use a Markov random field-based labeling framework. However, the specification of cliques and potential functions is a difficult problem which is further compounded when we have a vector of observed values. As a compromise, we have allowed limited interaction between the jump and crease edge observations. At each edge site, we replace the two likelihood ratios by the maximum value, allowing us to work with Markov random fields with scalar observations.

3.2 Clique Potential Functions

The cliques of interest depend on the type of neighborhood used, while the potential function values depend on the labeling problem at hand. We have used the second-order neighborhood here. Figure 6 shows all possible cliques for this neighborhood, assuming rotational invariance since cliques of size greater than four do not occur in a second-order neighborhood.

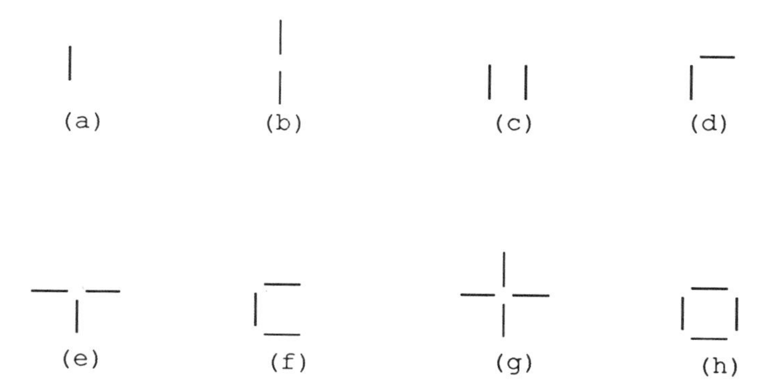

Figure 6: Cliques for the second-order neighborhood

Clique potentials are chosen based on the spatial structure or constraints that a clique symbolizes. For example, for the clique (b) in Figure 6, if both the edge sites are labeled EDGE, the clique potential should be assigned a small value, and if only one site is labeled EDGE, a high potential value should be assigned, since this indicates a break in the boundary. In other words, if the labels in the clique do not agree with a priori constraints of continuity and smoothness, then that labeling is discouraged by assigning a high potential value to the associated clique.

Different clique potential assignments have been suggested for $V_c(.)$ in the literature [5, 11, 16]. The HCF algorithm with clique potential functions given by Chou *et al.* [5] performed poorly in our initial segmentation experiments. A retrospective analysis showed that these parameters discouraged grouping of diagonal edges. Subsequent search for a better set of assignments has resulted in clique potentials that perform reasonably well for most of the test images in our experiments. The cliques used and the corresponding potential assignments are shown in Figure 7. Clique configurations not shown in the figure do not contribute to the total energy value. It should be emphasized here that these potential function values were chosen based on certain heuristics and empirical results.

The ICM algorithm described in Section 2 has been widely used for segmenting intensity images [9]. Therefore, in our initial experiments, we tried the ICM algorithm for the edge labeling problem, but the results were not satisfactory. One possible explanation is that the complex form of the cliques used here (cliques of size 3 and 4) introduce undesirable local minima because of interaction between cliques of different sizes, and ICM gets trapped in them easily. Since the HCF algorithm does not assign a label

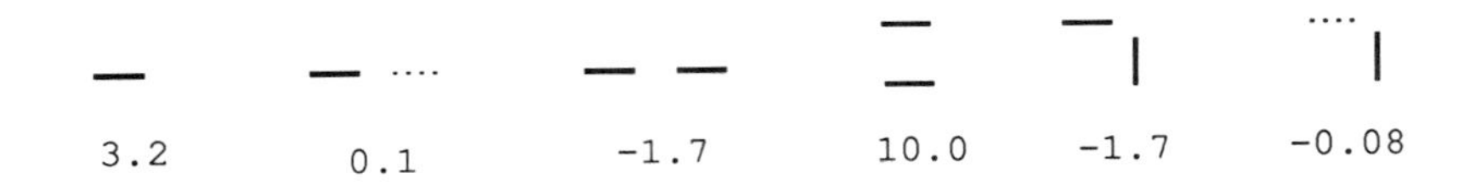

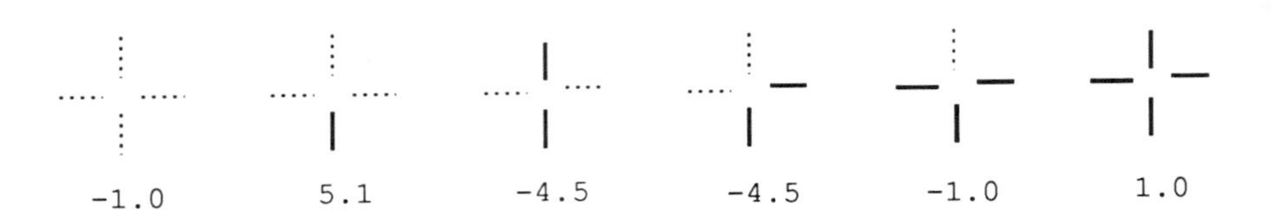

10.0

30.0

— : Edge site labeled EDGE

···· : Edge site labeled NON-EDGE

Figure 7: Cliques and their potential function values, $V_c(.)$

to a site until the confidence value of the site is maximum, it has a better chance of escaping the local minima. One of the limitations of the HCF algorithm is that it is inherently sequential. If a parallel implementation of the edge labeling problem is desired, then the MPM or the MAP algorithms are better candidates.

3.3 Edge Detection Algorithm

Our edge detection algorithm can be summarized as follows:

1. Compute the jump edge likelihood ratio, L_J for each edge site.

2. Compute the crease edge likelihood ratio, L_C for each edge site.
3. Edge Likelihood ratio, L_E for each edge site = max (L_J, L_C).
4. Use HCF algorithm to obtain the desired edge labeling.

The parameters to be specified in this algorithm are: neighborhood size, potential function values, and standard deviations of jump and crease edge statistics.

4 Segmentation Results

This section describes the segmentation experiments performed using range images obtained from the Technical Arts scanner in our laboratory. We report results on five different range images. All range images were scaled to have depth values between 0 and 255. This enables using table look up techniques for computation of likelihoods to reduce computation time. The standard deviations of the jump and crease edge statistics can be computed using an arbitrary test image (here *coffee cup* image was used). The estimated standard deviations were 1.05 for the jump edge statistic and 0.05 for the crease edge statistic. The same cliques and potential function values were used for all the images. We used a 1×4 window ($N = 4$) for computing the jump edge likelihoods and a 1×2 window ($N = 2$) for crease edge likelihoods. The segmentation results for these five images are presented in Figures 8–12. All these images are not of the same size because, in some of the images, background pixels were deleted to reduce unnecessary processing. Each figure shows the following details:

- Original range image is shown in part (a) of the Figure.
- Range image is shown as a shaded image in part (b) of the Figure. The shaded image highlights the edges and surfaces in the original image. In the shaded image the pixel value is proportional to the difference in angle between the viewing direction and the surface normal at that pixel.
- Jump boundaries obtained by thresholding the jump edge likelihood ratios (threshold = 9.0) are shown in part (c). This is a binary image.
- Crease boundaries obtained by thresholding the crease edge likelihood ratios (threshold = 9.0) are shown in part (d). This is a binary image.
- Superposition of images in (c) and (d) is shown in part (e). This is a binary image.

- Output of the HCF algorithm is shown in part (f). This is a binary image.

Figure 8 shows a 122×226 range image of a polyhedron. This is a fairly simple image to segment and as shown in the figure, our algorithm detects all the crease and jump edges present in the image. The boundary which separates the triangle shaped region from the rest of the object in (b) is a crease edge and therefore is not detected in (c) which shows only jump boundaries, but is present in (d). Observe that some jump boundaries are detected in the crease edge map as well. This is because jump boundaries affect surface normal computation to some extent and may result in an artificial discontinuity in surface normal at these edge sites. The crease edge is broken at a couple of places in (e), but these gaps are filled in by our algorithm as shown in (f).

Figure 9 shows the results for a 104×164 image of a coffee cup. The jaggedness of the second edge near the bottom of the cup (top, in the figure) is due to the proximity of the two edges, and does not cause much of a problem when the edge map is used for obtaining a surface segmentation, as shown in Figure 14. Our algorithm fails to connect the broken edges in the bottom left boundary; this is probably due to the large likelihoods for parallel edges resulting from very high depth gradient in this region. Figure 10 shows the results for a 194×218 image of a pair of wooden blocks. Again, all significant edges have been found correctly. Many spurious edges that appear in Figure 10(e) have been removed by the HCF algorithm in Figure 10(f). The second parallel edge near the head of the T-shaped block, although not quite visible in the image, is due to a small change in surface normals. The crease edge near the top corner is broken in the final segmentation indicating the need for postprocessing when segmentation into surface patches is desired. Figure 11 shows the results for an 83×156 image of a rotationally symmetric industrial object. The superior performance of our algorithm can be seen by comparing parts (e) and (f) in Figure 11. Figure 12 shows the results for a 240×240 image of an occluded scene of wooden blocks (also shown in Figure 1(a)). A comparison with the desired segmentation shown in Figure 1(c) reveals that, surprisingly, curvature edges in the image are also detected by our algorithm.

All the edges have been detected and localized quite accurately for all five test images used in our experiments. Table 1 gives the computation times for segmenting these five images on a Sun 4/330 workstation. The HCF labeling step accounts for a major portion (about 85%) of the run time. The time taken by this algorithm is comparable to other edge detection algorithms for range images [12].

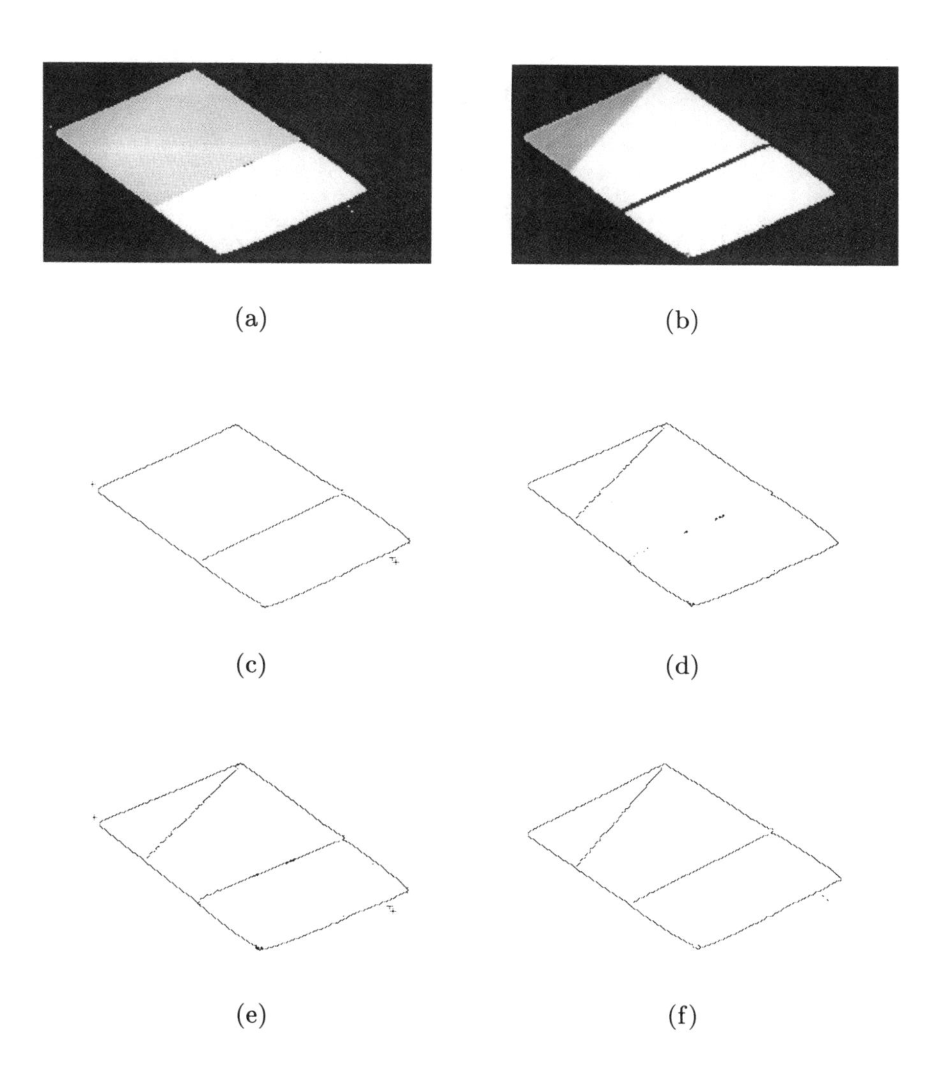

Figure 8: Segmentation results for the *block1* image: (a) range image; (b) shaded image; (c) jump edges; (d) crease edges; (e) combined edge map; (f) final segmentation.

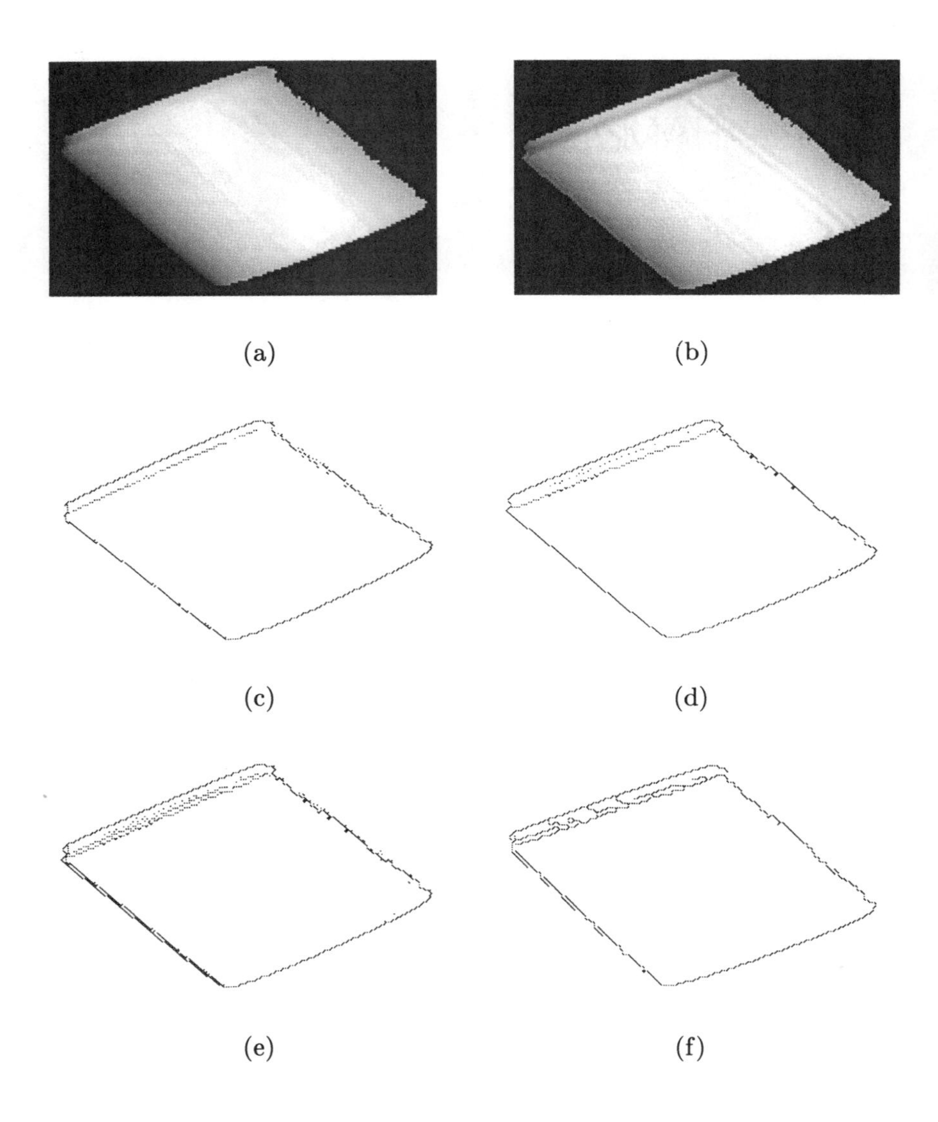

Figure 9: Segmentation results for the *coffee cup* image: (a) range image; (b) shaded image; (c) jump edges; (d) crease edges; (e) combined edge map; (f) final segmentation.

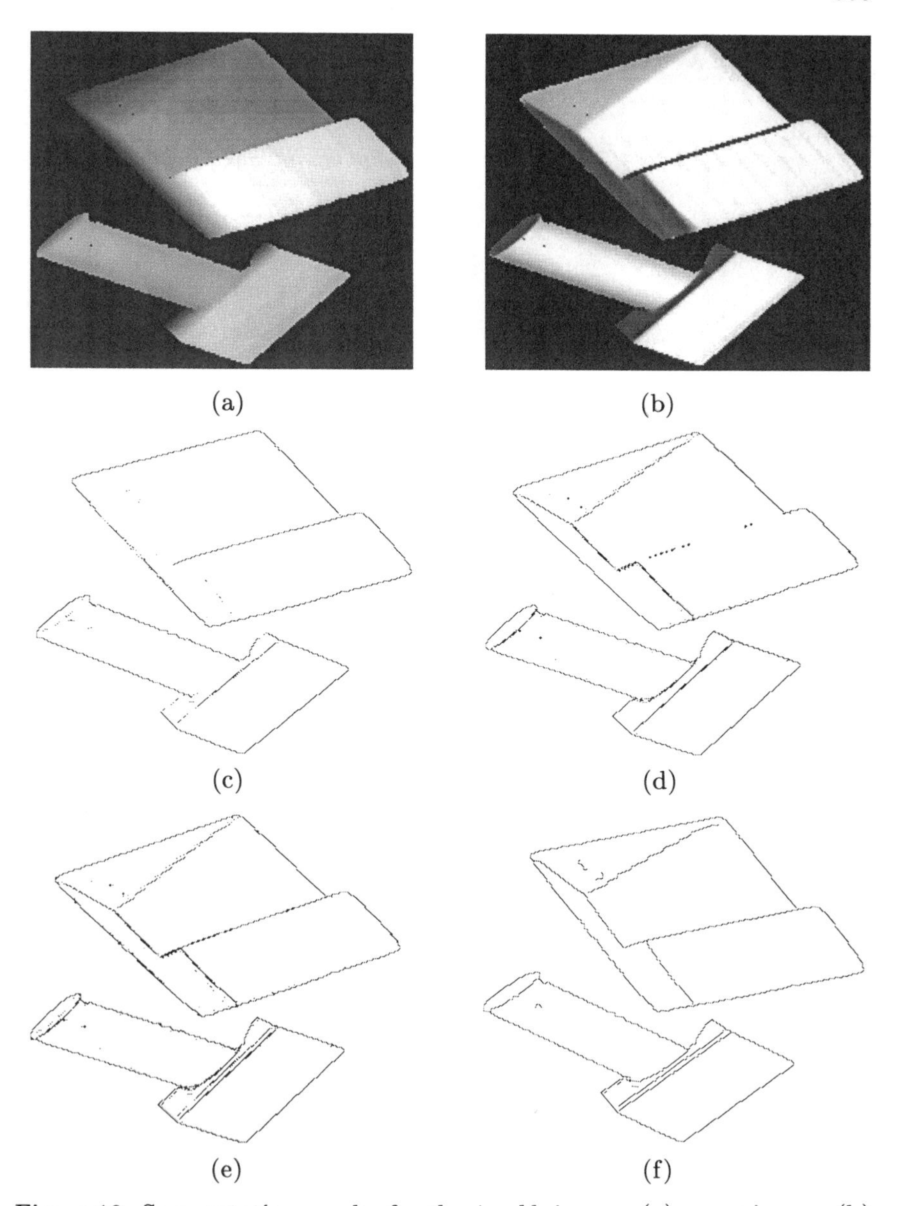

Figure 10: Segmentation results for the *jumble* image: (a) range image; (b) shaded image; (c) jump edges; (d) crease edges; (e) combined edge map; (f) final segmentation.

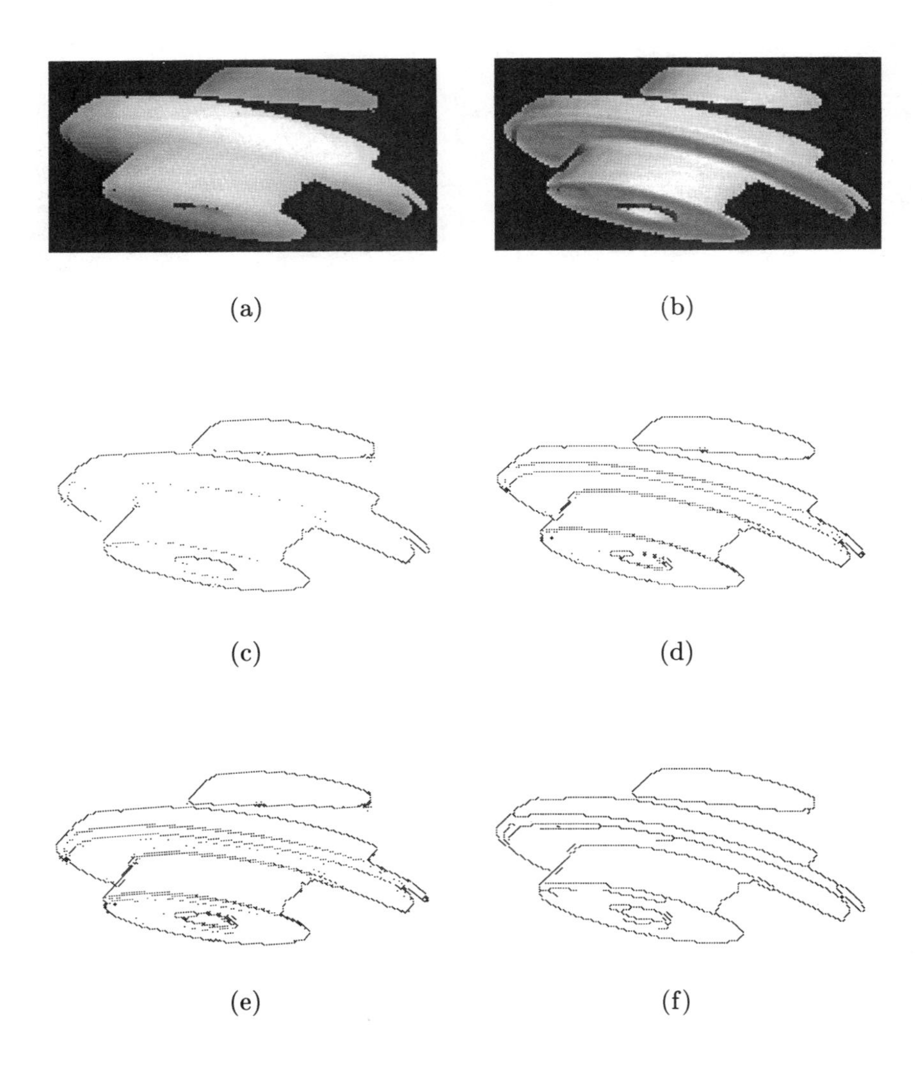

Figure 11: Segmentation results for the *rotsym* image: (a) range image; (b) shaded image; (c) jump edges; (d) crease edges; (e) combined edge map; (f) final segmentation.

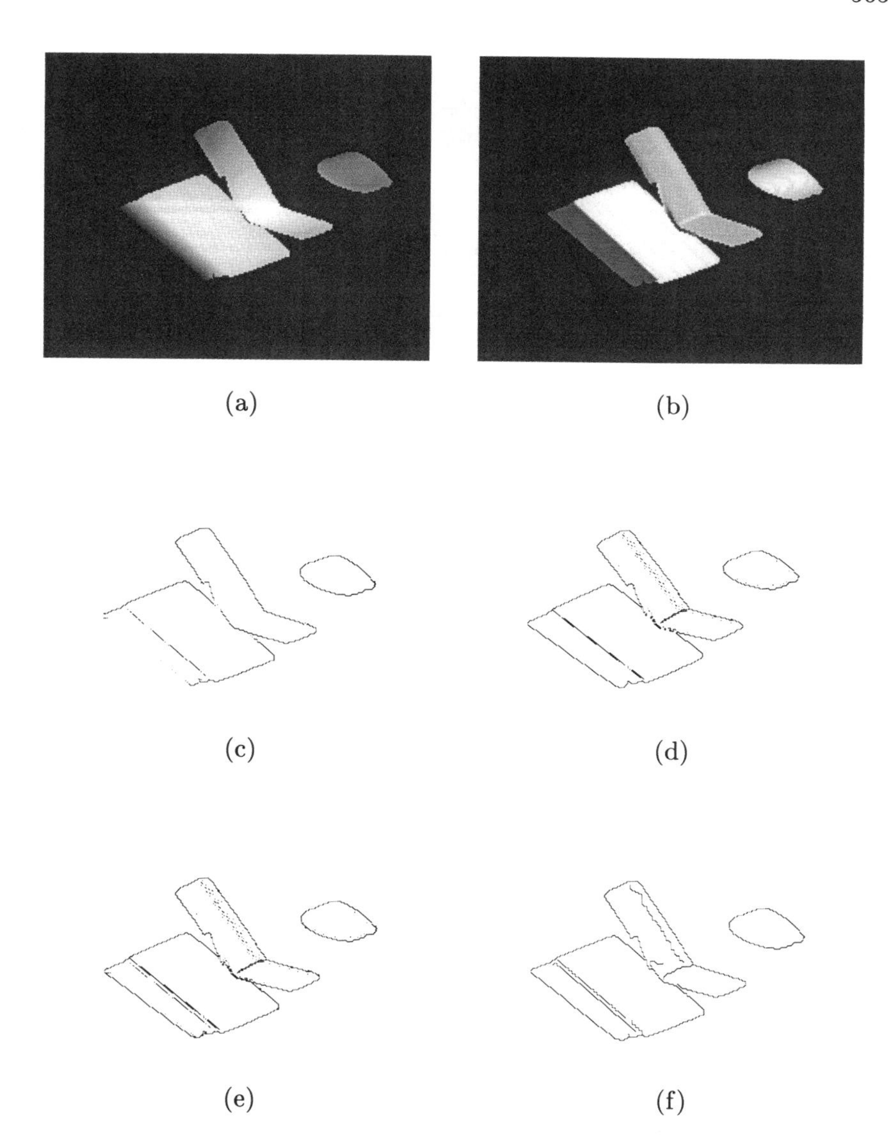

Figure 12: Segmentation results for the *occluded scene* image: (a) range image; (b) shaded image; (c) jump edges; (d) crease edges; (e) combined edge map; (f) final segmentation.

Image	Time
block1	69.7
coffee cup	47.3
jumble	102.5
rotsym	37.1
occluded scene	145.6

Table 1: Execution time in seconds

5 Hybrid Method for Segmentation

Our edge detection method, like all other edge-based segmentation methods, does not always result in closed boundaries. We need a postprocessing step to extract surface patches in the range image. An obvious solution is to apply some heuristics to convert the edge map with broken boundaries into closed contours which represent surface boundaries. This is the approach used by Fan *et al.* [12]. However, this reduces the robustness of the segmentation method significantly, because small spurious edges may get treated as boundaries if excessive relaxation is used to recover broken edges. To overcome this problem, we have developed a hybrid segmentation method which uses our MRF-based edge finding algorithm for discontinuity detection, and a region-based method for continuity detection. The outputs of the two algorithms are combined using a simple approach.

We have used the clustering-based segmentation method of Hoffman and Jain [13]. Their method used six features per pixel in clustering: 3 coordinate values and the 3 components of the surface normal at the pixel. Pixel coordinates incorporate spatial information and surface normals provide continuity and smoothness information. The clustering algorithm is unable to group all the pixels belonging to a single curved surface; therefore, Hoffman and Jain specify that the clustering algorithm generates a solution with a large number of clusters or surface patches. A merging step then merges adjacent patches which are of the same surface type and are not separated by a jump or crease edge.

Our hybrid segmentation algorithm works as follows: the region-based segmentation algorithm provides an initial segmentation. The oversegmented solution produced at this stage has the desirable characteristic that these surface patches are almost always contained in a single surface in the scene. Therefore, all true edges are preserved in the oversegmented solution to within a couple of pixels accuracy. The boundary segments in the oversegmented solution are validated using evidence from the edges found in the MRF-based edge detection algorithm.

Consider the boundary between an arbitrary pair of adjacent surface patches, p_1 and p_2. Let N_T be the total number of edge sites along this boundary. Let us, for the sake of clarity, refer to the edge sites on this boundary as candidate sites. For each candidate site, we verify whether there is a corresponding edge site in the edge-based segmentation which is marked as an EDGE. In order to allow for localization disparities between the two segmentations, matching is relaxed to include the two nearest sites parallel to the current site in the edge-based segmentation. Consider a candidate site, i, on the boundary of interest. We mark candidate site, i, to have a match in the edge-based segmentation if any one of the three corresponding edge sites mentioned above have been labeled as EDGE. Let N_M be the number of candidate sites that have a match in the edge-based segmentation. The surface patches p_1 and p_2 are merged if

$$(\frac{N_M}{N_T}) < t.$$

The segmentation results were the same for a wide range ($t \in (0.40, 0.80)$) of values of the threshold, t. We used $t = 0.60$ for the results reported in this chapter.

The segmentation results of the hybrid algorithm are shown in Figures 13–17. Part (a) of these figures shows the initial segmentation obtained by the region-based method, and part (b) shows the final segmentation. Comparing these figures with Figures 8–12, we see that in the output of the clustering procedure, surface patches do not cross over jump or crease edges. Also, broken edges in the edge-based segmentation are connected in the final segmentations. For example, the broken crease edge in the wedge of the polyhedron in Figure 10(f) is now connected in Figure 15(b). The robustness of the hybrid segmentation method is demonstrated by the result shown in Figure 13(b). In this case, even though an extra segment was detected near the crease edge, this segment is restricted to a very small area, illustrating the much desired *graceful degradation* property. The final segmentations for all the test images show that all but the very small surface patches have been grouped correctly and their shape retained; this is an important trait for a segmentation algorithm meant to be used in an object recognition system.

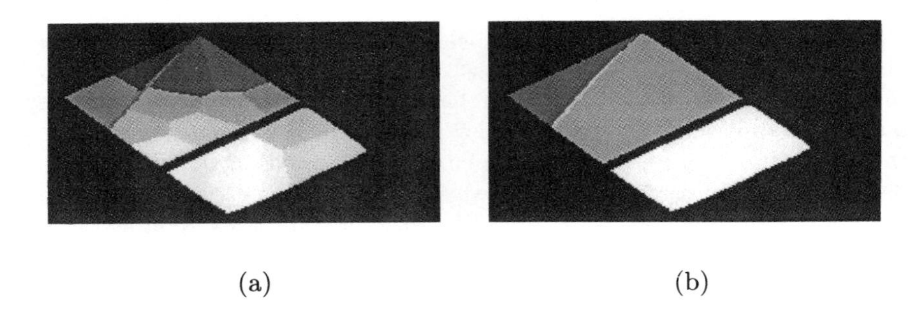

(a) (b)

Figure 13: Hybrid segmentation results for the *block1* image: (a) initial region-based segmentation; (b) result of combining region-based and edge-based segmentations.

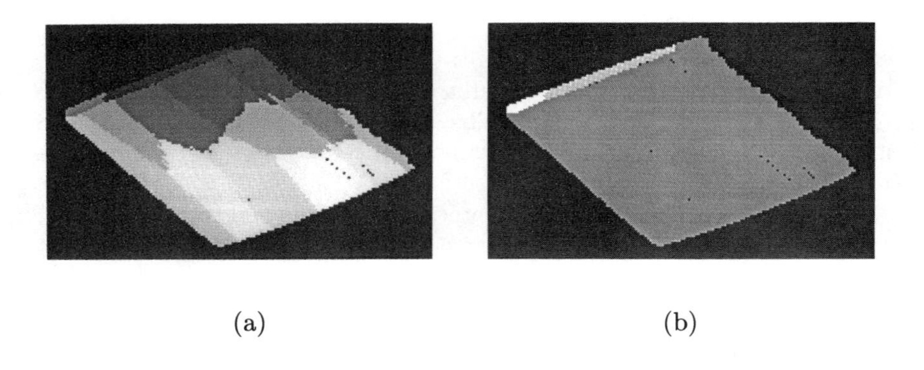

(a) (b)

Figure 14: Hybrid segmentation results for the *coffee cup* image: (a) initial region-based segmentation; (b) result of combining region-based and edge-based segmentations.

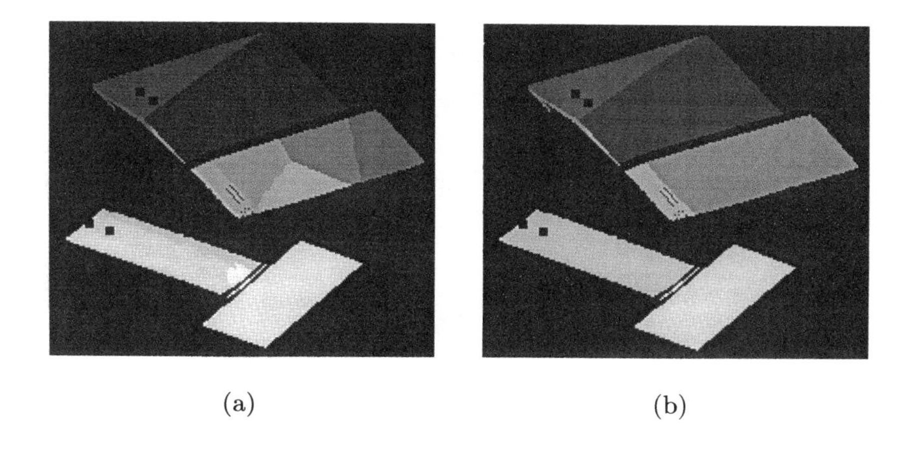

(a) (b)

Figure 15: Hybrid segmentation results for the *jumble* image: (a) initial region-based segmentation; (b) result of combining region-based and edge-based segmentations.

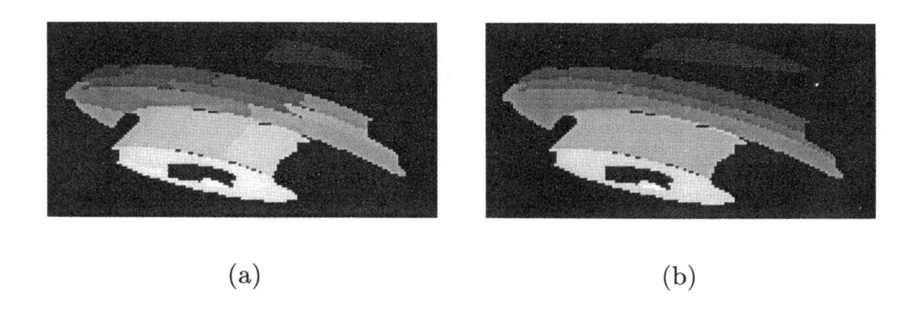

(a) (b)

Figure 16: Hybrid segmentation results for the *rotsym* image: (a) initial region-based segmentation; (b) result of combining region-based and edge-based segmentations.

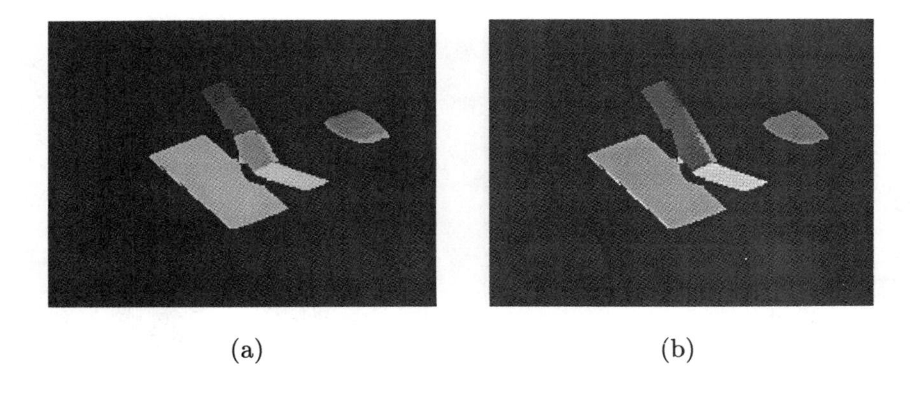

(a) (b)

Figure 17: Hybrid segmentation results for the *occluded jumble* image: (a) initial region-based segmentation; (b) result of combining region-based and edge-based segmentations.

6 Conclusions

We have developed an edge detection algorithm for range images which uses Markov random field models to handle uncertainty in edge labeling. Our algorithm handles both the jump edges and crease edges. Our methodology consists of computing jump and crease edge likelihoods at each edge site, using the larger of the two values as the edge likelihood. The edge likelihoods are then used in a Bayesian framework with MRF prior density on the edge labels to derive the posterior distribution of labels. An approximation to the maximum a posteriori estimate is obtained using the HCF estimate which was earlier shown to perform well for intensity images by Chou *et al.* [5]. This edge-based segmentation was integrated with a region-based segmentation using a simple technique. Segmentation results indicate that the algorithm performs quite well on range images acquired from the range scanner in our laboratory.

We believe that the clique potential values used here can be improved. It is, in general, difficult to choose potential function values which result in an optimal performance in the sense of being closest to the human grouping mechanism. Perhaps this is the weakest link in the use of MRF models to practical problems in computer vision. The quantification of various prior constraints and Gestalt principles in the form of clique potentials, and the selection of appropriate neighborhood size, are formidable problems. Nevertheless, as we have shown here, for specific problems it is possible to utilize contextual information and other constraints in the form of MRF models. The results of experiments using the hybrid segmentation algorithm showed that excellent surface segmentation results can be obtained by integrating edge-based and region-based approaches.

Bibliography

[1] Besag, J., "Spatial Interaction and the Statistical Analysis of Lattice Systems," *J. Royal Statistical Society B 36*, pp. 192–236, 1974.

[2] Besag, J., "On the Statistical Analysis of Dirty Pictures," *J. Royal Statistical Society B 48*, pp. 259–302, 1986.

[3] Besl, P. J. and Jain, R. C., "Segmentation Through Variable-Order Surface Fitting," *IEEE Trans. Pattern Anal. Machine Intel.*, Vol. PAMI-10, pp. 167–192, 1988.

[4] Blake, A. and Zisserman, A., *Visual Reconstruction*, MIT Press, Cambridge, MA, 1987.

[5] Chou, P., Brown, C. and Raman, R., "A Confidence-Based Approach to the Labeling Problem," *Proc. IEEE Workshop on Computer Vision*, pp. 51–56, Miami Beach, Florida, 1987.

[6] Chou, P., "The Theory and Practice of Bayesian Image Labeling," *Technical Report 258*, University of Rochester, 1988.

[7] Cohen, F. S. and Cooper, D. B., "Simple Parallel Hierarchical and Relaxation Algorithms for Segmenting Noncausal Markovian Random Fields," *IEEE Trans. Pattern Anal. Machine Intell.*, Vol. PAMI-9, pp. 195–219, 1987.

[8] Dubes, R. C. and Jain, A. K., "Random Field Models in Image Analysis," *J. Applied Statistics*, Vol. 16, 2, pp. 131–164, 1989.

[9] Dubes, R. C., Jain, A. K., Nadabar, S. G. and Chen, C. C., "MRF Model-Based Algorithms for Image Segmentation," *Proc. 10th Int. Conf. Pattern Recognition*, Atlantic City, Vol 2, pp. 808–814, June 1990.

[10] Flynn, P. J. and Jain, A. K., "On Reliable Curvature Estimation," *Proc. IEEE 1989 Conf. Computer Vision and Pattern Recognition*, San Diego, pp. 110–116, June 1989.

[11] Geman, S. and Geman, D., "Stochastic Relaxation, Gibbs Distributions, and the Bayesian Restoration of Images," *IEEE Trans. Pattern Anal. Machine Intell.*, Vol. PAMI-6, pp. 721–741, 1984.

[12] Fan, T. J., Medioni, G. and Nevatia, R., "Segmented Descriptions of 3-D Surfaces," *IEEE J. Robotics and Automation*, Vol. RA-3, pp. 527–538, 1987.

[13] Hoffman R. and Jain, A., "Segmentation and Classification of Range Images," *IEEE Trans. Pattern Anal. Machine Intell.*, Vol. PAMI-9, pp. 608–620, 1987.

[14] Inokuchi, S., Nita, T., Matsuda, F. and Sakurai, Y., "A Three Dimensional Edge-Region Operator for Range Pictures," *Proc. 6th Int. Conf. Pattern Recognition*, pp. 918–920, Munich, West Germany, 1982.

[15] Jain, R. C. and Jain, A. K., *Range Image Understanding,* Springer-Verlag, Berlin, 1989.

[16] Marroquin, J., Mitter, S. and Poggio, T., "Probabilistic Solution of Ill-Posed problems in Computational Vision," *J. American Statistical Association*, Vol. 82, pp. 76–89, 1987.

[17] Mitiche, A. and Aggarwal, J. K., "Detection of Edges Using Range Information," *IEEE Trans. Pattern Anal. Machine Intell.*, Vol. PAMI-5, pp. 174–178, 1983.

[18] Poggio, T., Torre, V. and Koch, C., "Computational Vision and Regularization Theory," *Nature 317*, pp. 314–319, 1985.

[19] Rimey, R. D. and Cohen, F. S., "A Maximum-Likelihood Approach to Segmenting Range Data," *IEEE J. Robotics and Automation*, Vol. RA-4, pp. 277–286, 1988.

[20] Yokoya, N. and Levine, M. D., "Range Image Segmentation Based on Differential Geometry: A Hybrid Approach," *IEEE Trans. Pattern Anal. Machine Intell.*, Vol. PAMI-11, pp. 643–649, 1989.

Index